EVIDENCE-BASED CLIMATE SCIENCE

EVIDENCE-BASED CLIMATE SCIENCE

DATA OPPOSING CO$_2$ EMISSIONS AS THE PRIMARY SOURCE OF GLOBAL WARMING

SECOND EDITION

Edited by

DON J. EASTERBROOK

ELSEVIER

AMSTERDAM • BOSTON • HEIDELBERG • LONDON • NEW YORK • OXFORD
PARIS • SAN DIEGO • SAN FRANCISCO • SINGAPORE • SYDNEY • TOKYO

Elsevier
Radarweg 29, PO Box 211, 1000 AE Amsterdam, Netherlands
The Boulevard, Langford Lane, Kidlington, Oxford OX5 1GB, United Kingdom
50 Hampshire Street, 5th Floor, Cambridge, MA 02139, United States

Notices
Knowledge and best practice in this field are constantly changing. As new research and experience broaden our understanding, changes in research methods, professional practices, or medical treatment may become necessary.

Practitioners and researchers must always rely on their own experience and knowledge in evaluating and using any information, methods, compounds, or experiments described herein. In using such information or methods they should be mindful of their own safety and the safety of others, including parties for whom they have a professional responsibility.

To the fullest extent of the law, neither the Publisher nor the authors, contributors, or editors, assume any liability for any injury and/or damage to persons or property as a matter of products liability, negligence or otherwise, or from any use or operation of any methods, products, instructions, or ideas contained in the material herein.

Library of Congress Cataloging-in-Publication Data
A catalog record for this book is available from the Library of Congress

British Library Cataloguing-in-Publication Data
A catalogue record for this book is available from the British Library

ISBN: 978-0-12-804588-6

For information on all Elsevier publications
visit our website at https://www.elsevier.com/

  Working together
to grow libraries in
developing countries

www.elsevier.com • www.bookaid.org

Publisher: Candice Janco
Acquisition Editor: Louisa Hutchins
Editorial Project Manager: Emily Thomson
Production Project Manager: Maria Bernard
Designer: Maria Ines Cruz

Typeset by TNQ Books and Journals

Contents

I CLIMATIC PERSPECTIVES

II TEMPERATURE MEASUREMENTS

III EXTREME WEATHER EVENTS

IV POLAR ICE

V CARBON DIOXIDE

VI OCEANS

VII SOLAR INFLUENCES ON CLIMATE

VIII CLIMATE MODELS

IX CLIMATE PREDICTIONS

List of Contributors

J.W. Abbot The Climate Modelling Laboratory, Noosa Heads, QLD, Australia; Institute of Public Affairs, Melbourne, VIC, Australia

H.I. Abdussamatov Pulkovo Observatory of the RAS, St. Petersburg, Russia

D. Archibald Rhaetian Management, City Beach, WA, Australia

Christopher Monckton of Benchley Science and Public Policy Institute, Washington, DC, United States

J.S. D'Aleo American Meteorological Society, Hudson, NH, United States

D.J. Easterbrook Western Washington University, Bellingham, WA, United States

D.M.W. Evans Science Speak, Perth, Australia

E.L. Fix Avionics Fix, Beavercreek, OH, United States

T. Heller Doctors for Disaster Preparedness

M. Khandekar Expert Reviewer IPCC 2007 Climate Change Documents, Toronto, ON, Canada

S. Lüning Independent Researcher, Lisbon, Portugal

J.J. Marohasy The Climate Modelling Laboratory, Noosa Heads, QLD, Australia; Institute of Public Affairs, Melbourne, VIC, Australia

P. Moore Ecosense Environmental Inc., Vancouver, BC, Canada; Frontier Centre for Public Policy

N.-A. Mörner Paleogeophysics & Geodynamics, Saltsjöbaden, Sweden

F. Vahrenholt German Wildlife Foundation, Hamburg, Germany

Preface

The question of "global warming" as a result of rising CO_2 has raised contentious issues. Arguments for this assertion in Intergovernmental Panel on Climate Change (IPCC) reports depend heavily on computer climate models and so-called "consensus among scientists" claims, whereas skeptics of this contention depend on application of the scientific method to actual measured data.

Plotting of the average of 102 IPCC climate models runs against measured temperatures over the same time period (Fig. 1, Spencer, 2015; Christy, 2016) shows that computer models have failed miserably, overestimating temperatures by large amounts.

In his February 2, 2016, testimony before the House Committee on Science, Space, and Technology, Dr. John Christy pointed out:

"Because this result challenges the current theory of greenhouse warming in relatively straightforward fashion, there have been several well-funded attacks on those of us who build and use such datasets and on the datasets themselves. As a climate scientist, I've found myself, along with fellow like-minded colleagues, tossed into a world more closely associated with character assassination and misdirection, found in Washington politics, for example, rather than objective dispassionate discourse commonly assumed for the scientific endeavor. Investigations of us by congress and the media are spurred by the idea that anyone who disagrees with the climate establishment's view of dangerous climate change must be on the payroll of scurrilous organizations or otherwise mentally deficient. Also thrust into this milieu is promotional material, i.e., propaganda, attempting to discredit these data (and researchers) with claims that amount to nothing." "It is clear that climate models fall short on some very basic issues of climate variablility, being unable to reproduce 'what' has happened regarding global temperature, and therefore not knowing 'why' any of it happened."

Christy (2016) comments on "the failure of the scientific community to objectively approach the study of climate and climate change." "Climate science is a murky science with large uncertainties on many critical components such as cloud distributions and surface heat exchanges." "Our science has also seen the move toward 'consensus' science where 'agreement' between people and groups is elevated above determined, objective investigation. The sad progression of events here has even led to congressional investigations designed to silence (with some success) those whose voices, including my own, have challenged the politically-correct views on climate." "When a contrarian proposal is submitted that seeks to discover other possible explanations besides greenhouse gases for the small changes we now see, or one that seeks to rigorously and objectively investigate climate model output, there is virtually no chance for funding." "The term 'consensus science' will often be appealed to regarding arguments about climate change to bolster an assertion. This is a form of 'argument from authority.' Consensus, however, is a political notion, not a scientific notion." "We do not have laboratory methods of testing our hypotheses as many other sciences do. As a result what passes for science includes, opinion, arguments—from—authority, dramatic press releases, and fuzzy notions of consensus generated by preselected groups. This is not science."

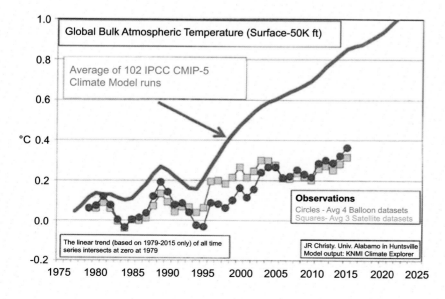

FIGURE 1 Average of 102 computer model runs plotted against observed temperatures from satellites and weather balloons. The computer models badly overestimate global temperatures (Christy, 2016).

In the end, data speaks far louder than rhetoric or "consensus." That is the underlying concept for this book, which brings together factual data on most of the topics related to global climate—hence the title "Evidence-based Climate Science." The data speaks for itself. "Dogma is an impediment to the free exercise of thought. It paralyses the intelligence. Conclusions based upon preconceived ideas are valueless. It is only the open mind that really thinks (Patricia Wentworth, 1949)."

CLIMATIC PERSPECTIVES

1

Climate Perspectives

D.J. Easterbrook

Western Washington University, Bellingham, WA, United States

OUTLINE

1. INTRODUCTION

The climatic changes that the Earth has experienced in the past several decades have led to an intense interest in their cause, with contentions by the United Nations (UN) Intergovernmental Panel on Climate Change (IPCC), activists, politicians, some climate scientists, and virtually all of the news media that catastrophic global warming and sea level rise due to increased atmospheric CO_2 will occur by the end of this century or before. However, many scientists point to data strongly suggesting that climate changes are a result of natural cycles, which have been occurring for thousands of years. Unfortunately, many nonscientist activists and the news media have entered the debate and their arguments have taken on political aspects with little or no scientific basis.

So what is the physical *evidence* for the cause of global warming and cooling? Proponents of CO_2-caused warming contend that the coincidence of global warming since 1978 with rising CO_2 means that CO_2 is the cause of the warming, and that 97% of all scientists agree that this will result in catastrophic events before the end of the century. However, this is not proof of anything—just because two things happen coincidently does not prove that one is the cause of the other. After 1945, CO_2 emissions soared for the next 30 years, but the climate cooled, rather than warmed, showing a total lack of correlation between CO_2 and climate. Then, in 1977, temperatures switched abruptly from cool to warm and the climate began to warm with no change in the rate of increase of CO_2.

2. THE "97%" MYTH

Every day, the news media, activists, politicians, and some climate scientists proclaim that 97% of all scientists agree that atmospheric CO_2 causes global warming and rising CO_2 will lead to global catastrophes. This claim has been echoed by the National Aeronautics and Space Administration (NASA), National Oceanic and Atmospheric Administration (NOAA), various scientific organizations, governments, President Obama, Secretary of State

Evidence-Based Climate Science, Second Edition
http://dx.doi.org/10.1016/B978-0-12-804588-6.00001-X

3

John Kerry, and many others. Where did the 97% number come from—was there some worldwide survey of all scientists? The 97% number is based on two publications—the first by Doran and Zimmerman (2009) and a later one by Cook et al. (2013).

The Doran and Zimmerman paper was a University of Illinois master's thesis by Maggie Zimmeran and her thesis advisor, Peter Doran, who claimed that "97% of climate scientists agree" that global warming is caused by rising CO_2. They sent an Internet survey to 10,257 people working at universities and government agencies and received 3146 replies. Of these, only 5% identified themselves as "climate scientists." Only two questions were asked: (1) "When compared with pre-1800 levels, do you think that global temperatures have generally risen, fallen, or remain relatively constant?" and (2) "Do you think human activity is a significant contributing factor in changing mean global temperature?" Of the 3146 replies, Doran and Zimmerman arbitrarily selected 79 responses, of whom 77 replied "yes." They divided 77 by 79 to get 97%, which was then elevated to "97% of *all* scientists" by various proponents of CO_2. The proper number should have been 77 divided by 3146, which equals 2%.

The Cook et al. (2013) paper was based on counting abstracts of climate papers. The authors contended that "Among [4014] abstracts expressing a position of AGW [Anthropogenic Global Warming], 97% endorsed the consensus position humans are causing global warming." However, Legates et al. (2013) point out that "the author's own analysis shows that only 0.5% of all 11,944 abstracts, and 1.6% of the 4014 abstracts expressing a position, endorsed anthropogenic warming as they had defined it."

Thus, the contention that "97% of all scientists agree that global warming is caused by CO_2" is simply not true, and those who continue to assert this are either uniformed or perpetuating a false statement. Legates et al. (2013) and Bast and Spencer (2014) conclude "The 97.1% consensus claimed by Cook et al. (2013) turns out upon inspection to be not 97.1% but 0.3%. Their claim of 97.1% consensus, therefore, is arguably one of the greatest items of misinformation in history."

On the other hand, the following statement has been signed by 31,478 American scientists, 9021 with PhDs, as part of the Global Warming Petition Project. All signers must have a degree in a scientific field.

> There is no convincing scientific evidence that human release of carbon dioxide, methane, or other greenhouse gases is causing or will, in the foreseeable future, cause catastrophic heating of the Earth's atmosphere and disruption of the Earth's climate.

Another relevant question here is whether or not even if there was a consensus, would it prove anything at all. As Feynman has pointed out, "The number of scientists who believe something is irrelevant to the validity of a concept." ("Consensus" means nothing.) As Galileo and other scientists have shown, it only takes one to prove a hypothesis wrong.

3. THE SCIENTIFIC METHOD VERSUS DOGMA

The scientific method is defined as "a method or procedure that has characterized natural science since the seventeenth century, consisting in systematic observation, measurement, and experiment, and the formulation, testing, and modification of hypotheses." Richard Feynman, Noble Prize winner in physics, has spoken and written eloquently about the methodology:

> Science is a method of finding things out by observation, experimentation, and testing, which is the ultimate judge of the truth of a concept. In general, we look for a new law by the following process: First we guess it; then we compute the consequences of the guess to see what would be implied if this law that we guessed is right; then we compare the result of the computation to nature, with experiment or experience, compare it directly with observation, to see if it works. If it disagrees with experiment, it is wrong. In that simple statement is the key to science. It does not make any difference how beautiful your guess is, it does not make any difference how smart you are, who made the guess, or what his name is—if it disagrees with experiment, it is wrong.
>
> *Richard Feynman*

If *any* exception to a concept can be proven by observation or experimentation, the concept is wrong. After Einstein published his Theory of Relativity, a group of 100 physicists was formed to attempt to prove that the theory was wrong. When asked about this, Einstein replied, I don't know why they need 100, it only takes one.

Pro-CO_2 activists have attempted to ridicule those who are skeptical of CO_2-caused global warming. However, as Einstein, Feynman, and others have pointed out, all scientists are skeptics—it is important to doubt in order to test concepts and look in new directions. Anyone who is not a skeptic is not a scientist.

The arguments for CO_2-caused global warming are not based on the scientific method, but on unsupported assertions and computer modeling (which has proven to be highly inaccurate). A high-level UN IPCC official, commenting on scientific proof of hypotheses, stated: "Proof is for mathematical theorems and alcoholic beverages. It's not for science. Science is all about "credible theories" and "best explanations." You don't need "proof" when you have "credible theories." In other words, the scientific method, practiced for 500 years, is now passé and all that is needed is enough followers to support a theory. Most pro-CO_2 assertions are therefore dogmas, principles laid down by an authority as incontrovertibly true and which may not be challenged by data, changed, or discarded.

4. COMPARISON OF COMPUTER MODELING OF CLIMATE WITH MEASURED TEMPERATURES

Because of the absence of any physical evidence that CO_2 causes global warming, the main argument for CO_2 as the cause of warming rests largely on computer modeling. Thus, the question becomes, how good are the computer models in predicting climate? We can test this by comparing global warming predicted by the IPCC models against actual climate change over the past two decades. Fig. 1.1 shows a comparison of predicted temperature made by 90 computer models with measured temperatures. The climate models failed miserably and didn't come anywhere near the temperatures that were later measured.

5. NO GLOBAL WARMING FOR 18 YEARS AND 8 MONTHS

A critical test of assertions that CO_2 causes global warming is that through December 2015, there has been no global warming at all (Fig. 1.2) even though CO_2 has continued to rise.

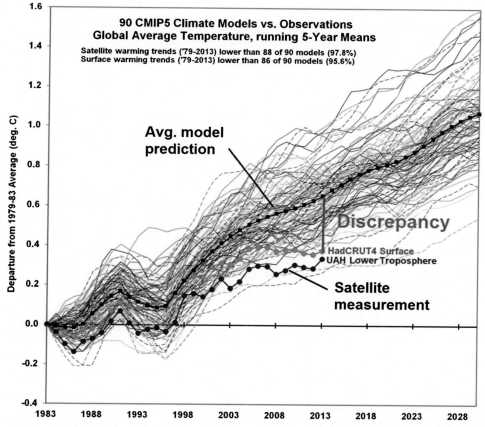

FIGURE 1.1 Comparison of computer model predictions and satellite and surface measurements. Computer climate models have failed dismally to predict temperature changes. *Modified from Spencer (2015), http://www.drroyspencer.com.*

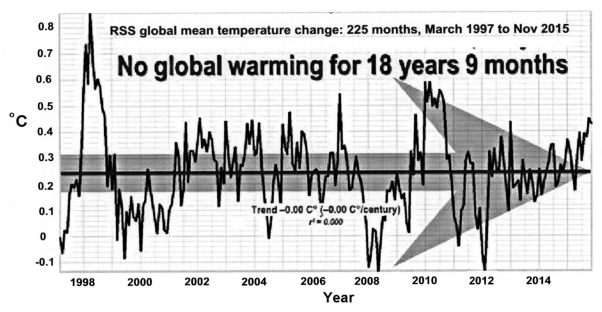

FIGURE 1.2   Remote Sensing Systems (RSS) satellite temperature measurements prove that there has been no global warming at all for 18 years and 8 months (Monckton, 2015, https://wattsupwiththat.com/2015/12/04/the-robust-pause-resists-a-robust-el-nio-still-no-global-warming-at-all-for-18-years-9-months/).

6. HOTTEST-YEAR-EVER CLAIMS

NASA and NOAA have made repeated claims that global temperatures for several years have been "the hottest ever recorded." However, these assertions are based on badly corrupted surface data that is contradicted by uncorrupted University of Alabama at Huntsville and Remote Sensing Systems satellite data (Fig. 1.3).

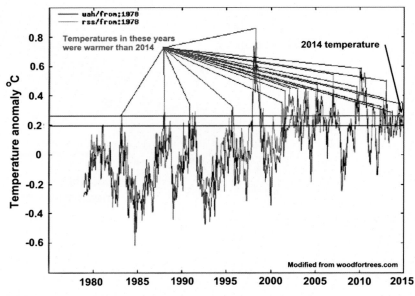

FIGURE 1.3   University of Alabama at Huntsville (UAH) and Remote Sensing Systems (RSS) satellite temperature data contradicting claims that 2014 and 2015 were the "hottest years ever." More than a dozen temperature peaks are higher than the 2014 temperature.

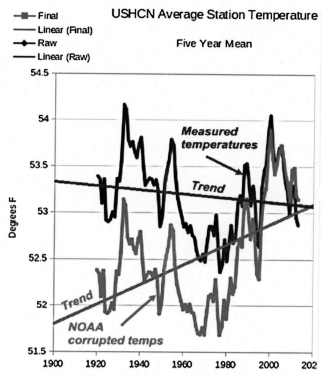

FIGURE 1.4   Temperature data corruption by NOAA. Measured temperatures are shown in blue, corrupted NOAA temperatures are shown in red. Note that NOAA decreased measured temperatures in the 1930s and 1940s by a full degree but made no changes to temperatures from about 1990 onward, thus changing a cooling trend into a warming trend (Heller, 2001, https://stevengoddard.wordpress.com/).

7. DATA CORRUPTION

Blatant temperature data corruption by NOAA and NASA is discussed in Chapter 2 of this volume. One example of this is shown in Fig. 1.4.

8. THE PAST IS THE KEY TO THE FUTURE

If CO_2 is incapable of explaining global warming, what natural possibilities exist? A vast amount of physical evidence of climate change over the past centuries and millennia has been gathered by scientists. Significant climate changes have clearly been going on for many thousands of years, long before the recent rise in atmospheric CO_2. In order to understand modern climate changes, we need to look at the past history of climate changes. The past is the key to the future—to know where we are headed in the future, we need to know where we have been in the past. This volume is intended to document past climate changes and present physical evidence for possible causes. It includes data related to the causes of global climate change by experts in meteorology, geology, atmospheric physics, solar physics, geophysics, climatology, and computer modeling.

Time and nature will be the final judge of the cause of global warming. The next decade should tell us the answer. If CO_2 is the cause of global warming and the computer models are correct, then warming of 2°F since 2000 should occur by 2040. If the climate continues to cool, then the computer models must be considered invalid, and we must look to other causes. As we enter the Grand Solar Minimum and cooling deepens in the next decade, as it did in 1790 and 1645, than a strong case can be made for solar variation as the main cause of climate change.

The reader is invited to toss aside all of the political rhetoric that has been introduced into the global warming debate, focus on the scientific evidence presented in the chapters in this volume, and draw his or her own conclusions. Dogma is an impediment to the free exercise of thought—it paralyses the intelligent. Conclusions based upon preconceived ideas are valueless—it is only the open mind that really thinks.

References

Bast, J.L., Spencer, R., May 2014. The myth of climate change '97%'. Wall Street Journal.

Cook, J., Nuccitelli, D., Green, S.A., Richardson, M., Winkler, B., Painting, R., Way, R., Jacobs, P., Kuce, A., 2013. Quantifying the consensus on anthropogenic global warming in the scientific literature. Environmental Research Letters 8, 2.

Doran, P.T., Zimmerman, M.K., 2009. Examining the scientific consensus on climate change. EOS 90, p22–23.

Legates, D.R., Soon, W., Briggs, W.M., Monckton, C., 2013. Climate Consensus and 'misinformation': A Rejoinder to Agnotology: Scientific Consensus, and the Teaching and Learning of Climate Change. Springer Science and Business Media, Dordrecht.

Spencer, 2015. http://www.drroyspencer.com.

Monckton, 2015. https://wattsupwiththat.com/2015/12/04/the-robust-pause-resists-a-robust-el-nio-still-no-global-warming-atall-for-18-years-9-months/.

TEMPERATURE MEASUREMENTS

2

A Critical Look at Surface Temperature Records

J.S. D'Aleo

American Meteorological Society, Hudson, NH, United States

1. INTRODUCTION

Although warming from 1979 to 1998 is well supported, major questions exist about long-term trends. Climategate-inspired investigations suggest global surface-station data are seriously compromised. The data suffer significant contamination by urbanization and other local factors such as land-use/land-cover changes and instrument siting that does not meet government standards. There was a major station dropout, which occurred suddenly around 1990, and a significant increase in missing monthly data in the stations that remained. There are also uncertainties in ocean temperatures—no small issue, as oceans cover 71% of the Earth's surface.

These factors all lead to significant uncertainty and in most cases a tendency for overestimation of century-scale temperature trends. Indeed, numerous peer-reviewed papers cataloged here have estimated that these local issues with the observing networks may account for **30%, 50%, or more** of the warming shown since 1880. After the data with all its issues are collected, further adjustments are made, each producing more warming.

"[W]hen data conflicts with models, a small coterie of scientists can be counted upon to modify the data to agree with models' projections," says MIT meteorologist Dr. Richard Lindzen.

In this paper, we look at some of the issues in depth and the recommendations made for a reassessment of global temperatures necessary to make sensible policy decisions.

2. THE GLOBAL DATA CENTERS

Five organizations publish global temperature data. Two—Remote Sensing Systems (RSS) and the University of Alabama at Huntsville (UAH)—are satellite data sets. The three terrestrial data sets provided by the institutions—NOAA's National Climatic Data Center (NCDC), NASA's Goddard Institute for Space Studies (GISS/GISTEMP), and the University of East Anglia's Climatic Research Unit (CRU)—all depend on data supplied by surface stations administered and disseminated by NOAA under the management of the National Climatic Data Center in Asheville, North Carolina. The Global Historical Climatology Network (GHCN) is the most commonly cited measure of global surface temperature for the last 100 years.

Around 1990, NOAA/NCDC's GHCN data set lost more than three-quarters of the climate measuring stations around the world. A study by Willmott et al. (1991) calculated a +0.2°C bias in the global average owing to pre-1990 station closures. Douglas Hoyt had estimated approximately the same value in 2001 due to station closures around 1990. A number of station closures can be attributed to Cold-War era military base closures, such as the DEW Line (the Distant Early Warning Line) in Canada and its counterpart in Russia.

The world's surface observing network had reached its golden era in the 1960s to 1980s, with more than 6000 stations providing valuable climate information. Now, there are fewer than 1500 remaining.

It is a fact that the three data centers each performed some final adjustments to the gathered data before producing their own final analysis. These adjustments are frequent and often poorly documented. The result was almost always to produce an enhanced warming even for stations which had a cooling trend in the raw data. The metadata, the information about precise location, station moves, and equipment changes were not well documented and shown frequently to be in error, which complicates the assignment to proper grid boxes and makes the efforts of the only organization that attempts to adjust for urbanization, NASA GISS, problematic.

As stated here relative to Hansen et al. (2001),[1] "the problem [accuracy of the latitude/longitude coordinates in the metadata] is, as they say, 'even worse than we thought.' One of the consumers of GHCN metadata is of course GISTEMP, and the implications of imprecise latitude/longitude for GISTEMP are now considerably greater, following the change in January 2010 to use of satellite-observed night light radiance to classify stations as rural or urban throughout the world, rather than just in the contiguous United States as was the case previously. As about a fifth of all GHCN stations changed classification as a result, this is certainly not a minor change."

Among some major players in the global temperature analyses, there is even disagreement about what the surface air temperature really is (see "The Elusive Absolute Surface Air Temperature (SAT)" by Dr. James Hansen.[2] Essex et al. questioned whether a global temperature existed here.[3])

Satellite measurements of the lower troposphere (around 600 mb) are clearly the better alternative. They provide full coverage and are not contaminated by local factors. Even NOAA had assumed satellites would be the future solution for climate monitoring. Some have claimed satellite measurements are subject to error. RSS and UAH in 2005 jointly agreed[4] that there was a small net cold bias of 0.03°C in their satellite-measured temperatures, and corrected the data for this small bias. In contrast, the traditional surface-station data we will show suffer from many warm biases that are orders of magnitude greater in size than the satellite data, yet that fact is often ignored by consumers of the data.

Some argue that satellites measure the lower atmosphere and that this is not the surface. This difference is real but it is irrelevant (CCSP[5]). The lower troposphere around 600 mb was chosen because it was above the mixing level and

[1] http://oneillp.wordpress.com/2010/03/13/ghcn-metadata/.

[2] http://data.giss.nasa.gov/gistemp/abs_temp.html.

[3] http://www.uoguelph.ca/~rmckitri/research/globaltemp/globaltemp.html.

[4] http://www.marshall.org/article.php?id=312.

[5] http://www.climatescience.gov/Library/sap/sap1-1/finalreport/.

so with polar orbiters the issues of the diurnal variations are eliminated. Also there is a high correlation between temperatures in the lower to middle troposphere and the surface.

Anomalies from satellite data and surface-station data have been increasing in the last three decades. When the satellites were first launched, their temperature readings were in better agreement with the surface-station data. There has been increasing divergence over time, which can be seen in Fig. 2.1 (derived from Klotzbach et al., 2009). In the first plot, we see the temperature anomalies as computed from the satellites and assessed by UAH and RSS and the station-based land surface anomalies from NOAA NCDC (Fig. 2.1).

The divergence is made clearer when the data are scaled such that the difference in 1979 is zero (Fig. 2.2).

The Klotzbach paper finds that the divergence between surface and lower-tropospheric trends is consistent, with evidence of a warm bias in the surface temperature record but not in the satellite data.

Klotzbach et al. described an "amplification" factor for the lower troposphere as suggested by Santer et al. (2005, 2008) due to greenhouse gas trapping relative to the warming at the surface. Santer refers to the effect as

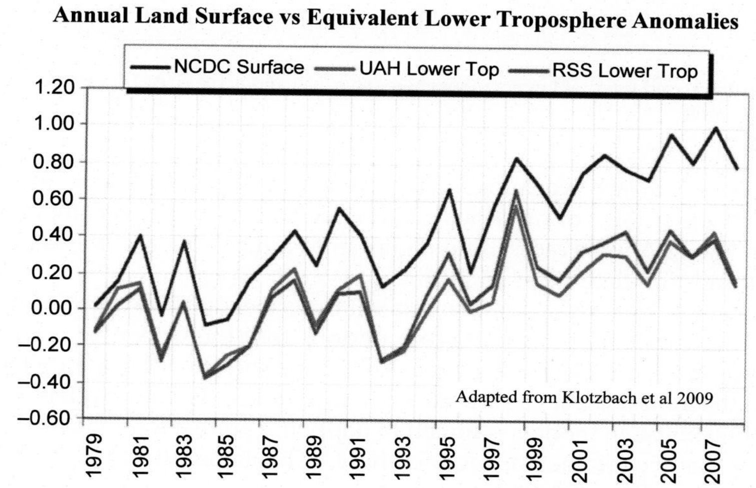

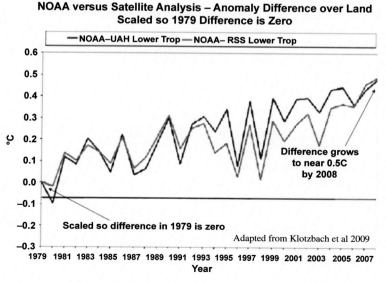

FIGURE 2.1 Annual land surface anomalies compared to UAH and RSS lower-tropospheric temperature anomalies since 1979. *Klotzbach, P.J., Pielke Sr, R.A., Pielke Jr, R.A., Christy, J.R., McNider, R.T., 2009. An alternative explanation for differential temperature trends at the surface and in the lower troposphere. Journal of Geophysical Research 114, D21102. doi:10.1029/2009JD011841, http://sciencepolicy.colorado.edu/admin/publication_files/resource-2792-2009.52.pdf.*

FIGURE 2.2 NOAA annual land temperatures minus annual UAH lower troposphere (*blue line*) and NOAA annual land temperatures minus annual RSS lower troposphere (*green line*) over the period from 1979 to 2008.

Annual Model Forecast versus Actual Lower Troposphere Anomalies

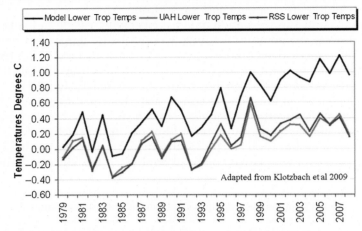

FIGURE 2.3   Model amplification-based forecast lower troposphere (*blue line*) and actual UAH (*green line*) and RSS lower troposphere (*purple line*) over the period from 1979 to 2008.

"tropospheric amplification of surface warming." This effect is a characteristic of all of the models used in the UNIPCC and the USGRCP "ensemble" of models by Karl et al. (2006), which was the source for Karl (1995), which in turn was relied upon by the EPA in its recent Endangerment Finding (Federal Register/Vol. 74, No. 239/Tuesday, December 15, 2009/Rules and Regulations at 66,510).

As Dr. John Christy, keeper of the UAH satellite data set describes it, "The amplification factor is a direct calculation from model simulations that show over 30-year periods that the upper air warms at a faster rate than the surface—generally 1.2 times faster for global averages. This is the so-called 'lapse rate feedback' in which the lapse rate seeks to move toward the moist adiabat as the surface temperature rises. In models, the convective adjustment is quite rigid, so this vertical response in models is forced to happen. The real world is much less rigid and has ways to allow heat to escape rather than be retained as models show." This latter effect has been documented by Chou and Lindzen (2005) and Lindzen and Choi (2009).

The amplification factor was calculated from the mean and median of the 19 GCMs that were in the CCSP SAP 1.1 report (Karl et al., 2006). A fuller discussion of how the amplification factor was calculated is available in the Klotzbach paper.[6]

The ensemble model forecast curve (upper curve) in Fig. 2.3 was calculated by multiplying the NOAA NCDC surface temperature for each year by the amplification factor, and thus is the model projected tropospheric temperature. The lower curves are the actual UAH and RSS lower-tropospheric satellite temperatures.

This strongly suggests that instead of atmospheric warming from greenhouse effects dominating, surface-based warming very likely due to uncorrected urbanization and land-use contamination is the biggest change. Since these surface changes are not fully adjusted for, trends from the surface networks are not reliable.

3. THE GOLDEN AGE OF SURFACE OBSERVATION

In this era of ever-improving technology and data systems, one would assume that measurements would be constantly improving. This is not the case with the global station observing network. The Golden Age of Observing was several decades ago. It is gone.

The Hadley Centre's CRU at East Anglia University is responsible for the CRU global data. NOAA's NCDC, in Asheville, NC, is the source of the GHCN and of the U.S. Historical Climate Network (USHCN). These two data sets are relied upon by NASA' GISS in New York City and by Hadley/CRU in England.

All three have experienced degradation in data quality in recent years.

---

[6] http://pielkeclimatesci.files.wordpress.com/2009/11/r-345.pdf.

Ian "Harry" Harris, a programmer at CRU, kept extensive notes of the defects he had found in the data and computer programs that the CRU uses in the compilation of its global mean surface temperature anomaly data set. These notes, some 15,000 lines in length, were stored in the text file labeled "Harry_Read_Me.txt," which was among the data released by the whistle-blower with the Climategate emails. This is just one of his comments:

> [The] hopeless state of their (CRU) database. No uniform data integrity, it's just a catalogue of issues that continues to grow as they're found...I am very sorry to report that the rest of the databases seem to be in nearly as poor a state as Australia was. There are hundreds if not thousands of pairs of dummy stations, one with no WMO and one with, usually overlapping and with the same station name and very similar coordinates. I know it could be old and new stations, but why such large overlaps if that's the case? Aarrggghhh! There truly is no end in sight...This whole project is SUCH A MESS. No wonder I needed therapy!!...I am seriously close to giving up, again. The history of this is so complex that I can't get far enough into it before by head hurts and I have to stop. Each parameter has a tortuous history of manual and semi-automated interventions that I simply cannot just go back to early versions and run the update prog. I could be throwing away all kinds of corrections—to lat/lons, to WMOs (yes!), and more. So what the hell can I do about all these duplicate stations?

According to Phil Jones, former director of CRU, 'there is some truth' to the charge that he failed to update and organize the raw data supporting the CRU temperature data set, on which the IPCC relies in its reports to make temperature projections, and that at least some of the original raw data were lost. This should raise questions about the quality of global data.

In the following email, CRU's Director at the time, Dr. Phil Jones, acknowledges that CRU mirrors the NOAA data:

> Almost all the data we have in the CRU archive is exactly the same as in the GHCN archive used by the NOAA National Climatic Data Center.

In the Russell inquiry into CRU's role in Climategate, they estimated at least 90% of the data were the same. Steve McIntyre's analysis showed 95.6% concordance. NASA uses the GHCN as the main data source for the NASA GISS data.

Dr. Roger Pielke Sr., in this post[7] on the three data sets, notes:

> The differences between the three global surface temperatures that occur are a result of the analysis methodology as used by each of the three groups. They are not 'completely independent.' Each of the three surface temperature analysis suffer from unresolved uncertainties and biases as we documented, for example, in our peer reviewed paper.[8]

Dr. Richard Anthes, President of the University Corporation for Atmospheric Research, in testimony to Congress[9] in March 2009, noted:

> The present federal agency paradigm with respect to NASA and NOAA is obsolete and nearly dysfunctional, in spite of best efforts by both agencies.

4. VANISHING STATIONS

More than 6000 stations were in the NOAA database for the mid-1970s, but just 1500 or fewer are used today. NOAA claims the real-time network includes 1200 stations, with 200 to 300 stations added after several months and included in the annual numbers. NOAA is said to be adding additional U.S. stations now that USHCN v2 is available, which will inflate this number, but make it disproportionately U.S stations.

There was a major disappearance of recording stations in the late 1980s to the early 1990s. Fig. 2.4 compares the number of global stations in 1900, the 1970s, and 1997, showing the increase and then decrease (Peterson and Vose[10]).

---

[7] http://pielkeclimatesci.wordpress.com/2009/11/25/an-erroneous-statement-made-by-phil-jonesto-the-media-on-the-independence-of-the-global-surface-temperature-trend-analyses-of-cru-gissand-ncdc/.

[8] http://pielkeclimatesci.files.wordpress.com/2009/10/r-321.pdf.

[9] http://www.ucar.edu/oga/pdf/Anthes%20CJS%20testimony%203-19-09.pdf.

[10] http://www.ncdc.noaa.gov/oa/climate/ghcn-monthly/images/ghcn_temp_overview.pdf.

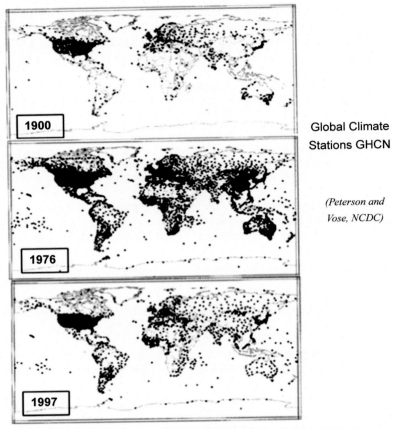

Global Climate
Stations GHCN

*(Peterson and
Vose, NCDC)*

FIGURE 2.4   Stations in 1900, 1976, and 1997 used in the global GHCN database. *Peterson, T.C., Vose, R.S., 1997. An overview of the global historical climatology network temperature database. Bulletin of the American Meteorological Society 78, 2837–2849.*

Dr. Kenji Matsuura and Dr. Cort J. Willmott at the University of Delaware have prepared an animation.[11] See the lights go out in 1990, especially in Asia.

Fig. 2.5 is a chart[12] of all GHCN stations, and the average annual temperature show the drop focused around 1990. In this plot, those stations with multiple locations over time are given separate numbers, which inflates the total number. While a straight average is not meaningful for global temperature calculation (because areas with more stations would have higher weighting), it illustrates that the disappearance of so many stations in an uneven fashion may have introduced a distribution bias (Fig. 2.5).

As can be seen in the figure, the straight average of all global stations does not fluctuate much until 1990, at which point the average temperature jumps up. This observational bias can influence the calculation of area-weighted averages to some extent. As previously noted, a study by Willmott, Robeson, and Feddema ("Influence of Spatially Variable Instrument Networks on Climatic Averages," 1991) calculated a +0.2°C bias in the global average owing to pre-1990 station closures. Others have attempted experiments (Mosher, Grant, Lilligren) that purport to show this does not necessarily translate into a warm bias given the "anomaly method" (using anomalies or departures from normal base period values instead of the actual temperatures). The effect may not be definitively known until a full data reconstruction can take place.

Global databases all compile data into latitude/longitude-based grid boxes and calculate temperatures inside the boxes using data from the stations within them or use the closest stations (weighted by distance) in nearby boxes.

The use of anomalies instead of mean temperatures greatly improve the chances of filling in some of the smaller holes (empty grid boxes) or not producing significant differences in areas where the station density is high, but they

[11] http://climate.geog.udel.edu/ ~climate/html_pages/Ghcn2_images/air_loc.mpg.

[12] http://www.uoguelph.ca/ ~rmckitri/research/nvst.html.

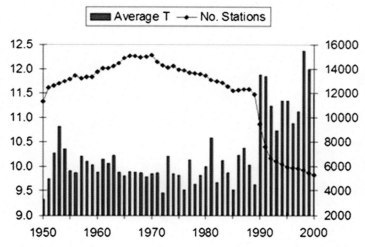

FIGURE 2.5   Plot of the number of total station IDs in each year since 1950 and the average temperatures of the stations in the given year.

can't be relied on to accurately estimate anomalies in the many large data-sparse areas (Canada, Greenland, Brazil, Africa, parts of Russia). To fill in these areas requires NOAA and NASA to reach out as far as 1200 km.

There are 8000 grid boxes globally (land and sea). If the Earth is 71% ocean, approximately 2320 grid boxes would be over land (the actual number will vary as some grid boxes will overlap or may just touch the coast).

With 1200 stations in the real-time GHCN network, that would be enough to have 51.7% of the land boxes with a station. However, since stations tend to cluster, that number is smaller. Our calculation is that the number is around 44%, or 1026 land grid boxes without a station.

For data in empty boxes, GHCN will look to surrounding areas as far away as 1200 km (in other words, using Atlanta, Georgia to estimate a monthly or annual anomaly in Chicago; using Birmingham, Alabama to estimate New York City; and using Los Angeles to estimate Jackson Hole, Wyoming).

Certainly an isolated vacant grid box surrounded by boxes with data in them may be able to yield a reasonably representative extrapolated anomaly value from the surrounding data.

But in data-sparse regions, such as is much of the Southern Hemisphere, when you have to extrapolate from more than one grid box away you are increasing the data uncertainty. If you bias it towards having to look towards more urbanized or airport regions or lower-elevation coastal locations as E.M. Smith has detected, you are adding potential warm bias to uncertainty. This has been the case in the north in the large countries bordering on the Arctic (Russia and Canada), where the greatest warming is shown in the data analyses; but also in Brazil, where fast-growing cities are used to estimate anomalies in the Amazon.

To ascertain whether a net bias exists, E.M. Smith has conducted first an analysis of mean temperatures for whatever stations existed by country or continent/subcontinent. He then applied a dT method,[13] which is a variation of "First Differences," as a means of examining temperature data anomalies independent of actual temperature. dT/year is the "average of the changes of temperature, month now vs. the same month that last had valid data, for each year." He then does a running total of those changes, or the total change, the "Delta Temperature," to date. He is doing this for every country (see footnote [14]). His next step will be to attempt to splice/blend the data into the grids.

Even then, uncertainty will remain, which only more complete data set usage would improve. Fig. 2.6 powerfully illustrates this was a factor even before the major dropout. Brohan et al. (2006) showed the degree of uncertainty in surface temperature sampling errors for 1969 (here for CRUTEM3). The degree of uncertainty exceeds the total global warming signal (Fig. 2.6).

---

[13] http://chiefio.wordpress.com/2010/02/28/last-delta-t-an-experimental-approach/.

[14] http://chiefio.wordpress.com/.

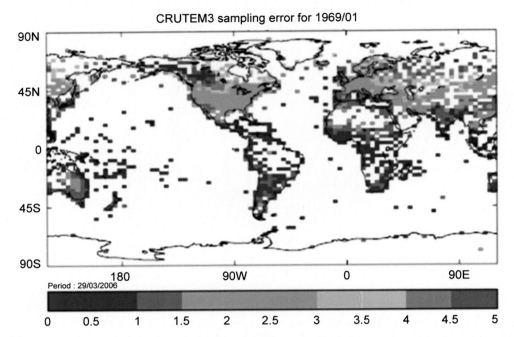

FIGURE 2.6 Temperature anomaly sampling errors (C) for January 1969 on the HadCM3 atmosphere grid. *Brohan, P., Kennedy, J.J., Harris, I., Tett, S.F.B., Jones, P.D., 2006. Uncertainty estimates in regional and global observed temperature changes: a new dataset from 1850. Journal of Geophysical Research 111, D12106. doi: 10.1029/2005JD006548.*

5. SEE FOR YOURSELF: THE DATA ARE A MESS

Look for yourself following these directions using the window into the NOAA GHCN data provided by NASA GISS.[15] Point to any location on the world map (say, central Canada). You will see a list of stations and approximate populations. Locations with less than 10,000 people are assumed to be rural (even though Oke has shown a town of 1000 can have an urban warming bias of 2.2°C).

You will see that the stations have a highly variable range of years with data. Try to find a few stations where the data extend to the current year. If you find some, you will likely see gaps in the graphs. To see how incomplete the data set is for that station, click in the bottom left of the graph *Download monthly data as text*. For many, many stations you will see the data set in a monthly tabular form has many missing data months, mostly after 1990 (designated by 999.9) (Fig. 2.7).

These facts suggest that the golden age of observations was in the 1950s to 1980s. Data sites before then were more scattered and did not take data at standardized times of day. After the 1980s, the network suffered from loss of stations and missing monthly data. To fill in these large holes, data were extrapolated from greater distances away.

Indeed this is more than just Russia. Forty percent of GHCN v2 stations have at least one missing month (Fig. 2.8). This is concentrated in the winter months, as analyst Verity Jones[16] has shown here:

> Much of the warming signal in the global average data can be traced to winter warming (lows are not as low). If we now have a series of cooler years, particularly cooler winter months with lower lows, my concern is that missing months, particularly winter months could lead to a warm bias.

NOAA tells us that by 2020, we will have as much data for the 1990 and 2000s as we had in the 1960 and 1970s. We are told that other private sources have been able to assemble more complete data sets in near real time (example: WeatherSource). Why can't our government with a budget far greater than these private sources do the same or better? This question has been asked by others in foreign nations.

---

[15] http://data.giss.nasa.gov/gistemp/station_data/.

[16] http://diggingintheclay.blogspot.com/2010/03/of-missing-temperatures-and-filled-in.html.

SVERDLOVSK, RUSSIA

	Jan	Feb	Mar	Apr	May	Jun	Jul	Aug	Sep	Oct	Nov	Dec
1987	-18.6	-8.3	-5.5	0.9	14.9	20	19.3	16.2	8.6	2.1	-11.2	-11.5
1988	-12.7	-9.6	-2.6	4.8	10.3	19.5	22.6	18.3	9.7	3.9	-7	-9.9
1989	-15.1	-9.6	-0.3	1.5	11.9	21.4	22.8	14.2	10.2	2.8	-4.2	-10.2
1990	-14.2	-6.8	0.1	5.3	9.6	999.9	19	16.4	999.9	1.2	-5	-8.3
1991	999.9	-12.6	-7.1	9.2	15.5	999.9	17.6	13.6	10.6	7.2	-3.6	-13.6
1992	-12	-9.2	-4.1	1.8	10.1	14	16.3	13.8	10.9	2.1	-5.3	999.9
1993	-8.9	-11.8	-6	3.2	10.8	17.9	18.6	16.5	5.8	2.4	-13.2	-9.9
1994	-10.2	-17.2	999.9	999.9	11.4	999.9	999.9	14.9	11	6.6	-7	-14
1995	-10.2	-4.2	-0.6	10.7	13	17	999.9	16.9	999.9	3.9	-3.7	-12.7
1996	-14.1	-11.2	-3.9	0.6	12.2	19.1	19.2	999.9	7.1	1.9	-2.3	-10.2
1997	-18.4	-9.4	-2.1	6.2	12	16.7	15.9	14	11.2	5.9	-7.3	-14.7
1998	-11	-14.8	-3.3	-1.4	11.8	18.5	21.6	17.6	8.3	3.5	-12.7	-7.1
1999	-12.6	-7.8	-8.6	4.8	9	15.1	20.2	15.6	9.3	7	-10.4	-6.4
2000	-12.9	-6.9	-1.7	7.2	8.3	19.1	20.5	999.9	8.9	2.3	-6.5	-12.2
2001	-12.1	-14.9	-3.4	6.8	13	14.6	17.9	999.9	10.7	0.6	-4.6	-12.3
2002	-9.2	-4.2	-1	3	9.3	14	19	13.1	11.1	2.1	-3.7	-18.5

FIGURE 2.7   The monthly average temperatures in degrees Celsius 1987 to 2002. The 999.9 values are missing months. These require estimation from surrounding sites.

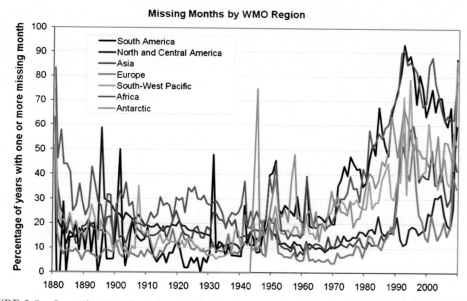

FIGURE 2.8   Quantification of missing months in annual station data. *Analysis and graph: Andrew Chantrill.*

6. STATION DROPOUT WAS NOT TOTALLY RANDOM

6.1. Canada

After 1990, just one thermometer remains in the database for everything north of the 65th parallel. That station is Eureka, which has been described as the "garden spot of the Arctic" thanks to the flora and fauna abundant around the Eureka area, moreso than anywhere else in the High Arctic. Winters are frigid but summers are slightly warmer than at other places in the Canadian Arctic.

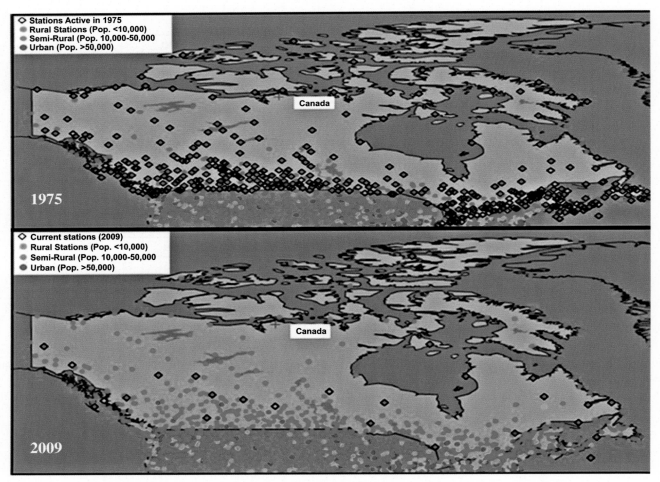

FIGURE 2.9   Canadian stations used in annual analyses in 1975 and 2009. *Verity Jones from GHCN.*

NOAA GHCN used only 35 of the 600 Canadian stations in 2009, down from 47 in 2008. A case study by Tim Ball confirmed Environment Canada claims that weather data are available elsewhere from airports across Canada, and indeed hourly readings can be found on the internet for many places in Canada (and Russia) not included in the global databases. Environment Canada reported in the National Post[17] that there are 1400 stations in Canada with 100 north of the Arctic Circle, where NOAA uses just one. See E.M. Smith's analysis in footnote [18].

Verity Jones plotted the stations from the full network rural, semirural, and urban for Canada and the northern United States both in 1975 and again in 2009. She also marked with diamonds the stations used in the given year. Notice the good coverage in 1975 and very poor, virtually all in the south in 2009. Notice the lack of station coverage in the higher-latitude Canadian region and Arctic in 2009 (Fig. 2.9).

6.2. New Zealand and Australia

Smith found that in New Zealand the only stations remaining had the words "water" or "warm" in the descriptor code. Some 84% of the sites are at airports, with the highest percentage in southern cold latitudes.

In Australia, Torok et al. (2001) [19] observed that in European and North American cities urban—rural temperature differences scale linearly with the logarithms of city populations. They also learned that Australian city heat islands

---

[17] http://www.nationalpost.com/news/story.html?id=2465231#ixzz0dY7ZaoIN.

[18] http://chiefio.wordpress.com/2009/11/13/ghcn-oh-canada-rockies-we-dont-need-no-rockies/.

[19] http://www.co2science.org/articles/V5/N20/C3.php.

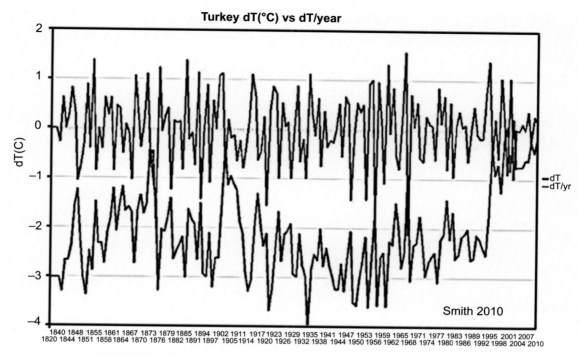

FIGURE 2.10   Smith analysis of Turkey temperatures using First Differences.

are generally smaller than those in European cities of similar size, which in turn are smaller than those in North American cities. The regression lines for all three continents converge in the vicinity of a population of 1000 people, where the urban—rural temperature difference is approximately $2.2 \pm 0.2°C$, essentially the same as what Oke (1973) had reported two decades earlier.

Smith finds the Australian dropout[20] was mainly among higher-latitude, cooler stations after 1990, with the percentage of city airports increasing to 71%, further enhancing apparent warming. The trend in "island Pacific without Australia and without New Zealand" is dead flat. The Pacific Ocean islands are NOT participating in "global" warming. Changes of thermometers in Australia and New Zealand are the source of any change.

6.3. Turkey

Turkey had one of the densest networks of stations of any country. E.M. Smith calculated anomaly process similar to First Differences. Then dT is the running total of those changes, or the total change, the Delta Temperature, to date. Note the step-up after 1990 cumulative change in temperature and the change per year for Turkey.[21]

His dT method[22] is a variation of First Differences as a means of examining temperature data anomalies independent of actual temperature. dT/year is the "average of the changes of temperature, month now vs. the same month that last had valid data, for each year" (Fig. 2.10).

Despite that apparent warming, the Turkish Met Service finds evidence for cooling. A peer-reviewed paper (Murat Turke, Utku M. Sumer, Gonul Kilic, State Meteorological Service, Department of Research, Climate Change Unit, 06,120 Kalaba-Ankara, Turkey) concludes:

Considering the results of the statistical tests applied to the 71 individual stations data, it could be concluded that annual mean temperatures are generally dominated by a cooling tendency in Turkey.

[20] http://chiefio.wordpress.com/2009/10/23/gistemp-aussy-fair-go-and-far-gone/.

[21] http://chiefio.wordpress.com/2010/03/10/lets-talk-turkey/.

[22] http://chiefio.wordpress.com/2010/02/28/last-delta-t-an-experimental-approach/.

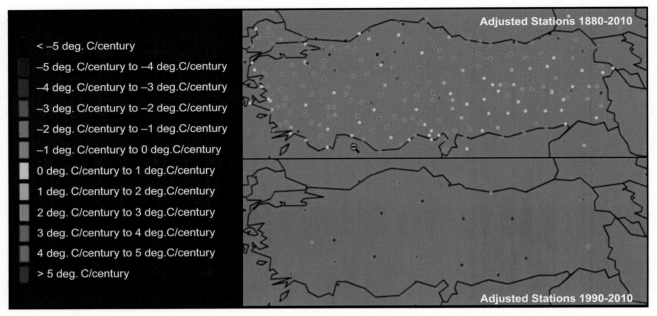

FIGURE 2.11   Verity Jones maps showing station temperature trends for (top) all stations active from 1880–2010 and (bottom) for stations active after 1990. The result is that Turkey is shown to be warming when the data show cooling.

See, in Verity Jones website Digging in the Clay,[23] the dropout of stations from nearly 250 to 39, leaving behind warming stations. Twenty-five of the 39 stations are shown as the other stations did not have complete enough data to determine a reliable trend (less than 10 years without missing months) (Fig. 2.11).

7. INSTRUMENT CHANGES AND SITING

The World Meteorological Organization (WMO), a specialized agency of the United Nations,[24] grew out of the International Meteorological Organization (IMO), which was founded in 1873. Established in 1950, the WMO became the specialized agency of the United Nations (in 1951) for meteorology, weather, climate, operational hydrology, and related geophysical sciences.

According to the WMO's own criteria, followed by NOAA's National Weather Service, temperature sensors should be located on the instrument tower at 1.5 m (5 feet) above the surface of the ground. The tower should be on flat, horizontal ground surrounded by a clear surface, over grass or low vegetation kept less than 4 inches high. The tower should be at least 100 m (110 yards) from tall trees, or artificial heating or reflecting surfaces, such as buildings, concrete surfaces, and parking lots.

Very few stations meet these criteria.

8. ALONG COMES "MODERNIZATION"

The modernization of weather stations in the United States replaced many human observers with instruments that initially had major errors, had "warm biases" (HO-83) (Jones and Young, 1995), or were designed for aviation and were not suitable for precise climate trend detection [Automated Surface Observing Systems (ASOS) and the Automated Weather Observing System (AWOS)]. Also, the new instrumentation was increasingly installed on unsuitable sites that did not meet WMO criteria.

[23] http://diggingintheclay.blogspot.com/2010/03/no-more-cold-turkey.html.

[24] http://www.unsystem.org/en/frames.alphabetic.index.en.htm#w.

FIGURE 2.12 USHCN climate station in Bainbridge, GA, showing the MMTS pole sensor in the foreground near the parking space, building, and air conditioner heat exchanger, with the older Stevenson Screen in the background located in the grassy area (surfacestations.org).

During recent decades there has been a migration away from old instruments read by trained observers. These instruments were generally in shelters that were properly located over grassy surfaces and away from obstacles to ventilation and heat sources. Today we have many more automated sensors (he MMTS) located on poles cabled to the electronic display in the observer's home or office or at airports near the runway where the primary mission is aviation safety.

The installers of the MMTS instruments were often equipped with nothing more than a shovel. They were on a tight schedule and with little budget. They often encountered paved driveways or roads between the old sites and the buildings. They were in many cases forced to settle for installing the instruments close to the buildings, violating the government specifications in this or other ways (Fig. 2.12).

Davey and Pielke (2005) found a majority of stations, including climate stations in eastern Colorado, did not meet WMO requirements for proper siting. They extensively documented poor siting and land-use change issues in numerous peer-reviewed papers, many summarized in the landmark paper "Unresolved issues with the assessment of multidecadal global land surface temperature trends" (2007).[25]

In a volunteer survey project first reported on in Fall et al. (2011), Anthony Watts (Watts, 2009) and his more than 650 volunteers (www.surfacestations.org) found that over 900 of the first 1067 stations surveyed in the 1221-station U.S. climate network did not come close to meeting the specifications. Only about 3% met the ideal specification for siting. They found stations located next to the exhaust fans of air-conditioning units, surrounded by asphalt parking lots and roads, on blistering-hot rooftops, and near sidewalks and buildings that absorb and radiate heat. They found 68 stations located at wastewater treatment plants, where the process of waste digestion causes temperatures to be higher than in surrounding areas. In fact, they found that 90% of the stations fail to meet the National Weather Service's own siting requirements that stations must be 30 m (about 100 feet) or more away from an artificial heating or reflecting source.

The average warm bias for inappropriately sited stations exceeded 1°C using the National Weather Service's own criteria, with which the vast majority of stations did not comply.

[25] http://pielkeclimatesci.files.wordpress.com/2009/10/r-321.pdf.

USHCN weather station at Hopkinsville, KY (Pielke et al., 2006). The station is sited too close to a building, too close to a large area of tarmac, and directly above a barbecue.

USHCN station at Tucson, AZ, in a parking lot on pavement. (Photo by Warren Meyer, courtesy of surfacestations.org.)

FIGURE 2.13   USHCN siting issues at Hopkington, Kentucky and Tucson, Arizona.

Numerous sensors are located at waste treatment plants. An infrared image of the scene shows the output of heat from the waste treatment beds right next to the sensor. (Photos by Anthony Watts, surfacestations.org.)

FIGURE 2.14   One of many waste treatment plants serving as stations in USHCN.

A report from last spring with some of the earlier findings can be found in footnote [26]. Some examples from these sources (Figs. 2.13 and 2.14):

As of October 25, 2009, 1067 of the 1221 stations (87.4%) had been evaluated by the surfacestations.org volunteers and evaluated using the Climate Reference Network (CRN) criteria.[27] 90% were sited in ways that result in errors exceeding 1°C according to the CRN handbook.

[26] http://wattsupwiththat.files.wordpress.com/2009/05/surfacestationsreport_spring09.pdf.

[27] http://www1.ncdc.noaa.gov/pub/data/uscrn/documentation/program/X030FullDocumentD0.pdf.

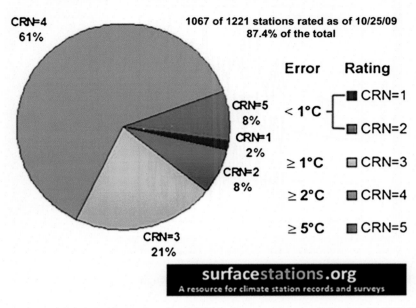

FIGURE 2.15   Surfacestations.org quality rating by stations for 1067 U.S. climate stations as of October 25, 2009. Only 10% meet minimal CRN ranking (CRN 1 or 2).

This siting issue remains true even by the older "100-foot rule" criteria for COOP stations, specified by NOAA[28] for the U.S. Cooperative Observer network, where they specify "The sensor should be at least 100 feet (~30 m) from any paved or concrete surface" (Fig. 2.15).

Dr. Vincent Gray, IPPC reviewer for AR1 through IV published on some issues related to temperature measurements.[29]

In 2008, Joe D'Aleo asked NCDC's Tom Karl about the problems with siting and about the plans for a higher-quality Climate Reference Network (CRN—at that time called NERON). He said he had presented a case for a more complete CRN network to NOAA, but NOAA said it was unnecessary *because they had satellite monitoring*. The CRN was capped at 114 stations and would not provide meaningful trend assessment for about 10 years. NOAA has since reconsidered and now plans to upgrade about 1000 climate stations, but meaningful results will be even further in the future.

In monthly press releases no satellite measurements are ever mentioned, although NOAA claimed that was the future of observations.

9. ADJUSTMENTS NOT MADE, OR MADE BADLY

The Climategate whistle-blower proved what those of us dealing with data for decades already knew. The data were not merely degrading in quantity and quality: they were also being manipulated. This is done by a variety of postmeasurement processing methods and algorithms. The IPCC and the scientists supporting it have worked to remove the pesky Medieval Warm Period, the Little Ice Age, and the period e-mailer Tom Wigley referred to as the "warm 1940s blip." There are no adjustments in NOAA and Hadley data for urban contamination. The adjustments and nonadjustments instead increased the warmth in the recent warm cycle that ended in 2001 and/or inexplicably cooled many locations in the early record, both of which augmented the apparent trend.

[28] http://www.nws.noaa.gov/om/coop/standard.htm.

[29] http://icecap.us/images/uploads/Gray.pdf.

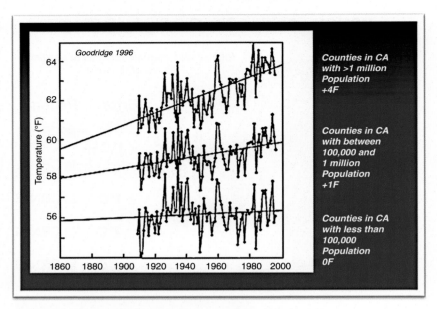

FIGURE 2.16   Jim Goodrich analysis of warming in California counties by population 1910–1995.

10. HEAT FROM POPULATION GROWTH AND LAND-USE CHANGES

10.1. Urban Heat Island

Weather data from cities as collected by meteorological stations are indisputably contaminated by urban heat-island bias and land-use changes (Landsberg, 1981). This contamination has to be removed or adjusted for in order to accurately identify true background climatic changes or trends. In cities, vertical walls, steel, and concrete absorb the sun's heat and are slow to cool at night. More and more of the world is urbanized (population increased from 1.5 billion in 1900 to 6.8 billion in 2010).

The urban heat-island effect occurs not only for big cities but also for towns. Oke (who won the 2008 American Meteorological Society's Helmut Landsberg award for his pioneer work on the effect of urbanization on local microclimates) had a formula for the warming that is tied to population. Oke (1973) found that the urban heat-island (in °C) increases according to the formula:

$$\text{Urban heat} - \text{island warming} = 0.317 \ln P, \text{where} P = \text{population.}$$

Thus a village with a population of 10 has a warm bias of 0.73°C. A village with 100 has a warm bias of 1.46°C, and a town with a population of 1000 people has a warm bias of 2.2°C. A large city with a million people has a warm bias of 4.4°C.

Goodrich (1996) showed the importance of urbanization to temperatures in his study of California counties in 1996. He found for counties with a million or more population, the warming from 1910 to 1995 was 4°F; for counties with 100,000 to 1 million, it was 1°F, and for counties with less than 100,000 there was no change (0.1°F) (Fig. 2.16).

11. U.S. CLIMATE DATA

Compared to the GHCN global database, the USHCN database is more stable (Fig. 2.17).

When first implemented in 1990 as version 1, USHCN employed 1221 stations across the United States. In 1999, NASA's James Hansen published a graph of USHCN v.1 annual mean temperature (Fig. 2.18). Hansen correctly noted:

The US has warmed during the past century, but the warming hardly exceeds year-to-year variability. Indeed, in the US the warmest decade was the 1930s and the warmest year was 1934.

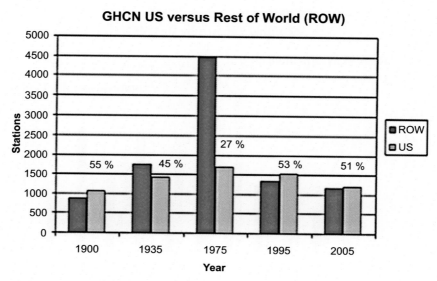

FIGURE 2.17   Comparison of number of GHCN temperature stations in the United States versus the rest of the world (ROW). *From http://www. appinsys.com/GlobalWarming/ClimateData.htm.*

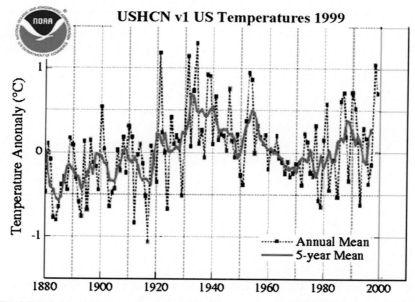

FIGURE 2.18   NOAA NCDC USHCN version 1 annual U.S. temperatures as of 1999.

USHCN was generally accepted as the world's best database of temperatures. The stations were the most continuous and stable and had adjustments made for time of observation, urbanization, known station moves, or land-use changes around sites, as well as instrumentation changes.

Note how well the original USHCN agreed with the state record high temperatures.

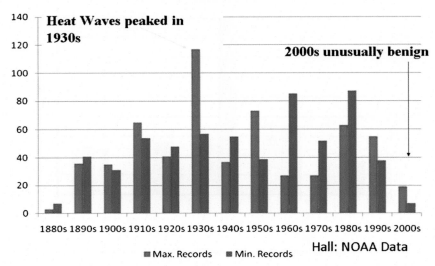

FIGURE 2.19   United States all-time monthly record lows and highs by decade. *Compiled by Hall from NOAA NCDC data.*

12. U.S. STATE HEAT RECORDS SUGGEST RECENT DECADES ARE NOT THE WARMEST

The 1930s were, by far, the hottest period for the time-frame. In absolute terms, the 1930s had a much higher frequency of maximum temperature extremes than the 1990s or 2000s or the combination of the last two decades. This was shown by Bruce Hall and Dr. Richard Keen,[30] also covering Canada (Fig. 2.19).

NCDC's Karl et al. (1988) employed an urban adjustment scheme for the first USHCN database (released in 1990). He noted that the national climate network formerly consisted of predominantly rural or small towns with populations below 25,000 (as of 1980 census) and yet that an urban heat-island effect was clearly evident.

Tom Karl et al.'s adjustments were smaller than Oke had found (0.22°C annually on a town of 10,000, 1.81°C on a city of 1 million, and 3.73°C for a city of 5 million). Karl observed that in smaller towns and rural areas the net urban heat-island contamination was relatively small, but that significant anomalies showed up in rapidly growing population centers.

13. MAJOR CHANGES TO USHCN IN 2007

NOAA had to constantly explain why their global data sets which had no such adjustment were showing warming, and the United States not so much. NOAA began reducing the UHI around 2000 (noticed by state climatologists and seen in this analysis of New York City's Central Park data at http://icecap.us/index.php/go/new-and-cool/central_park_temperatures_still_a_mystery/), and then in USHCN version 2 released for the U.S. stations in 2009, the urban heat-island adjustment was eliminated, which resulted in an increase of 0.3°F in warming trend since the 1930s. See the animating GIF at http://stevengoddard.files.wordpress.com/2010/12/1998uschanges.gif.

In 2007 the NCDC, in its version 2 of USHCN, inexplicably removed the Karl urban heat-island adjustment and substituted a change-point algorithm that looks for sudden shifts (discontinuities). This is best suited for finding site moves or local land-use changes (like paving a road or building next to sensors or shelters), but not the slow ramp-up of temperature characteristic of a growing town or city (Fig. 2.20).

David Easterling, Chief of the Scientific Services Division at NOAA, in one of the NASA FOIA emails, noted: "One other fly in the ointment, we have a new adjustment scheme for USHCN (V2) that appears to **adjust out** some, if not most, of the "local" trend that includes land-use change and urban warming."

[30] http://icecap.us/index.php/go/new-and-cool/more_critique_of_ncar_cherry_picking_temperature_record_study/.

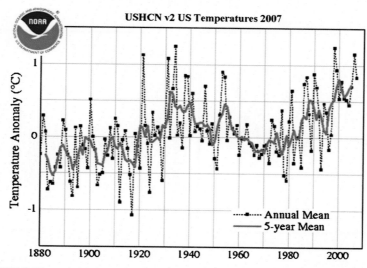

FIGURE 2.20   NOAA NCDC USHCN version 2 annual mean temperatures as of 2007.

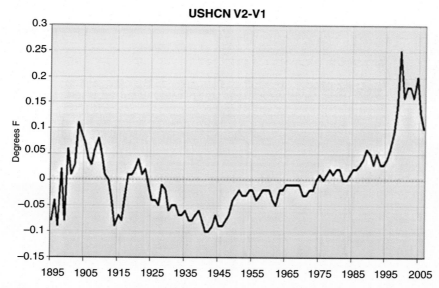

FIGURE 2.21   NOAA NCDC USHCN version 2 minus version 1 annual mean temperatures.

The difference between the old and new is shown here. Note the significant post-1995 warming and mid-20th-century cooling owing to de-urbanization of the database (Fig. 2.21).

The change can be seen clearly in this animation[31] and in blink charts for Wisconsin[32] and Illinois.[33] Here are two example stations with USHCN version 1 and version 2 superimposed (thanks to Mike McMillan). Notice the clear tendency to cool off the early record and leave the current levels near recently reported levels or to increase them. The net result is either reduced cooling or enhanced warming not found in the raw data (Fig. 2.22).

The new algorithms are supposed to correct for urbanization and changes in siting and instrumentation by recognizing sudden shifts in the temperatures (Fig. 2.23). It should catch the kind of change shown above in Tahoe City,

[31] http://climate-skeptic.typepad.com/.a/6a00e54eeb9dc18834010535ef5d49970b-pi.

[32] http://www.rockyhigh66.org/stuff/USHCN_revisions_wisconsin.htm.

[33] http://www.rockyhigh66.org/stuff/USHCN_revisions.htm.

USHCN v1 Versus v2

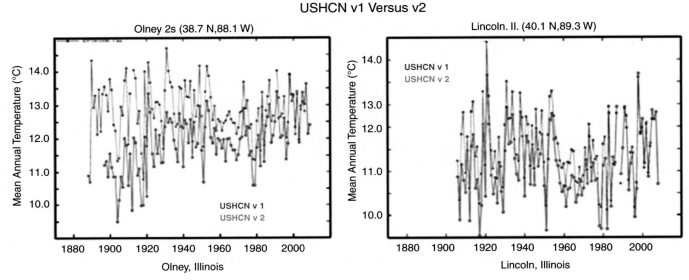

FIGURE 2.22   NOAA USJCN version 1 versus version 2 for Olney and Lincoln, Illinois.

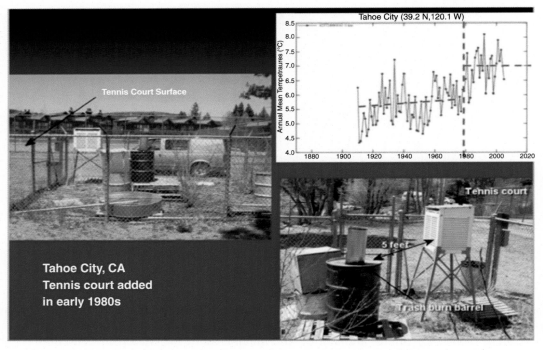

FIGURE 2.23   Tahoe City, California data and photos. *Courtesy of Anthony Watts, surfacestations.org.*

California (Fig. 2.23). It is unlikely to catch the slow warming associated with the growth of cities and towns over many years, as in Sacramento, California, in Fig. 2.24.

In a conversation during Anthony Watts' invited presentation about the surface stations projects to NCDC, on April 24, 2008, he was briefed on USHCN2's algorithms and how they operated by Matt Menne, lead author of the USHCN2 project. While Mr. Watts noted improvements in the algorithm can catch some previously undetected events like undocumented station moves, he also noted that the USHCN2 algorithm had no provision for long-term filtering of signals that can be induced by gradual local urbanization, or by long-term changes in the siting environment, such as weathering/coloring of shelters or wind blocking due to growth of shrubbery/trees.

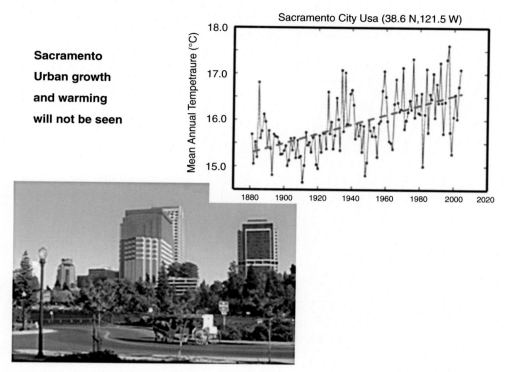

Sacramento Urban growth and warming will not be seen

FIGURE 2.24   Sacramento, California data and photos. *Courtesy of Anthony Watts, surfacestations.org.*

When Mr. Menne was asked by Mr. Watts if this lack of detection of such long-term changes was in fact a weakness of the USHCN algorithm, he replied "Yes, that is correct." Essentially USHCN2 is a short-period filter only, and cannot account for long-term changes to the temperature record, such as UHI, making such signals indistinguishable from the climate-change signal that is sought.

See some other examples of urban versus nearby rural.[34] Doug Hoyt, who worked at NOAA, NCAR, Sacramento Peak Observatory, the World Radiation Center, Research and Data Systems, and Raytheon, where he was a Senior Scientist, did this analysis[35] of the urban heat island. Read beyond the references for interesting further thoughts (Fig. 2.25).

Even before the version 2, Balling and Idso (2002) [36] found that the adjustments being made to the raw USHCN temperature data were "producing a statistically significant, but spurious, warming trend" that "approximates the widely-publicized 0.50°C increase in global temperatures over the past century." There was actually a linear trend of progressive cooling of older dates between 1930 and 1995. "It would thus appear that in this particular case of 'adjustments,' the *cure* was much worse than the *disease*. In fact, it would appear that the cure may actually *be* the disease."

It should be noted that even with the changes to the USHCN, the correlations with CO_2 are intermittent, with just 44 years warming while CO_2 increased and 62 years cooling even as CO_2 rose, not a convincing story for greenhouse CO_2 climate dominance at least with the U.S. data, even with all its warts, generally accepted as the most complete and stable data sets in the world (Fig. 2.26).

[34] http://www.appinsys.com/GlobalWarming/GW_Part3_UrbanHeat.htm.

[35] http://www.warwickhughes.com/hoyt/uhi.htm.

[36] http://www.co2science.org/articles/V12/N50/C1.php.

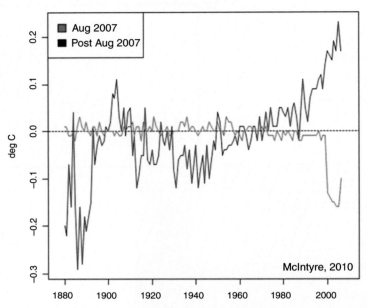

FIGURE 2.25   Adjustments to U.S. data August 2007 and then post-August 2007. *McIntyre, 2010.*

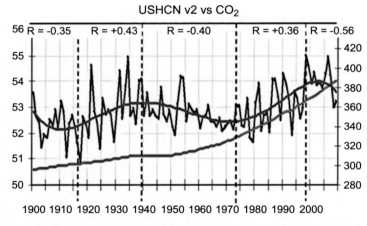

FIGURE 2.26   USHCN version 2 annual temperatures versus ERSL CO_2 annual concentrations ppm. Pearson coefficient shown for the warming and cooling intervals.

14. HADLEY AND NOAA

No real urbanization adjustment is made for either NOAA or CRU global data. Jones et al. (1990: Hadley/CRU) concluded that urban heat-island bias in gridded data could be capped at 0.05°C/century. Jones used data by Wang, which Keenan[37] has shown was fabricated. Peterson (2006) agreed with the conclusions of Jones, Easterling et al. (1997) that urban effects on twentieth century globally and hemispherically averaged land air temperature time-series do not exceed about 0.05°C from 1900 to 1990.

[37] http://www.informath.org/WCWF07a.pdf.

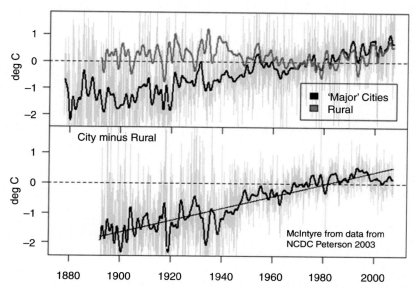

FIGURE 2.27   GHCN version 2 annual temperatures for stations identified by Peterson in 2003, separated by rural and major cities with the city minus rural. *McIntyre, 2007.*

Peterson (2003) and Parker (2004) argue urban adjustment is not really necessary. Yet Oke (1973) showed a town of 1000 could produce a 2.2°C (3.4°F warming). The UK Met Office (UKMO) has said[38] future heat waves could be especially deadly in urban areas, where the temperatures could be 9°C or more above today's, according to the Met Office's Vicky Pope. NASA summer land surface temperature of cities in the Northeast were an average of 7–9°C (13–16°F) warmer than surrounding rural areas over a three year period, new NASA research shows. It appears, the warmers want to have it both ways. They argue that the urban heat-island effect is insignificant, but also argue future heat waves will be most severe in the urban areas. This is especially incongruous given that greenhouse theory has the warming greatest in winters and at night.

The most recent exposition of CRU methodology is Brohan et al. (2006), which included an allowance of 0.1°C/century for urban heat-island effects in the uncertainty but did not describe any adjustment to the reported average temperature. To make an urbanization assessment for all the stations used in the HadCRUT data set would require suitable metadata (population, siting, location, instrumentation, etc.) for each station for the whole period since 1850. No such complete metadata are available.

The homepage for the NOAA temperature index[39] cites Smith and Reynolds (2005) as authority. Smith and Reynolds in turn state that they use the same procedure as CRU: ie, they make an allowance in the error-bars but do not **correct** the temperature index itself. The population of the world went from 1.5 to 6.7 billion in the twentieth century, yet NOAA and CRU ignore population growth in the database with only a 0.05–0.1°C uncertainty adjustment.

Steve McIntyre challenged Peterson (2003), who had said, "Contrary to generally accepted wisdom, no statistically significant impact of urbanization could be found in annual temperatures,"[40] by showing that the difference between urban and rural temperatures for Peterson's station set was 0.7°C and between temperatures in large cities and rural areas 2°C. He has done the same for Parker (2004) (Fig. 2.27).[41]

Runnalls and Oke (2006) concluded that:

Gradual changes in the immediate environment over time, such as vegetation growth or encroachment by built features such as paths, roads, runways, fences, parking lots, and buildings into the vicinity of the instrument site, typically lead to trends in the series...Distinct

[38] http://icecap.us/index.php/go/joes-blog/cities_to_sizzle_as_islands_of_heat/.

[39] http://www.ncdc.noaa.gov/oa/climate/research/anomalies/anomalies.html.

[40] http://climateaudit.org/2007/08/04/1859/.

[41] http://climateaudit.org/2007/06/14/parker-2006-an-urban-myth/.

régime transitions can be caused by seemingly minor instrument relocations (such as from one side of the airport to another or even within the same instrument enclosure) or due to vegetation clearance. This contradicts the view that only substantial station moves involving significant changes in elevation and/or exposure are detectable in temperature data.

Numerous other peer-reviewed papers and other studies have found that the lack of adequate urban heat-island and local land-use change adjustments could account for up to half of all apparent warming in the terrestrial temperature record since 1900.

Siberia is one of the areas of greatest apparent warming in the record. Besides station dropout and a 10-fold increase in missing monthly data, numerous problems exist with prior temperatures in the Soviet era. City and town temperatures determined allocations for funds and fuel from the Supreme Soviet, so it is believed that cold temperatures were exaggerated in the past. This exaggeration in turn led to an apparent warming when more honest measurements began to be made. Anthony Watts has found that in many Russian towns and cities uninsulated heating pipes[42] are in the open. Any sensors near these pipes would be affected. The pipes also contribute more waste heat to the city over a wide area.

The physical discomfort and danger to observers in extreme environments led to some estimations or fabrications being made in place of real observations, especially in the brutal Siberian winter. See this report.[43] This was said to be true also in Canada along the DEW Line, where radars were set up to detect incoming Soviet bombers during the Cold War.

McKitrick and Michaels (2004, 2007) gathered weather station records from 93 countries and regressed the spatial pattern of trends on a matrix of local climatic variables and socioeconomic indicators such as income, education, and energy use. Some of the nonclimatic variables yielded significant coefficients, indicating a significant contamination of the temperature record by nonclimatic influences, including poor data quality.

The two authors repeated the analysis on the IPCC gridded data covering the same locations. They found that approximately the same coefficients emerged. Though the discrepancies were smaller, many individual indicators remained significant. On this basis they were able to rule out the hypothesis that there are no significant nonclimatic biases in the data. Both de Laat and Maurellis and McKitrick and Michaels concluded that nonclimatic influences add up to a substantial warming bias in measured mean global surface temperature trends.

Ren et al. (2007), in the abstract of a paper on the urban heat-island effect in China, published in *Geophysical Research Letters*, noted that "annual and seasonal urbanization-induced warming for the two periods at Beijing and Wuhan stations is also generally significant, with the annual urban warming accounting for about 65—80% of the overall warming in 1961—2000 and about 40—61% of the overall warming in 1981—2000."

This result, along with the previous mentioned research results, indicates a need to pay more attention to the urbanization-induced bias that appears to exist in the current surface air temperature records.

Numerous studies show the effects of urban anthropogenic warming on local and regional temperatures in many diverse, even remote, locations. Jáuregui (2005) discussed the UHI in Mexico, Torok et al. (2001) in southeast Australian cities. Block et al. (2004) showed effects across central Europe. Zhou et al. (2004), Jones et al. (2008), He et al. (2006) and Li et al. (2004) across China, Velazquez-Lozada et al. (2006) across San Juan, Puerto Rico, and Hinkel et al. (2003) even in the village of Barrow, Alaska. In all cases, the warming was greatest at night, and in higher latitudes, chiefly in winter.

Kalnay and Cai (2003) found regional differences in U.S. data but overall very little change, and if anything a slight decrease in daily maximum temperatures for two separate 20-year periods (1980—1999 and 1960—1979), and a slight increase in nighttime readings. They found these changes consistent with both urbanization and land-use changes from irrigation and agriculture.

Christy et al. (2006) showed that temperature trends in California's Central Valley had significant nocturnal warming and daytime cooling over the period of record. The conclusion is that, as a result of increases in irrigated land, daytime temperatures are suppressed owing to evaporative cooling and nighttime temperatures are warmed in part owing to increased heat capacity from water in soils and vegetation. Mahmood et al. (2006) also found similar results for irrigated and nonirrigated areas of the Northern Great Plains.

Two Dutch meteorologists, Jos de Laat and Ahilleas Maurellis, showed in 2006 that climate models predict there should be no correlation between the spatial pattern of warming in climate data and the spatial pattern of industrial

---

[42] http://wattsupwiththat.com/2008/11/15/giss-noaa-ghcn-and-the-odd-russian-temperatureanomaly-its-all-pipes.

[43] http://wattsupwiththat.com/2008/07/17/fabricating-temperatures-on-the-dew-line/.

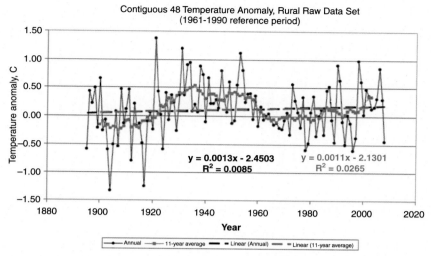

FIGURE 2.28 Edward long analysis of rural raw stations for the lower 48 states, USHCN version 2. Note the very small trend (0.12°C/century) in this data set and the significant peak in the 1930s.

development. But they found that this correlation does exist and is statistically significant. They also concluded it adds a large upward bias to the measured global warming trend.

Ross McKitrick and Patrick Michaels, in 2007, showed a strong correlation between urbanization indicators and the "urban adjusted" temperatures and that the adjustments are inadequate. Their conclusion: "Fully correcting the surface temperature data for non-climatic effects reduce the estimated 1980−2002 global average temperature trend over land by about half."

As Pielke et al. (2007a,b) also notes:

Changnon and Kunkel (2006) examined discontinuities in the weather records for Urbana, Illinois, a site with exceptional metadata and concurrent records when important changes occurred. They identified a cooling of 0.17°C caused by a non-standard height shelter of 3 m from 1898 to 1948. After that there was a gradual warming of 0.9°C as the University of Illinois campus grew around the site from 1900 to 1983. This was followed by an immediate 0.8°C cooling when the site moved 2.2 km to a more rural setting in 1984. A 0.3°C cooling took place with a shift in 1988 to Maximum-Minimum Temperature systems, which now represent over 60% of all USHCN stations. The experience at the Urbana site reflects the kind of subtle changes described by Runnalls and Oke (2006) and underscores the challenge of making adjustments to a gradually changing site.

A 2008 paper[44] by Hadley's Jones et al. has shown a considerable contamination in China, amounting to 1°C/century. This is an order of magnitude greater than the amount previously assumed (0.05−0.1°C/century uncertainty).

In a 2009 article,[45] Brian Stone of Georgia Tech wrote:

Across the US as a whole, approximately 50 percent of the warming that has occurred since 1950 is due to land use changes (usually in the form of clearing forest for crops or cities) rather than to the emission of greenhouse gases. Most large US cities, including Atlanta, are warming at more than twice the rate of the planet as a whole. This is a rate that is mostly attributable to land use change.

In a paper posted on SPPI,[46] Dr. Edward Long summarized his findings as follows: Both raw and adjusted data from the NCDC have been examined for a selected contiguous U.S. set of rural and urban stations, 48 one per state. The raw data provide 0.13 and 0.79°C/century temperature increase for the rural and urban environments (Figs. 2.28 and 2.29).

[44] http://www.warwickhughes.com/blog/?p=204.

[45] http://www.gatech.edu/newsroom/release.html?nid=47354.

[46] http://scienceandpublicpolicy.org/images/stories/papers/originals/Rate_of_Temp_Change_Raw_.and_ Adjusted_NCDC_ Data.pdf.

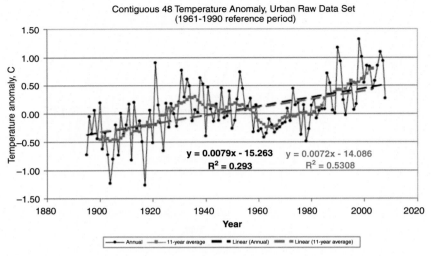

FIGURE 2.29　Edward Long urban annual temperatures and trend from USHCN version 2 annual temperatures for the lower 48 states Note the trend of 0.79°C for this data set with the 1930 peak but with the second recent peak higher.

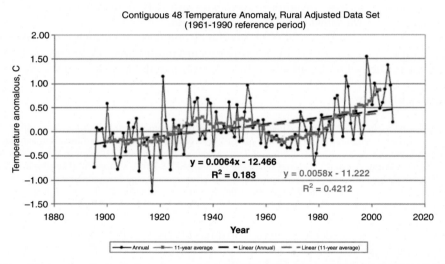

FIGURE 2.30　Edward Long plot of adjusted rural annual temperatures. Note the trend has increased to 0.64°C for this data set.

One would expect the urban would be adjusted to match the uncontaminated rural data. Instead the rural is adjusted to look more like the urban with the warming since 1895 increased over half a degree from just 0.13°C to 0.64°C while the urban trend decreased an insignificant 0.02°C (Fig. 2.30).

The adjusted data provide 0.64 and 0.77°C/century, respectively. Comparison of the adjusted data for the rural set to that of the raw data shows a systematic treatment that causes the rural adjusted set's temperature rate of increase to be five-fold more than that of the raw data. This suggests the consequence of the NCDC's protocol for adjusting the data is to cause historical data to take on the time-line characteristics of urban data. The consequence, intended or not, is to report a false rate of temperature increase for the contiguous United States.

15. FINAL ADJUSTMENTS: HOMOGENIZATION

Dr. William Briggs, in a five part series on the NOAA/NASA process of homogenization on his blog,[47] noted the following:

> At a loosely determined geographical spot over time, the data instrumentation might have changed, the locations of instruments could be different, there could be more than one source of data, or there could be other changes. The main point is that there are lots of pieces of data that some desire to stitch together to make one whole. Why? I mean that seriously. Why stitch the data together when it is perfectly useful if it is kept separate? By stitching, you introduce error, and if you aren't careful to carry that error forward, the end result will be that you are far too certain of yourself. And that condition— unwarranted certainty— is where we find ourselves today.

It has been said by NCDC in Menne et al. "On the reliability of the US surface temperature record" (in press) and in the June 2009[48] "Talking Points: related to Is the US. Surface Temperature Record Reliable?" that station siting errors do not matter. However, the way NCDC conducted the analysis gives a false impression because of the homogenization process used.

Here's a way to visualize the homogenization process. Think of it like measuring water pollution. Fig. 2.31 gives a simple visual table of CRN station quality ratings and what they might look like as water pollution turbidity levels, rated as 1 to 5 from best to worst turbidity (Fig. 2.31).

In homogenization, the data are weighted against the nearby neighbors within a radius. And so a station might start out as a "1" data-wise, might end up getting polluted with the data of nearby stations and end up as a new value, say weighted at "2.5". Even single stations can affect many other stations in the GISS and NOAA data homogenization methods carried out on U.S. surface temperature data (Fig. 2.32).[49,50]

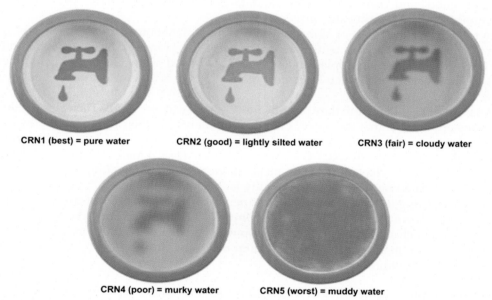

FIGURE 2.31 Simple visual table of CRN station quality ratings and what they might look like as water pollution turbidity levels, rated as one to five from best to worst turbidity.

In the map in Fig. 2.32, applying a homogenization smoothing, weighting stations by distance nearby the stations with question marks, what would you imagine the values (of turbidity) of them would be? And, how close would these two values be for the East Coast station in question and the West Coast station in question? Each would be closer to a smoothed center average value based on the neighboring stations.

---

[47] http://wmbriggs.com/blog/?p=1459.

[48] www.ncdc.noaa.gov/oa/about/response-v2.pdf.

[49] http://wattsupwiththat.com/2009/07/20/and-now-the-most-influential-station-in-the-giss-record-is/.

[50] http://wattsupwiththat.com/2008/09/23/adjusting-pristine-data/.

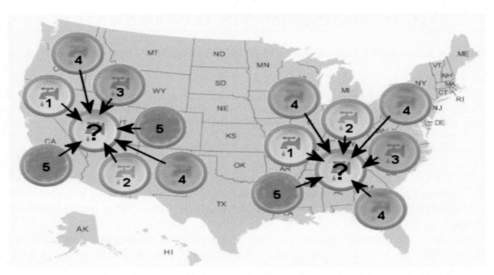

FIGURE 2.32   In homogenization the data are weighted against the nearby neighbors within a radius. And so a station might start out as a "1" data-wise, and might end up getting polluted with the data of nearby stations and end up as a new value, say weighted at "2.5."

Essentially, NCDC is comparing *homogenized data* to *homogenized data*, and thus there would *not* likely be any large difference between "good" and "bad" stations in that data. All the differences have been smoothed out by homogenization (pollution) from neighboring stations!

The best way to compare the effect of siting between groups of stations is to use the "raw" data, before it has passed through the multitude of adjustments that NCDC performs. However, NCDC is apparently using homogenized data. So instead of comparing apples and oranges (poor-sited vs. well-sited stations) they essentially just compare apples (Granny Smith vs. Golden Delicious) of which there is little visual difference beyond a slight color change.

They cite 60 years of data in the graph they present, ignoring the warmer 1930s. They also use an early and incomplete surfacestations.org data set that was never intended for analysis in their rush to rebut the issues raised. However, our survey most certainly cannot account for changes to the station locations or station siting quality any further back than about 30 years. By NCDC's own admission (see "Quality Control of pre-1948 Cooperative Observer Network Data"[51]) they have little or no metadata posted on station siting much further back than about 1948 on their MMS *meta*-database. Clearly, siting quality is dynamic over time.

The other issue about siting that NCDC does not address is that it is a significant contributor to extreme temperature records. By NOAA's own admission in "PCU6—Unit No. 2 Factors Affecting the Accuracy and Continuity of Climate Observations,"[52] such siting issues as the rooftop weather station in Baltimore contributed many erroneous high temperature records, so many in fact that the station had to be closed.

NOAA wrote about the Baltimore station:

> A combination of the rooftop and downtown urban siting explain the regular occurrence of extremely warm temperatures. Compared to nearby ground-level instruments and nearby airports and surrounding COOPs, it is clear that a strong warm bias exists, partially because of the rooftop location. Maximum and minimum temperatures are elevated, especially in the summer. The number of 80 plus minimum temperatures during the one-year of data overlap was 13 on the roof and zero at three surrounding LCD airports, the close by ground-based inner Baltimore harbor site, and all 10 COOPs in the same NCDC climate zone. Eighty-degree minimum are luckily, an extremely rare occurrence in the mid-Atlantic region at standard ground-based stations, urban or otherwise.

Clearly, siting does matter, and siting errors have contributed to the temperature records of the United States, and likely the world GHCN network. Catching such issues isn't always as easy as NOAA demonstrated in Baltimore (Fig. 2.33).

[51] http://ams.confex.com/ams/pdfpapers/68379.pdf.

[52] http://www.weather.gov/om/csd/pds/PCU6/IC6_2/tutorial1/PCU6-Unit2.pdf.

*Baltimore
USHCN station
circa 1990's
(photo courtesy NOAA)*

FIGURE 2.33   Baltimore USHCN rooftop station around 1999.

There is even some evidence that the change-point algorithm does not catch some site changes it should catch and that homogenization doesn't help. Take, for example, Lampasas, Texas, as identified by Anthony Watts (Fig. 2.34). The site at Lampasas moved close to a building and a street from a more appropriate grassy site after 2001. Note even with the GISS "homogeneity" adjustment (red) applied to the NOAA adjusted data, this artificial warming remains although the old data (blue) is cooled to accentuate warming even further (Fig. 2.35).

The net result is to make the recent warm cycle maximum more important relative to the earlier maximum in the 1930s, and note the sudden warm blip after the station move remains.

Other examples (and there are many, many such examples) include (Fig. 2.36):

Adjustments to the raw data are responsible for the New Zealand warming trend shown by the National Institute of Water and Atmospheric Research (NIWA). New Zealand Climate Science Coalition (NZCSC) publicly called on NIWA http://icecap.us/index.php/go/political-climate/high_court_asked_to_invalidate_niwas_official_nz_temperature_record1/to admit no valid statistical justification for its claims of a 0.91°C rise in New Zealand's average temperature last century for the Seven Station Series (7SS) (Fig. 2.37).

For the globe, the final adjusted data set is then used to populate a global grid, interpolating up to 1200 km (745 miles) to grid boxes that had become now vacant by the elimination of stations.

FIGURE 2.34   Lampasas, Texas relocated station. *Photograph by Julie K. Stacy, surfacestations.org.*

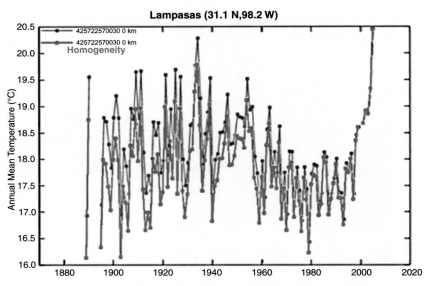

FIGURE 2.35   Lampasas, Texas relocated station before (*blue*) and after (*red*) homogenization. Note the cooling of the old data but no correction for the station move in 2001.

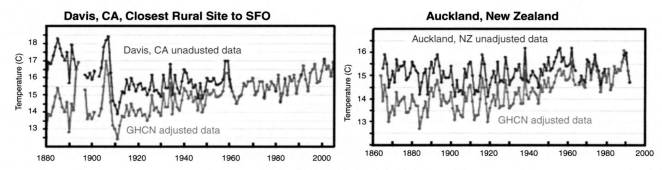

FIGURE 2.36   GHCN raw versus adjusted for Davis, California and Auckland, New Zealand.

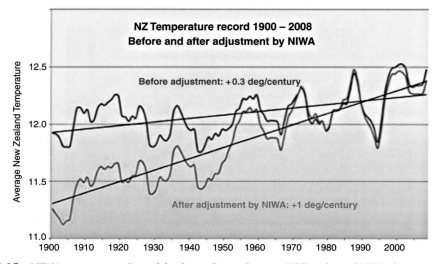

FIGURE 2.37   NIWA raw versus adjusted for Seven Sisters Stations (7SS). Adjusted NIWA becomes GHCN raw.

The data are then used for estimating the global average temperature and anomaly and for initializing or validating climate models.

After the Menne et al. (2010) paper, NCDC recognized their position on station siting was untenable and requested $100 million to upgrade the siting of 1000 climate stations in the 1220 station network.

The NASA/NOAA homogenization process has been shown to significantly alter the trends in many stations where the siting and rural nature suggest the data are reliable. In fact, adjustments account for virtually all the trends in the data (multi-author paper accepted 2011).

16. PROBLEMS WITH SEA SURFACE TEMPERATURE MEASUREMENTS

The world is 71% ocean. The Hadley Centre only trusts data from British merchant ships, mainly plying Northern Hemisphere routes. Hadley has virtually no data from the Southern Hemisphere's oceans, which cover four-fifths of the hemisphere's surface. NOAA and NASA use ship data reconstructions (Smith and Reynolds, 2004; Smith et al., 2008). The gradual change from taking water in canvas buckets to taking it from engine intakes introduces uncertainties in temperature measurement. Different sampling levels will make results slightly different. How to adjust for this introduced difference and get a reliable data set has yet to be resolved adequately, especially since the transition occurred over many decades. A chart, taken from Kent et al. (2007), shows how methods of ocean-temperature sampling have changed over the past 40 years (Fig. 2.38).

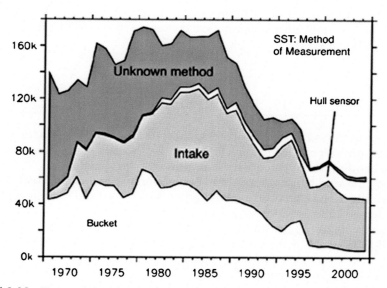

FIGURE 2.38   Kent et al. (2007) depicted methods of measurement for sea surface temperatures.

We have reanalysis data based on reconstructions from ships, from buoys (which also have problems with changing methodology), and, in recent decades, from satellites. The oceans offer some opportunity for mischief, as the emails released by the Climategate whistle-blower showed clearly.

This report[53] analyzed climate model (Barnet et al., 2001) forecasts of ocean temperatures from 1955 to 2000 vs. actual changes. It found models greatly overstated the warming, especially at the surface, where the actual change was just about 0.1°C over that period.

There is another data set that may better resolve this discrepancy with time, the ARGO buoys. The Argo network,[54] which may eventually overcome many of the prior problems, became operational in mid-2003.

Before Argo, starting in the early 1960s, ocean temperatures were measured with bathythermographs (XBTs). They are expendable probes fired into the water by a gun, that transmit data back along a thin wire. They were nearly

[53] http://www.worldclimatereport.com/archive/previous_issues/vol6/v6n16/feature1.htm#http://www.world climatereport.com/archive/previous_issues/vol6/v6n16/feature1.htm.

[54] http://www.argo.ucsd.edu/About_Argo.html.

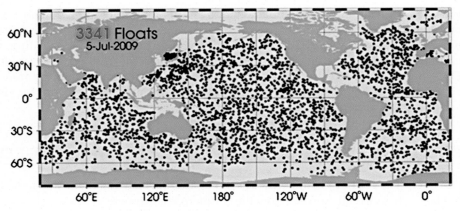

FIGURE 2.39   ARGO float network as of July 2009.

all launched from ships along the main commercial shipping lanes, so geographical coverage of the world's oceans was poor—for example, the huge southern oceans were not monitored. XBTs do not go as deep as Argo floats, and their data are much *less* accurate (Met Office,[55] Argo[56]) (Fig. 2.39).

Early results showed a cooling (Loehle, 2009), but some issues may exist with the quality control of the early measurements, and the strong El Niño in 2009 to 2010 produced a brief pop-up, now reversing. We believe in the future this data set may give us the best indication of ocean heat content, which could be the most robust and reliable indication of climate trends (Pielke, 2008[57]).

17. LONG-TERM TRENDS

Just as the Medieval Warm Period was an obstacle to those trying to suggest that today's temperature is exceptional, and the UN and its supporters tried to abolish it with the "hockey-stick" graph, the warmer temperatures in the 1930 and 1940s were another inconvenient fact that needed to be "fixed."

In each of the databases, the land temperatures from that period were simply adjusted downward, making it look as though the rate of warming in the 20th century was higher than it was, and making it look as though today's temperatures were unprecedented in at least 150 years (Figs. 2.40–2.44).

Wigley[58] even went so far as to suggest that sea surface temperatures for the period should likewise be "corrected" downward by 0.15°C, making the 20th-century warming trend look greater but still plausible. This is obvious data doctoring.

In the Climategate emails, Wigley also noted[59]:

> Land warming since 1980 has been twice the ocean warming—and skeptics might claim that this proves that urban warming is real and important.

NOAA, then, is squarely in the frame. First, the unexplained major station dropout with a bias towards warmth in remaining stations. Next, the removal of the urbanization adjustment and lack of oversight and quality control in the siting of new instrumentation in the United States database degrades what once was the world's best data set, USHCNv1. Then, ignoring a large body of peer review research demonstrating the importance of urbanization and land-use changes, NOAA chooses not to include any urban adjustment for the global data set, GHCN.

---

[55] http://www.metoffice.gov.uk/weather/marine/observations/gathering_data/argo.html.

[56] http://wwlw.argo.ucsd.edu/Novel_argo.html.

[57] http://pielkeclimatesci.files.wordpress.com/2009/10/r-334.pdf.

[58] http://www.eastangliaemails.com/emails.php?eid=1016&filename=1254108338.txt.

[59] http://www.eastangliaemails.com/emails.php?eid=1067&filename=1257546975.txt.

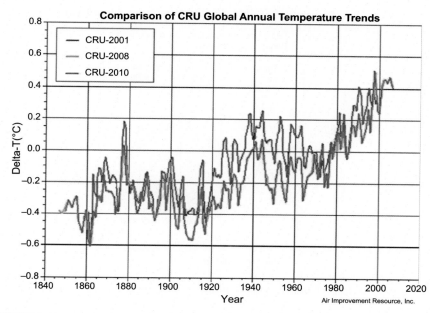

FIGURE 2.40   Comparing Hadley CRU 2001 versus 2008 and 2010 annual mean temperatures.

GISS Temperatures Change Yearly				
	2006	*2007*	*2008*	*2009*
1996	−0.18	−0.16	−0.16	−0.06
1997	0.05	0.04	0.04	0.14
1998	1.24	1.24	1.24	1.31
1999	0.94	0.94	0.94	1.07
2000	0.65	0.54	0.54	0.69
2001	0.89	0.78	0.78	0.92
2002	0.67	0.55	0.55	0.69
2003	0.65	0.53	0.53	0.69
2004	0.54	0.46	0.46	0.61
2005	0.99	0.71	0.71	0.92
2006	*	1.15	1.15	1.31
2007	*	*	0.84	0.88
2008	*	*	*	0.12

FIGURE 2.41   Comparing NASA GISS values for recent years as reported in the year shown. Note the shift down in 2007 after correction for the millennium bug identified by McIntyre and then the shift up again in 2009.

As shown, these and other changes that have been made alter the historical record and mask cyclical changes that could be readily explained by natural factors like multidecadal ocean and solar changes.[60]

The CRU data have seen changes even in the last decade with a cooling of the early and middle parts of the twentieth century and dramatic post-1990 warming when most of the issues discussed emerged (Fig. 2.40). The green is the 2001 global temperature plot and the red that in 2010 (with data through 2009).

Is NASA in the clear? No. It works with the same GHCN/USHCN base data (plus the SCAR data from Antarctica). To its credit, as we have shown, its U.S. database includes an urban adjustment that is reasonable, but as Steve McIntyre showed,[61] NASA uses population data and adjusted GHCN temperature records for cities

[60] http://icecap.us/images/uploads/ATMOSPHERIC_CIRCULATION.doc.

[61] http://icecap.us/images/uploads/US_AND_GLOBAL_TEMP_ISSUES.pdf.

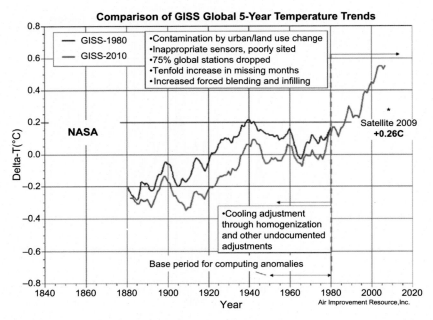

FIGURE 2.42   Comparing NASA GISS global values from 1980 to 2010. Note the string cooling prior to 1980. Warming post 1980 was due to many issues unaccounted for. Compare to UAH value for 2009.

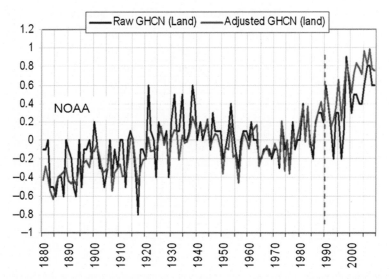

FIGURE 2.43   Comparing NOAA GHCN version 2 raw versus adjusted data. A similar cooling in the early record and recent warming is clearly shown. Raw data show little warming from peak in 1930s to 1940s to 1990s and 2000s.

in a warming direction as often as they do in a cooling direction. This we have seen is due to very poor metadata from GHCN, which GISS uses to match with satellite night light to define a station as urban, suburban or rural.

And their homogenization process and other nondocumented final adjustments result in an increase in apparent warming, often by cooling the early record, as can be seen in several case studies that follow.

NASA also seems to constantly rehash the surface data. John Goetz[62] showed that 20% of the historical record was **modified 16 times** in the 2½ years ending in 2007. 1998 and 1934 ping-pong regularly between first and second warmest year as the fiddling with old data continues.

---

[62] http://wattsupwiththat.com/2008/04/08/rewriting-history-time-and-time-again/.

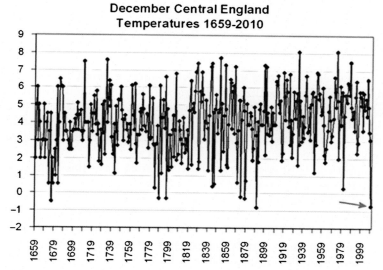

FIGURE 2.44   Central England Temperatures (CETs) for December from 1659 to 2010 s only 1890.

In 2007, NASA adjusted post-2000 data,[63] when Steve McIntyre found a bug in the USHCN data, down by 0.12—0.15°C. Note how the data were adjusted up again in 2009 (USHCN V2) (see Fig. 2.41).

Earlier version of NASA data was extracted from an earlier paper by Hansen in 1980 and is compared in Fig. 2.42. In 1987 (green on the graph in Fig. 2.42), the GISS temperatures were modified down in the middle part of the century from the 1980 version (blue), which enhanced the apparent warming in time for Dr. Hansen's testimony in front of Congress.

Cooling before 1980 is dramatic. Warming after 1990 was due to the myriad of issues with the data in this period, as we have identified above.

E-mail messages obtained by a Freedom of Information Act request reveal that NASA concluded that its own climate findings were inferior to those maintained by both the University of East Anglia's CRU —the scandalized source of the leaked Climategate e-mails—and the National Oceanic and Atmospheric Administration's National Climatic Data Center.

The e-mails from 2007 reveal that when a *USA Today* reporter asked if NASA's data "was more accurate" than other climate-change data sets, NASA's Dr. Reto A. Ruedy replied with an unequivocal no. He said[64] "the National Climatic Data Center's procedure of only using the best stations is more accurate," admitting that some of his own procedures led to less accurate readings.

"My recommendation to you is to continue using NCDC's data for the US means and [East Anglia] data for the global means," Ruedy told the reporter.

A similar tale is seen with NOAA GHCN version 2 before and after adjustment (Fig. 2.43).

The longest history of unaltered data is the Central England Temperature set, established during the Little Ice Age in 1659. Note how December 2010 was the second coldest December in the record (just 0.1°C above 1890). Long-term warming is seen coming out of the LIA but no acceleration upwards can be detected (Fig. 2.44).

18. SUMMARY

Climategate has sparked a flurry of examinations of the global data sets not only at CRU, NASA, and NOAA, but in various countries throughout the world. Though the Hadley Centre implied their data were in agreement with other data sets and were thus trustworthy, the truth is that other data centers and the individual countries involved were forced to work with degraded data and appear to be each involved in data manipulation.

[63] http://data.giss.nasa.gov/gistemp/updates/200708.html.

[64] http://pajamasmedia.com/files/2010/03/GISS-says-CRU-Better0001.pdf.

Kevin Trenberth, IPCC lead author, NCAR and CRU associate said:

> It's very clear we do not have a climate observing system…This may be a shock to many people who assume that we do know adequately what's going on with the climate, but we don't.

Essex et al. (2007), question even whether a global average surface temperature is meaningful.

Climate change is real. There has been localized warming due to population growth and land-use changes. There are cooling and warming periods that can be shown to correlate well with solar and ocean cycles. You can trust in the data that show that there has been warming from 1979 to 1998, just as there was warming around 1920 to 1940. But there has been cooling from 1940 to the late 1970s and since 2001. The long-term trend on which this cyclical pattern is superimposed has been exaggerated.

As shown, record highs and rural temperatures in North America show the cyclical pattern but suggest the 1930s to 1940 peak was higher than the recent peak around 1998. Recent ranking was very likely exaggerated by the numerous data issues discussed. Given these data issues and the inconvenient truths in the Climategate e-mails, the claim that 2010 was the warmest year, and that the 2000s were the warmest decade in the record—or as some claim in a millennium or two—is not credible.

These factors all lead to significant uncertainty and a tendency for overestimation of century-scale temperature trends. An obvious conclusion from all findings above is that the global databases are seriously flawed and can no longer be trusted to assess climate trends. And, consequently, such surface data should not be used for decision making.

We enthusiastically support Roger Pielke Sr. who, after exchanges with Phil Jones over data sets, called for[65]:

> an inclusive assessment of the surface temperature record of CRU, GISS and NCDC. We need to focus on the science issues. This necessarily should involve all research investigators who are working on this topic, with formal assessments chaired and paneled by mutually agreed to climate scientists who do not have a vested interest in the outcome of the evaluations.

Georgia Tech's Dr. Judith Curry's comments on Roger Pielke Jr.'s blog also support such an effort:

> In my opinion, there needs to be a new independent effort to produce a global historical surface temperature dataset that is transparent and that includes expertise in statistics and computational science…The public has lost confidence in the data sets…Some efforts are underway in the blogosphere to examine the historical land surface data (e.g. such as GHCN), but even the GHCN data base has numerous inadequacies.

Judith is part of the newly announced Berkeley Earth Surface Temperature (BEST) Project, which aims to develop an independent analysis of the data from land stations, which would include many more stations than had been considered by the Global Historic Climatology Network. We trust they will include scientists who understand the issues we have raised and will make the reconstructed data sets available for independent review and analysis.

It should be noted that replication is required by the Data Quality Act according to the government's own Office of Management and Budget. Though such an effort can be done locally through tedious research and analysis in the United States, the status of the publicly available global databases (GHCN, GISS, CRU) makes that extremely difficult or impossible currently. Until then, satellite data is the only trustworthy data set.

Acknowledgments

I wish to thank Anthony Watts who provided invaluable analysis, and considerable constructive feedback and suggestions for this analysis. I wish to also thank Roger Pielke Sr., Steve McIntyre, E.M. Smith, and Verity Jones and many others cited in this compilation study for their tireless efforts with regard to issues with temperature measurements.

References

Balling, R.C., Idso, C.D., 2002. Analysis of adjustments to the United States historical climatology network (USHCN) temperature database. Geophysical Research Letters. http://dx.doi.org/10.1029/2002GL014825.

Barnett, T.P., Pierce, D.W., Schnur, R., 2001. Detection of anthropogenic climate change in the world's oceans. Science 292, 270–274.

Block, A., Keuler, K., Schaller, E., 2004. Impacts of anthropogenic heat on regional climate patterns. Geophysical Research Letters 31, L12211. http://dx.doi.org/10.1029/2004GL019852.

---

[65] http://wattsupwiththat.com/2010/01/14/pielke-senior-correspondence-with-phil-jones-onklotzbach-et-al/.

Brohan, P., Kennedy, J.J., Harris, I., Tett, S.F.B., Jones, P.D., 2006. Uncertainty estimates in regional and global observed temperature changes: a new dataset from 1850. Journal of Geophysical Research 111, D12106. http://dx.doi.org/10.1029/2005JD006548.

Changnon, S.A., Kunkel, K.E., 2006. Changes in instruments and sites affecting historical weather records: A case study. Journal of Atmospheric Oceanic Technology 23, 825—828.

Christy, J.R., Norris, W.B., Redmond, K., Gallo, K.P., 2006. Methodology and results of calculating central California surface temperature trends: Evidence of human induced climate change? Journal of Climate 19, 548—563.

Chou, Lindzen, 2005. Comments on examination of the decadal tropical mean ERBS nonscanner radiation data for the Iris hypothesis. Journal of Climate 18, 2123—2127.

Davey, C.A., Pielke Sr., R.A., 2005. Microclimate exposures of surface-based weather stations — implications for the assessment of long-term temperature trends. Bulletin of the American Meteorological Society 86 (4), 497—504.

de Laat, A.T.J., Maurellis, A.N., 2006. Evidence for influence of anthropogenic surface processes on lower tropospheric and surface temperature trends. International Journal of Climatology 26, 897—913.

Essex, C., Andresen, B., McKitrick, R.R., 2007. Does a global temperature exist? Journal of Nonequilibrium Thermodynamics 32 (1).

Easterling, D.R., Horton, B., Jones, P.D., Peterson, T.C., Karl, T.R., Parker, D.E., Salinger, M.J., Razuvayev, V., Plummer, N., Jamason, P., Folland, C.K., 1997. Maximum and minimum temperature trends for the globe. Science 277, 364—367.

Fall, S., Watts, A., Nielsen-Gammon, J., Jones, E., Niyogi, D., Christy, J., Pielke Sr., R.A., 2011. Analysis of the impacts of station exposure on the U.S. Historical Climatology Network temperatures and temperature trends. Journal of Geophysical Research 116, D14120. http://dx.doi.org/10.1029/2010JD015146.

Goodridge, J.D., 1996. Comments on regional simulations of greenhouse warming including natural variability. Bulletin of the American Meteorological Society 77, 1588—1599.

Hansen, J., Ruedy, R., Sato, M., Imhoff, M., Lawrence, W., Easterling, D., Peterson, T., Karl, T., 2001. A closer look at United States and global surface temperature change. Journal of Geophysical Research 106 (D20), 23947—23963.

He, Y., Lu, A., Zhang, Z., Pang, H., Zhao, J., 2006. Seasonal variation in the regional structure of warming across China in the past half century. Climate Research 28, 213—219.

Hinkel, K.M., Nelson, F.E., Klene, S.E., Bell, J.H., 2003. The urban heat island in winter at Barrow, Alaska. International Journal of Climatology 23, 1889—1905. http://dx.doi.org/10.1002/joc.971.

Jáuregui, E., 2005. Possible impact of urbanization on the thermal climate of some large cities in Mexico. Atmosfera 18, 249—252.

Jones, C.G., Young, K.C., May 1995. An Investigation of temperature discontinuities introduced by the installation of the HO-83 thermometer. Journal of Climate 8 (5), 1394.

Jones, P.D., Groisman, P.Y., Coughlan, M., Plummer, N., Wangl, W.C., Karl, T.R., 1990. Assessment of urbanization effects in time series of surface air temperatures over land. Nature 347, 169—172.

Jones, P.D., Lister, D.H., Li, Q., 2008. Urbanization effects in large-scale temperature records, with an emphasis on China. Journal of Geophysical Research 113, D16122. http://dx.doi.org/10.1029/2008JD009916.

Kalnay, E., Cai, M., 2003. Impacts of urbanization and land-use change on climate. Nature 423, 528—531.

Karl, T.R., Diaz, H.F., Kukla, G., 1988. Urbanization: its detection and effect in the United States climate record. Journal of Climate 1, 1099—1123.

Karl, T.R., 1995. Critical issues for long-term climate monitoring. Climate Change 31, 185.

Karl, T.R., Hassol, S.J., Miller, C.D., Murray, W.L. (Eds.), 2006. Temperature Trends in the Lower Atmosphere: Steps for Understanding and Reconciling Differences. A Report by the Climate Change Science Program and the Subcommittee on Global Change Research, Washington, DC.

Kent, E.C., Woodruff, S.D., Berry, D.I., 2007. Metadata from WMO Publication No. 47 and an assessment of voluntary observing ship observation heights in ICOADS. Journal of Atmospheric and Oceanic Technology 24 (2), 214—234.

Klotzbach, P.J., Pielke Sr., R.A., Pielke Jr., R.A., Christy, J.R., McNider, R.T., 2009. An alternative explanation for differential temperature trends at the surface and in the lower troposphere. Journal of Geophysical Research 114, D21102. http://dx.doi.org/10.1029/2009JD011841. http://sciencepolicy.colorado.edu/admin/publication_files/resource-2792-2009.52.pdf.

Landsberg, H.E., 1981. The Urban Climate. Academic Press.

Li, Q., et al., 2004. Urban heat island effect on annual mean temperatures during the last 50 years in China. Theoretical and Applied Climatology 79, 165—174.

Lindzen, R.S., Choi, Y.S., 2009. On the determination of climate feedbacks from ERBE data. Geophysical Research Letters 36, L16705. http://dx.doi.org/10.1029/ 2009GL039628.

Loehle, C., 2009. Cooling of the global ocean since 2003. Energy & Environment 20 (No. 1&2), 101—104 (4).

Mahmood, R., Foster, S.A., Keeling, T., Hubbard, K.G., Carlson, C., Leeper, R., 2006. Impacts of irrigation on 20th century temperature in the Northern Great Plains. Global Planetary Change 54, 1—18.

McKitrick, R.R., Michaels, P.J., 2007. Quantifying the influence of anthropogenic surface processes and inhomogeneities on gridded global climate data. Journal of Geophysical Research 112, D24S09. http://dx.doi.org/10.1029/2007JD008465.

McKitrick, R., Michaels, P.J., 2004. A test of corrections for extraneous signals in gridded surface temperature data. Climate Research 26 (2), 159—173. "Erratum," Climate Research 27(3), 265—268.

Menne, M.J., Williams, C.N., Palecki, M.A., 2010. On the reliability of the U.S. surface temperature record. Journal of Geophysical Research 115, D11108. http://dx.doi.org/10.1029/2009 JD013094.

Oke, T.R., 1973. City size and the urban heat island. Atmospheric Environment 7, 769—779.

Parker, D.E., November 18, 2004. Climate: large-scale warming is not urban. Nature 432, 290. http://dx.doi.org/10.1038/432290a.

Peterson, T.C., Vose, R.S., 1997. An overview of the global historical climatology network temperature database. Bulletin of the American Meteorological Society 78, 2837—2849.

Peterson, T.C., 2003. Assessment of urban versus rural in situ surface temperatures in the contiguous United States: no difference found. Journal of Climate 16 (18), 2941—2959.

Peterson, 2006. Examination of potential biases in air temperature caused by poor station locations. Bulletin of the American Meteorological Society 87, 1073—1089.

Pielke Sr., R.A., 2008. A broader view of the role of humans in the climate system. Physics Today 11, 53—55.

Pielke Sr., R.A., Nielsen-Gammon, J., Davey, C., Angel, J., Bliss, O., Doesken, N., Cai, M., Fall, S., Niyogi, D., Gallo, K., Hale, R., Hubbard, K.G., Lin, X., Li, H., Raman, S., 2007a. Documentation of uncertainties and biases associated with surface temperature measurement sites for climate change assessment. Bulletin of the American Meteorological Society 88 (6), 913–928.

Pielke Sr., R.A., Davey, C., Niyogi, D., Fall, S., Steinweg-Woods, J., Hubbard, K., Lin, X., Cai, M., Lim, Y.-K., Li, H., Nielsen-Gammon, J., Gallo, K., Hale, R., Mahmood, R., Foster, S., McNider, R.T., Blanken, P., 2007b. Unresolved issues with the assessment of multi-decadal global land surface temperature trends. Journal of Geophysical Research 112, D24S08. http://dx.doi.org/10.1029/2006JD008229. http://www.climatesci.org/publications/pdf/R-321.pdf.

Ren, G.Y., Chu, Z.Y., Chen, Z.H., Ren, Y.Y., 2007. Implications of temporal change in urban heat island intensity observed at Beijing and Wuhan stations. Geophysical Research Letters 34, L05711. http://dx.doi.org/10.1029/2006GL027927.

Runnalls, K.E., Oke, T.R., 2006. A technique to detect microclimatic inhomogeneities in historical records of screen-level air temperature. Journal of Climate 19, 959–978.

Santer, B.D., Wigley, T.M.L., Mears, C., Wentz, F.J., Klein, S.A., Seidel, D.J., Taylor, K.E., Thorne, P.W., Wehner, M.F., Gleckler, P.J., Boyle, J.S., Collins, W.D., Dixon, K.W., Doutriaux, C., Free, M., Fu, Q., Hansen, J.E., Jones, G.S., Ruedy, R., Karl, T.R., Lanzante, J.R., Meehl, G.A., Ramaswamy, V., Russell, G., Schmidt, G.A., 2005. Amplification of surface temperature trends and variability in the tropical atmosphere. Science 309, 1551–1556.

Santer, B.D., et al., 2008. Consistency of modeled and observed temperature trends in the tropical troposphere. International Journal of Climatology 28, 1703–1722. http://dx.doi.org/10.1002/joc.1756.

Smith, T.M., Reynolds, R.W., 2004. Improved extended reconstruction of SST (1854–1997). Journal of Climate 17, 2466–2477.

Smith, T.M., Reynolds, R.W., 2005. A global merged land air and sea surface temperature reconstruction based on historical observations (1880–1997). Journal of Climate 18, 2021–2036.

Smith, T.A., Reynolds, R.W., Petersen, T.C., Lawrimore, J., 2008. Improvements to NOAA's historical merged land–ocean surface temperature analysis (1880–2006). Journal of Climate 21, 2283–2296.

Torok, S.J., Morris, C.J.G., Skinner, C., Plummer, N., 2001. Urban heat island features of southeast Australian towns. Australian Meteorological Magazine 50, 1–13.

Velazquez-Lozada, A.V., Gonzalez, J.E., Winter, A., 2006. Urban heat island effect analysis for San Juan, Puerto Rico. Atmospheric Environment 40, 1731–1741.

Watts, A., 2009. Is the U.S. Surface Temperature Record Reliable? The Heartland Institute, Chicago, IL. ISBN:10.1-934791-26-6. http://wattsupwiththat.files.wordpress.com/2009/05/surfacestationsreport_spring09.pdf.

Willmott, Robeson, Feddema, December 1991. Influence of spatially variable instrument networks on climatic averages. Geophysical Research Letters 18 (No. 12), 2249–2251.

Zhou, L., Dickinson, R., Tian, Y., Fang, J., Qingziang, L., Kaufman, R., Myneni, R., Tucker, C., June 29, 2004. Rapid Urbanization warming China's climate faster than other areas. Proceedings of the National Academy of Sciences of the United States of America.

CHAPTER

3

Is the NASA Surface Temperature Record an Accurate Representation?

T. Heller

Doctors for Disaster Preparedness

We have all grown used to seeing graphs like Fig. 3.1 from NASA, showing nearly continuous global warming over the last 135 years, with a flat period between 1940 and 1980, and 1.2°C warming from 1880 through 2000. Much of climate science, journalism, and public policy is based around the belief that these NASA graphs are an accurate representation of the temperature record, and that the apparent warming that is shown in the graphs is due to an increase in atmospheric CO_2. Understanding the accuracy, consistency, and integrity of these graphs is therefore very important.

However, if we look at earlier versions of the same graph, we see something very different. Fig. 3.2, published by NASA in 2001, showed 1975 as barely warmer than 1880 with less than 0.6°C warming from 1880 to 2000. The 2001 version showed only half as much warming from 1880 to 2000 as the 2016 version of the same graph in Fig. 3.1.

The recent increase of 0.6°C in NASA's reported warming over the 1880–2000 time period is particularly troubling because the recent alterations extend far outside their own error bars. Fig. 3.3 shows the 2001 version and the 2016 versions at the same scale, normalized to the most recent common decade. The blue lines represent the 2001 error bars, and the green lines represent the 2016 error bars. Note how the alterations are about three times as large as NASA's reported error. This indicates a scientific process that is completely broken.

Fig. 3.4 shows the changes that have been made to the NASA 1880–2000 temperature trend since 2001, a total of more than half a degree. The alterations have been almost as large as the entire trend reported in 2001, indicating a

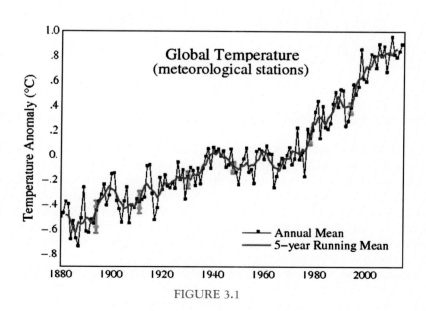

FIGURE 3.1

Evidence-Based Climate Science, Second Edition
http://dx.doi.org/10.1016/B978-0-12-804588-6.00003-3

49

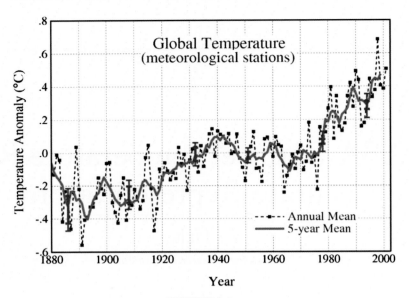

FIGURE 3.2

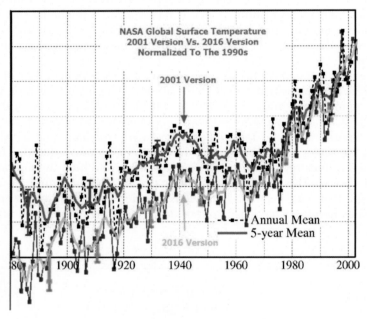

FIGURE 3.3

signal-to-noise ratio of close to zero. In other words, the NASA temperature graph is meaningless from a scientific point of view.

Going farther back in time with published NASA temperature records, we see even larger discrepancies. NASA's James Hansen in 1981 (Fig. 3.5) showed a spike in temperatures around 1940, followed by nearly 0.3°C cooling until the late 1960s. This spike and cooling has been almost completely erased from the NASA temperature record.

Fig. 3.6 overlays Hansen's 1981 graph on the current NASA surface temperature record and shows how pre-1980 temperatures have been cooled. By cooling the past, NASA increased the total amount of warming, removed the 1940s spike, and removed the post 1940s cooling. These alterations did not occur all at once but rather have occurred in successive stages, where the past is made cooler to create the appearance of more warming. The amount of warming from 1880 to 1980 has been doubled since Hansen 1981 (Fig. 3.7).

Before NASA was involved, the National Academy of Sciences and NCAR both showed even more post-1940s cooling, and that Northern Hemisphere temperatures were no warmer in the late 1960s than they were at the beginning of the 19th century. These graphs are completely different from the current NASA graph.

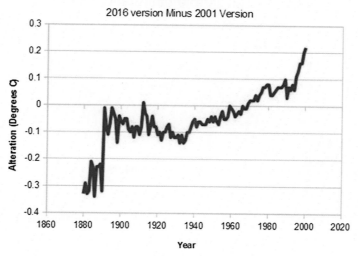

FIGURE 3.4

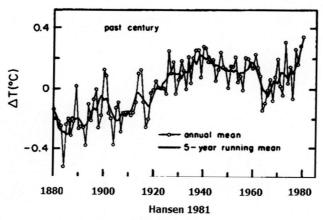

FIGURE 3.5

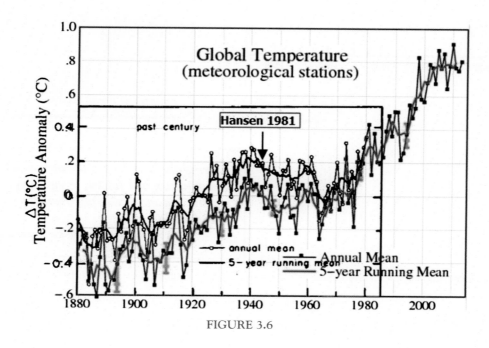

FIGURE 3.6

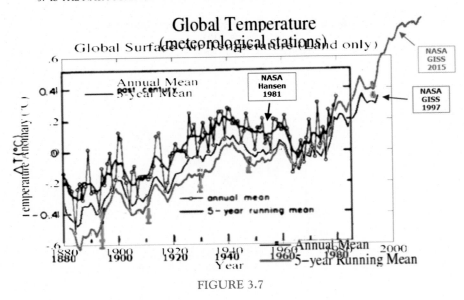

FIGURE 3.7

In 1974, the US Science Board reported a sharp cooling since World War II (WWII). Fig. 3.9 is from the November 1976 issue of *National Geographic*.

That same issue of *National Geographic* included the graph in Fig. 3.9, and the following text: "Downward trend of temperature since 1938 has come nearly halfway back to the chill of the Little Ice Age 300 year ago." This downward trend has been nearly erased by NASA and NOAA.

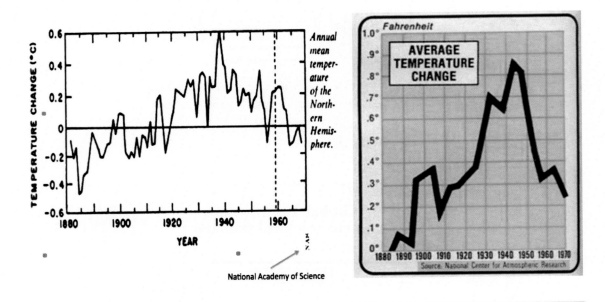

FIGURE 3.8   National Academy of Sciences Report 1975, National Center For Atmospheric Research 1974, National Science Board 1974

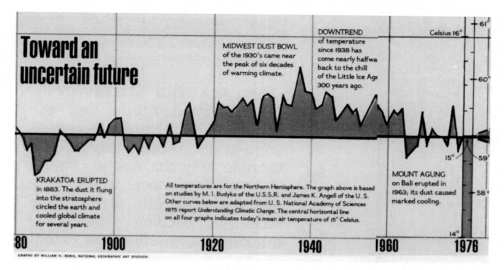

FIGURE 3.9

The Corpus ...aller-Times (Corpus Christi, Texas) · 4 Mar 1974, Mon

To consider some of the urgent questions that a changing climate poses for the world, several of America's most prominent meteorologists gathered here last week for discussions at the annual meeting of the American Association for the Advancement of Science. What emerged was a grim forecast for the future.

It was generally agreed that the extraordinary warmth of the early 20th century is drawing to a close, and that a global cooling trend has been under way since about 1940. But the scientists cannot explain why.

Dr. James D. Hays of Columbia University, leader of the scientific team, called

But a moderate cooling trend has already begun, Dr. Hays said, adding:

"If you project the relationship between the orbits and the climate in the future, this cooling trend should continue for on the order of 20,000 years. In that length of time I think there is not much doubt that we will build substantial ice on the Northern Hemisphere continents."

The New York Times
Published: November 30, 1976
Copyright © The New York Times

FIGURE 3.10

Lawrence Journal-World - Mar 11, 1979

One thing is indisputable: The world has been cooling off since World War II, something like one degree Fahrenheit. But that may be only a temporary swing in the climate.

Dr. J. Murray Mitchell, of the National Oceanic and Atmospheric Administration says the world has been cooling off in the long run.

"On an average it's cooled down by something like one degree Fahrenheit or half-a-degree Celsius, and that cooling began around World War II.

The New York Times
MONDAY, JANUARY 30, 1961

SCIENTISTS AGREE WORLD IS COLDER

But Climate Experts Meeting Here Fail to Agree on Reasons for Change

By WALTER SULLIVAN

After a week of discussions on the causes of climate change, an assembly of specialists from several continents seems to have reached unanimous agreement on only one point: it is getting colder.

FIGURE 3.11

The Monroe News-Star (Monroe, Louisiana) · 9 Jun 1976, Wed · Pa

A most gloomy view of world weather trends has emerged from the Central Intelligence Agency. The report was made public, with CIA permission, by the House Agriculture Committee to force decision makers in government to come to grips with what some committee members feel is a serious problem.

Says one, "we must no longer accept with blind faith the notion that food production will continue to increase and climates will remain stable."

The CIA "working paper" contends that the global climate is cooling after 50 years of the most favorable farming weather since the eleventh century. The economic and political impact, the report asserts, "is almost beyond comprehension."

The New York Times

SATURDAY, JULY 18, 1970

U.S. and Soviet Press Studies of a Colder Arctic

By WALTER SULLIVAN

The United States and the Soviet Union are mounting large-scale investigations to determine why the Arctic climate is becoming more frigid, why parts of the Arctic sea ice have recently become ominously thicker and whether the extent of that ice cover contributes to the onset of ice ages.

FIGURE 3.12

Ukiah Daily Journal (Ukiah, California) · 20 Nov 1974, Wed · Page 17 ⓘ

The cooling trend heralds the start of another ice age, of a duration that could last from 200 years to several millennia.

Quoting various authorities around the world, he said the current trend could lead to either a "little ice age" of 200 to 400 years, or to a "great ice age" that could last 10,000 years "and send rivers of solid ice again as far south as Yosemite

the present and future effects of overpopulation.

Sixty theories have been advanced, he said, to explain the global cooling period.

The Bering Straits dam Ponte estimated would cost $100 billion which could be shared between Russia and the United States with the possible help of other countries such as Japan.

FIGURE 3.13

During the 1970s, the American Association for the Advancement of Science reported a global cooling trend since 1940, as did Columbia University (Fig. 3.10).

In 1961 there was "unanimous agreement" among climate experts that Earth was getting colder. In 1979 the cooling was considered "indisputable" (Fig. 3.11) and NOAA reported about 1°F cooling since WWII.

The CIA warned that the cooling trend threatened political stability. The United States and Russia mounted "large-scale investigations to determine why the Arctic climate is becoming more frigid" and the ice was getting thicker (Fig. 3.12).

Sixty theories were put forward to explain the cooling (Fig. 3.13).

Further evidence for the cooling was that during the late 1950s and 1960s, glaciers were advancing for the first time in over a century (Fig. 3.14).

This cooling was a big change from the previous decades. Glaciers were disappearing in the early part of the century, and many had vanished by 1923 (Fig. 3.15).

The Canberra Times (ACT : 1926 - 1995) (about) ◄ Thursday 18 July 1963

Glaciers Grow In Norway

OSLO (A.A.P.-Reuter). — Norway's glaciers are in the process of becoming thicker again after a period of 200 years of gradually melting down, according to glaciologist, Mr. Olav Liestol.

Last year nearly all glaciers increased by more than one metre — approximately four feet.

The New York Times

SUNDAY, OCTOBER 16, 1955

Northwest Glaciers Again Advance

Glaciers in the Pacific Northwest mountain ranges are advancing substantially for the first time in about one hundred years, finds Richard Hubley, project leader of an expedition sent to the Olympic and Cascade Mountains. Mr. Hubley, a graduate student in meteorology at the University of Washington, conducted the project with Ed La Chapelle, snow physicist for the United States Forest Service in Alta, Utah, and Michael W. Hane, a university physics student.

The team spent July and August studying Blue Glacier on Mount Olympus in the Olympic National Park and surveyed from the air forty-eight glaciers in the Cascade Mountains in mid-September. In both areas a substantial forward motion of glaciers, ranging from ten feet to 350 feet annually, was detected. It is possible that this movement is a forerunner of major glacial activity in temperate regions in other parts of the world. Similar advances have occurred at intervals during the past several hundred years. The last was in the nineteenth century. Since that time, glaciers have been retreating.

glacier as related to weather conditions.

Blue Glacier is 3¼ miles long and covers an area of 2¾ miles on the slopes of Mount Olympus from the 7,800-foot level to 3,900 feet. Its flat, broad snow fields at the higher reaches terminate in a long valley tongue. Mr. Hubley and his associates found that the terminus had advanced ten feet in the past year, and that ice in the middle portions of the glacier moved as rapidly as one foot a day.

Photographs Compared

In the second phase of the research, aerial photos were taken of forty-eight glaciers in the Cascade Mountains, which had not been photographed aerially since 1950. Comparative photographs showed that thirty-three of the forty-eight glaciers had advanced. Most of the other fifteen could not be measured because they were heavily covered with snow, although the pictures were taken in early autumn. The thick snow that remained throughout the summer indicates these glaciers also are growing.

FIGURE 3.14

The Brooklyn Daily Eagle (Brooklyn, New York) • 1 Aug 1907 Thu •

Glaciers Disappearing.

Remarkable shrinkings have been going on among the Swiss glaciers, especially during the past two or three years. According to measurements, the great Rhone glacier, one of the sights of Switzerland—at least, it used to be—has lost in the past two years no less than 88,250 square feet, and others have been shortened by anywhere from 20 to 40 feet. Among

Medford Mail Tribune (Medford, Oregon) • 29 Dec 1923, Sat

CINCINNATI, Dec. 29.—(By the Associated Press)—The hot dry seasons of the past few years have caused rapid disintegration of glaciers in Glacier National park, Montana, professor W. G. Waterman of Northwestern university declared in an address today before the Geological section of the American Association for the Advancement of Science.

Sperry Glacier, studied by Professor Waterman, has lost one-quarter, or perhaps one-third of its ice in the past 18 years, he said. If this rapid retreat should continue, the professor added, the glacier would almost disappear in another 25 years, but he expressed the opinion that the long dry seasons of the past few years is over with probabilities of a lessening in the retreat.

Daily Mercury (Mackay, Qld. : 1906 - 1954) (about) ◄ Saturday 7 April 1923 ►

NORTH POLE MELTING.

MANY GLACIERS VANISHED.

Is the North Pole going to melt entirely? Are the Artic regions warming up, with the prospect of a great climatic change in that part of the world?

Science is asking these questions (says "Popular Science Siftings"). Reports from fishermen, seal hunters, and explorers who sail the seas around Spitzbergen and the eastern Arctic all point to a radical change in climatic conditions, with hitherto unheard-of high temperatures on that part of the earth's surface

The New York Times

SUNDAY, DECEMBER 21, 1930

WORD comes from Switzerland that the Alpine glaciers are in full retreat. Out of 102 glaciers observed by Professor P. L. Mercanton of the University of Lausanne and his associates more than two-thirds have been found to be shrinking. Does this mean the approach of a warmer climate, such as swept over our globe thousands of years ago? Will palms, cypresses, magnolias, myrtles and olive trees thrive at the feet of the Adirondacks, as they did in those distant days?

FIGURE 3.15

During the 1930s and 1940s, the glaciers of Norway and Greenland were rapidly melting and faced "catastrophic collapse." By 1952 the glaciers of Norway had lost half of their mass (Fig. 3.16).

National Geographic showed images of rapid glacial retreat in the Alps by 1940 (Fig. 3.17).

Evidence for the earlier warmth and post WWII cooling is very compelling. So why does it no longer appear in the NASA temperature record? NASA shows the period during the 1960s when glaciers were growing as warm, and the earlier period when they were melting as cold (Fig. 3.18). The NASA temperature record does not appear consistent with the evidence on the ground, whereas the 1975 National Academy of Sciences graph is consistent.

This Climategate email provides a big hint as to what has happened to the data since 1975. Climate scientists wanted to get rid of the "1940s blip":

From: Tom Wigley <wigley@ucar.edu>
To: Phil Jones <p.jones@uea.ac.uk>
Subject: 1940s
Date: Sun, 27 Sep 2009 23:25:38 -0600
Cc: Ben Santer <santer1@llnl.gov>

FIGURE 3.16

FIGURE 3.17

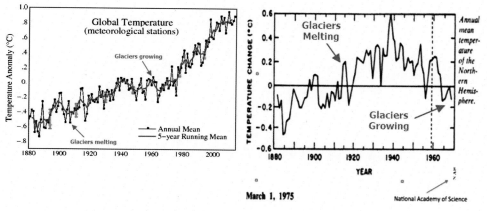

FIGURE 3.18

So, if we could reduce the ocean blip by, say, 0.15 degC, then this would be significant for the global mean – but we'd still have to explain the land blip.

It would be good to remove at least part of the 1940s blip, but we are still left with "why the blip."

Removing the 1940s blip is exactly what NASA did. This is particularly evident in Greenland, Iceland, and United States temperature data (Fig. 3.19).

In 1999 NASA's James Hansen reported that the United States cooled half a degree since the 1930s:

in the US there has been little temperature change in the past 50 years, the time of rapidly increasing greenhouse gases—in fact, there was a slight cooling throughout much of the country (Fig. 3.20).

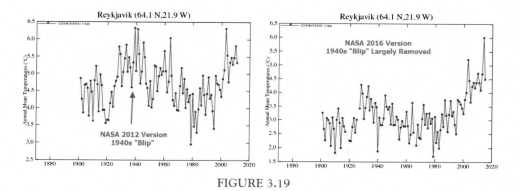

FIGURE 3.19

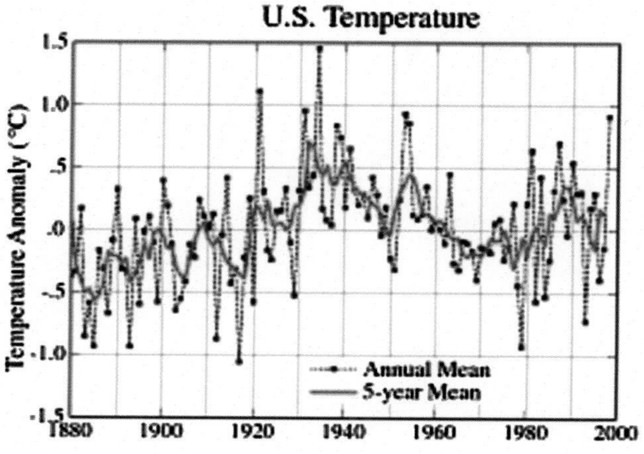

FIGURE 3.20

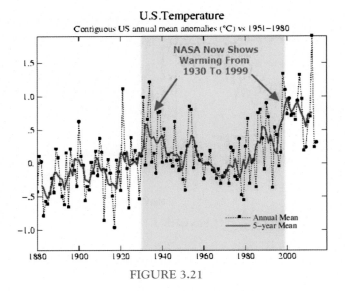

FIGURE 3.21

NASA now shows warming during that same period. NASA US temperature data is based on NOAA United States Historical Climatology Network (USHCN) data, which is massively altered to create the appearance of warming (Fig. 3.21).

Fig. 3.22 shows how the NOAA data is altered. The blue line shows the averaged measured temperature at all USHCN stations, and the red line shows the average adjusted temperature. The warming trend over the last century is entirely due to "adjustments" by NOAA. Note how the "1940s blip" has been largely removed.

Fig. 3.23 shows the adjustments made by NOAA to the US temperature record—a total of 1.6°F.

A large portion of these adjustments is due to NOAA simply making up data. If they are missing data at a particular station one month, they use a computer model to fabricate the temperature for that month. Since 1970, the percentage of fabricated data has increased from 10% to almost half of the data (Fig. 3.24). The US temperature record from NASA and NOAA is thus a completely meaningless fabrication, which bears no resemblance to the thermometer data from which it is derived.

In summary, the NASA global and US temperature records are neither accurate nor credible representations of reality.

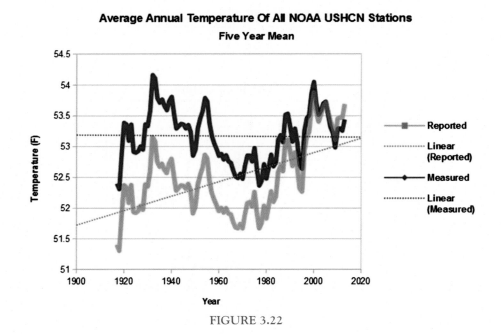

FIGURE 3.22

3. WAS 2014 "THE WARMEST YEAR ON RECORD"? AND DOES IT MATTER?

63

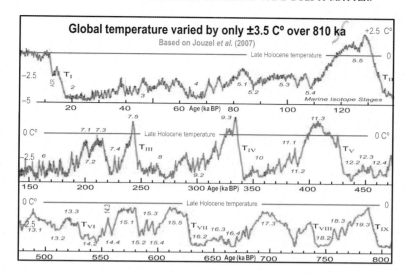

- At present we are about 1°C above that long-run mean, but each of the previous three interglacial warm periods was warmer than the present.
- The most recent such warm period was 2.5°C warmer than the present. So there is nothing "unprecedented" about present-day temperatures.

3. WAS 2014 "THE WARMEST YEAR ON RECORD"? AND DOES IT MATTER?

Mr. Varley says:

> The WMO confirmed that globally 2014 was the warmest year on record, with 14 of the 15 hottest years occurring this century.

Yet again we are given only part of the story. The following are among the relevant considerations omitted or erroneously presented:

- When one talks of "the warmest year on record," the "record" only goes back to 1850 (HadCRUT4), or 1880 (NASA, GISS, and NCDC), or 1979 (RSS and UAH). The first three depend on the same historical climate data network. They all show 2014 as the warmest year since 1850 (or 1880). So what?
- The RSS and UAH satellite data sets do not show 2014 as the warmest year. It would have been fairer if this fact had been mentioned.
- The Mediaeval Optimum was warmer than today by up to 3°C in some places. The Roman, Minoan, and Old Kingdom climate optima were also warmer. The Holocene Climate Optimum was warmer than today for 4000 years.
- According to the two satellite data sets, there has been no global warming at all for more than 18 years. The trend is zero. In a briefing on global warming, you should surely have mentioned that fact.

Mr. Varley says:

> Closer to home, Met Office statistics show that in 2014 the UK as a whole experienced its warmest year on record with the eight warmest years in this series all occurring since 2002. It was also the warmest year on record in the Central England Temperature series which extends back to 1659.

Yet again, balancing considerations are omitted.

- Cherry-picking an individual year, or a selection of individual years, is not how statistical trends on time series are determined.
- The Central England Temperature Record is cited, but with no mention of the fact that from 1693 to 1733 the rate of global warming was twice that which occurred over any period of 15 years or more during the 20th century.

4. WHAT IS THE IDEAL UNITED KINGDOM AND GLOBAL TEMPERATURE?

Mr. Varley says:

> The UK mean temperature for 2014 was 9.9°C, 1.1°C above the 1981—2010 long-term average and the warmest year in the UK series ...

The balancing considerations are the following:

- Cherry-picking an individual year in an individual territory tells us nothing about the global temperature trend, which has been statistically indistinguishable from zero for at least a quarter of a century on the satellite measures, notwithstanding record increases in CO_2 concentration.
- In a cold country like Britain, what problems would be caused by the temperature warming up a little? The human body works best at about 19°C, yet the United Kingdom average temperature in 2014 was less than 10°C.
- And what, in any event, is the ideal global (or, for that matter, UK) temperature? Unless we are told that, we cannot be at all sure that an increase of 1°C in UK surface temperature is anything other than welcome.

5. STRETCHING THE VERTICAL AXIS OF THE TEMPERATURE GRAPH

Next, Mr. Varley reproduces a graph of global temperature change since 1850 published by the Intergovernmental Panel on Climate Change (IPCC) in its 2013 *Fifth Assessment Report*. The IPCC stretched the graph along its vertical axis (and Mr. Varley has stretched it a little further):

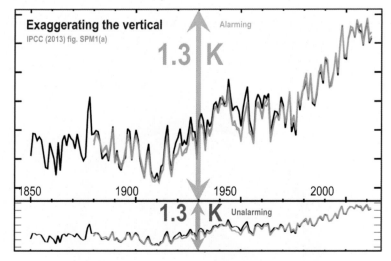

- Such alterations of the aspect ratio by accentuating the vertical axis are calculated to make the actually rather small change in global temperature over the period seem bigger than it was.

- In the upper panel of the image, we have restored the aspect ratio of the IPCC's original graph. In the lower panel, we have reduced the vertical emphasis to show that the apparent steepness of the temperature increase in the IPCC's graph is merely an artefact of the choice of aspect ratio. Vertical exaggeration, now commonplace in climate science, is a rather less-than-honest graphical technique.

6. IMAGINED EFFECTS OF GLOBAL WARMING

Mr. Varley says:

In addition to the Earth's surface temperature, many tens of other climate variables are measured, extending from high in the atmosphere to the depths of the oceans. These are analysed at academic and research centres around the world, with thousands of scientists pooling their findings and expertise to build a picture of past and current climate. Emerging from these observations is evidence of change: global and regional temperatures are increasing; Arctic sea ice, mountain glaciers and snow cover are shrinking; warming oceanic waters are expanding, leading to sea level rise; atmospheric humidity is rising as a warmer atmosphere's capacity to hold water increases; the frequency of rainfall and temperature extremes has increased. These changes are already impacting on natural and human systems.

Now for the balance:

- Of course there is "evidence of change": the climate has been changing for 4.5 billion years, and it will continue to change.
- No one denies that the climate changes. The question is whether man has had or may yet have a significant effect, and whether that effect, if significant, will be beneficial or detrimental.
- Mr. Varley makes no mention of the real difficulties in distinguishing between natural and anthropogenic climate change.
- Mr. Varley's statement that global and regional temperatures are increasing is scientifically meaningless in the absence of a stated start date.
- Global temperature has not increased for more than 18 years, and has not increased significantly in the quarter of a century since the IPCC's *First Assessment Report* in 1990.
- Some regions, such as Antarctica and central Africa, have scarcely warmed, if at all.
- Global temperatures are lower than during previous climate optima during the Holocene.
- Global temperatures are also lower than in each of the previous four interglacial warm periods.
- Mr. Varley says Arctic sea ice, mountain glaciers, and snow cover are shrinking, but he is silent on the fact that Antarctic sea ice has grown; that mountain glaciers in the Himalayas, in Greenland, and in Antarctica show a long-established and unalarming pattern of advance and retreat; and that winter snow cover in the Northern Hemisphere shows no particular trend:

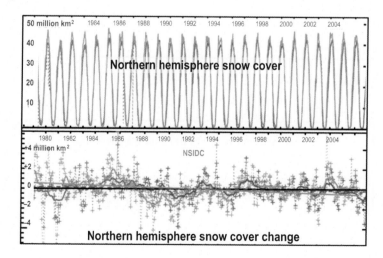

- Mr. Varley says warming oceanic waters are expanding, leading to sea level rise, but is careful not to quantify this. According to the GRACE gravitational-recovery satellites, sea level actually fell from 2003 to 2008:

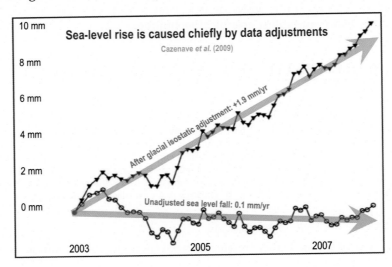

- According to the ENVISAT sea level satellite, sea level rose from 2004 to 2012 at a mean rate equivalent to just 1.3 inches per century:

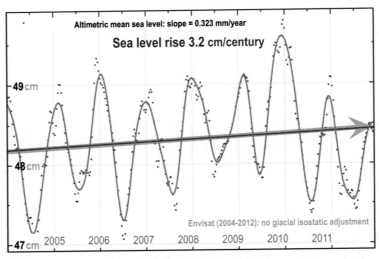

- The intercalibration errors between the series of laser-altimetry satellites from which the "official" sea level record is obtained are greater than the sea level rise they purport to show.

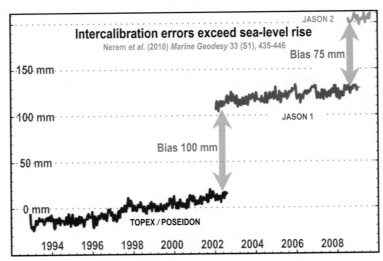

- Tide gauges and benchmarks show very little sea level rise. And why should there be much sea level rise? The ARGO bathythermographs show that in the first 11 years of the record, 2004–14, the ocean to a depth of 1900 m warmed at a rate equivalent to 0.23°C per century:

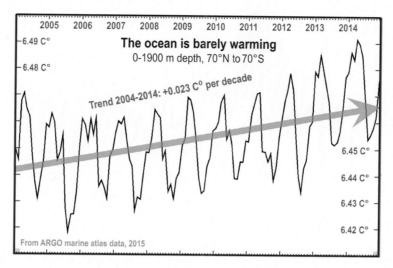

- Mr. Varley says that atmospheric humidity is increasing: but not all records show this, as the following chart of column water vapor demonstrates:

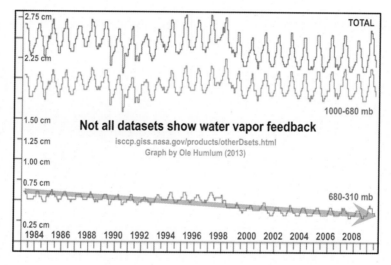

- Mr. Varley says the frequency of temperature extremes has increased. However, the weather is like the cricket: new records are set somewhere in the system all the time. It is the nature of the object.
- Significantly, there have been just about as many cold-weather records as hot-weather records set in recent years, even though theory would lead us to suspect fewer cold-weather records in a rapidly warming world (though it is not warming by much). And more all-time high-temperature records were set in the 1930s than in any decade since:

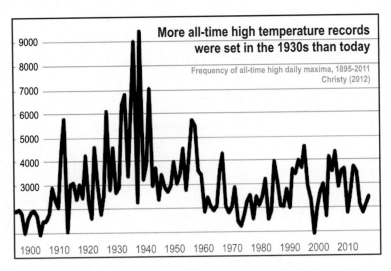

- Mr. Varley says the frequency of rainfall extremes has increased. Yet the IPCC, both in its 2012 report on extreme weather and in its 2013 *Fifth Assessment Report* draws no such conclusion.
- On the contrary, the IPCC says there is little or no evidence that such rainfall changes as have occurred are anthropogenic.
- Mr. Varley's own Met Office records tell a story different from Mr. Varley's. For instance, the longest annual national rainfall record, the quarter of a millennium in England and Wales, shows little change.
- There is also very little change in US annual rainfall over the 48 mainland states:

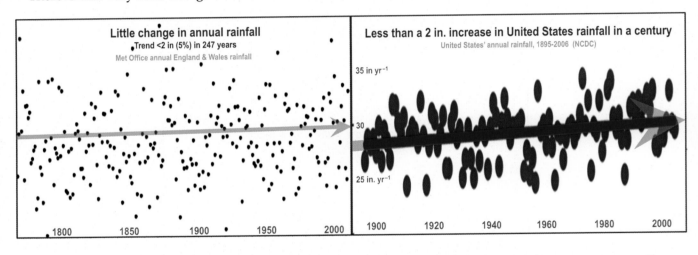

In fact, as the Global Warming Policy Foundation recently concluded that deaths from extreme weather are currently at an all-time low, notwithstanding record increases both in greenhouse gas concentrations and in global population over the period covered by its graph:

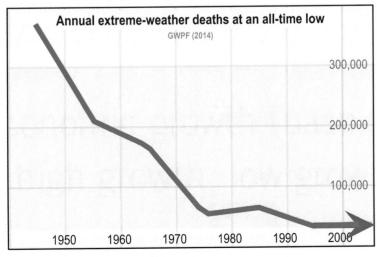

Next, Mr. Varley shows a graph of September sea ice extent in the Arctic similar to the following:

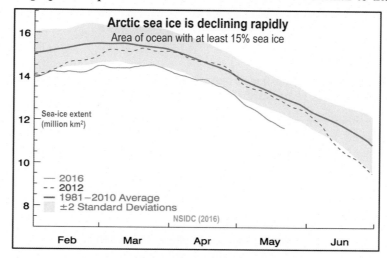

What he does not show is the Antarctic sea ice extent. It has increased somewhat:

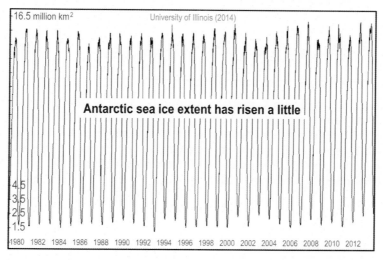

Moreover, in the Arctic as in the Antarctic, the amplitude of the seasonal variation dwarfs the relatively small changes in sea ice extent:

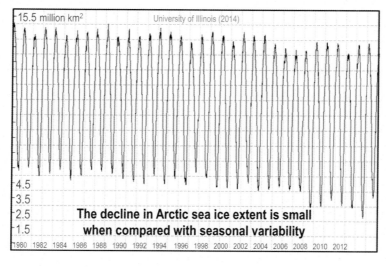

Mr. Varley says:

> Observations of the climate to current day show that the poles have warmed at twice the global average, and computer model predictions suggest this trend will continue.

Here is a more complete picture:

- In recent decades the south polar region has shown little or no warming, as even the IPCC concedes.
- The extent of global sea ice shows remarkably little change over the past 35 years:

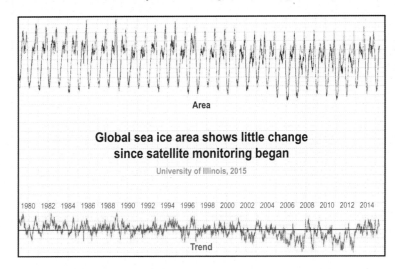

- Computer model predictions of 1.4–4.5°C global warming per CO_2 doubling have remained unchanged for 36 years: yet the IPCC's medium-term global warming predictions made 25 years ago have proven exaggerated by a factor of two. Would it not have been appropriate at least to mention the models' continuing exaggerations?
- Likewise, there is now a substantial list of papers in the reviewed journals suggesting that climate sensitivity could be as little as one-fifth of the IPCC's current central estimate of about 3°C per CO_2 doubling. Mr. Varley has given only one side of the climate-sensitivity case.

8. RISING CO_2 CONCENTRATION AND ITS EFFECT ON GLOBAL TEMPERATURE

Mr. Varley says:

> Since pre-industrial times, the atmospheric concentration of CO_2 has risen by 40% to a level unprecedented in at least 800,000 years … This has led to an enhanced greenhouse effect.

> Scientists have calculated that more than half of the observed warming since the mid 20th century was caused by the increase in man-made greenhouse gases.

Once again, there is plenty of balancing evidence:
Today's CO_2 concentration may be unprecedented in 800,000 years, but, notwithstanding the increase in anthropogenic emissions, to the nearest tenth of 1% there is no CO_2 in the air at all.

- In the Neoproterozoic era, 750 million years ago, the atmosphere was 30% CO_2 and the planet did not fry.
- In the Cambrian era 550 million years ago the concentration was 20–25 times today's.
- In the Jurassic era it was 12–15 times today's. Yet here we all are.
- Mr. Varley has mentioned only the theoretical harm that he imagines warmer weather may cause, without mentioning the many benefits of increased CO_2 concentration, not least in increasing the net primary productivity of trees and plants worldwide by 2% per decade; increasing the yield of staple crops by up to 40% per CO_2 doubling; and increasing the resistance of plants and crops to drought.

- Also, cold is a far worse killer than warmth. It is no accident that 90% of all species live in the tropics and fewer than 1% at the poles.
- Mr. Varley says, "Scientists have calculated that more than half of the observed warming since the mid-20th century was caused by the increase in man-made greenhouse gases." Certainly that is what the IPCC has long maintained. However, in this respect the IPCC is not honoring its obligation to reflect the peer-reviewed scientific literature.
- Though propagandists have sought to maintain that there is a "97% consensus" to the effect that recent global warming is mostly man made, the truth—given in Legates et al. (2015)—is that only 0.3% of climate science papers published in the 21 years 1991–2011 stated that recent global warming was mostly man made:

0.3% consensus, not 97.1%

'The scientific consensus that human activity is very likely causing <u>most</u> of the <u>current</u> GW (anthropogenic global warming, or AGW)' – Cook _et al._ (2013)

11944	abstracts (1991-2011) were reviewed by Cook _et al._	100%
7930	were arbitrarily excluded for expressing no opinion	66.4%
3896	were marked as agreeing we cause <u>some</u> warming	32.6%
64	were marked as stating we caused <u>most</u> warming	0.5%
41	actually stated we caused <u>most</u> warming since 1950	0.3%
0	were marked as endorsing manmade catastrophe	0.0%

- The truth is that at present we are unable to distinguish between the respective magnitudes of the anthropogenic and natural components in the global warming that unaccountably stopped more than 18 years ago.
- In one sense, however, it might legitimately be said that global warming is man made. For the terrestrial temperature records have been relentlessly and unidirectionally altered to make early 20th-century temperatures cooler and later temperatures warmer, in a manner calculated falsely and perhaps substantially to overstate the true warming rate in the 20th century:

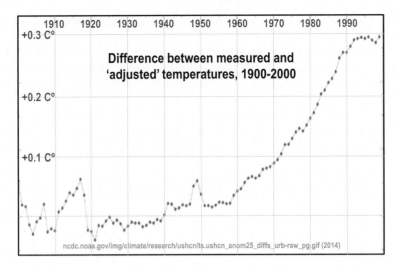

The NCDC's adjustments are influential, because all three of the longest-standing terrestrial temperature records rely on its historical climate network for the compilation of their data sets. The changes made by the NCDC to the historical climate network data in just eight years are shown here:

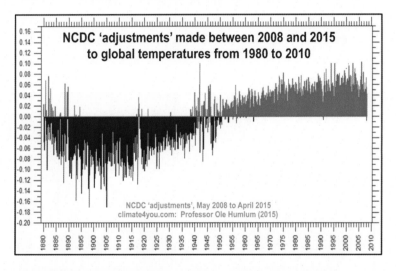

- The tampering over the past seven years shows how earlier temperatures have been pushed ever lower and later temperatures pushed ever higher. There may or may not be legitimate reasons for this tampering, which always appears to go in the direction of amplifying man's influence on climate (the equivalent GISS "adjustment" is even larger than for NCDC), but it introduces an additional uncertainty to temperature measurements that Mr. Varley's article fails to reflect:

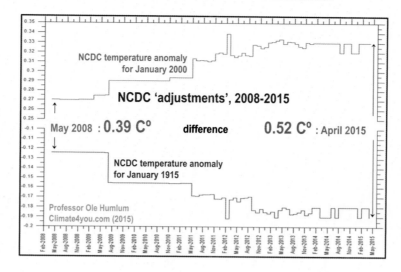

9. ARE THE COMPUTER MODELS OF CLIMATE RELIABLE?

Mr. Varley says:

It is through models that we predict future climate, but not until they have been tried and tested to see how well they reproduce historic climate. Simulations of the future point towards further warming, and changes in all components of the climate system, including means and extremes of temperature, more intense and frequent rainfall events over many land areas, increases in sea level, and further ice melt.

Again, here are some of the balancing considerations omitted by Mr. Varley:

- First, the models have been "tried and tested." They have failed. Anyone can retune them to match past climate. The real test is whether they can predict future climate. They cannot. In 1990 the IPCC predicted that the rate of global warming would be twice what has occurred in the 25 years since then:

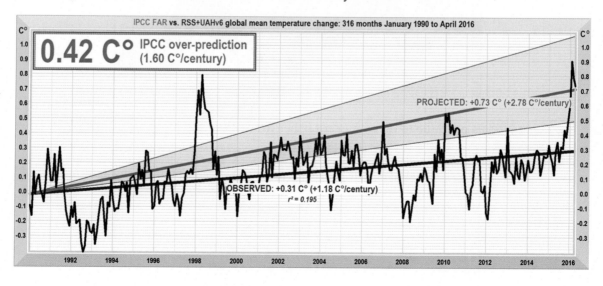

- Second, the IPCC predicted in 2007 that there should have been significant global warming in the decade since 2005. However, there has been hardly any:

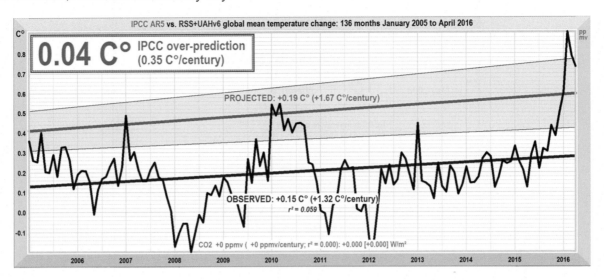

- Third, the IPCC in 2007 and again in 2013 predicted short-term global warming, relying on the CMIP3 and CMIP5 computer models, but again the models' predictions have proven excessive:

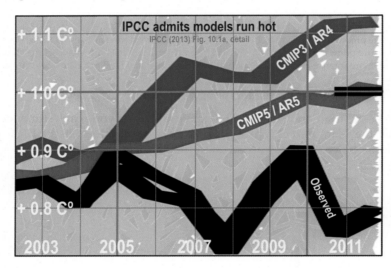

- Fourth, the latest models continue to diverge ever farther from observation:

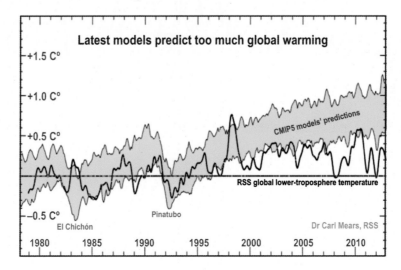

- Fifth, models have also overpredicted regional warming, for instance in the tropics:

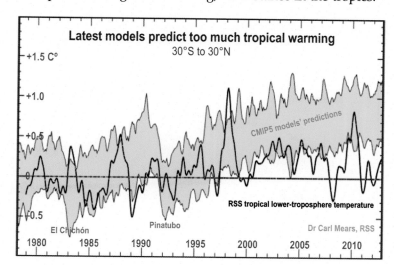

- Sixth, warming at the North Pole has also been somewhat overpredicted:

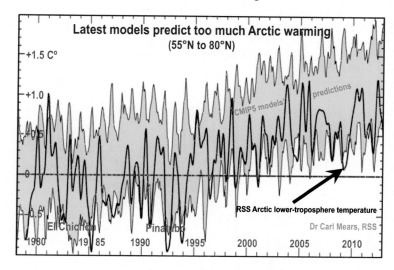

- Seventh, models can be tuned to fit the past but cannot predict the future:

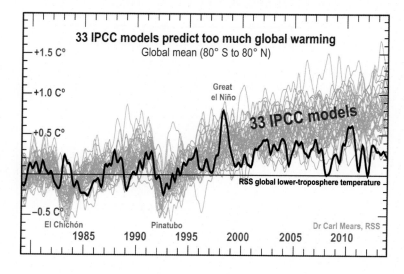

- Eighth, even the oldest weather organization in the world gets it wrong:

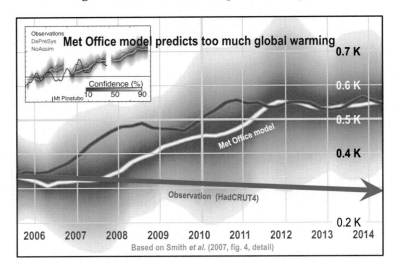

- Ninth, even the earliest predictions were exaggerated, and it was on the basis of these exaggerations that urgent action on climate was demanded:

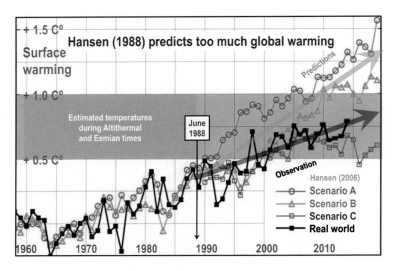

- Tenth, models also overpredict ocean warming, and by a wide margin:

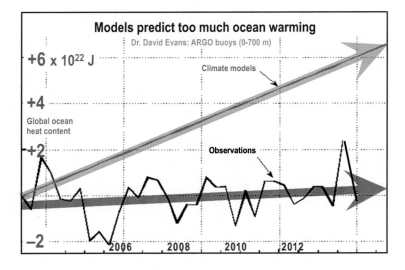

- Eleventh, the prefinal draft of the IPCC's 2013 *Fifth Assessment Report* showed a graph demonstrating that all four previous *Assessment Reports* had flagrantly exaggerated their predictions of the increase in methane concentration. The observed trend falls below the prediction interval in all four previous *Assessment Reports*. However, at the insistence of Hungary and Germany, the graph was removed from the final report not because it was incorrect but because "it might give ammunition to sceptics."

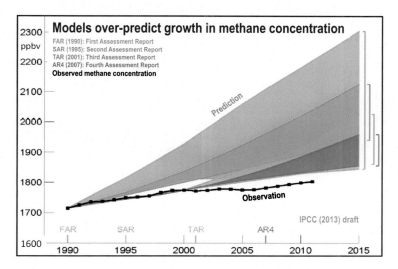

- Twelfth, the models have been particularly bad at predicting temperature trends in the crucial tropical mid-troposphere about six miles up, and the IPCC's self-confidence in models' predictive skill increases as the gap between the models' exaggerations and observed reality widens:

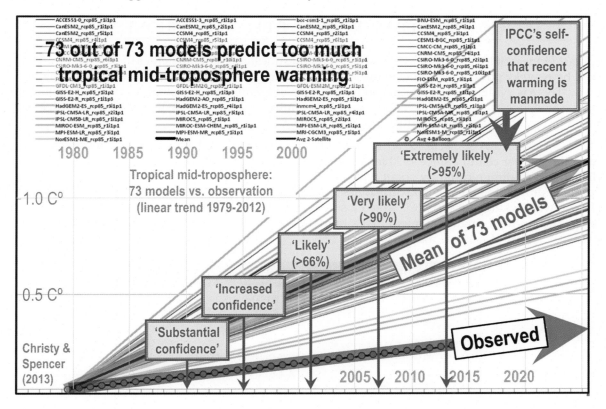

The IPCC has itself conceded that the models in which it once imprudently placed absolute faith have proven defective, at least in the medium term. In 2013 the IPCC admitted that 111 of 114 models had run hot, explicitly abandoned them, and substituted what it called its "expert assessment" for their output. The effect was dramatic: the IPCC all but halved its predictions of near-term warming:

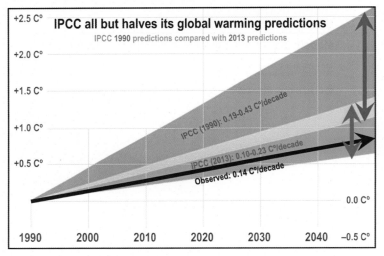

The models, then, have not proven reliable. Though the chief executive of the Met Office might be forgiven for not mentioning as many failures by the models as have been illustrated here, no account of the supposed threat from global warming will be balanced unless it records that the models have erred, and have very nearly always erred on the side of considerable exaggeration of what may not after all be a threat.

10. HAS CLIMATE SCIENCE BECOME DISHONEST?

The one-sidedness of Mr. Varley's article raises legitimate questions about whether those who profit from the vast sums paid by panicky governments to climate science are acting not only in a fair and balanced way but also in an honest way. There is evidence of unethical conduct by a small number of influential scientists promoting alarm that, based on the evidence, is unjustifiable. Some examples of outright dishonesty will now be given. Mr. Varley's article contains no hint of this unethical conduct.

- The IPCC's 1990 *First Assessment Report* had stated its "substantial confidence" in the models: yet the warming trend in the quarter century since 1990 falls substantially below the entire interval that the IPCC had then predicted. Overclaiming certainty about what is uncertain is the central dishonesty in climate.
- The prefinal draft of the IPCC's 1995 *Second Assessment Report* stated five times that evidence for human influence on climate was lacking, but the IPCC asked a single scientist to rewrite the report to remove all five references, replacing them with a single statement that a human influence on global climate was now discernible. The "consensus" is that of just one man:

IPCC (1995): Scientists' final draft	As published
"None of the studies cited above has shown clear evidence that we can attribute the observed [climate] changes to the specific cause of increases in greenhouse gases."	
"No study to date has positively attributed all or part [of observed climate change] to anthropogenic causes."	"The body of ... evidence now points to a discernible human influence on global climate."
"While none of these studies has specifically considered the attribution issue, they often draw some attribution conclusions, for which there is little justification."	
"Claims of positive detection of significant climate change are likely to remain controversial until uncertainties in the total natural variability of the climate system are reduced."	
"When will an anthropogenic effect on climate be identified? It is not surprising that the best answer to this question is, 'We do not know.'"	Dr Ben Santer, L.L.N.L. (consensus of *one*)

- The front cover of the World Meteorological Organization's (WMO) 1999 *State of the Global Climate* report showed three attempted reconstructions of 1000 years' temperature changes derived from tree rings (below left): however, from 1960 onward the true tree ring data (white inset panel, below right) did not reflect the observed warming. The true data for one of the three (green, below right) showed a decline where thermometers showed an increase. To conceal these divergences, the WMO graph spliced the last 50 years' measured warming (black, below right) onto the tree ring data, but without disclosing that that was what had been done. All three tree ring records were tampered with to conceal the splicing, making it appear that they matched real temperatures. In particular, the pronounced decline in the green tree ring dataset, a decline that was directly contrary to measured global temperatures and hence establishing that tree rings are not a suitable way to reconstruct past global temperature change, was eradicated:

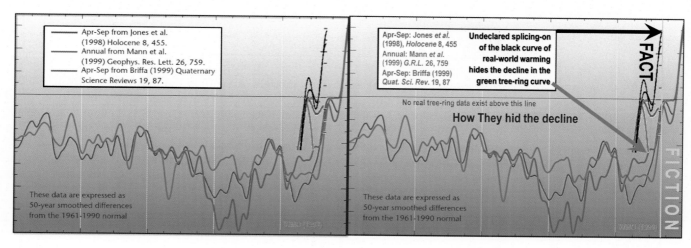

- The "hockey stick" reconstruction of the past 1000 years' global temperatures in the IPCC's 2001 *Third Assessment Report* (below right), visibly similar to the WMO's graph, also purported to abolish the medieval warm period. The IPCC adopted this doctored graph as its logo until independent research showed it was a fabrication. Subsequently, the author of one of the three tree ring data sets and of the hockey stick graph denied under oath in court that he was an author of the WMO's 1999 graph: yet his name appeared on the graph and in the acknowledgments on the inside front cover of the publication:

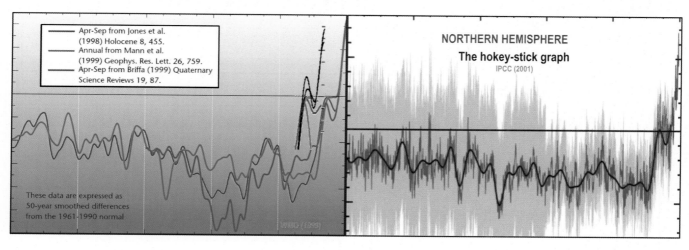

- The hockey stick graph is inconsistent with empirical reconstructions in some 450 peer-reviewed papers. It is also inconsistent with the sea level record and with the previous understanding of the past 1000 years' temperature change:

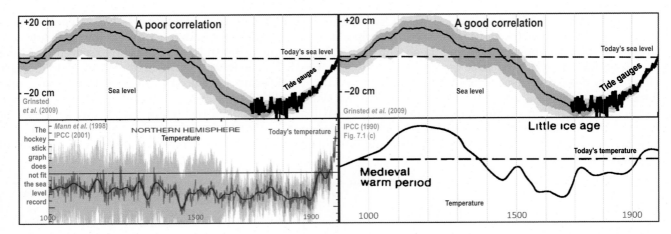

- The IPCC has refused to correct a major statistical error in its models. An influential graph in the IPCC's 2007 *Fourth Assessment Report* showed the temperature record of the previous 155 years overlain by four linear trend lines starting at 150, 100, 50, and 25 years previously (below). Each successive trend line was steeper than the last. The offending graph was displayed twice in the report, each time with the conclusion that the rate of global warming was accelerating and that humans were to blame. In fact, the slope of the 25-year trend line had two previous precedents in the temperature data (yellow arrows, below right), so there had been no acceleration in the warming rate. The fact that there has been no global warming for more than 18 years confirms the absence of any acceleration. The IPCC had used a false statistical technique:

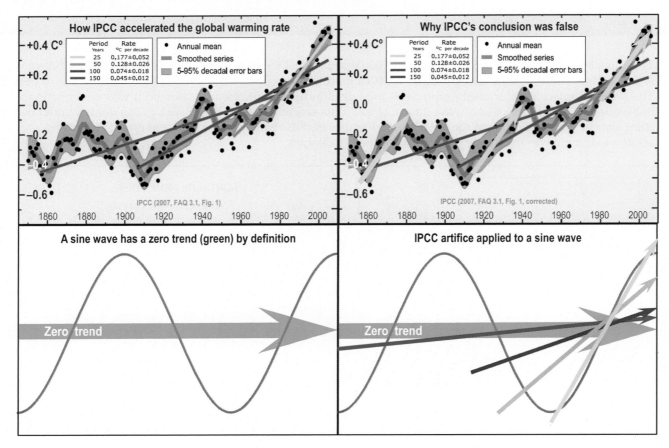

- A sine wave (above left) has a zero trend by definition, but the same false technique can be made to show that it has an apparently accelerating uptrend (above right). The IPCC refused to correct the error when one of its own reviewers complained about it.

- As noted earlier, there has been systemic tampering with the terrestrial temperature records so as to arbitrarily to depress the true temperatures in the early 20th century and to elevate them at the end of the century in a manner calculated artificially to increase the apparent rate of global warming. The satellite records, however, are not so easily tampered with. Since they began in 1979, they show appreciably less warming than the much-altered terrestrial records. In April 2015, Professor Terence Kealey, former vice-chancellor of Buckingham University, announced an independent inquiry into the tampered terrestrial records.

11. ARE FLOODS AND DROUGHTS WORSENING AND CROPS FAILING AS PREDICTED?

Mr. Varley says there are:

... implications for flooding, drought, crop production, inundation of coastal communities, and threats to ecosystems unable to adapt quickly enough to the rapid rate of climate change.

The balancing facts:

- CO_2 is good for crops. Crop yields have grown with particular rapidity since 1950. There are three main reasons: improved agricultural practices, warmer weather, and CO_2 fertilization (nearly all anthropogenic CO_2 was emitted after 1950):

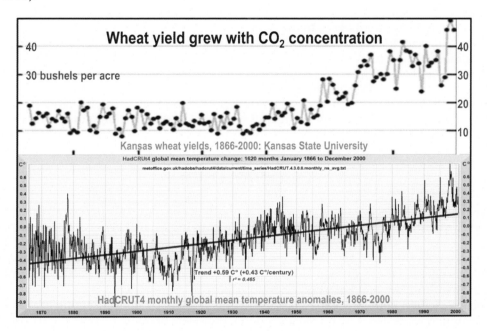

- Rising CO_2 in the air causes a very rapid increase in combined crop yield:

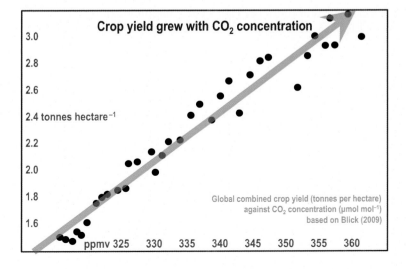

- As for droughts, the area of the globe suffering drought conditions has declined throughout the past 35 years:

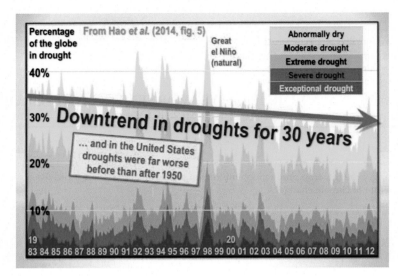

- As for "inundation of coastal communities," sea level is not rising fast enough to make much difference globally; nor, with present trends, is it at all likely to do so.
- As for adaptation of species to "rapid climate change," the conspicuous feature of the past two decades is the very slow—indeed, almost nonexistent—rate of global warming. There is no evidence that the rate of global warming seen in the terrestrial and ocean data sets—equivalent to 0.25°C per century—will accelerate tenfold by 2100, as it would have to do to reach the IPCC's central estimate of 21st-century global warming.
- Temperatures vary by as much as 100F° between midnight and midday in some places. The notion that species cannot adapt to a warming one-twentieth as big as this diurnal variation is ill founded, and the evidence to date indicates that no warming at that rate is at all likely.

12. CONCLUSIONS

The Met Office makes very large sums every year out of climate change. It is part of an international network of governmental and corporate interests that benefit greatly from giving a narrowly one-sided view of global warming science.

The wider range of scientific facts and results than those that Mr. Varley chooses to put forward surely demonstrate that—at the very least—there are two sides to the climate question. And it is equally surely the duty of the Met Office to take a neutral, fair, and balanced scientific stance.

On the evidence presented here, Mr. Varley has misled his readers by not presenting a balanced account of the state of global warming science. He is by no means unique. Profiteers of doom all over the world have taken advantage of the near-universal ignorance of science among politicians, press, and public. That ignorance is costly, not only in treasure but also in lives. It is too often falsely claimed that climate change harms the poor. There has not been enough change to harm anyone, nor will there be. However, misguided policies to make the rich richer by addressing the nonproblem that was global warming are already making the poor poorer still.

References

Jouzel, J., Masson-Delmotte, V., Cattani, O., et al., 2007. Orbital and millennial Antarctic climate variability over the past 800,000 years. Science 317, 793–796.

Legates, D.R., Soon, W., Briggs, W.M., 2015. Christopher Monckton of Brenchley. 2015. Climate Consensus and 'Misinformation': A Rejoinder to Agnotology, Scientific Consensus, and the Teaching and Learning of Climate Change. Science and Education 24, 299.

Ljungqvist, F.C., 2010. A new reconstruction of temperature variability in the extra-tropical Northern Hemisphere during the last two millennia. Geografiska Annaler: Physical Geography 92 A (3), 339–351. September 2010.

C H A P T E R

5

Southeast Australian Maximum Temperature Trends, 1887–2013: An Evidence-Based Reappraisal

J.J. Marohasy, J.W. Abbot

The Climate Modelling Laboratory, Noosa Heads, QLD, Australia; Institute of Public Affairs, Melbourne, VIC, Australia

O U T L I N E

1. INTRODUCTION

Climate change is generally reported as an annual mean increase in land surface temperatures of 0.8°C globally since 1880 (eg, Lawrimore et al., 2011). In Australia the trend is typically reported as 0.9°C since 1910, with maximum temperatures warming at a rate of 0.8°C per century and minimum temperatures at a rate of 1.1°C per century (Bureau of Meteorology and CSIRO, 2014). These trends are derived from the compilation of adjusted, weighted, and then gridded, surface temperature recordings from weather stations. A high degree of consistency has been reported in national and global trends (Brohan et al., 2006; Fawcett et al., 2012). The adjustment process is called homogenization, and consistent with the meaning of the word, ensures that temperature trends within and between regions, across nations, and around the globe are somewhat uniform (Peterson et al., 1998; Fawcett et al., 2012).

The first homogenized temperature data set for Australia was created by Torok and Nicholls (1996), based on the detection of discontinuities in unique temperature series when compared with neighboring stations. The homogenization technique employed by Torok and Nicholls (1996) was derived from Easterling and Peterson (1995), and

Evidence-Based Climate Science, Second Edition
http://dx.doi.org/10.1016/B978-0-12-804588-6.00005-7

details the combination of regression analysis and nonparametric statistics used to achieve "relative" homogeneity and to detect undocumented discontinuities relative to neighboring stations. The same principles are now applied in the development of national and global temperature databases, for the most part through automated algorithms which, under some supervision, can simultaneously and/or sequentially compare data from neighboring weather stations.

Recently published studies of the effect of homogenization on regional and national temperature trends have found that such adjustments can result in an artificial warming bias (Zhang et al., 2014; De Freitas et al., 2014; Oyler et al., 2015; Stockwell and Stewart, 2012; Boretti, 2013). Homogenized temperature series are also increasingly contested on popular weblogs (eg, Nova, 2015) and in the mainstream media (eg, Booker, 2014).

Adjustments made by the Australian Bureau of Meteorology to temperatures as measured at an agricultural research station near the town of Rutherglen in southeast Australia are arguably the most contentious in the Australian Climate Observations Reference Network–Surface Air Temperature data set. This is the data set used to report official national and regional trends. Through the application of a percentile-matching algorithm, the Bureau has changed a slight cooling trend of 0.35°C per century in the *minimum* temperature series as measured at Rutherglen, into dramatic warming of 1.73°C per century in the official ACORN-SAT record for Australia (Lloyd, 2014a, 2014b). The adjustments were made by the Bureau independently of any documented changes in equipment, site location, or recording time, but ostensibly on the basis that comparative sites showed warming (Bureau of Meteorology, 2014a).

A recent analysis of the actual measured temperatures from the listed comparative sites for Rutherglen that have long and homogenous records found that, contrary to the general advice from the Bureau (2014a), there is statistically significant cooling in the temperatures record from 1913 to the present (Marohasy and Abbot, 2015a). This cooling is attributed to changes in local land use associated with the building of the Snowy Mountain's hydroelectricity scheme, and over 800,000 ha (nearly 2 million acres) of land newly under irrigation in the immediate vicinity of Rutherglen, specifically in a region known as the Riverina (Marohasy and Abbot, 2015a).

Here we extend our study of minimum temperatures at Rutherglen, to show how the longest continuous *maximum* temperature series for the Riverina is also remodeled by the Bureau, again changing an overall cooling trend into warming. Specifically, the temperature series for Deniliquin is truncated and then homogenized by the Bureau to show dramatic warming of +1.2°C per century. We made this discovery as we searched for the longest continuous temperature series from southeast Australia with the objective of creating a realistic, replicable, historical temperature profile for the region. To the extent possible, we wanted to understand the history of climate variability and change in southeast Australia, before data were subjected to the Bureau's percentile-matching algorithm (Trewin, 2013).

For the purposes of this study, southeast Australia is defined as the landmass east of 140° east longitude and south of 35° south latitude, including the entire state of Victoria. A schematic representation of this region is shown in Fig. 5.1, excluding the island of Tasmania.

Global warming is commonly reported as an average increase in mean temperatures. A problem with this approach, however, is that the mean, by its very nature is not a real physical measurement. The mean is a statistic calculated as the average of the maximum and minimum, which may not always trend in the same direction. For example, during a period of drought, while maximum temperatures may increase, minimum temperatures may fall because of clear skies at night. Arguably, measures of maxima give a better indication of regional climate change because of higher rates of turbulent mixing of the lower atmosphere in the daytime.

2. MATERIALS AND METHODS

2.1 Choice of Locations and Start Date

The southeast corner of Australia includes the state of Victoria. The Bureau of Meteorology has 11 locations in its ACORN-SAT data set from Victoria, which are used to generate the official temperature profile for this state. In this study we used only three of these, Cape Otway lighthouse, Wilsons Promontory lighthouse, and Melbourne Regional Office.

We wanted records that began on, or before, 1887 with very little, if any, missing data. The year 1887 was chosen as it predates the notorious Federation drought (1895–1902), and potentially provides 126 years of data, with an equal number of years before and after 1950; a year possibly marking a climate shift in the Australian temperature record.

Of the other locations used in ACORN-SAT, Kerang has one of the longest records, but it is not continuous, with many months of data missing in 1908 and through the 1940s, 1950s, 1960s, and 1990s. The Mildura record does not

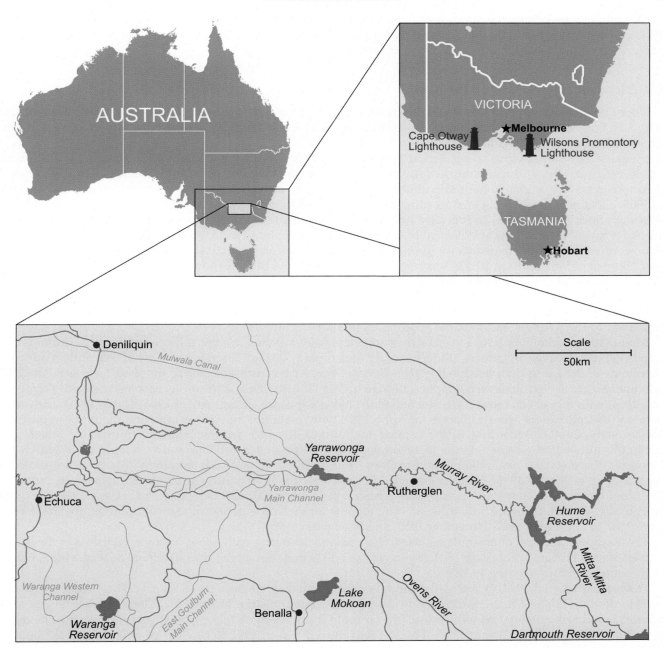

FIGURE 5.1   Locations of Melbourne, Hobart, Wilsons Promontory Lighthouse, Cape Otway Lighthouse, Echuca and Deniliquin in South East Australia.

begin until about 1890, and there is a significant site change in about 1950. The Nhill record does not begin until about 1895, and ends in 2010. The record for Sale begins in about 1895, and there was a significant site move in 1945. Orbost, Laverton, and Rutherglen are the other sites used by the Bureau to generate a temperature profile for Victoria, but the records for these centers do not begin until about 1940, 1941, and 1913, respectively.

From our study of the 27 sites listed by the Bureau as comparative sites for Rutherglen (Bureau of Meteorology, 2014a), we identified that only Echuca and Deniliquin had long continuous records predating 1887 (Marohasy and Abbot, 2015a).

In this study, we also consider the temperature record for Hobart, the capital of the island state Tasmania, which has a record that is almost as long and continuous as the record for Melbourne. Hobart, however, was not included in the final calculation of the weighted mean for southeast Australia.

Monthly raw maximum temperatures can be downloaded from the Bureau website for each of these six locations. The early data were manually read from liquid-in-glass thermometers. All except the measurements from Echuca, however, are now made by automatic weather stations.

2.2 Metadata and Regional Histories

Cape Otway and Wilsons Promontory lighthouses are located at the extreme southeast of the Australian continent (Fig. 5.1), both with exceptionally long and continuous temperature records unlikely to be effected by changes in local land use or urban heat islands.

Mercury thermometers were installed at Wilsons Promontory (Bureau no. 85096) in November 1872 (Bureau of Meteorology, 2014b), and temperature data are available for downloading as monthly values from 1878. How the thermometers were originally housed is unknown. According to Torok (1996), a Stevenson screen was installed prior to 1905, and a new screen in 1933. According to the ACORN-SAT station catalog (Bureau of Meteorology, 2012), the area was affected by bushfires in February 1951, but the instrument enclosure was not directly affected. According to Torok (1996), all instruments at Wilsons Promontory had to be replaced in 1959 following a storm. However, this is not recorded in either the metadata for the station (Bureau of Meteorology, 2014b) or in the official ACORN-SAT catalog (Bureau of Meteorology, 2012). According to the ACORN-SAT catalog and the metadata, an automatic weather station was installed in September 2000.

The maximum temperature series from 1887 to 2013 is complete, except for six missing months in 1951, eight missing months in 1954, and one missing month in 1976. Of some concern to us, solar panels were installed in close proximity to the weather station at the Wilsons Promontory lighthouse in 1993.

Manually read, mercury liquid-in-glass thermometers were installed in a thermometer shed in the grounds of the Cape Otway lighthouse in January 1865 (Bureau of Meteorology, 2014c). Alterations were subsequently made to this thermometer shed, and completed in July 1898 (Torok, 1996). A Stevenson screen was supplied "at least before 1902," and then a new Stevenson screen supplied in March 1908 (Torok, 1996). A third Stevenson screen was supplied in July 1954, with this screen moved to a position with "better exposure" in September 1966 (Torok, 1996). The original mercury thermometers were replaced in April 1994 with platinum resistance temperature probes (Bureau of Meteorology, 2014c). According to the catalog for all ACORN-SAT weather stations (Bureau of Meteorology, 2012), this automated weather station installed in April 1994 proved unreliable, and was replaced in March 1995. However, the metadata available online for Otway lighthouse (Bureau of Meteorology, 2014c) do not record the replacement in March 1995. These metadata indicate that this temperature probe was replaced in November 2010, and that the mercury thermometer was replaced in October 2002 and December 2012. Cape Otway is Bureau site number 090015, there are only three missing months for the entire period of the maximum temperature record from 1887 to 2013: May 1957, December 1958, and September 1995.

The European settlement of southeast Australia began in Tasmania with a first colony established in the location of present-day Hobart in February 1804. It was not until 1834 that the current site of Melbourne was settled, with the first immigrant ships arriving in 1839.

Hobart (Bureau no. 094029) is now the capital of the island state of Tasmania (Fig. 5.1). The weather station has been located on Ellerslie Road, which is now an inner city suburb, since March 1881. The thermometers were first housed in a thermometer shed with a double iron roof (Torok, 1996). A Stevenson screen is thought to have been installed soon after 1895 (Torok, 1996). The site was moved 5 m in 1992 to be clear of a newly erected building. At the same time, June 29, 1992, an automatic weather station was installed (Bureau of Meteorology, 2012) and became the primary instrument for measurement on November 1, 1996. Data are missing for 13 months (02/1881, 08/1881–03/1882, 03–06/1892) from an otherwise complete record for the period March 1881 to December 2013.

Melbourne is located on Port Phillip Bay in the far southeast of mainland Australia (Fig. 5.1). Melbourne was officially declared a city by Queen Victoria in 1847 and became the capital of the newly founded colony of Victoria in 1851. During the Victorian gold rush of the 1850s, it was transformed into one of the world's largest and wealthiest cities, with a population of 490,000 in 1890. After the federation of Australia in 1901, Melbourne was the nation's interim seat of government until 1927. The population of Melbourne has continued to grow, reaching 4.4 million in 2014.

The Melbourne weather station is at a site with the official name "Melbourne Regional Office." However, this weather station has always been in the Melbourne central business district adjacent to two major roads (Bureau no. 086071). Temperatures have been recorded in the general vicinity of this site since June 1855. The site is known to be affected by an urban heat island with increasing levels of road traffic and general urbanization over the period of the record. The site is surrounded by tall buildings, with the large apartment building constructed in 2006 thought to have had a particularly significantly impact on the local climate. The construction of this building coincided with the time that the automatic weather station, installed in August 1986, became the primary instrument in November 1996. When a Stevenson screen was installed at this site is unknown. The maximum temperature series for Melbourne is complete for the entire period of the record from May 1855.

Echuca is a town located on the banks of the Murray and Campaspe Rivers in Victoria (Fig. 5.1). Temperatures were first recorded in a thermometer shed in the grounds of a police station in June 1881 (Torok, 1996). When a Stevenson screen was installed is unknown, but probably between 1906 and 1908 (Nicholls et al., 1996). The site was moved to the Echuca post office in 1939, moved again in 1967, and then to the airport in 1985. During the entire period of this record (06/1881—12/2013), the site retained the same Bureau no. 80015. Only three months of data (07/1902, 08/1902, 02/2012) are missing for the entire period.

Deniliquin was established on the convergence of major stock routes between the colonies of Queensland, New South Wales, and Victoria, with the post office opening in January 1850. Overstocking, combined with plagues of rabbits, and then drought resulted in the collapse of the local pastoral industry in the late 1800s. Irrigation in this region began in the early 1900s, and had a measurable effect on stream flows from the 1920s (Wen, 2009). The Hume reservoir (Fig. 5.1) was completed in 1936. The Snowy Mountains Scheme, which channels water which once flowed east to the Pacific Ocean, west of the Great Dividing Range to the Riverina, took 25 years to build. This scheme now comprises 16 major dams, including the Darmouth Reservoir (Fig. 5.1). The Yarrawonga weir was completed in 1939 and is the point of greatest diversion of water from the scheme. The Mulwala canal and Yarrawonga main channel (Fig. 5.1) distribute water from the Yarrawonga weir to over 800,000 ha (nearly 2 million acres) of irrigated farmland in the immediate vicinity of both Deniliquin and Echuca (Murray Darling Basin Authority, 2005).

Temperatures were measured at Wilkinson Street, Deniliquin, from February 1867 to June 2003 (Bureau no. 074128). This site was closed, and temperatures were recorded from an automatic weather station at the Deniliquin airport from June 1997 (Bureau no. 74258). There are only three missing months of data for this entire record: November 1958, March 1959, and February 1967.

2.3 Quality Control

In modern climate science, the intrinsic quality of output from a weather station is always assessed relative to output from other weather stations. In modern manufacturing, however, quality control is not dependent on comparisons with output from production lines at other factories. Control charts are used extensively in manufacturing industries for assessment of the production process through analysis of the variance in one or more important characteristics of the final product over a specific time period (Ryan, 1989; Taylor, 1991). The relatively simple transparent statistical technique also has application for the quality control of temperature time series.

In this study, temperatures are output from mercury thermometers or automatic weather stations deployed with the specific purpose of producing accurate representations of surface air temperatures. This type of temperature data always displays diurnal and seasonal variation, which can be accommodated when using monthly data in a control chart by focusing on the annual mean as the subgroup mean and treating each of the 12 months within a calendar year as a measure of variation in the process. The mean monthly temperature, as a measure of within-sample variation, can thus also be used to calculate upper and lower control limits, set at three standard deviations from the actual process average. I-M/R—R/S Control Charts, when deployed using the statistical software Minitab (version 17), can also provide a visual representation of changes in the standard deviation of the annual mean for the period of each series, as well as the moving range of the annual mean.

A main purpose of homogenization in modern climate science is to correct for discontinuities created by equipment and site changes (Hansen et al., 2001). As control charts are a measure of the within- and between-year variation, major discontinuities should be evident as step-change, resulting in exceedance of the upper and/or lower control limit; and minor discontinuities as step-change in the position of the subgroup mean (ie, annual mean) above or below the center line. Each of the six temperature series considered in this study was thus subject to assessment using an I-M/R—R/S Control Chart.

2.4 Adjustments to the Deniliquin Record

In September 2014, following articles in *The Australian* newspaper, the Bureau published a list of adjustments made in the creation of individual ACORN-SAT series and the individual comparative sites used in the detection of discontinuity (Bureau of Meteorology, 2014d). More information on the statistical technique is available in Trewin (2013).

TABLE 5.1   Adjustments Made to Maximum Temperatures as Recorded at Wilkinson Street, Deniliquin

Series	Date	Adjustment	Reason
ACORN-SAT	27/09/1984	+0.04	"Move"
	13/08/1971	−0.16	"Move"
	01/01/1960	−0.54	"Statistical"
	01/01/1950	+0.56	"Site environment"
	01/01/1943	−0.38	"Statistical"
Marohasy and Abbot	01/1908	−1.93	Discontinuity detected using control chart, likely associated with installation of Stevenson screen

Adjustment Applied to all data prior to the date shown. The ACORN-SAT record begins in January 1910. The Marohasy and Abbot series begins in 1887.

Adjustments made to the Deniliquin maximum temperature record between 1910 and 1984, as detailed in this document, are shown in Table 5.1.

While five discontinuities have been identified by the Bureau in the Deniliquin maximum temperature series (Table 5.1), we found no discontinuities using I-M/R–R/S control charts in the same period of record (Fig. 5.2). The exceedance of the moving range of the subgroup means (Fig. 5.2, middle chart) in 1957 is acknowledged, but considered real and climatic, due to exceptionally heavy rainfall and flooding in the previous year in this region

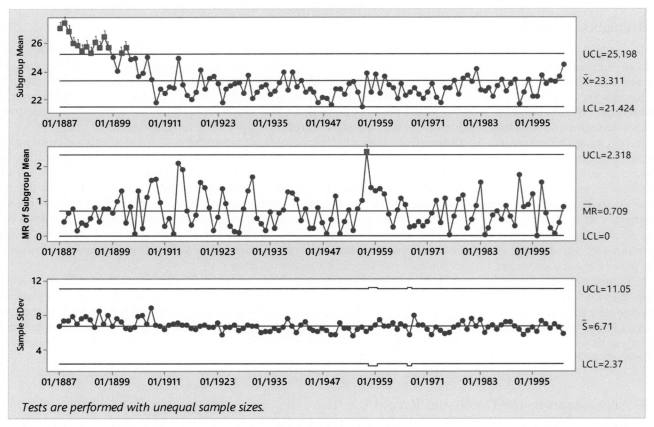

Tests are performed with unequal sample sizes.

FIGURE 5.2   I-MR–R/S Control Chart showing raw maximum temperatures as recorded at Wilkinson Street, Deniliquin (1887–2002). The subgroup mean shown (top chart) represents the means for each year eliminating the within-subgroup component of variation (ie, the seasons), thus tracking process location (ie, annual temperature change). The moving range (MR) of the subgroup mean (middle chart) tracks between year variation. The sample standard deviation (bottom chart) plots process variation using the within-subgroup (ie, within year) component of variation.

(Close, 1990; Condon, 2002). We identified discontinuity in the pre-1910 record (Fig. 5.2), determining it was likely due to the installation of a Stevenson screen in 1908 (Fig. 5.2).

The adjustment of −1.93°C to all data before 1908 was determined through calculation of the average difference in the two methods of producing a measure of surface temperatures before and after 1908 (ie, with, and without, a Stevenson screen). Specifically, this difference was calculated by subtracting the mean of the annual means for the years 1887−1907, from the mean of the annual means for the years 1909−29. This difference is 2.61°C. We then subtracted 0.17°C for each decade (the extent of cooling at the nearby location of Echuca, Table 5.2) for this approximately four-decade period, to account for the cooling over the same period. We thus determined that all measurements before 1908 at Deniliquin could be corrected through the subtraction of 1.93°C (ie, 2.61−(0.17*4)); Table 5.1.

The raw temperature record for the Deniliquin Wilkinson Street site (Bureau no. 74128) ends in June 2003. Data are available for a period of overlap with the Deniliquin airport (Bureau no. 74258) from June 1997 to June 2003 when both weather stations were operational. We performed a two-sampled t-test of paired values to determine whether the mean of the values measured at Wilkinson Street (23.5) was different from the mean of the values measured at the airport (23.3) at alpha 0.05 (Minitab version 17). This test determined that the mean of the values from Wilkinson Street was statistically different from the mean of the values from the airport ($p < 0.005$). A plot of the paired values showed that the airport was cooler in winter and warmer in summer, so any adjustment would need to be seasonally specific. This difference, however, would be within the standard error of the mean when calculated for the entire

TABLE 5.2 Temperature Trends as Rate per Decade Based on Annual Mean Temperatures for the Period 1887−2013, and Discrete Periods Therein, for Raw and Homogenized Temperature Series, Including Composites

Location and/or series	1887−2013	1887−1949	1911−40	1921−50	1950−2013	1951−80
Cape Otway						
Raw, no infill	**−0.09**	**−0.53**	−0.04	**−0.19**	**+0.10**	**+0.15**
Raw, infilled, series 1	**−0.09**	n/a	−0.04	**−0.19**	n/a	n/a
Homogenized, series 9	**+0.06**	n/a	−0.04	**−0.19**	n/a	n/a
Homogenized, ACORN-SAT	n/a	n/a	−0.04	**−0.19**	n/a	n/a
Wilsons Promontory						
Raw, no infill	**+0.03**	**−0.10**	+0.03	−0.03	**+0.17**	**+0.27**
Raw, infilled, series 12	**+0.03**	−0.09	+0.03	−0.03	+0.17	n/a
Homogenized, ACORN-SAT	n/a	n/a	+0.03	−0.01	n/a	n/a
Echuca						
Raw, no infill	+0.01	**−0.17**	−0.13	+0.02	**+0.19**	+0.22
Deniliquin						
Raw, no infill (Wilkinson St)	n/a	**−0.65**	+0.03	**−0.26**	n/a	+0.04
Homo. and composite (M&A)	−0.02	**−0.24**	+0.03	**−0.26**	**+0.16**	+0.04
Homogenized, ACORN-SAT	n/a	n/a	0.00	−0.15	**+0.15**	**+0.32**
Hobart						
Raw, no infill	**+0.06**	**−0.06**	+0.05	−0.05	**+0.17**	**+0.26**
Melbourne						
Raw, no infill	**+0.10**	+0.05	+0.23	−0.04	**+0.21**	−0.01
SE Australia						
Homogenized, weighted ACORN-SAT	n/a	n/a	+0.05	−0.14	**+0.20**	**+0.23**
Homo. & composite, weighted (Marohasy & Abbot)	+0.03	**−0.15**	0.00	−0.11	**+0.19**	+0.21

Bold values are statistically significant based on analysis of variance for regression of annual temperatures as a fitted line plot ($p < 0.05$). The numbered series are as described in Marohasy and Abbot (2015b).

Wilkinson Street record. No adjustment was thus made to the Deniliquin record for the site change from Wilkinson Street to the airport.

2.5 Calculating the Weighted Mean for Southeast Australia

The Bureau publishes maximum temperature anomalies for each state (Bureau of Meteorology, 2015a). This anomaly is based on the application of a complex weighting system to each of the individual series, specifically, "Temperature anomalies…are based on daily and monthly gridded data with more than one station contributing towards values at each grid point. Unlike simpler methods such as Thiessen polygons, there is no specific set of weights attached to these. The effective contributions change on a daily or monthly basis, depending on which station did or did not report on any given day or month" (Bureau of Meteorology, 2015b).

We wanted a much simpler and more transparent system. The city of Melbourne only covers about 0.02% of the landmass of Victoria. However, other places in southeast Australia are also likely to be affected by urban heat islands. The temperatures as recorded at the lighthouses are likely to be indicative of temperatures in coastal and near-coastal parts of southeast Australia. The temperatures at Echuca and Deniliquin are likely to be indicative of inland regions, particularly the irrigation areas of the Riverina, and also other irrigation areas including Sunraysia. After studying a map of topography and land use in southeast Australia east of 140° east longitude and south of 35° south latitude, we made some arbitrary decisions about relative weightings. In particular, we determined that temperatures at Echuca and Deniliquin would be representative of 54% of the landmass of southeast Australia, the lighthouses 45%, and Melbourne 1%. We applied these weightings accordingly to develop a single temperature series for southeast Australia from 1887 to 2013.

2.6 Calculating Trends

Climate change is typically reported as the slope of a linear regression fitted to mean annual values (eg, Bureau of Meteorology and CSIRO, 2014). While the resulting calculation gives a rate per year, this is typically reported as a rate per century in the popular press and also increasingly in technical papers. In the context of this study—where we were not only interested in change over the entire period of the record (1887–2013) but also change for discrete intervals within this period—we report the rate of change per decade in Table 5.2. For some of the locations, we also show temperature trends for ACORN-SAT, and for series previously calculated for the lighthouses (Table 5.2). In the previous study of the lighthouses (Marohasy and Abbot, 2015b), it was evident that the cooling in the first half of the record is most obvious for the period 1921 to 1950, and so this discrete interval is used again in this study (Table 5.2).

3. RESULTS AND DISCUSSION

3.1 Cape Otway Lighthouse

Cape Otway has one of the longest temperature records of anywhere in the Southern Hemisphere and probably the longest continuous record of anywhere in Australia unaffected by an urban heat island. In a previous study (Marohasy and Abbot, 2015b), we used I-M/R—R/S charts to shows two discontinuities associated with changes to the housing of the thermometers at this lighthouse. Correcting for these discontinuities through seasonally specific reductions of 1.46 and 1.31°C to all temperatures before 1898 and 1908, respectively, resulted in the creation of the series shown in Fig. 5.3. For the period of the record from January 1887 to December 2013, the subgroup mean, which is the annual mean, fluctuates around the overall mean of 16.7°C (Fig. 5.4, top chart). There is a single exceedance of the lower control limit in 1949, and of the upper control limit in 2013, which are the coldest and hottest years in the record, respectively (Fig. 5.3). The moving range of the subgroup mean (middle chart), and standard deviation of the annual means (bottom chart), fluctuate within their respective control limits without showing significant change over the period of the record (Fig. 5.3).

No discontinuity is evident in the record at 1994 (Fig. 5.3). This is the year an automatic weather station was installed at Cape Otway. Yet according to the Bureau's ACORN-SAT adjustment summary (2014d), this equipment change caused a disruption to the record, and as a consequence, in the ACORN-SAT record for Cape Otway all temperatures before 1994 are lowered by 0.5°C. This discontinuity was apparently detected by the Bureau after comparisons were made using the percentile-matching algorithm, specifically using 1996—2000 as the reference period and 10 towns, cities, and airports in the southeast of Australia as comparative stations (Bureau of Meteorology, 2014d).

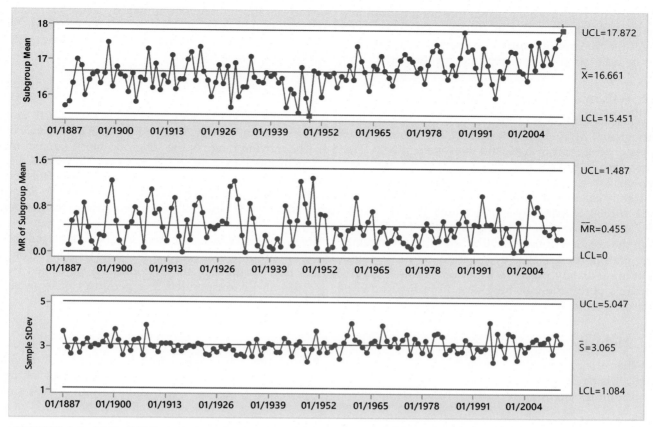

FIGURE 5.3   I-MR–R/S Control Chart showing homogenized maximum temperatures as recorded at Cape Otway (1887–2013).

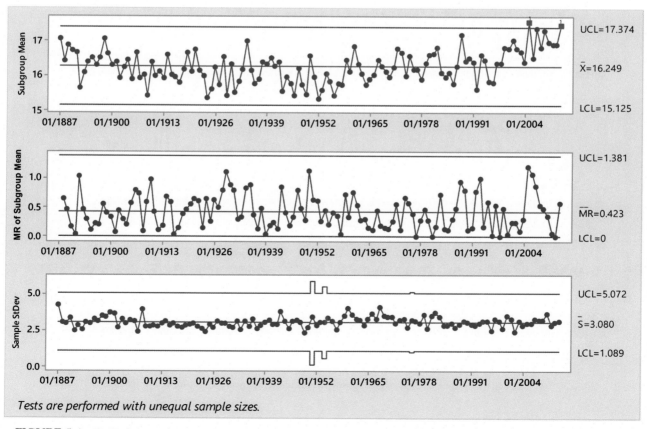

Tests are performed with unequal sample sizes.

FIGURE 5.4   I-MR–R/S Control Chart showing raw maximum temperatures as recorded at Wilsons Promontory (1887–2013).

Of concern to us, however, is that most of the locations have records shorter than one climate cycle (30 years), and in the case of Colac (Bureau no. 090174), data are available only for two of the four reference years (1996 and 1997).

All the maximum temperature series for Cape Otway lighthouse show significant cooling, equivalent to 0.19°C per decade (almost 2° per century) for the period from 1921 to 1950 (Table 5.2).

3.2 Wilson's Promontory Lighthouse

When monthly values for Wilson's Promontory lighthouse for the period January 1887 to December 2013 are run through an I-M/R–R/S control chart, the record appears homogenous until about 2004, with the first exceedance of the upper control chart in 2005, and then again in 2013 (Fig. 5.4). The moving range of the subgroup mean (annual mean) shows steady decline from this year (Fig. 5.4, middle chart), with the annual mean temperatures showing a distinct step-up. There is no documented equipment change or site move at this time. No adjustments were made to this record.

Trends calculated from these raw values indicate dramatic warming of nearly 3°C per century (Table 5.2) from 1951 to 1980. Statistically significant cooling occurred from 1887 to 1947 (Table 5.2). The overall trend for the period 1887 to 2003 is one of warming at a rate of 0.3°C per century.

In the creation of the ACORN-SAT maximum temperature series for Wilsons Promontory, the Bureau reduces all temperatures before 1950 by −0.26°C. Yet no discontinuity is apparent in the record at this time (Fig. 5.4). The year 1950 immediately precedes a period of missing values. It is noteworthy that the discontinuity perceived by the Bureau's percentile-matching algorithm for Rutherglen in 1966 also corresponded to a period of missing values (Marohasy and Abbot, 2015a).

3.3 Hobart

When the raw maximum monthly values for Ellerslie Road, Hobart are analyzed using an I-M/R–R/S control chart, it is evident that the series from 1887 to 2013 is in control until the very end, with exceedance of the upper control limit beginning in 2010 and then repeated in 2012 and 2013 (Fig. 5.5). The overall pattern is very similar

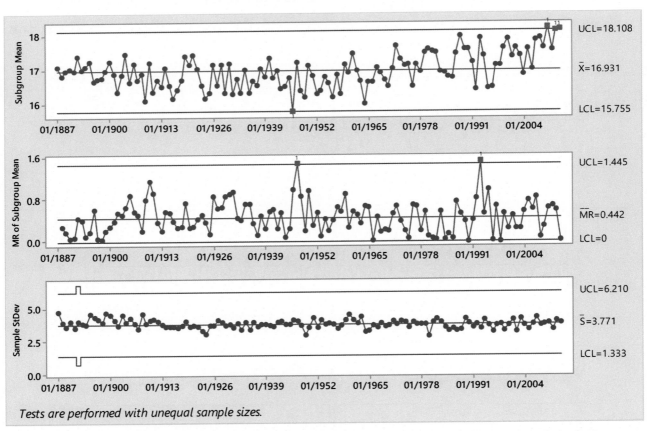

FIGURE 5.5   I-MR–R/S Control Chart showing raw maximum temperatures as recorded at Hobart (1887–2013).

to the series for Wilsons Promontory lighthouse (Fig. 5.4), with cooling to approximately 1950 and then dramatic warming. The warming for the 30 years from 1951 to 1980 is a statistically significant 0.26°C per century (Table 5.2).

3.4 Melbourne

The overall pattern at Melbourne (Fig. 5.6) is one of more sustained warming for the entire period of the record (Fig. 5.6). The first exceedance of the upper control limit by the subgroup mean is in 1961. There then appears to be a step-up in the location of the annual means from 1996 (Fig. 5.6, top graph). The position of the annual/subgroup mean showed some incremental increase relative to the overall mean (19.9°C) until 1996 (Fig. 5.6). In 1996, there appears to be a shift in the mean that corresponds with the transition to recording from the automatic weather station and also construction of the new building (Bureau of Meteorology, 2012).

This step-change is also evident in the minimum temperature series, and, interestingly, the Bureau made an adjustment at this point by adding 0.5°C to all values before 1996 in the creation of the homogeneous series for ACORN-SAT (Bureau of Meteorology, 2014d). The only adjustment that the Bureau makes to the maximum temperature series for Melbourne, however, is to add 0.41 to all values before 1990 (Bureau of Meteorology, 2014d). Yet there is no discontinuity in the Melbourne maximum temperature record at 1990 (Fig. 5.6). Standard deviations (bottom graph), moving range of the annual mean (middle chart), and the subgroup mean appear to be close to overall mean values at this point. Maximum temperatures at Melbourne show an overall increase of 0.1°C per decade for the overall period of the record (Table 5.2). There is no warming in the Melbourne record from 1951 to 1980 (Table 5.2). Almost all the warming occurs after 1980 (Table 5.2).

While the effects of the growth of a city may create a gradual increase in minimum and maximum temperatures, rather than any obvious discontinuity, in the case of Melbourne, the 1996 large building appears to have created an obvious discontinuity in the record (Fig. 5.6). Until 1996, the annual mean maximum temperature was fluctuating above and below the center line, but after 1996, a step-up appears in the temperature series away from the center line, with exceedance of the upper control limit from 2005 (Fig. 5.6).

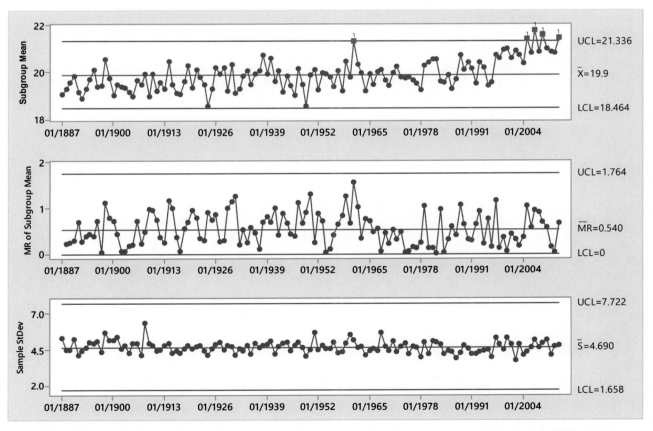

FIGURE 5.6   I-MR—R/S Control Chart showing raw maximum temperatures as recorded at Melbourne (1887—2013).

3.5 Echuca

Echuca is located on the banks of the main channel of Australia's longest river, the Murray River (Fig. 5.1). When raw values are run through an I-M/R–R/S control chart, the years with missing monthly values (1902 and 2012) display major discontinuities in the record, with 1902 exceptionally hot and 2012 exceptionally cold. All months for these years were thus removed and the data run a second time through a control chart (Fig. 5.7). This chart shows exceedance of the upper control limit in 1901 and 1914, and of the lower control limit in 1924 for annual means (Fig. 5.7, top chart). The moving range exceeds the upper control limit in 1915, which immediately follows the very hot year of 1914 (Fig. 5.7, middle chart).

Over the period of the raw maximum temperature record, seven peaks are evident in the annual values, and these peaks correspond with periods of exceptionally low rainfall and/or low river flow (Fig. 5.7, top graph). An extensive literature describes how low rainfall in this region is often accompanied by anomalously high air temperatures (eg, Lockart et al., 2009).

Distinct peaks in 1901, 1914, 1940, 1982, and 2007 correspond directly with periods of exceptionally low rainfall in the state of Victoria (Fig. 5.8). The peak in 1961 (Fig. 5.7) does not show in the rainfall chart as exceptionally dry, but does correspond to a period of low river flow (Close, 1990). Notwithstanding these significant shorter cycles, there is statistically significant cooling for the first 62 years of the record, from 1887 to 1949, followed by significant warming for the following 63 years, from 1950 to 2013 (Table 5.2).

Droughts in 1902 and 1914 predated the construction of major water infrastructure in this region. According to the oral history of the region and early photographs, the Murray River dried up completely during the drought in this region in 1914–15 (Marohasy, 2012). The recent temperature spike in 2007 (Fig. 5.6), which also corresponds to exceptionally low rainfall, occurred at a time when there were good flows along the Murray River with continual discharge from Hume Dam (Marohasy, 2012), and temperatures were not as high as they were in 1914 (Fig. 5.7). Significant regrowth of the forests have occurred since 1887 (Marohasy, 2005), which is the year before the

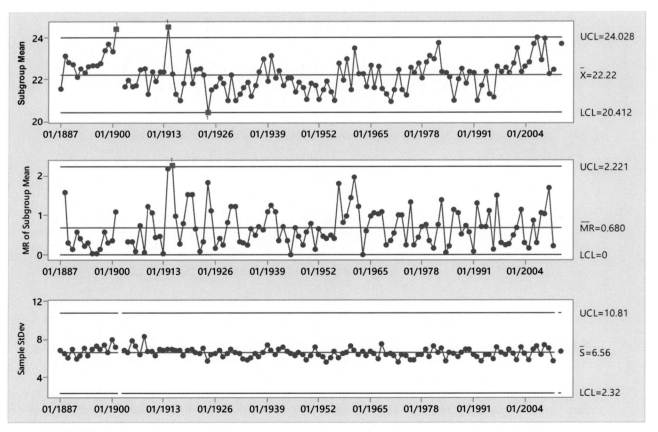

FIGURE 5.7   I-MR–R/S Control Chart showing raw maximum temperatures as recorded at Echuca (1887–2013).

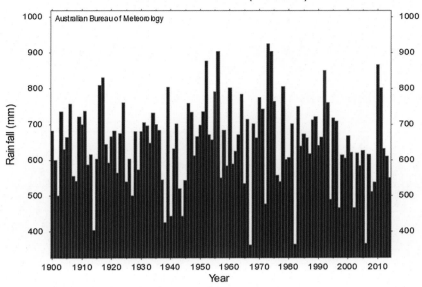

FIGURE 5.8   Annual rainfall Victoria (1900–2014), Australian Bureau of Meteorology.

Conservator of Forests was appointed in 1888. This appointment was to protect the forest from further overcutting, control overgrazing, introduce silviculture treatments, and protect the forest from fire.

3.6 Deniliquin

The raw maximum temperature data for Deniliquin show major exceedance of the upper control limit in the early record (Fig. 5.2). We hypothesize that this is due to an equipment change in 1908. The maximum temperature series for Deniliquin, when corrected for the discontinuity at 1908, and with the addition of the series from the airport, mostly cycles within the upper and lower control limits (Fig. 5.9).

Exceedance of the upper control limit in 1887, 1888, and 1914 corresponds to periods of El Niño warming, and occurs before the development of significant irrigated agriculture in the region. Exceedance of the moving range of the subgroup mean (Fig. 5.9, middle chart) occurs in 1957, which immediate follows an exceptionally wet year for eastern Australia, accompanied by significant flooding (Condon, 2002; Close, 1990).

The discontinuities identified by the Bureau in the Deniliquin record (Table 5.1) are not evident in the raw record for Deniliquin (Fig. 5.2), or the record with the adjustment for the installation of the Stevenson screen in 1908 (Fig. 5.9). After applying the adjustments as detailed in Table 5.1 for ACORN-SAT to the raw series for Deniliquin from 1910, an artificial warming trend is generated of 1.2°C per century (Fig. 5.10).

The period of the ACORN-SAT record for Deniliquin from 1910 to 1949 is represented by the red line in Fig. 5.10. The peak in 1914 is reduced by 0.5°C in the ACORN-SAT record, so it is no longer the hottest year (Fig. 5.10). The very hottest year of 1888 in the raw temperature series for Deniliquin, and the very hot years of 1887 and 1897, are completely removed from the ACORN-SAT record, which is the official record for Australia.

The maximum temperature series for Deniliquin from 1887 to 1949 shows dramatic cooling of 2.4°C per century (Fig. 5.10, Table 5.2). Ignoring this period of cooling, the overall simple linear trend is almost flat with slight cooling of −0.0019°C per year, which is equivalent to approximately −0.02 per decade (Table 5.2), and −0.2 per century.

Clearly shortening the record for Deniliquin so it only begins in 1910, and also adjusting the series using the percentile-matching algorithm with comparison sites, can create an impression that there has been consistent warming, particularly if only a simple linear regression is then applied to the data (Fig. 5.10, red straight dashed line).

The types of adjustments detailed here for Deniliquin in the ACORN-SAT record for Australia (Table 5.1) have been reported for other parts of the world on blogs (eg, Homewood, 2015) and in the popular press (Booker, 2015). The difference in the global mean from such adjustments made by the U.S. National Climate Data Center is on the order of 0.5°C (Humlum, 2015), with the past cooled relative to the present.

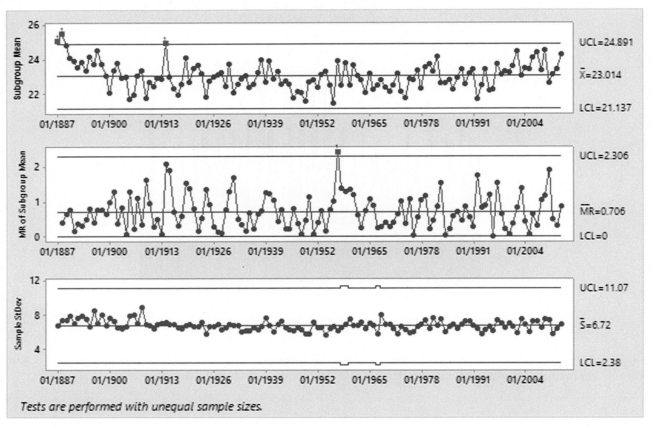

FIGURE 5.9   I-MR—R/S Control Chart showing homogenized maximum temperatures as recorded at Deniliquin (1887—2013).

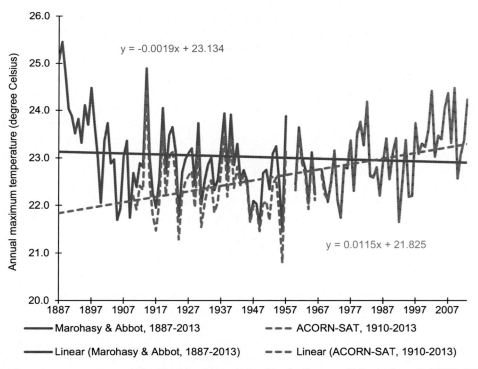

FIGURE 5.10   Annual maximum temperatures Deniliquin as homogenized by the Bureau of Meteorology (ACORN-SAT, 1910—2013) and by Marohasy and Abbot (1887—2013).

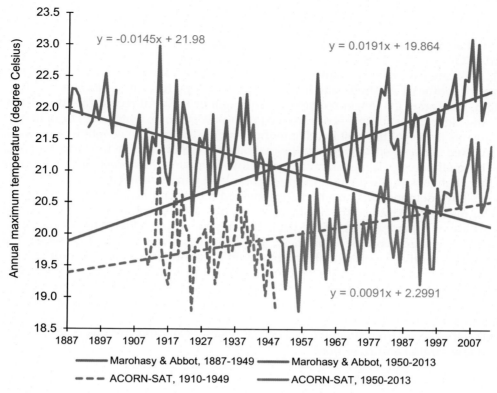

FIGURE 5.11   Mean annual maximum temperatures Southeast Australia, ACORN-SAT (1910–2013) and Marohasy and Abbot (1887–2013).

3.7 Mean Maximum Temperatures for Southeast Australia

When the series as described above in Figs. 5.3, 5.4, 5.6, 5.7, and 5.9 are combined applying the weighting system as described in Section 2.5, a temperature profile surprisingly similar to the Bureau's official ACORN-SAT mean maximum temperature series for Victoria is created (Fig. 5.11). Because the Marohasy and Abbot series shows a distinct cooling trend to 1949, and then warming to 2013, these two periods are shown in different colors in Fig. 5.11, as blue and green lines, respectively.

The overall mean is 1.5°C cooler for the ACORN-SAT reconstruction (20.1°C ± 0.06) relative to the Marohasy and Abbot series (21.6°C ± 0.06). This is not unexpected, given that the ACORN-SAT series describes the mean maximum for the state of Victoria, while the Marohasy and Abbot series is for a region that extends some distance farther north and also west (Fig. 5.1). Notwithstanding the overall cooler trend for ACORN-SAT Victoria, and that the ACORN-SAT series only starts in 1910, there is a high level of synchrony in both the inter-annual variability and overall trends, Fig. 5.11.

The extent of the warming from 1950 to the present exceeds the known capacity of carbon dioxide as a forcing agent, given measured concentrations in the atmosphere over this period (Long and Collins, 2013; Florides and Christodoulides, 2009; Laubereau and Iglev, 2013). The even higher rates of warming reported of 0.21°C and 0.20°C respectively, by Melbourne and Hobart, would be a consequence of the urban heat island effect (Wu and Yan, 2013). There may also be an urban heat island component in the record for Wilsons Promontory from the solar panels (Nemet, 2009) recently installed in close proximity to the weather station.

The cooling from 1887 to 1949 in southeast Australia is likely to be in part a consequence of the large-scale water infrastructure development along the Murray River. However, there is also statistically significant cooling in the record for Cape Otway and cooling evident in the records for Wilsons Promontory, Hobart, and also Melbourne from 1921 to 1950 (Table 5.2).

4. CONCLUSIONS

Claims that the Earth is heating up because of human-influenced global warming, with July 2015 the hottest month on record (NOAA, 2015), 2014 the hottest year globally (Kahn, 2015), and 2013 the hottest year in Australia (Marohasy et al., 2014) are based on a weighted subset of substantially remodeled surface air temperature measurements. In the case of southeast Australia, the Bureau of Meteorology relies on 11 locations with records of various quality and length, truncating the longer records so they begin in 1910. All records are then adjusted based on the percentile-matching algorithm and time series from comparative stations, which may be of low quality. The net effect of the adjustments is to cool the past relative to the present. The specific weightings subsequently assigned to each of the 11 adjusted series used to generate the overall temperature profile for Victoria change on a daily and monthly basis. This methodology is intrinsically unscientific, involving the remodeling of actual physical measurements in ways that change both the direction and magnitude of temperature change.

In this study, we have demonstrated an alternative method for generating a single temperature profile for southeast Australia. Our method is based on the five longest continuous temperature series available for this region from 1887. Adjustments are only made to two of the temperature series (Cape Otway and Deniliquin), and correspond with documented changes in recording equipment. The weightings assigned to each series in the calculation of the overall temperature profile are consistent, and based on local topography and land use.

Our results indicate that statistically significant cooling in southeast Australia of 1.5°C per century occurred from 1887 to 1949, and then statistically significant warming of 1.9°C per century occurred from 1950 to 2013. This overall cooling and then warming trend cannot be explained by current anthropogenic global warming theory. The significant cooling and then warming is superimposed on interannual variability. Shorter cycles of warming and cooling are most evident in the record from the town of Echuca, where peaks and troughs in the record correspond with years of high and low rainfall and river flow.

Acknowledgments

This research was funded by the B. Macfie Family Foundation.

The Australian Bureau of Meteorology continues to make raw and homogenized temperature series for locations across Australia publicly available and increasingly issues advice clarifying the comparative stations used in the homogenization of individual ACORN-SAT temperature series.

Public interest in Australia in the homogenization of temperature trends was stimulated by a series of newspaper article by journalist Graham Lloyd. A panel of statisticians (including Michael Martin, Professor of Statistics, Australian National University; Patty Solomon, Professor of Statistical Bioinformatics, University of Adelaide; and Terry Speed Professor of Bioinformatics, Monash University) was subsequently appointed by the government to review methodologies at the Bureau of Meteorology. Recommendations in their first report have influenced our approach to the quality control of temperature series.

Jennifer Marohasy is grateful to Anthony Cox, Bill Johnston, Bob Fernley–Jones, Case Smit, Chris Gilham, David Stockwell, Dennis Jensen, Geoff Sherrington, Joanne Nova, John Nicol, John Sayers, Ken Stewart, Lance Pidgeon, Malcolm Roberts, Marc Hendrickx, Merrick Thomson, Phillip Goode, Roger Underwood, Tom Quirk, and other thoughtful Australians, for helpful discussion and the free exchange of valuable information over the last three years.

References

Booker, C., 2014. How Much Warming Is Just Fiddled Data? Herald Sun, News Ltd., UK.

Booker, C., 2015. The Fiddling with Temperature Data Is the Biggest Science Scandal Ever. The Telegraph, News Ltd., UK.

Boretti, A., 2013. Statistical analysis of the temperature records for the Northern Territory of Australia. Theoretical and Applied Climatology 114, 567–573.

Brohan, P., Kennedy, J.J., Harris, I., Tett, S.F.B., Jones, P.D., 2006. Uncertainty estimates in regional and global observed temperature changes: a new data set from 1850. Journal of Geophysical Research 11, D12106.

Bureau of Meteorology, 2012. The Australian Climate Observations Reference Network — Surface Air Temperature (ACORN-SAT). Station Catalogue, Commonwealth of Australia, Melbourne.

Bureau of Meteorology, 2014a. ACORN-SAT Station Adjustment Summary — Rutherglen (As at 24 September 2014). Commonwealth of Australia, Melbourne.

Bureau of Meteorology, 2014b. Basic Climatological Station Metadata. Wilsons Promontory Lighthouse, Commonwealth of Australia, Melbourne.

Bureau of Meteorology, 2014c. Basic Climatological Station Metadata. Cape Otway Lighthouse, Commonwealth of Australia, Melbourne.

Bureau of Meteorology, 2014d. ACORN-SAT Station Adjustment Summary. Commonwealth of Australia, Melbourne.

Bureau of Meteorology, 2015a. Annual Maximum Temperature Anomaly — Victoria (1910–2014).

Bureau of Meteorology, 2015b. Australian Climate Observations Reference Network — Surface Air Temperature (ACORN-SAT). Response to the Recommendations of the Technical Advisory Forum, Commonwealth of Australia, Melbourne.

Bureau of Meteorology and CSIRO, 2014. State of the Climate 2014. Commonwealth of Australia, Melbourne.

Close, A., 1990. The impact of man on the natural flow regime. In: Mackay, N., Eastburn, D. (Eds.), The Murray. Murray Darling Basin Commission, Canberra, pp. 61–76.

Condon, D., 2002. Out of the West: Historical Perspectives on the Western Division of New South Wales. Octopus Productions, Sydney.

De Freitas, C.R., Dedekind, M.O., Brill, B.E., 2014. A reanalysis of long-term surface air temperature trends in New Zealand. Environmental Modeling & Assessment. http://dx.doi.org/10.1007/s10666-014-9429-z.

Easterling, D.R., Peterson, T.C., 1995. A new method for detecting undocumented discontinuities in climatological time series. International Journal of Climatology 15, 369–377.

Fawcett, R.J.B., Trewin, B.C., Braganza, K., Smalley, R.J., Jovanovic, B., Jones, D.A., 2012. On the Sensitivity of Australian Temperature Trends and Variability to Analysis Methods and Observation Networks, CAWCR Technical Report No. 050, the Centre for Australian Weather and Climate Research. Bureau of Meteorology, Melbourne.

Florides, G.A., Christodoulides, P., 2009. Global warming and carbon dioxide through sciences. Environment International 35, 390–401.

Hansen, J., et al., 2001. A closer look at United States and global surface temperature change. Journal of Geophysical Research 106 (2), 23,947–23,962.

Homewood, P., 2015. All of Paraquay's Temperature Record has been Tampered with. https://notalotofpeopleknowthat.wordpress.com/2015/01/26/all-of-paraguays-temperature-record-has-been-tampered-with/.

Humlum, O., 2015. Climate4You Update July 2015. http://www.climate4you.com/Text/Climate4you_July_2015.pdf.

Kahn, B., 2015. 2014 Officially Hottest Year on Record. Scientific America.

Laubereau, A., Iglev, H., 2013. On the direct impact of the CO_2 concentration rise to the global warming. EPL 104, 29001.

Lawrimore, J.H., et al., 2011. An overview of the Global Historical Climatology Network monthly mean temperature data set. Journal of Geophysical Research 116. http://dx.doi.org/10.1029/2011JD016187.

Lloyd, G., 2014a. Climate Records Contradict Bureau of Meteorology. The Australian, News Ltd.

Lloyd, G., 2014b. 'More Time' to Find Rutherglen Temperature Record. The Australian, News Ltd.

Lockart, N., Kavetski, D., Franks, S.W., 2009. On the recent warming in the Murray-Darling Basin: land surface interactions misunderstood. Geophy. Res. Lett. 36, L24405. http://dx.doi.org/10.1029/2009GL040598.

Long, D.J., Collins, M., 2013. Quantifying global climate feedbacks, responses and forcing under abrupt and gradual CO_2 forcing. Climate Dynamics 41, 2471–2479.

Marohasy, J., 2005. Australia's Environment Undergoing Renewal, not collapse. International Journal of Energy and Environment 16 (3), 457–480.

Marohasy, J., 2012. Plugging the Murray River's Mouth: The Interrupted Evolution of a Barrier Estuary. National Library of Australia. http://pandora.nla.gov.au/pan/66941/20150206-0852/jennifermarohasy.com/wp-content/uploads/2012/02/Plugging-the-Murray-Rivers-Mouth-120212.pdf.

Marohasy, J., Abbot, J., Stewart, K., Jensen, D., 2014. Modelling Australian and Global Temperatures: What's Wrong? Bourke and Amberley as Case Studies. Sydney Institute, Occasional Paper.

Marohasy, J., Abbot, J., 2015a. Quantifying Uncertainty in Measured and Homogenized Minimum Temperature Series from Rutherglen, Australia (1913 to 2014), In preparation.

Marohasy, J., Abbot, J., 2015b. Assessing the quality of eight different maximum temperature time series as input when using artificial neural networks to forecast monthly rainfall at Cape Otway, Australia. Atmospheric Research 166, 141–149.

Murray Darling Basin Authority, 2005. Yarrawonga Weir. http://www.mdba.gov.au/what-we-do/managing-rivers/river-murray-system/dams-weirs/yarrawonga-weir.

Nemet, F.G., 2009. Net radiative forcing from widespread deployment of photovoltaics. Environmental Science & Technology 43, 2173–2178.

Nicholls, N., Tapp, R., Burrows, K., Richards, D., 1996. Historical thermometer exposures in Australia. International Journal of Climatology 16, 705–710.

Nova, J., 2015. The BOM: Homogenizing the Heck Out of Australian Temperature Records. http://joannenova.com.au/2015/08/the-bom-homogenizing-the-heck-out-of-australian-temperature-records/.

NOAA, 2015. State of the Climate: Global Analysis for July 2015.

Oyler, J.W., Dobrowski, S.Z., Ballantyne, A.P., Klene, A.E., Running, S.W., 2015. Artificial amplification of warming trends across the mountains of the western United States. Geophysical Research Letters 42, 153–161.

Peterson, T.C., et al., 1998. Homogeneity adjustments of in situ atmospheric climate data: a review. International Journal of Climatology 18, 1493–1517.

Ryan, T.P., 1989. Statistical Methods for Quality Improvement, fourth ed. John Wiley & Sons, New York.

Stockwell, D.R.B., Stewart, K., 2012. Biases in the Australian high quality temperature Network. International Journal of Energy and Environment 23, 1273–1294.

Taylor, W.A., 1991. Optimization and Variation Reduction in Quality. McGraw-Hill, New York.

Torok, S.J., 1996. The Development of a High Quality Historical Temperature Data Base for Australia (PhD thesis). University of Melbourne.

Torok, S.J., Nicholls, N., 1996. A historical annual temperature dataset for Australia. Australian Meteorological Magazine 45, 251–260.

Trewin, B., 2013. A daily homogenized temperature data set for Australia. International Journal of Climatology 33, 1510–1529.

Wen, L., 2009. Reconstruction natural flow in a regulated system, the Murrumbidgee River, Australia, using time series analysis. Journal of Hydrology 364, 216–226.

Wu, K., Yan, X., 2013. Urbanization and heterogenous surface warming in eastern China. Chinese Science Bulletin 58 (12), 1363–1373.

Zhang, L., et al., 2014. Effect of data homogenization on estimate of temperature trend: a case of Huairou station in Beijing Municipality. Theoretical and Applied Climatology 115, 365–373.

EXTREME WEATHER EVENTS

Weather Extremes

J.S. D'Aleo[1], M. Khandekar[2]

[1]American Meteorological Society, Hudson, NH, United States; [2]Expert Reviewer IPCC 2007 Climate Change Documents, Toronto, ON, Canada

1. INTRODUCTION

The debate on the anthropogenic global warming/extreme weather (AGW/EW) link continues unabated in media and popular magazines and also in scientific journals. This debate appears to become louder during the Northern Hemisphere summer, in particular when EW events, such as heat waves, floods, hurricanes, and rainstorms, make headlines with strong suggestions that these events were sparked by AGW and associated climate change. In the last five years, several incidences of such EW events have repeatedly ignited this debate. Among some of the notable events of last five years were (1) June 2015 heat wave in Pakistan and parts of India as the monsoon onset was disrupted by a strong El Niño; (2) extensive flooding in Kashmir Valley in the Himalayan foothills during the first week of September 2014, which led to over 500 deaths; (3) a severe heat wave in parts of China during the summer of 2013; (4) a month-long heat wave in the U.S. Midwest during the summer of 2012, accompanied by drought in the corn belt; (5) a month-long heat wave in Russia in 2010, which led to over 50,000 deaths; and (6) extensive flooding during the 2010 monsoon season in Pakistan, which affected over 20% of land area and claimed close to 2000 fatalities.

Understandably, such EW events garnered worldwide media attention, with long articles suggesting "how extreme weather has now become a norm." Several prominent climate scientists, mainly associated with the Intergovernmental Panel on Climate Change (IPCC) (EOS, 2010), claimed that "such weather extremes are due to global

Evidence-Based Climate Science, Second Edition
http://dx.doi.org/10.1016/B978-0-12-804588-6.00006-9

warming caused by increasing concentration of human-added CO_2 in the atmosphere and that such EW events would be on the rise in future." The World Meteorological Organization (WMO) in Geneva, Switzerland issued a statement in August 2010 that "the weather related cataclysm of July and August fit the pattern predicted by climate scientists (and climate models)."

The United States Climate Change Science Program (USCCSP) 3.3 (2008) focused on extremes: "Changes in extreme weather and climate events have significant impacts and are among the most serious challenges to society in coping with a changing climate. Many extremes and their associated impacts are now changing. For example, in recent decades most of North America has been experiencing more unusually hot days and nights, fewer unusually cold days and nights, and fewer frost days. Heavy downpours have become more frequent and intense. Droughts are becoming more severe in some regions, though there are no clear trends for North America as a whole. The power and frequency of Atlantic hurricanes have increased substantially in recent decades, though North American mainland land-falling hurricanes do not appear to have increased over the past century. Outside the tropics, storm tracks are shifting northward and the strongest storms are becoming even stronger."

What was missing in such declarations and scientific commentaries was a careful assessment of past EW events and whether or not such events have indeed increased in recent years. If so, is human CO_2-induced climate change causing such increase or are these EW events a result of natural variability?

2. EXTREME WEATHER EVENTS AND THE EARTH'S CLIMATE

It is important to realize that EW events are an integral part of the Earth's climate system, and that such events are triggered by large-scale atmosphere—ocean circulation systems and their complex interaction with local and regional weather patterns. Throughout the recorded history of the Earth's climate, EW events (hot or cold extremes) have always occurred somewhere. For example, an ENSO (El Niño—Southern Oscillation) event in equatorial eastern Pacific can produce significant temperature and precipitation anomalies in different parts of the world (Kiladis and Diaz, 1989), and these anomalies can lead to EW events, depending on the phase and intensity of a given ENSO event. Even without an ENSO event, the large-scale atmospheric flow patterns often give rise to violent interaction of higher-latitude colder air with lower-latitude warmer air, producing weather events that can develop into an EW event. The question that is currently extensively debated is: Are EW events of recent years more frequent and are they linked to recent (human CO_2-induced) climate change?

The IPCC (2001, 2007) suggested a definite link between EW events and the recent climate warming. However, the IPCC does not provide a physical basis for such a link, but a simplistic suggestion that "a warmer climate holding more moisture can lead to excessive rains and floods in some regions, while droughts and reduced rains in some other regions." This simple suggestion has now morphed into the hypothesis that a warmer climate will bring "more rain, floods and also droughts and heat waves."

Most atmospheric general circulation models (GCMs) used by the IPCC project a mean temperature increase of 2—5°C for the next 50—100 years and a commensurate increase in atmospheric moisture, leading to more EW events. The IPCC also projects fewer future cold weather extremes, while the last few years have witnessed increasing cold weather extremes worldwide! What's more, satellite data has shown no global warming for over 18½ years, which should invalidate any claims that any recent extremes are the result of global warming.

3. HEAT EXTREMES

Heat waves are a characteristic of summers in North America in particular. When these heat waves become anchored by a persistent high-pressure ridge and amplified by dry ground feedback, they become protracted and newsworthy.

A National Aeronautics and Space Administration (NASA) scientist, James Hansen, published a 2012 op-ed in *The Washington Post* entitled "Climate change is here—and worse than we thought." His thesis: The worst hot spells of the past decade were "a consequence of climate change" and have "virtually no explanation other than climate change." However, as we discuss below, the heat extremes of last few years are part of natural climate variability, and linking these extremes to human CO_2 emissions is without any merit.

3.1 European Heat Wave of Summer 2003

The summer 2003 was exceptionally warm and dry in large parts of Europe, particularly in most of France, where a large number of mainly older people died in July. This extreme weather event garnered many newspapers report as well as scientific commentaries and papers. A paper by Stott et al. (2004) argued that the primary cause of this heat wave was the increasing atmospheric CO_2 concentration and the resulting climate warming. A recent study by Weisheimer et al. (2011) analyzed climate model simulations based on the seasonal ensemble forecasting system of the European Centre for Medium-Range Weather Forecasts and concluded that dry soil conditions persisting for several months before the summer 2003 were a key factor in triggering this exceptional heat wave. Weisheimer et al. discount increasing atmospheric CO_2 concentration as the cause of the 2003 heat wave.

3.2 Russian Heat Wave of July 2010

The unusually hot summer of July 2010, especially in Moscow and vicinity, led to widespread forest and peat fires and several thousand deaths (Dole et al., 2011). This heat wave was certainly an EW event and captured headlines of many European and North American news media. During the last two years, a number of peer-reviewed papers analyzed the sequence of events that led to this EW event. According to Dole et al. (2011), "internal atmospheric dynamic processes that produced and maintained intense and long-lived (atmospheric) blocking" caused this heat wave.

3.3 U.S. Summer 2012 Heat Wave

The two-month long heat wave in the conterminous United States, which was accompanied by a severe drought primarily in the "corn belt states," was extensively debated in newspapers and TV media. This drought caused a significant reduction of the 2012 U.S. corn harvest that impacted world corn and grain prices. The main theme of the related printed and broadcasted news was that extreme weather has now become a norm. This theme was echoed by Ms. Connie Hedegaard (European Union Commissioner for Climate Action) in an article of *The European Voice* on September 11, 2012, entitled "Weather Extremes: The New Normal." Interestingly, most TV and print media in North America ignored similar past EW events, such as the recurring droughts and heat waves on the American/Canadian prairies during the Dust Bowl years of the 1920s and 1930s. An extensive literature is available on the anomalous North American climate of the 1920s and the 1930s.

Fig. 6.1 illustrates that the decade of the 1930s was the all-time hottest decade in the conterminous United States, while the subsequent decades and especially the 1990s and 2000s appear to have been rather mild compared to the 1930s. It is noteworthy that in Canada the hottest ever recorded temperature at Yellow Grass, Saskatchewan on the Canadian Prairies reached 45°C (113°F) on July 5, 1937. Also, the deadliest heat wave in Canada occurred from July 5–12, 1936, when over 1000 people died of heat exhaustion in Manitoba and southern Ontario (Khandekar, 2002), and Toronto, Canada's largest city, recorded the highest temperature of 41°C (106°F) for three days in a row.

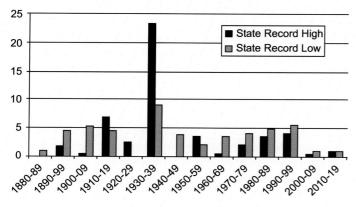

Number of State Record High & Low Temperatures by Decade

FIGURE 6.1 Highest and lowest decadal temperature records for various states in conterminous United States. The decade of the 1930s was the hottest decade in the United States. *Data from NOAA compiled by Dr. John Christy from NCDC data, http://www.ncdc.noaa.gov/extremes/scec/records.*

The heat wave of summer 2012 in the United States was by no means unprecedented when compared to the heat waves of the Dust Bowl years. Further, the recurrent heat waves and droughts during 1952 to 1954 and 1980 (Namias, 1955, 1982) indicate that heat waves are not uncommon in the U.S.—Canadian plains region.

Globally, the continental heat records were all set many decades ago (http://wmo.asu.edu/#continental). Cold records at the same link were generally more recent.

3.4 Pakistan/India Heat Wave of June 2015

A severe heat wave with temperatures as high as 49°C (120°F) struck southern Pakistan in June 2015. It caused the deaths of about 2000 people from dehydration and heat stroke, mostly in Sindh province and its capital city, Karachi. Widespread failures of the electrical grid left many locations without working air-conditioners, fans, or water pumps, adding further to the death toll. The heat wave coincided with the month of Ramadan, in which Muslims observe fasting with no drinking from dawn till dark. This religious ritual may also have led to more deaths during this heat wave.

4. COLD EXTREMES

The IPCC (2007) and (2013) documents categorically state that as the Earth's climate continues to warm, winters in the future will become milder and shorter with less snow accumulation over the Northern Hemisphere. Many IPCC scientists also speculated about warmer, shorter winters and decreasing snow accumulation in the future. The reality of climate change appears to be at odds with the IPCC projection, as shown by observations of extreme cold winters over Europe, North America, and parts of China and Japan in the past 10 years or so. Has the Earth's climate entered into a new phase since the new millennium? Let us consider observational evidence as reported in various news media reports and also in many scientific papers published in the new millennium.

1. The winter of 2002—2003 was the first of the five extreme cold winters in Europe and parts of the U.S. mainland and Canada since the new millennium. The 2002—2003 winter saw several dozen extreme cold episodes from eastern Canada (Kodera, 2003) to western Europe to as far south as Vietnam and Bangladesh, where several hundred people died of long exposure to month-long low temperatures. The Gulf of Finland was frozen completely for the first time since 1947.

2. The winter of 2005—2006 was exceptionally cold over most of western and northeastern Europe (Petukhov and Semenov, 2010). In Poland, 23 people were reported to have died as a result of extreme cold, and heavy snow blanketed the Baltic coast city of Gdansk. Heavy snow was also reported in parts of Hungary to Germany to the Yorkshire region in the UK during the last week of December 2005. In northern England, a low temperature of −12°C (10.4°F) was accompanied by high winds. Elsewhere, parts of China were very cold and in Japan, Tokyo and a few other cities received record-breaking snowfall.

3. The cold winter of 2007—2008 produced all-time record snow accumulation in the conterminous United States from Alaska to Oregon, Utah, and Colorado, and across the upper Midwest to northern New England. Winter 2007 was also cold in parts of South America, and snow fell in Buenos Aires in July 2007 for the first time since 1918.

4. The winters of 2008—2009 and 2009—2010 were both exceptionally cold and snowy over parts of Western Europe and also in the U.S. mid-Atlantic states (Cattiaux et al., 2010; Jung et al., 2011; Seager et al., 2010). Record-breaking snow was reported for Baltimore, Philadelphia, and Washington, DC, with snow amounts of about 190 cm (~75″) and above. In Europe, Scotland suffered some of the coldest winter months in almost 100 years, while Siberia witnessed perhaps the coldest winter (2009—2010) ever, according to Russian climatologists. In Argentina and Chile, June and July 2010 were very cold, leading to several dozen deaths.

5. December 2010 was the second-coldest December in the long-running, Central England temperature data set since records began during the Little Ice Age in 1659.

6. Winters of 2011—2012 and 2012—2013 were once again very cold and snowy in parts of central and eastern Europe and also in parts of North America. In eastern Europe, an exceptionally cold spell was reported for about three weeks in February 2012, when some locales in the Czech Republic reported low temperatures of −40°C (−40°F) and below for a week. Several dozen people were reported to have died in the Czech Republic as a result of extreme cold. The Danube River remained frozen until late May 2012. In North America, very heavy snow accumulation occurred in the Canadian Rockies and Alaska. In the Canadian Rockies, record-breaking snow

accumulation of over 900 cm was registered at Sunshine Village, a popular ski resort. At Valdez, Alaska, more than 350 cm of snow fell in the first two weeks of January 2012. The winter of December 2012 to March 2013 brought several large snowstorms across Europe and North America. March 2013 saw heavy snowfalls in northern France, Germany, and Belgium, with snowfall amounts in excess of 25 cm in many places. In north India, the months of December 2012 and January 2013 were significantly colder than normal leading to several hundred deaths, as reported in various media in India. The reported deaths in India were primarily elderly people living in poorly built housing with no heating or insulation (Khandekar, 2013)

7. The winter of 2013—2014 was one of the coldest, snowiest, and longest winters in North America in 40 years. A total of 7689 low-temperature records were broken or tied during the two coldest months in the winters of 2013—2014 (January) and 2014—2015 (February) (Fig. 6.2A and B). Heavy snow accumulation and poor visibility led to several thousand flight cancellations in Canada and the conterminous United States in January 2014. The 2013—2014 winter was also significantly colder in parts of north Japan and India. In northwest India, snow accumulation of up to 75 cm was reported in January and February 2014; such heavy snow accumulation is rare in that region. Japan witnessed two major snowstorms in February 2014, causing traffic chaos, several deaths, and cancellation of hundreds of flights. In South America, July 2013 saw snow in several locales in southern Brazil, where TV footage showed snow hanging from palm trees.

The recent winter of 2014—2015 was, once again, exceptionally cold in the northeastern United States and in Atlantic Canada. Boston reported record-breaking snow close to 300 cm (~110″). Most of the northeast United States and Canadian Atlantic Provinces saw several heavy snowfalls and blizzards throughout the winter season with snow amounts of 25 cm or more on several occasions; further, the northeastern U.S. states witnessed the coldest winter since 1895 (Fig. 6.3). February 2015 was one of the coldest months in most of southern Ontario (Canada), and Toronto (Canada's largest city) recorded low temperature of −35°C (−31°F) on several days.

Fig. 6.4 illustrates how winter months (January to March) globally have shown no warming in 20 years. Instead, a cooling of 0.9°C (1.5°F)/decade has been identified in the northeastern United States. Cooling of a lesser magnitude has been shown for the lower 48 U.S. states and for winters in the UK for the last 20 years.

Elsewhere, heavy snow avalanches in the Panjshir Valley of Afghanistan killed more than 200 people in February 2015. The Kashmir Valley region in the Himalayan foothills recorded several snowfalls during the months of February and March 2015. The winter months of January and February 2015 were once again significantly colder than normal in north India, causing transportation delays due to heavy fog and affecting lives of over 300 million people living in major north Indian cities and towns.

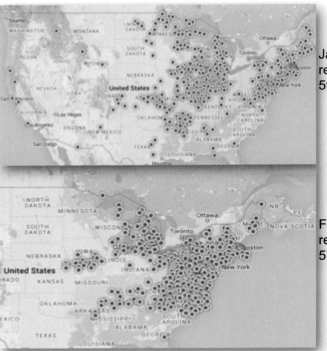

January 2014 had 2945 record lows broken and 599 tied – a total of 3544

February 2015 had 3573 record lows broken and 572 tied – a total of 4145

FIGURE 6.2 Plots of daily low temperature records broken and tied during the cold months of (A) January 2014 (total of 3544) and (B) February 2015 (total of 4145).

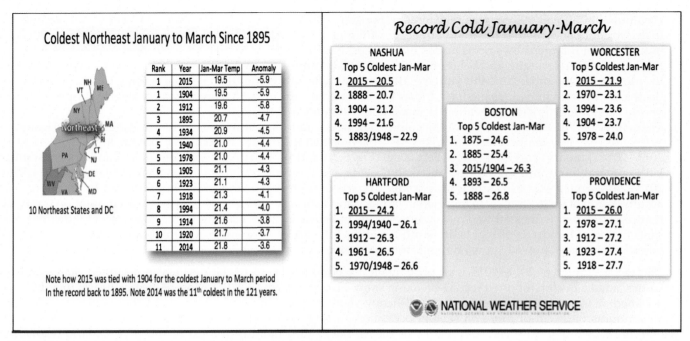

FIGURE 6.3   United States northeast region ranking of the coldest January to March periods since 1895. The year 2015 tied with 1904 as the coldest such period in the 120-year record.

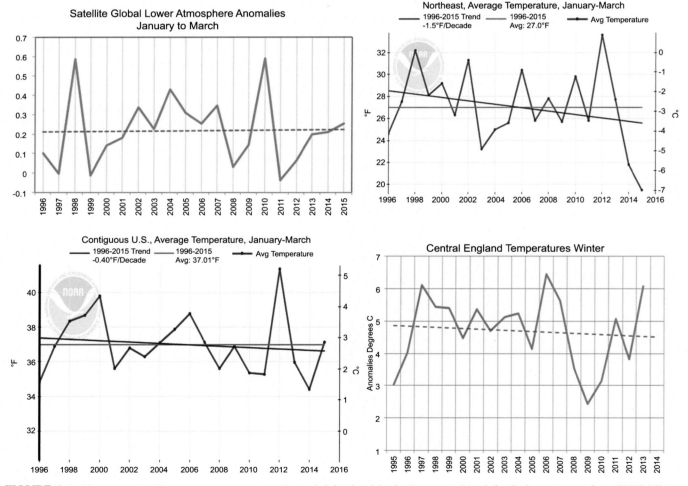

FIGURE 6.4   Twenty-year cold season temperature trends: top left for the globe for January to March for the lower troposphere *(UAH 6.0)*, top right for the 10-state northeast region for January to March, bottom left for the same period for the lower 48 states, and bottom right for winter in the UK *(Central England Temperature data set, established 1659)*.

For additional incidences of cold weather extremes, refer to the NIPPC (Non-Governmental International Panel on Climate Change) 2013 document Climate Change Reconsidered II (Idso et al., 2013).

The IPCC has steadfastly refused to refer to any cold weather extremes in their recent documents (IPCC, 2013, 2007) despite growing incidences of such events worldwide. In its 2007 documents (IPCC AR4 II, chapter on "Assessments of Observed Changes and Responses in Natural and Managed Systems"), the IPCC provides an extensive discussion of the "2003 summer heat wave in Europe" while totally ignoring any reference to the severely cold winter of 2002–2003, just six months earlier! In its documents published in September 2013, the IPCC once again makes no reference to any cold weather events or to many studies on recent colder winters appearing in peer-reviewed journals in last six years (eg, Benestad (2010) and Lockwood et al. (2010)). As summarized by van Geel and Ziegler (2013) the IPCC underestimates the sun's role in climate change. This negatively impacts their credibility.

5. FLOODS AND DROUGHTS

5.1 Conterminous USA and Southern Canada

In general, floods in the conterminous United States have not increased. Fig. 6.5 shows that losses due to floods as a percentage of the GDP (gross domestic product) for the United States have actually dropped by about 75% since 1940. (Pielke Jr.)

The Dust Bowl years of the 1920s and the 1930s were the most drought-prone years on the U.S.–Canadian prairies. Among some of the notable drought years were 1919–1921, 1929–1931, and 1936–1937 (Khandekar, 2004). The droughts of 1999–2001 were reminiscent of the Dust Bowl droughts on the Canadian prairies in particular. Since about 2003, no severe droughts have occurred, either in U.S. or in Canadian prairies (Garnett and Khandekar, 2010). Fig. 6.6 for the conterminous United States shows dry/wet areas (in percentage) since 1900 with no long-term trend.

The recent drought in Texas reached severe levels in 2011–2012, but it has been brief relative to the long seven-year drought of the 1950s. California turned dry in 2013, and the drought has persisted through the 2014–2015 wet season. However, like Texas, the California annual precipitation shows no clear long-term trend.

5.2 Droughts and Floods Globally

5.2.1 Pakistan Floods of 2010

Floods in Pakistan began in late July 2010, from heavy monsoon rains that affected the Indus River basin. Approximately one-fifth of Pakistan's total land area was underwater, approximately 796,095 km^2 (307,374 sq mi). According to Pakistani government data, the floods directly affected about 20 million people, mostly by destruction of property, livelihood, and infrastructure, with a death toll of close to 2000. It was the 65th greatest flood in the last

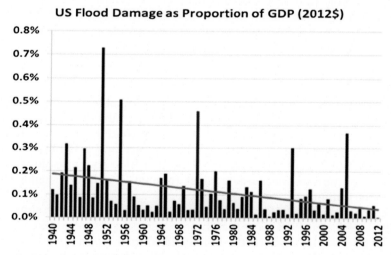

FIGURE 6.5   Annual flood losses have decreased from about 0.2% of U.S. GDP to <0.05% since 1940. *Flood loss data from NOAA Hydrologic Information Center.*

FIGURE 6.6 NOAA percent area of the United States wet or dry, 1900–2012, based on the Palmer Drought Index. Major multiyear droughts can be seen in the 1930s and 1950s, with long-period dryness in the late 1980s and the early 2000s. Wet periods are seen in the early 1900s, the early 1980s, and the 1990s. No clear long-term trends are apparent.

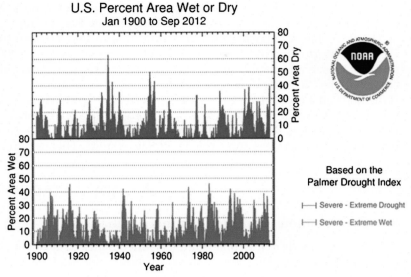

National Climatic Data Center / NESDIS / NOAA

1200 years. The worst flood on record occurred in 1931 in China. Flooding on all the major river systems was responsible for an estimated 3.7 million deaths worldwide.

Note that the monsoon season in Pakistan is almost in tandem with the Indian monsoon, which is primarily driven by regional as well as global scale features like ENSO (El Niño–Southern Oscillation), Eurasian snow cover during the (previous) winter season, and the QBO (Quasi-Biennial Oscillation) phase of the equatorial stratospheric wind oscillation (see eg, Khandekar, 1996). The monsoon season in Pakistan is generally of a shorter duration, from about July 1 until about the third week of September. Pakistan as a whole receives just about 50 cm (~20″) of monsoon rains, compared to about 85 cm for the whole of India during the June–September season.

Major droughts and floods in the Indian monsoon have been linked to El Niño/La Niña events in several studies published in the last 30 years (see, for example, Khandekar and Neralla, 1984). The 2010 Pakistan floods appear to have been triggered by a rapid transition of the ENSO phase from El Niño to La Niña between spring and summer of 2010. The La Niña phase sparked convective activity in the Bay of Bengal, where several low-pressure systems developed in late July 2010. These low-pressure systems while traveling northwestward carried sufficient moisture into northwest India and vicinity, causing heavy flooding in parts of northwest India and over a large area of Pakistan. An examination of monsoon climatology reveals that the 2010 Pakistan floods, although seemingly unprecedented, were well within natural variability of monsoonal climate over the Indian subcontinent (Khandekar, 2010). Finally, Fig. 6.7 shows droughts and floods in the Indian monsoon, which are strongly linked to El Niño and La Niña events in the equatorial eastern and central Pacific. Note here that major droughts and floods in the Indian monsoon have occurred irregularly with no linkage to global warming or climate change.

5.2.2 Bangladesh Flooding

Bangladesh is often mentioned as a country that is especially vulnerable to "global warming"–induced sea level rise because it is located on the Ganges Delta, which is highly vulnerable to floods due to the confluence of three rivers: Ganga, Brahmaputra, and Meghna. These rivers frequently flood during the summer monsoon when heavy upstream rains generate huge runoffs that flow through the country and eventually drain into the Bay of Bengal through a narrow passage. About 75% of Bangladesh is less than 10 m above sea level, while 80% of its land area is flood plain. Developing appropriate measures against frequent floods is a major challenge for this country with a population of over 160 million.

A comprehensive paper by Mirza (2003) documents several floods in Bangladesh since 1955 and further discusses three extreme floods in 1987, 1988, and 1998. Note that these extreme floods were linked to the La Niña events, which in general are conducive to large-scale floods on the Indian subcontinent (see Fig. 6.7). Outside of the monsoon season (May–September), Bangladesh also experiences severe flooding during landfall of tropical cyclones in the Bay of Bengal, which normally invade the country during October and November. Among the most disastrous floods was

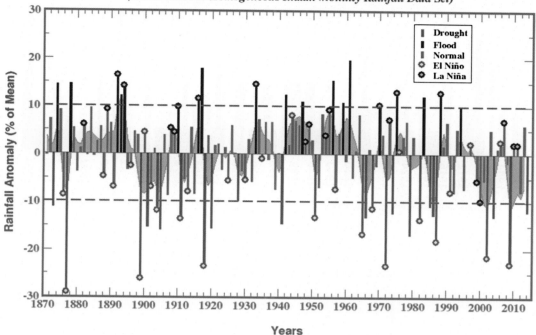

FIGURE 6.7   All-India summer monsoon rainfall, 1871—2014. *Courtesy of Indian Institute of Tropical Meteorology, http://tropmet.res.in/~kolli/MOL/.*

one associated with the landfall of a tropical cyclone on November 12—13, 1970, when an estimated one quarter million people were drowned (see Appendix A). Note here that sea level rise due to recent warming of the earth's climate is not of major concern for Bangladesh today (Morner, 2010).

6. TROPICAL CYCLONES AND TORNADOES

Tropical cyclones and hurricanes are a major concern for the U.S. Gulf Coast and the Atlantic seaboard states, where about two to three landfall hurricanes occur annually. A large body of literature on U.S. hurricanes and their climatology over the past 100 years or more is available. Two papers on climate change impact on hurricane intensity are worth noting. Webster et al. (2005) suggested significant increase in hurricane frequency and strength in a warming climate. In contrast to Webster's study, Klotzbach and Landsea (2015) found that the global frequency of category four and five hurricanes has shown a small, insignificant, downward trend while the percentage of category four and five hurricanes has shown a small, insignificant, upward trend between 1990 and 2014. An important dynamic parameter called Accumulated Cyclone Energy experienced a large and significant downward trend during the same period, 1990—2014 (Maue, 2011).

NOAA has found that hurricane frequency has not increased in the United States since at least 1900.

Hurricane damages in the United States (adjusted for changes in population, wealth, and consumer price index) have not changed since at least 1900 (Figs. 6.8 and 6.9).

No significant trend (up or down) in global tropical cyclone landfalls has occurred since 1970 (Fig. 6.10).

Finally, it is of interest to note some historical facts on U.S. hurricanes: (1) five hurricanes made landfall in New York City from 1815 to 1821 during the Little Ice Age Dalton Solar Minimum, and (2) eight major, devastating hurricanes made landfall on the U.S. east coast from 1938 to 1960.

6.1 Tornadoes

The central United States is the most tornado-prone region in the world. Best available data suggest that tornadoes in the United States have not increased in frequency, intensity, or normalized damage since 1950, and

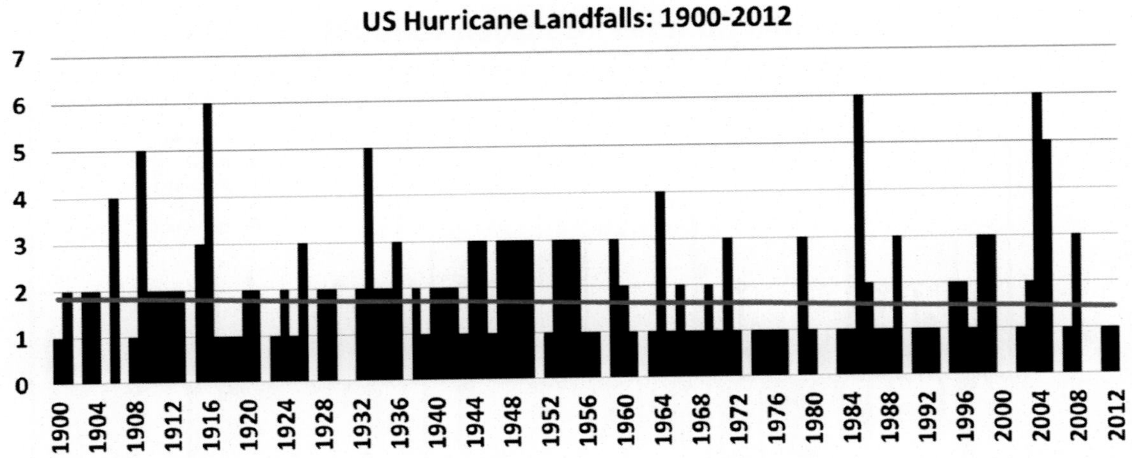

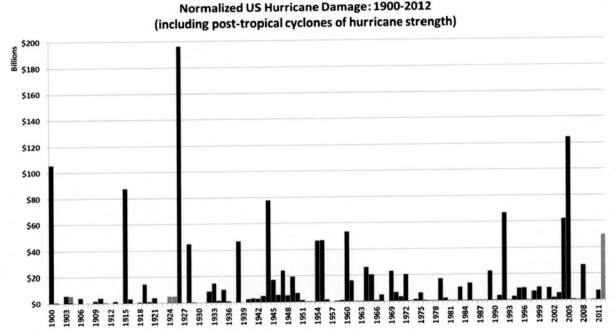

FIGURE 6.8   U.S. hurricane landfalls, 1900–2012. The red line shows the linear trend, exhibiting a decrease from about 2 to 1.5 landfalls per year since 1900. *NOAA.*

FIGURE 6.9   Normalized U.S. hurricane damage, 1900–2012.

some evidence suggests they have actually declined. This is true for all tornadoes and strong-to-violent ones (Fig. 6.11).

Normalized tornado damage shows some tendency for a downward trend (Fig. 6.12). The three years after the 2011 La Niña spike are among the quietest three-year period on record.

7. ECONOMIC LOSSES DUE TO EXTREME WEATHER

Escalating economic losses due to weather extremes has now become the linchpin in the present debate on climate change and an alleged urgent need to reduce human CO_2 emissions. In a recent article, Wuebbles et al. (2014) identified over 150 weather disasters in the United States since 1980, each inflicting damage worth U.S. $1 billion or more (relative to the 2013 U.S. Consumer Price Index). They identify 2011 and 2012 as the two most "extreme weather"

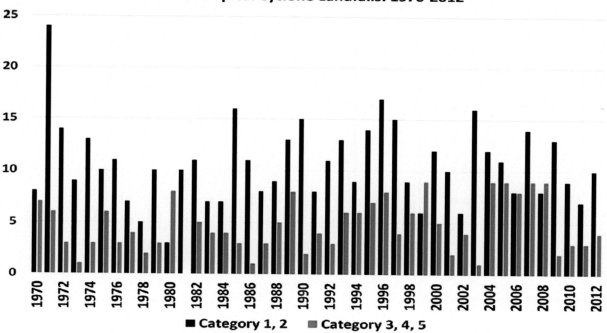

FIGURE 6.10   Global tropical cyclone landfalls, 1970–2010. The black lines show categories 1 and 2, and the red lines show categories 3 to 5.

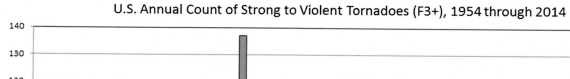

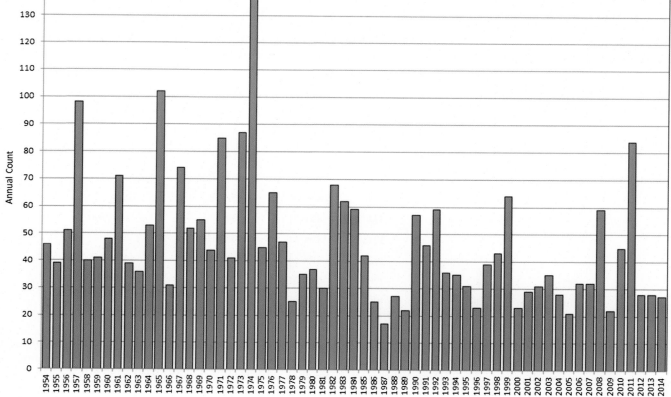

FIGURE 6.11   U.S. annual number of strong to violent (EF3+) tornadoes, 1954–2014. *NOAA SPC.*

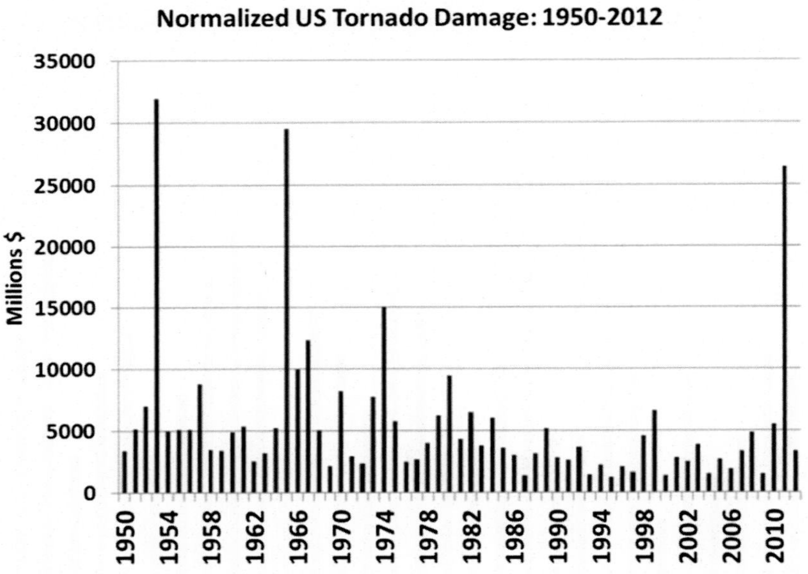

Normalized US Tornado Damage: 1950-2012

FIGURE 6.12   Normalized U.S. Tornado Damage, 1950–2012. Estimated total damage if tornadoes of past years occurred with 2012 levels of development, updating Simmons et al. (2012).

years for the conterminous United States, with 14 and 11 strong weather events, respectively. Notable among such EW events were: (1) several incidences of tornado outbreaks in the south-central United States in 2011, resulting in the deaths of over 500 people and damage to infrastructure of several tens of billions of dollars; and (2) major economic losses in 2012 due to a heat wave in the central plains of the United States, accompanied by a severe drought that inflicted economic losses of about 33 billion dollars, and the landfall of Hurricane Sandy, causing extensive damage to heavily industrialized New Jersey and vicinity.

Hurricane damage today is greatly magnified by the increased number of people and property at risk, as well as monetary inflation. NOAA/ESRI compiled a list of the top 10 hurricanes ranked in order of likely damage if they occurred today. Only Katrina and Andrew are within the last quarter century in the study. Notable is that estimated losses from Hurricane Sandy are $68.5 billion, which would have ranked fourth.

Rank	Storm	Year	Damage	Damage Today	Deaths
1	Great Miami	1926	$105 M	$157 B	372
2	Katrina	2005	$81 B	$81 B	1836
3	Galveston	1900	$20 M	$78 B	8000
4	Galvestion II	1915	$56 M	$61.7 B	275
5	Andrew	1992	$26.5 B	$57.7 B	65
6	Hurricane of '38	1938	$308 M	$39.2 B	600–800
7	Florida	1944	$100 M	$38.7 B	318
8	Lake Okeechobee	1928	$100 M	$33.6 B	4078
9	Donna	1960	$387 M	$29.6 B	163
10	Camille	1969	$1.42 B	$21.2 B	256

Among other EW events that inflicted heavy losses were flooding in southeastern United States, mostly due to hurricane landfall and snow storms in the northeastern United States. A similar report on worldwide economic losses from extreme weather suggests 2011 was the most expensive, with total losses at about $380 billion. The report further projects disaster losses to exceed over $500 billion by 2015 and beyond. These and similar other reports include only the WWE as identified by the IPCC.

Economic losses due to extreme cold weather have NOT been estimated in any study so far. In view of the fact that such events have been on the rise in recent years, it is instructive to analyze how cold weather events impact the socioeconomics of various regions. Below, we attempt to estimate economic losses over North America during the winters of 2013–2014 and 2014–2015.

Both the 2013–2014 and 2014–2015 winters inflicted significant economic losses in the United States and Canada. The losses were primarily due to disruption in transportation, delay and cancelation of thousands of airline flights, delays in construction and engineering, and in loss of revenue to businesses due to poor sales during extreme cold weather.

An economic slowdown in the first quarter (January–March) of 2014 and 2015 was reported by both the United States and Canada, part of the slowdown being attributed to extreme cold winter. The most recent report on the world economy prepared by the Organization for Economic Cooperation and Development, a Paris-based international economic organization, has blamed 2015 winter blizzards and heavy snow in northeastern United States and eastern Canada for an economic drawdown in the first quarter of 2015.

Using representative numbers for Canadian and U.S. GDP, we estimate economic losses of about $30 billion for Canada and about $50 billion for the conterminous U.S. for each of the winter seasons 2013–2014 and 2014–2015. These numbers may be considered as conservative, since indirect losses (like increased energy need, impact on human health, etc.) are not taken into account. In reality, total economic losses for North America (excluding Mexico) could be as high as about $100 billion for each of the winters of 2013–2014 and 2014–2015.

A similar calculation for Europe, which witnessed extreme winter conditions in various regions during the 2008–2009 to 2012–2013 winters, can be made. Accounting for different socioeconomic states for eastern and western Europe, we obtain estimates of anywhere from $25 to $50 billion as direct economic losses over Europe as a whole. Once again, these numbers must be considered conservative, because many factors like indirect losses due to health impact of extreme cold winters, increased energy costs, etc. are not taken into account here. Elsewhere, Japan and China also experienced economic losses in 2013 and 2014 due to extreme winter conditions, which are estimated at $10 to $25 billion.

These calculations on economic losses due to extreme winters are illustrative of how present climate change can inflict damage to various regions and countries, depending on their socioeconomic structure. A report by Lord Nicholas Stern (UK) on economic impacts of global warming and climate change was published in 2006. This review is often referred to as an authoritative report for guidance in dealing with future climate change and its impacts on human societies. The Stern review emphasizes the need to reduce human CO_2 emissions, even at an enormous cost of several trillion dollars, as an "insurance" against huge economic losses in the future due to EW events. The Stern report relies heavily on the IPCC and makes no reference to future economic losses due to extreme cold weather. As the above discussion suggests, economic losses due to cold weather extremes accompanied by increased snow accumulation (Fig. 6.13) are now becoming comparable to similar losses due to warm weather extremes!

In the absence of any reference to increasing economic losses due to colder future climate, the utility of the Stern report for estimating future climate-change mitigation strategy remains questionable. Pielke (2014), using loss data from Munich Re, has shown how global weather-related losses due to weather extremes have declined since 1990 as a proportion of global GDP (Fig. 6.14).

8. HUMAN FATALITIES AND HEAT AND COLD EXTREMES

While acknowledging the unfortunate deaths of several thousand people in the European and Russian heat waves of 2003 and 2010 and more recent heat wave fatalities (eg, more than1500 in India in May 2015), it is important to realize here that many more deaths have occurred in various regions of the world due to extreme cold weather and continue to occur even now. The large number of deaths due to extreme cold winters in Europe and elsewhere during the 1960s and 1970s has been well documented in news media and documents. Since 2010, several hundred deaths have occurred during cold winters in India and parts of Europe. During the Little Ice Age (~1650–1915 AD), several million Europeans died in extremely cold winters of the Maunder and the Dalton Solar Minimums.

Of all the weather extremes popularly attributed to climate change, heat waves allow for the most rapid and effective adaptation (http://www.ncbi.nlm.nih.gov/pubmed/14594620). Davis et al. (2003) found that as summer air temperatures increase (whether due to the expansion of urban heat islands, global warming, or both), U.S. heat-related mortality has declined. U.S. cities such as Tampa and Phoenix, with the most frequent hot weather, have the lowest heat-related mortality.

FIGURE 6.13 Northern Hemisphere snow extent in million square km for the winter season (December to March).

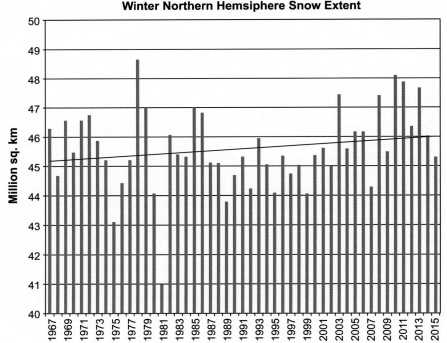

FIGURE 6.13 Northern Hemisphere snow extent in million square km for the winter season (December to March).

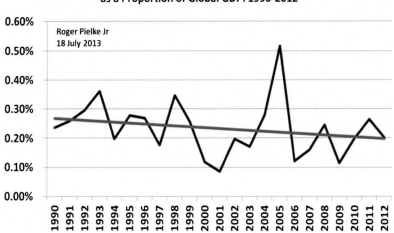

FIGURE 6.14 Global weather-related disaster losses as proportion of GDP, 1990–2012 (Pielke Jr.). *Source of loss data, Munich Re. Source of GDP data, UN.*

Gasparrini et al. (2015) examined over 38 million deaths in various regions of the world during the 1985–2012 period and concluded that **cold weather kills 20 times more than hot weather**. The study analyzed 74,225,200 deaths between 1985 and 2012 in 13 countries with a wide range of climates, from cold to subtropical.

Among many other studies of human fatalities, a study by Goklany (2011) is worth noting. Goklany analyzes deaths due to various extreme weather events and documents a significantly larger number of deaths during the winter season in most mid-latitude regions of the world. For the conterminous United States, Goklany documents over 100,000 more deaths in the winter season over all other seasons. More importantly, Goklany documents that significant reduction in deaths from extreme weather events is primarily due to improved early warning systems and better resilience to such EW events. For the UK, a study by Donaldson et al. (2001) documents more deaths during winter than during summer (Fig. 6.12). These and several other studies on human mortality clearly show more deaths during the winter/colder season than during the summer/hotter season. Ironically, the present debate on global warming and climate change seems to portray a totally opposite view, namely more human fatalities due to heat waves, droughts, floods etc. Once again, the current debate on climate change has created misconceptions of more human fatalities during warm weather events, which is contrary to reality (Figs. 6.15 and 6.16).

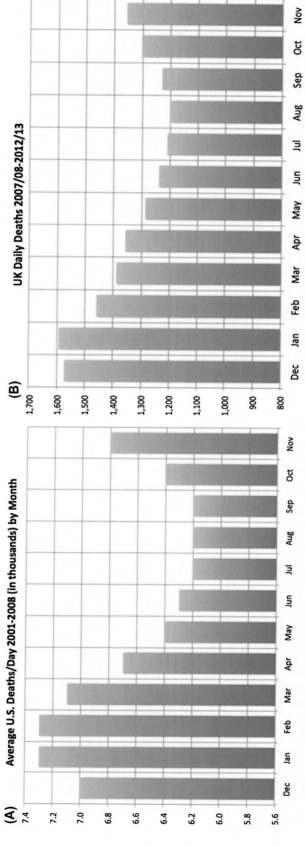

FIGURE 6.15 (A) Average U.S. deaths by month, 2001–2008. (*Indur M. Goklany, data from the U.S. National Center for Health Statistics.*) (B) UK daily deaths by month for 2007–2008 to 2012–2013. (*UK Office for National Statistics.*)

FIGURE 6.16 Global death and death rates due to extreme weather events, 1900–2008. *http://www.jpands.org/vol14no4/goklany.pdf.*

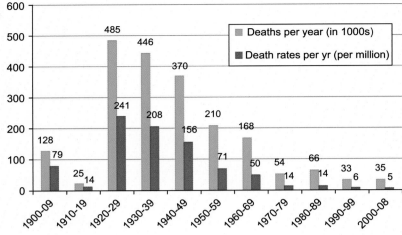

TABLE 6.1   Death Rates Per Year, 1900–1989 and 1990–2008, for Droughts, Floods, Storms, Extreme Temperatures, and Wildfires

	Deaths Per Year		Death Rates Per Year (Per Million People)	
	1900–1989	1990–2008	1900–1989	1990–2008
Droughts	130,044	225	58.19	0.04
Floods	75,169	7676	31.87	1.28
Storms	11,018	20,079	4	3.35
Mass movement—wet	441	780	0.15	0.13
Extreme temperatures	124	5144	0.03	0.82
Wildfires	22	69	0.01	0.01
TOTAL	216,819	33,973	94.24	5.63

http://www.jpands.org/vol14no4/goklany.pdf.

Human fatalities from extreme weather events have declined for the past nine decades. This is true for all types of extreme weather (Table 6.1).

9. SUMMARY AND CONCLUSIONS

The discussion above allows us to summarize several points as follows:

- Extreme weather events are an inherent aspect of the Earth's climate system.
- Extreme weather events have occurred throughout the recorded history of the Earth's climate.
- The Earth's climate warmed quickly during the first half of the 20th century. In North America, the decades of the 1920 and 1930s, known as the Dust Bowl years, witnessed extremes of climate, with recurring droughts and heat waves.
- During the period 1945–1977, when the mean temperature of the Earth declined by about 0.25°C, a number of notable (and tragic) extreme weather events occurred (Appendix A). Most climate scientists attribute these extreme weather events to natural climate variability.
- Many climate scientists and environmentalists have attributed recent extreme weather events to the warming of the Earth's climate. However, this attribution is not substantiated by data. A careful assessment of many well-publicized extreme weather events of the last 10 years suggests that they are due to natural climate variability.
- Hurricanes and tropical storms do not show increasing trends in frequency or in intensity.
- When examined closely, no increase is apparent in extreme weather events in recent years compared to the period 1945–1977, when the Earth's mean temperature was declining. The global warming/extreme weather link is more a perception than reality (Khandekar et al., 2005). The purported link has been fostered by extensive and

uncritical media attention to recent EW events. The latest IPCC documents appear to deemphasize the warming/extreme weather link by suggesting "low confidence" in linking some of the events to recent warming of the climate. The majority of the U.S. and Canadian heat records and the worst multiyear droughts occurred in the 1930s and again in the 1950s. Cold weather extremes have definitely increased in recent years in Europe (2012—2013, 2011—2012, 2009—2010, 2005—2006) and North America (2014—2015, 2013—2014, 2012—2013, 2007—2008), rather than decreased, as claimed by the IPCC and NOAA reports. Colder winters in parts of Asia (2012—2013, 2002—2003) and South America (2007, 2010, and 2013) have occurred.

Economic losses due to worldwide EW events have declined, when compared with the GDP of major countries like the United States, Canada, and most of the countries in the European Union. Economic losses due to cold weather extremes appear to be on the rise in the United States and Canada.

- The Earth's climate may witness cold as well as warm weather extremes in the future. The best way to cope with present and future extreme weather events is to develop improved seasonal (long-range) climate forecasting capability so as to minimize the adverse impacts of such events.

Recent reports (eg, Goklany, 2011) show that human fatalities from extreme weather have declined significantly during the global warming era of the 20th century (1978—2000). The report points out that improved food productivity and increased wealth in developing nations has enabled these countries to cope with such extreme weather events and to reduce damage to property and the loss of human life.

A final comment: global warming and extreme weather pose no serious threat to humanity, either at present or in the foreseeable future.

References

Benestad, R.E., 2010. Low solar activity is blamed for winter chill over Europe. Environmental Research Letters 5. http://dx.doi.org/10.1088/1748—9326/5/2/021001.

Cattiaux, J., et al., 2010. Winter 2010 in Europe: a cold extreme in a warming climate. Geophysical Research Letters 37, L20704. http://dx.doi.org/10.1029/2010GL044613.

Davis, R.E., Knappenberger, P.C., Michaels, P.J., Novicoff, W.M., November 2003. Changing heat-related mortality in the United States. Environmental Health Perspectives 111 (14), 1712—1718, 14594620.

Dole, R., et al., 2011. Was there a basis for anticipating the 2010 Russian heat wave? Geophysical Research Letters 38, L06702. http://dx.doi.org/10.1029/2010GL046582.

Donaldson, G.C., et al., 2001. Heat and cold related mortality and morbidity and climate change. In: Maynard, R.L. (Ed.), Health Effects of Climate Change in the UK. Report to the Department of Health, UK, pp. 70—80.

EOS, 2010. Transactions. American Geophysical Union 91 (30), 255.

Garnett, R., Khandekar, M., 2010. Summer 2010: Wettest on the Canadian prairies in 60 years! CMOS (Canadian Meteorological and Oceanographic Society) Bulletin 38, 204—208.

Gasparrini, A., et al., 2015. Mortality rates attributable to high and low ambient temperature: a multi-country observational study. The Lancet. May 2015.

Goklany, I.M., 2011. Weather and Safety: The Amazing Decline in Deaths from Extreme Weather in an Era of Global Warming, 1910—2010. Report by Reason Foundation, US, 24 p.

van Geel, B., Ziegler, P.A., 2013. IPCC underestimates the sun's role in climate change. Energy and Environment 24, 431—451.

Idso, C.D., Carter, R.M., Singer, F. (Eds.), 2013. Climate Change Reconsidered II Physical Science. Published by the Heartland Institute, Chicago, USA, p. 993.

IPCC, 2001. Climate change 2001; the scientific basis. In: Houghton, J., et al. (Eds.), Contribution to the Working Group I to the Third Assessment Report of the Intergovernmental Panel on Climate Change. Cambridge University Press, 881 p.

IPCC, 2007. Climate change 2007; the scientific basis. In: Solomon, S., et al. (Eds.), Contribution to the Working Group I to the Fourth Assessment Report of the Intergovernmental Panel on Climate Change. Cambridge University Press, 725 p.

IPCC, 2013. Climate change 2013; the scientific basis. In: Stocker, T.F., et al. (Eds.), Contribution to the Working Group I to the Fifth Assessment Report of the Intergovernmental Panel on Climate Change. Cambridge University Press.

Jung, T., et al., 2011. Origin and predictability of the extreme negative NAO winter of 2009/10. Geophysical Research Letters 38, L07701. http://dx.doi.org/10.1029/2011GL046786.

Khandekar, M.L., 1996. El Nino/Southern Oscillation, Indian monsoon and world grain yields-A synthesis. In: El-Sabh, M., et al. (Eds.), Land-based Marine Hazards. Kluwer Pub, Netherlands, pp. 79—95.

Khandekar, M.L., 2002. Trends and changes in extreme weather events: an assessment with focus on Alberta and Canadian Prairies. Report prepared for Alberta Environment Edmonton Alberta, p. 56.

Khandekar, M.L., 2004. Canadian Prairie Drought: A Climatological Analysis. Report prepared for Alberta Environment 37 p. Edmonton Alberta CANADA.

Khandekar, M.L., Murty, T.S., Chittibabu, P., 2005. The global warming debate: a review of the state of science. Pure and Applied Geophysics 162, 1557—1586.

Khandekar, M.L., 2010. 2010 Pakistan floods: climate change or natural variability? Canadian Meteorological and Oceanographic Society CMOS Bulletin 38 (5), 165—167.

Khandekar, M.L., 2013. Are extreme weather events on the rise? Energy and Environment 24, 537—549.

Khandekar, M.L., Neralla, V.R., 1984. On the relationship between the sea surface temperature in the equatorial Pacific and the Indian monsoon rainfall. Geophysical Research Letters 11, 1137—1140.

Kiladis, G.N., Diaz, H.F., 1989. Global climatic anomalies associated with extremes of the Southern Oscillation. Journal of Climate 2, 1069—1090.

Klotzbach, P.J., Landsea, C.W., 2015. Extremely intense hurricanes: revisiting Webster et al. (2005) after 10 years. Journal of Climate 28, 7621—7629.

Kodera, K., 2003. Solar influence on the spatial structure of the NAO during the winter 1900—1999. Geophysical Research Letters 30 (4), 1175. http://dx.doi.org/10.1029/2002GL016584.

Lockwood, M., et al., 2010. Are cold winters in Europe associated with low solar activity? Environmental Research Letters 5. http://dx.doi.org/10.1088/1748—9326/5/2/024001.

Maue, R.N., 2011. Recent historically low global tropical cyclone activity. Geophysical Research Letters 38, L14803. http://dx.doi.org/10.1029/2011GL047711.

Mirza, M., 2003. Three recent extreme floods in Bangladesh: a hydro-meteorological analysis. Natural Hazards 28, 35—64.

Morner, N.-C., 2010. Sea level changes in Bangladesh: observational facts. Energy and Environment 21, 235—249.

Namias, J., September 1955. 1955: some meteorological aspects of droughts: with special reference to the summer of 1952—54 over the United States. Monthly Weather Review 199—205.

Namias, J., 1982. Anatomy of Great Plains protracted heatwaves (especially the 1980 US summer drought). Monthly Weather Review 110, 824—838.

Pielke Jr., R.A., 2014. The Rightful Place of Science: Disasters and Climate Change. Consortium for Science, Policy and Outcomes (Published November 1, 2014).

Petukhov, V., Semenov, V.A., 2010. A link between reduced Barents-Kara sea ice and cold winter extremes over northern continents. Journal of Geophysical Research 115, D21111. http://dx.doi.org/10.1029/2009JD013568.

Seager, R., et al., 2010. Northern Hemisphere winter snow anomalies: ENSO, NAO and the winter of 2009/10. Geophysical Research Letters 37, L14703. http://dx.doi.org/10.1029/2010GL043830.

Simmons, K.M., Sutter, D., Pielke Jr., R., 2012. Normalized tornado damage in the United States: 1951—2011. Environmental Hazards. http://dx.doi.org/10.1080/17477891.2012.738642.

Stott, P.A., Stone, D.A., Allan, M.R., 2004. Human contribution to the European heat wave of 2003. Nature 432, 610—614.

Webster, P.J., Holland, G.J., Curry, J.A., Chang, H.-R., 2005. Changes in tropical cyclone number and intensity in a warming environment. Science 309, 1844—1846. http://dx.doi.org/10.1126/science.1116448.

Weisheimer, A., et al., 2011. On the predictability of the extreme summer 2003 over Europe. Geophysical Research Letters 38, L05704. http://dx.doi.org/10.1029/2010GL046455.

Wuebbles, D.J., et al., 2014. Severe weather in the United States under a changing climate. EOS, Transactions American Geophysical Union 95 (18), 149—150.

APPENDIX A: NOTABLE EXTREME WEATHER EVENTS DURING THE 1945—1977 COOLING PERIOD

1. 1952—1954: An extended drought in conterminous United States, with October 1952 being the driest month ever recorded, according to the U.S. Weather Bureau.

2. October 1954: Deadliest flood in Canada, with over 80 people drowned in southern Ontario due to flooding by remnants of Hurricane Hazel.

3. Summers of 1954 and 1955: Major landfall Hurricanes along the East Coast, including Carol, Edna, and Hazel in 1954 and Connie and Dianne in 1955.

4. Summer 1961: India-wide most disastrous floods with heaviest summer monsoon rains June—September; several hundred people died in floods due to overflowing dams, rivers, etc.

5. 1962—1963: Exceptionally cold winter in most of Europe, with average mean daily temperature anomaly of −4.00°C.

6. May 1968: A cyclone in the Bay of Bengal struck the coast of Myanmar (former Burma), killing over 1000 people; nearby Bangladesh recorded up to 5 m of storm surge (with possible deaths).

7. July 1969: Hurricane Camille (category 4 or 5) slammed into the U.S. Gulf Coast, killing over 250 people.

8. November 13, 1970: A tropical cyclone in the Bay of Bengal struck Bangladesh (former East Pakistan), killing over 250,000 people, the largest number ever of human fatalities in one single extreme weather event.

9. Summer 1972: Severe drought during an Indian summer monsoon (June—September), leading to sharply reduced rice and other crop harvests; as a consequence, India had to import grain for several years.

10. April 1974: Largest outbreak of tornadoes in the U.S. Midwest in one day (April 3, 1974); over 250 people died, with property damage in excess of U.S. $1 billion.

11. December 1976—January 1977: One of the most severe winters in the United States and Canada; Buffalo, New York recorded several feet of snow. Also the southeastern United States recorded extreme cold for several days in January 1977.

PART IV

POLAR ICE

CHAPTER

7

Evidence That Antarctica Is Cooling, Not Warming

D.J. Easterbrook

Western Washington University, Bellingham, WA, United States

1. INTRODUCTION

Antarctica (Figs. 7.1 and 7.2) is the fifth-largest continent, covering 5.4 million square miles (14 million square km). It is the coldest, driest, and windiest continent. The average winter temperature in Antarctica is −81°F (−63°C) and has reached −128.6°F (−89.2°C) at Vostock in east Antarctica. The South Pole receives less than 4 in. (10 cm) of precipitation per year.

About 98% of the Antarctic continent is covered by glacial ice (Figs. 7.2 and 7.3) that averages about 1.2 miles (1.9 km) in thickness. The total volume of ice in Antarctica is about 6.4 million cubic miles (26.5 million cubic km), comprising about 90% of the world's ice and about 70% of the world's fresh water. The average height of the ice sheet is approximately 10,000 ft above sea level (Fig. 7.4). At its thickest point, the ice is 15,700 ft (4776 m) thick. The weight of the ice is estimated to have depressed the Earth's crust 3000 to 5000 ft, depending on the ice thickness.

Antarctica is divided by the Transantarctic Mountains, separating the East Antarctic Ice Sheet from the West Antarctic Ice Sheet (Fig. 7.5). The East Antarctic Ice Sheet is by far the biggest, making up 92% of the total glacial ice, while the West Antarctic Ice Sheet makes up only about 8%.

Evidence-Based Climate Science, Second Edition
http://dx.doi.org/10.1016/B978-0-12-804588-6.00007-0

FIGURE 7.1   Antarctica.

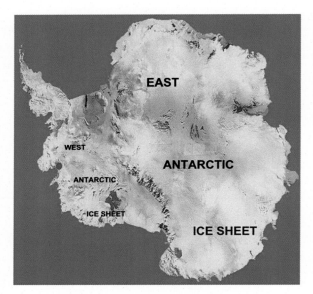

FIGURE 7.2   East and West Antarctic ice sheets.

FIGURE 7.3   Antarctic ice. *From NASA.*

IV. POLAR ICE

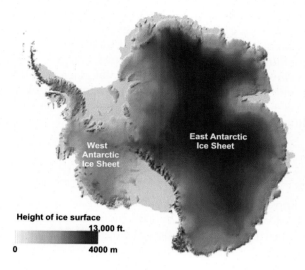

FIGURE 7.4   Height of the Antarctic ice sheet above sea level.

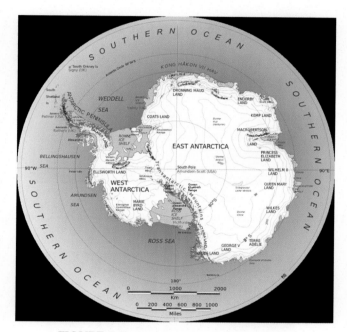

FIGURE 7.5   Map of Antarctica. *From NASA.*

2. ARE ANTARCTIC GLACIERS MELTING AT AN ACCELERATING RATE?

Every day the news media carry headline stories about "accelerated warming in Antarctica causing more rapid melting of the Antarctic ice sheet," and warn of "drowning major cities of the world by rising sea level." NOAA warned of 6.6 ft (2 m) of sea level rise by 2100 as a result of melting of polar ice. The U.S. Army Corps of Engineers recommended that planners consider a sea level rise of five feet "by 2070, and 150 million people in the world's large port cities will be at risk from coastal flooding, along with $35 trillion worth of property." "Inexorably rising oceans will gradually inundate low-lying areas." "By the next century, if not sooner, large numbers of people will have to abandon coastal areas in Florida and other parts of the world." "With seas four feet higher than they are today—a distinct possibility by 2100—about two-thirds of southeastern (SE) Florida is inundated. The Florida Keys have almost vanished. Miami is an island." NASA says "Ice melt in part of Antarctica appears unstoppable" and "the region has enough ice to raise global sea levels by 4 feet." The UN's most recent climate change report estimated that sea level rise "could displace tens of millions of people from coastal areas around the world."

All of these catastrophic scenarios are built upon the assumption of accelerating warming and melting of the Antarctic ice sheet. Is Antarctica really warming at an accelerating rate and is the Antarctic ice sheet melting? These are important allegations, so what do real physical data say about them?

The East Antarctic Ice Sheet contains 92% of the ice in Antarctica, is higher (average 10,000 ft), and colder than the West Antarctic Ice Sheet, which is lower, not as cold, and has outlet glaciers that terminate in the sea. Thus, the East Antarctic Ice Sheet is dominant with respect to the question of melting of polar ice and its effect on sea level. The average annual temperature is −58°F, so to melt any significant amount of ice requires warming to the melting point of ice (58 + 32 = 90°F), plus additional degrees for melting. Warming of 100°F in Antarctica is not remotely plausible.

3. IS ANTARCTICA WARMING OR COOLING?

Previous studies showed cooling across all of the much larger East Antarctic Ice Sheet (Fig. 7.6) and warming limited to the Antarctic Peninsula of west Antarctica. In 2009, Steig et al. published a controversial paper, "Warming of the Antarctic Ice-Sheet Surface Since the 1957 International Geophysical Year," in *Nature* (Fig. 7.7), contending that *all*

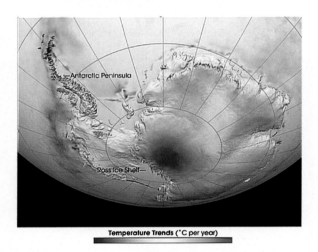

Temperature Trends (°C per year)

FIGURE 7.6   Antarctic temperatures. (Blue = cooling, red = warming.)

FIGURE 7.7   Cover of *Nature* alleging that all of Antarctica is warming.

of Antarctica was warming. Steig et al. stated that the "Antarctic is warming up and is contributing significantly to sea level rise; and that there is strong potential for a greater contribution to sea level rise from Antarctica in the future." Because of very limited surface temperature measurements from the huge East Antarctic Ice Sheet, Steig et al. projected temperatures from the Antarctic Peninsula and a few short records from coastal east Antarctica over all of the East Antarctic Ice Sheet. The Steig et al. analysis spread warming across the entire continent (Fig. 7.7). However, O'Donnell et al. (2011) showed that Steig's methodology was badly flawed, and using the same data as Steig et al., but with better technology, they produced a map similar to previous versions, ie, displaying cooling dominating most of the East Antarctic Ice Sheet with warming primarily constrained to the much smaller (8% of ice) Antarctica Peninsula. Thus, the Steig et al. view of warming of all of Antarctica has been totally discredited.

4. PHYSICAL DATA SHOW LACK OF ANTARCTIC WARMING

Measured satellite and surface temperatures confirm the lack of warming over most of Antarctica. The UAH and RSS satellite records (Figs. 7.8, 7.9, and 7.10) are the most comprehensive because of their extensive coverage of

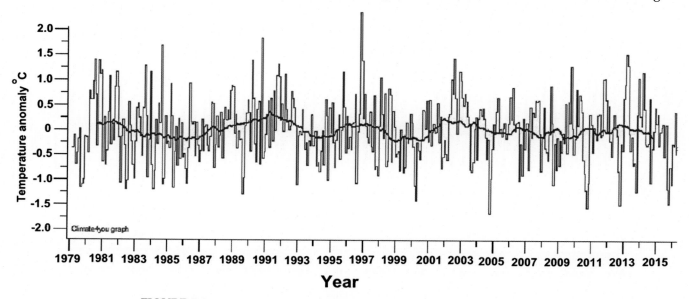

FIGURE 7.8   UAH Antarctic satellite temperatures show no warming for 37 years.

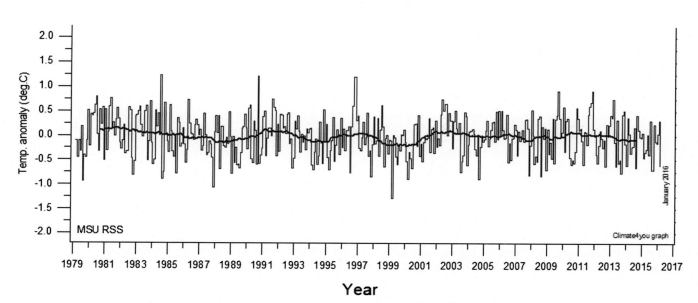

FIGURE 7.9   RSS Antarctic satellite temperatures show no warming for 37 years.

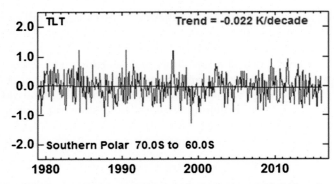

FIGURE 7.10   RSS Antarctic temperature record for the Southern Polar area, showing −0.02°C/decade cooling since 1979. (UAH temperatures are similar, with a drop of −0.01°C.)

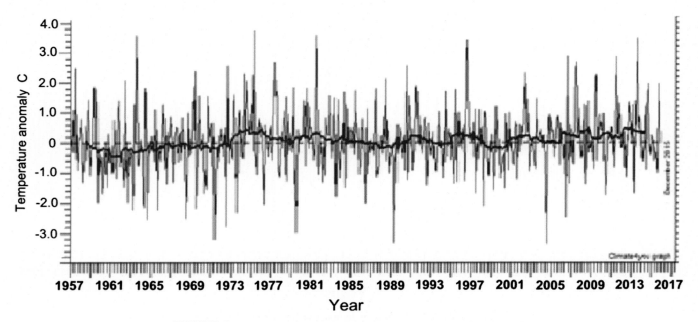

FIGURE 7.11   Antarctic surface temperatures since 1957. (HADCRUT.)

Antarctica. They show the same lack of warming as the surface temperature records (Fig. 7.11). The main conclusion to be drawn from these data is that at least 92% of glacial ice in Antarctica is increasing, not melting.

5. COOLING OF THE SOUTHERN OCEAN AROUND ANTARCTICA

The Southern Ocean around Antarctica has been getting markedly colder since 2006 (Fig. 7.12). Sea ice has increased substantially, especially since 2012.

6. WEST ANTARCTIC ICE SHEET

The West Antarctic Ice Sheet occupies a deep basin west of the main East Antarctic Ice Sheet (Fig. 7.13). It comprises only about 8% of glacial ice in Antarctica.

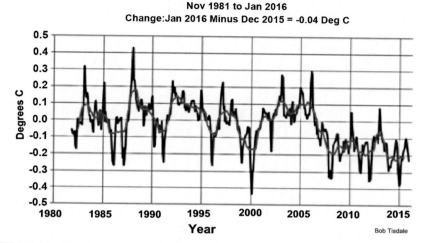

FIGURE 7.12   Temperature anomalies of the Southern Ocean showing sharp cooling since 2006.

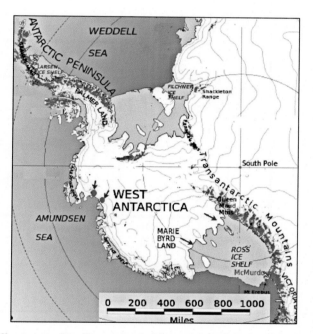

FIGURE 7.13   West Antarctic Ice Sheet west of the Transantarctic Mountains. Red dots are the Pine Island and Thwaites outlet glaciers.

The Antarctic Peninsula has been cooling sharply since 2006. Ocean temperatures have been plummeting since about 2007, sea ice has reached all-time highs, and surface temperatures at 13 stations on or near the Antarctic Peninsula have been cooling since 2000 (Fig. 7.14). Fig. 7.14 is a plot of temperature anomalies at 13 Antarctic stations on or near the Antarctic Peninsula. These data show that the Antarctic Peninsula was warming up until 2000 but has been cooling dramatically since then.

The Larsen Ice Shelf Station has been cooling at an astonishing rate of 1.8°C per decade (18°C per century) since 1995 (Fig. 7.15). Nearby Butler Island records even faster cooling at 1.9°C/decade. Sea ice around Antarctica is increasing because ocean temperature from the surface to 100 m dropped below the freezing point in 2008 and has stayed there since.

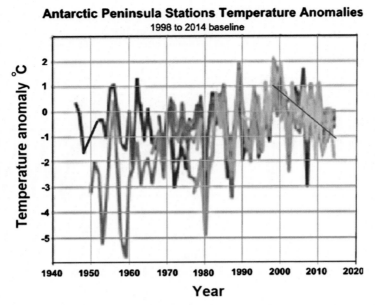

FIGURE 7.14   Temperature anomalies at 13 Antarctic stations on or near the Antarctic Peninsula, showing that the Antarctic Peninsula was warming up until 2000 but has been cooling dramatically since then. *From GISTemp.*

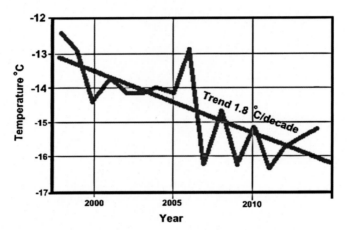

FIGURE 7.15   Annual average temperature at the Larsen Ice Shelf shows sharp cooling (1.8°C/decade). *From GISTemp.*

7. CONCLUSIONS

Satellite and surface temperature records and sea surface temperatures show that both the East Antarctic Ice Sheet and the West Antarctic Ice Sheet are cooling, not warming, invalidating the Steig et al. and news media assertion of warming of all of Antarctica.

- Satellite and surface temperature measurements show that the East Antarctic Ice Sheet is cooling, not warming, and glacial ice is increasing, not melting.
- Satellite and surface temperature measurements of the southern polar area show no warming over the past 37 years.
- Growth of the Antarctic ice sheets means sea level rise is not being caused by melting of polar ice and, in fact, is slightly lowering the rate of rise.
- Satellite Antarctic temperature records show 0.02°C/decade cooling since 1979.
- The Southern Ocean around Antarctica has been getting sharply colder since 2006.
- Antarctic sea ice is increasing, reaching all-time highs.
- Surface temperatures at 13 stations show the Antarctic Peninsula has been sharply cooling since 2000.
- These data demonstrate that the Steig et al. contention that all of Antarctica is warming is clearly false.

8. EVIDENCE OF STABILITY OF THE WEST ANTARCTIC ICE SHEET

In May, 2015, a *New York Times* headline stated "Scientists Warn of Rising Oceans from Polar Melt" and goes on to say: "A large section of the mighty West Antarctica ice sheet has begun falling apart and its continued melting now appears to be unstoppable" and "the melting could destabilize neighboring parts of the ice sheet and a rise in sea level of 10 feet or more may be unavoidable in coming centuries." Virtually every newspaper and TV news show went ballistic with dire predictions of the "unstoppable" catastrophe about to unfold. Two papers on the Pine Island and Thwaites glaciers in West Antarctica triggered this renewed outburst of catastrophic predictions.

A *Washington Post* headline stated "Research casts alarming light on decline of West Antarctic Glaciers" and goes on to say that "a rapidly melting section of the West Antarctic ice sheet appears to be in an irreversible state of decline, with nothing to stop the glaciers in this area from melting into the sea" and "the region's mile-thick ice sheet could collapse and raise sea levels as much as 11 feet…The consequences of such an amount of sea-level rise for the United States—or for any other coastal region—are staggering to contemplate…12.8 million Americans live on land less than 10 feet above their local high-tide line…$2.4 trillion worth of property is occupying this land, excluding Hawaii and Alaska…The cities that would be most affected include Miami, New Orleans, and New York…Within 100–200 years, one-third of West Antarctica could be gone…The effects of climate change are outpacing scientific predictions, driven in part, scientists say, by soaring levels of greenhouse gases in the atmosphere."

In a paper titled "Widespread, Rapid Grounding Line Retreat of Pine Island, Thwaites, Smith, and Kohler glaciers, West Antarctica from 1992 to 2011," Rignot et al. (2014) contend that increased flow velocity of several small outlet glaciers of the West Antarctic Ice Sheet as a result of increased rates of calving into the sea will lead to "unstoppable collapse" of the entire West Antarctic Ice Sheet and raising of sea level by 4 ft, which will displace tens of millions of people from coastal areas around the world. According to Rigot, an electrical engineer, "Warm ocean currents and geographic peculiarities have helped kick off a chain reaction at the Amundsen Sea-area glaciers, melting them faster than previously realized and pushing them 'past the point of no return'…The system [becomes] a chain reaction that is unstoppable, [with] every process of retreat feeding the next one…The glacial retreat there appears unstoppable."

Curiously, Rignot asserts that "heat makes the grounding line retreat inland, leaving a less massive ice shelf above. When ice shelves lose mass, they can't hold back inland glaciers from flowing toward the sea." Apparently he believes that the terminal area of the glacier acts like a dam, "holding back" the rest of the glacier, and if it is removed, the glacier will essentially slide into the sea. That's a false premise—every glaciologist knows that where a glacier terminates is determined by its mass balance between the amount of accumulation of new ice every year and the amount of ice loss by melting or calving. Thus, an important factor for the Rignot "unstoppable collapse" of the West Antarctic Ice Sheet is based on a false premise.

In a paper titled "Marine Ice Sheet Collapse Potentially Underway for the Thwaites Glacier Basin, West Antarctica," Joughin et al. (2014) also infer that the entire West Antarctic Ice Sheet will soon disappear, resulting in a sea level rise of up to 10 ft. The authors contend that recent retreat of the Pine Island and Thwaites glaciers has occurred because warm ocean water has caused melting of ice on the underside of the glaciers, causing them to thin and calve more rapidly. Because the base of most of the West Antarctic Ice Sheet lies below sea level, the authors contend that ocean water will melt its way up several small embayments under the ice sheet, which is more than 1000 miles across, and cause it to collapse abruptly. They refer to this as "unstoppable" because the glacier base is below sea level and they claim that there is nowhere that the glacier can ground, resulting in total collapse of the ice sheet into the sea.

To get a perspective of what is happening now and what might or might not happen in the future requires a look at the overall geologic setting and the scale of the size and thickness of the West Antarctic Ice Sheet relative to the Pine Island and Thwaites outlet glaciers. The West Antarctic Ice Sheet makes up only about 8½ % of Antarctic ice, and the Pine Island glacier makes up about 10% of the West Antarctic Ice Sheet. Most of the West Antarctic Ice Sheet lies SE of the Pine Island glacier and at its SW margin is about 1000 miles from the Pine Island and Thwaites outlet glaciers. Ice in the SE region flows into the Ross Sea, making the Ross Ice Shelf, and has little if anything to do with the part of the ice sheet that flows through the Pine Island and Thwaites outlet glaciers far to the north beyond the ice divide. The Pine Island and Thwaites glaciers are not independent glaciers—they are ice streams from the NW part of the West Antarctic Ice Sheet flowing through narrow embayments bounded by mountains. Their termini calve into the Amundsen Sea, but the rest of the ice sheet is grounded and all of the southwestern part discharges into the Ross Sea. The entire western and southern margins of the West Antarctic Ice Sheet are separated from the ocean by mountains, so these are virtually the only outlets for the ice. The total width of the Pine Island and Thwaites outlet glaciers makes up only about 60 miles of the 2500 miles of coastline along the western and southern margins of the

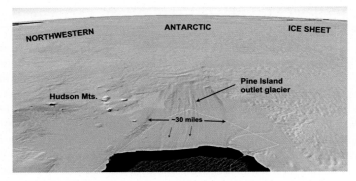

FIGURE 7.16   Pine Island outlet glacier and the northwestern part of the West Antarctic Ice Sheet.

ice sheet. The major ice discharge from the SW margin into the Ross Ice Shelf is not affected by what goes on in the northern part of the ice sheet. Scale is important—only when looking a map of the size of the West Antarctic Ice Sheet does it become apparent just how tiny the Pine Island and Thwaites outlet glaciers are relative to the size of the West Antarctic Ice Sheet (Figs. 7.13 and 7.16).

The rate of glacial retreat is estimated at 10–23 ft per year. The West Antarctic Ice Sheet is roughly 800–1000 miles across, depending on where you measure it. So melting at 10 ft per year would take 528,000 years and at 23 ft per year would take 229,565 years.

9. CREDIBILITY OF THE "UNSTOPPABLE COLLAPSE OF THE WEST ANTARCTIC ICE SHEET"

The base of most of the West Antarctic Ice Sheet lies below sea level (Figs. 7.17 and 7.18) and it is because of this that the West Antarctic Ice Sheet is predicted to collapse. The deepest parts of the subglacial basin are mostly about 10,000 ft (3300 m) deep and lie beneath the central portion of the ice sheet where the ice is the thickest (Fig. 7.19). More important than just depth below sea level is how thick the ice is relative to the depth below sea level. If the ice is thicker than the depth of its base below sea level, the ice will not float.

9.1 Thickness of the West Antarctic Ice Sheet

Fig. 7.19 shows the thickness of the West Antarctic Ice Sheet. Most of the ice sheet is more than 6000 ft (2000 m) thick and in places, reaches up to 10,000 ft (3000 m) thick. The importance of ice thickness is that virtually all of the ice sheet is considerably thicker than the depth below sea level to bedrock, so the ice is grounded and will not float.

Also important is the source area of the outlet glaciers. Fig. 7.20 shows ice divides and ice drainage areas. The Pine Island outlet glacier drains only a relative small portion of the West Antarctic Ice Sheet, so it is difficult to see how events there could result in collapse of the entire Antarctic Ice Sheet.

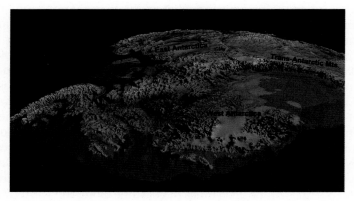

FIGURE 7.17   Subglacial topography of the West Antarctic Ice Sheet.

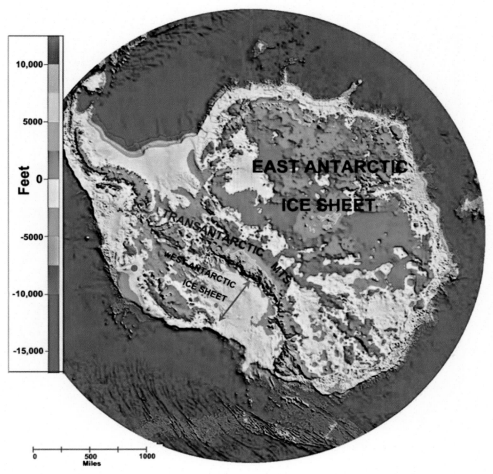

FIGURE 7.18 Subglacial topography in Antarctica. Most of the West Antarctic Ice Sheet lies below sea level, shown in dark and light blue. *Modified from Wikipedia.*

The authors assert that "we find no major bed obstacle that would prevent the glaciers from further retreat and draw down of the entire basin." But that is contrary to what is shown in Fig. 7.21, which is a profile of the West Antarctic Ice Sheet from the east coast to the Transantarctic Mountains, showing thickness of the ice sheet, sea level, and the subglacial floor. At its deepest part, the subglacial floor is 2000 m (6500 ft) below sea level, but almost all of the subglacial floor in this profile is less than 1000 m (3300 ft) below sea level. The ice is mostly more than 2500 m (8000 ft) thick, so basic physics tells us it will not float in 1000 m (3300 ft) of water nor will sea water melt its way under the ice. At 200 km (125 miles) up-ice from the terminus, the ice sheet is about 1600 m (5200 ft) thick and the subglacial floor is above sea level. At 300 km from the terminus, the subglacial floor is 1000 m (3300 ft) above sea level. About 700 km from the terminus, the ice is about 1700 m (5500 ft) thick and the subglacial floor is near sea level. About 1050 to 1150 km (650–700 miles) from the terminus, bedrock occurs at sea level. Because the depth of the subglacial floor below sea level is substantially less than the thickness of ice, it will not float and collapse!

10. ISOSTATIC REBOUND

The weight of thick glaciers resting on the Earth's surface causes depression of the area beneath the ice. The amount of depression is roughly proportional to the density of ice to the density of underlying rocks. If the ice melts away, the land will rebound to its original position. Thus, if the West Antarctic Ice Sheet melts away, where the ice is 10,000 ft thick, the land beneath will rebound several thousand feet, bringing most of the subglacial bed above sea level.

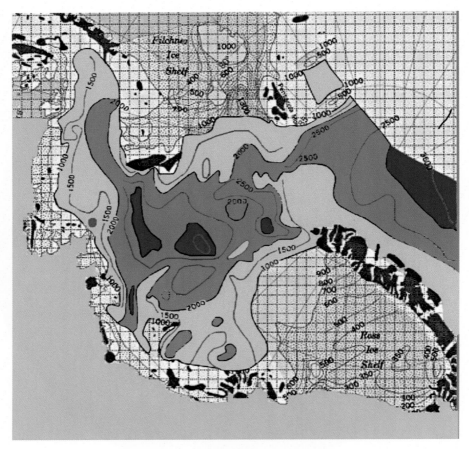

FIGURE 7.19 Thickness of the West Antarctic Ice Sheet. Deep blue = ice > 10,000 ft thick; medium blue = ice > 6500 ft thick; light blue = ice > 3300 ft thick.

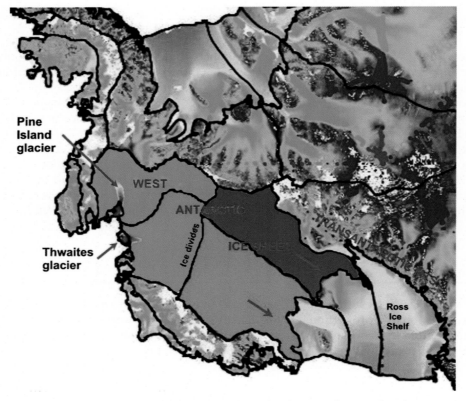

FIGURE 7.20 Ice divides and ice drainages in the West Antarctic Ice Sheet. Light green is the area of ice draining into the Pine Island glacier; dark green is ice draining into the Thwaites glacier; light and dark blue is ice draining into the Ross Sea.

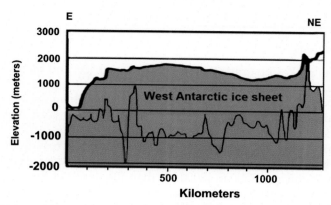

FIGURE 7.21   Profile through the West Antarctic Ice Sheet from the Amundsen Sea to the Transantarctic Mountains. *Modified from http:// antarcticglacier.org.*

10.1 Geothermal Heat Flow Under the Ice Sheet

All of the forecasters of "unstoppable collapse of the West Antarctic Ice Sheet" assume that recent accelerated ice flow and calving has been caused by melting from warm sea water beneath the ice. They did not consider geothermal heat as a possible cause.

In a paper titled "Evidence for Elevated and Spatially Variable Geothermal Flux Beneath the West Antarctic Ice Sheet," Schroeder (2014) presented evidence of elevated geothermal heat flow beneath the Thwaites glacier that *"is likely a significant factor in local, regional, and continental-scale ice sheet stability. Thwaites Glacier is one of the West Antarctica's most prominent, rapidly evolving, and potentially unstable contributors to global sea level rise. Uncertainty in the amount and spatial pattern of geothermal flux and melting beneath this glacier is a major limitation in predicting its future behavior and sea level contribution.*

Geothermal flux is one of the most dynamically critical ice sheet boundary conditions. We show that the Thwaites Glacier catchment has a minimum average geothermal flux of $\sim 114 \pm 10$ mW/m^2 with areas of high flux exceeding 200 mW/m^2 consistent with hypothesized rift-associated magmatic migration and volcanism. These areas of highest geothermal flux include the westernmost tributary of Thwaites Glacier adjacent to the subaerial Mount Takahe volcano and the upper reaches of the central tributary near the West Antarctic Ice Sheet. Large areas at the base of Thwaites Glacier are actively melting in response to geothermal flux consistent with rift-associated magma migration and volcanism. This supports the hypothesis that heterogeneous geothermal flux and local magmatic processes could be critical factors in determining the future behavior of the West Antarctic Ice Sheet.

The distribution of melt and geothermal flux (Fig. 7.10) includes several regions with high melt that are closely related to rift structure and associated volcanism. These include the entire westernmost tributary that flanks Mount Takahe (Fig. 7.10, location A), a subaerial volcano active in the Quaternary and several high-flux areas across the catchment adjacent to topographic features that are hypothesized to be volcanic in origin We also observe high geothermal flux in the upper reaches of the central tributaries that are relatively close to the site of the West Antarctic ice sheet Divide ice core (Fig. 7.10, location B), where unexpectedly high melt and geothermal flux have been estimated.

Our results produce high melt values adjacent to known volcanoes and structures that are morphologically suggestive of volcanic origin. We believe that both the magnitude and spatial pattern of geothermal flux we present reflect the geologic and glaciological reality of the Thwaites Glacier bed and that contrary to previous modeling, our results show regions of high geothermal flux that are in substantial agreement with levels inferred from the ice core drilling site near the ice divide for the Thwaites catchment.

Our results further suggest that the subglacial water system of Thwaites Glacier may be responding to heterogeneous and temporally variable basal melting driven by the evolution of rift-associated volcanism and support the hypothesis that both heterogeneous geothermal flux and local magmatic processes could be critical factors in determining the future behavior of the WAIS."

Other studies (Blankenship et al., 1993; Behrendt, et al., 1998; Bindschadler, 2011) have also found evidence of subglacial geothermal heat flow, especially along zones of former volcanic activity (Corr and Vaughan, 2008).

11. CONCLUSIONS

The evidence above shows that:

1. The West Antarctic Ice Sheet makes up only about 8% of Antarctic ice, and the Pine Island glacier makes up only about 10% of the West Antarctic Ice Sheet. Most of the West Antarctic Ice Sheet lies SE of the Pine Island glacier and at its SW margin is about 1000 miles from the Pine Island and Thwaites outlet glaciers.
2. The Pine Island and Thwaites outlet glaciers drain less than half of the West Antarctic Ice Sheet, so it is not plausible that they could cause collapse of the entire ice sheet.
3. The Pine Island and Thwaites outlet glaciers are only about 30 miles across, so draining 2.2 million km^3 of ice through their narrow channels or sending sea water 1000 miles under the ice sheet is not plausible.
4. Most of the ice sheet is much thicker (2500 m [8000 ft]) than the depth of the subglacial floor below sea level (1000 m [3300 ft]), so the ice will not float and sea water will not extend under the ice.
5. Almost all of the West Antarctic Ice Sheet subglacial floor is less than 1000 m (3300 ft) below sea level. The ice is mostly more than 2500 m (8000 ft) thick, so the ice sheet will not float in 1000 m (3300 ft) of water nor will sea water melt its way under the ice.
6. Studies of subglacial geothermal heat flow show that the area under the Thwaites glacier is unusually high and is the most likely cause of subglacial melting, rather than ocean water.
7. The West Antarctic Ice Sheet is NOT collapsing, the retreat of these small glaciers is NOT caused by global warming, and sea level is NOT going to rise 10 ft.

References

Behrendt, J.C., Finn, C.A., Blankenship, D.D., Bell, R.E., 1998. Aeromagnetic evidence for a volcanic caldera (?) complex beneath the divide of the West Antarctic ice sheet. Geophysics Research Letters 5, 4385–4388.

Bindschadler, R.A., 2011. Variability of basal melt beneath the Pine Island glacier ice shelf, West Antarctica. Journal of Glaciology 57, 581–595.

Blankenship, D.D., et al., 1993. Active volcanism beneath the West Antarctic ice sheet and implications for ice-sheet stability. Nature 361, 526–529.

Corr, H.F., Vaughan, D.G., 2008. A recent volcanic eruption beneath the West Antarctic Ice Sheet. Nature Geoscience 1, 122–125.

Joughin, I., Smith, B.E., Medley, B., 2014. Marine ice sheet collapse potentially under way for the Thwaites Glacier Basin, West Antarctica. Science 344, 735–738.

O'Donnell, R., et al., 2011. Improved methods for PCA-based reconstructions: case study using the Steig, et al. (2009) Antarctic temperature reconstruction. Journal of Climate 24, 2099–2115.

Rignot, E., Mouginot, J., Morlighem, M., Seroussi, H., Scheuchl, B., 2014. Widespread, rapid grounding line retreat of Pine Island, Thwaites, Smith and Kohler glaciers, West Antarctica from 1992 to 2011. Geophysics Research Letters 41, 3502–3509.

Schroeder, D.M., 2014. Evidence for elevated and spatially variable geothermal flux beneath the West Antarctic ice sheet. Proceedings of the National Academy of Sciences of the United States of America 111, 9070–9072.

Steig, E.J., et al., 2009. Warming of the antarctic ice-sheet surface since the 1957 International Geophysical Year. Nature 457, 459–462.

CHAPTER

8

Temperature Fluctuations in Greenland and the Arctic

D.J. Easterbrook

Western Washington University, Bellingham, WA, United States

OUTLINE

1. GREENLAND

The Greenland Ice Sheet (Figs. 8.1 and 8.2) is the second largest ice body in the world, covering 1,710,000 km² (660,000 square miles). The ice sheet is about 2400 km (1500 miles) long and 1100 km (680 miles) wide, covering about 80% of Greenland. The mean altitude of the ice is 2135 m (7000 ft), reaching more than 3000 m (10,000 ft) in places. The ice at the base of the ice sheet is 110,000 years old, but glacial ice has covered Greenland for at least the last 15 million years. The weight of the ice has depressed the subglacial bedrock floor of central Greenland into an elongate oval basin (Fig. 8.3) that extends below sea level, rimmed by mountains through which narrow outlet glaciers must flow to reach the Atlantic Ocean (Figs. 8.4 and 8.5). The basin in central Greenland is below sea level.

The mean annual temperature in Greenland is below freezing and melting is possible only around the periphery. The average annual temperature of Greenland is about 1.5°F (−17°C). Every month of the year has an average temperature below freezing. The lowest mean annual temperatures are about −24°F (−31°C).

Evidence-Based Climate Science, Second Edition
http://dx.doi.org/10.1016/B978-0-12-804588-6.00008-2

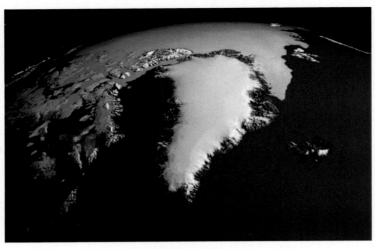

FIGURE 8.1   Greenland and Arctic sea ice. *NASA image.*

FIGURE 8.2   Greenland Ice Sheet.

Temperatures in Greenland were warmer in the earlier part of the last 4000 years, including centuries that were about 2°F (1°C) warmer than the decade of 2001–2010 (Chylek et al., 2004, 2006; Easterbrook, 2011). High−resolution Greenland ice core temperature records of the past 1000 years show the Medieval Warm Period (900–1300 AD), the Little Ice Age (1300–1977), and recent warming (1978–2000) (Grootes and Stuiver, 1997; Kobashi et al., 2011). Greenland was warmer from 1919 to 1932 than recently (Chylek et al., 2006). 1936–1946 was the warmest decade of the last 106 years in west Greenland.

1.1 Climate Changes in the Oxygen Isotope Record of Greenland Ice Cores

Variation of oxygen isotopes in ice cores can be used as a measure of temperature fluctuations. The Greenland ice cores have proven to be a great source of climatic data from the geologic past. Ancient temperatures can be measured using oxygen isotopes in the ice, and ages can be determined from annual dust accumulation layers in the ice. The oxygen isotope ratios of thousands of the Greenland Ice Sheet Project (GISP2) ice core samples were measured by Minze Stuiver and Peter Grootes at the University of Washington (Stuiver et al., 1995; Grootes and Stuiver, 1997) and these data have become a world standard.

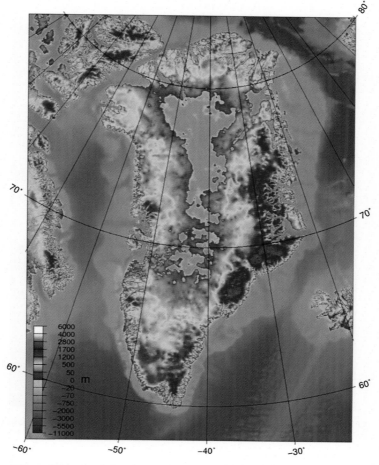

FIGURE 8.3 Subglacial bedrock floor of Greenland, forming an elongate oval basin. The blue area in central Greenland is below sea level. The dark reddish color consists of mountains through which narrow outlet glaciers must flow to reach the Atlantic Ocean. *Adapted from Wikipedia.*

FIGURE 8.4 Outlet glacier fed by the Greenland Ice Sheet.

FIGURE 8.5   Outlet glacier fed by the Greenland Ice Sheet. *Photo by Austin Post.*

Most atmospheric oxygen consists of ^{16}O but a small amount consists of ^{18}O, an isotope of oxygen that is somewhat heavier. When water vapor (H_2O) condenses from the atmosphere as snow, it contains a ratio of $^{16}O/^{18}O$ ($\delta^{18}O$) that reflects the temperature at the time. When the snow falls on a glacier and is converted to ice, it retains an isotopic "fingerprint" of the temperature conditions at the time of condensation. Measurement of the $^{16}O/^{18}O$ ratios in glacial ice hundreds or thousands of years old allows reconstruction of past temperature conditions. What makes these measurements so useful is the accuracy of dating of the samples accomplished by counting annual layers of dust that accumulated in the ice during each melt season on the glacier, giving a dating accuracy of a few years over thousands of years.

The oxygen isotopic composition of a sample is expressed as a departure of the $^{18}O/^{16}O$ ratio from an arbitrary standard:

$$\delta^{18}O = \frac{(^{18}O/^{16}O)_{sample} - (^{18}O/^{16}O) \times 10^3}{(^{18}O/^{16}O)_{standard}}$$

where $\delta^{18}O$ is the of ratio $^{18}O/^{16}O$ expressed in per mil (0/00) units.

The GISP2 temperature data includes two types: (1) oxygen isotope measurements ($\delta^{18}O$) that reflect temperatures at the time of snow accumulation, and (2) borehole temperature measurements that allow reconstruction of temperatures in degrees.

Although the GISP2 ice core data is site specific (Greenland), it has been well correlated with global glacial fluctuations and a wide range of other climate proxies and has become the "gold standard" among global climate reconstructions. However, keep in mind that temperature variations are latitude specific so temperatures from the GISP2 cores show a higher range of values than global data.

The age of each sample is accurately known from annual dust layers in the ice core. The top of the core is 1987. The $\delta^{18}O$ data clearly show remarkable swings in climate over the past 100,000 years. In just the past 500 years, Greenland warming/cooling temperatures fluctuated back and forth about 40 times, with changes every 25–30 years (27 years on the average). None of these changes could have been caused by changes in atmospheric CO_2, because they predate the large CO_2 emissions that began about 1945. Nor can the warming of

1915–1945 be related to CO$_2$, because it predates CO$_2$ emissions that soared after 1945. Thirty years of global cooling (1945–1977) occurred during the big post-1945 increase in CO$_2$.

1.1.1 Late Pleistocene Abrupt Climate Changes

The GSP2 ice core penetrated approximately 10,000 ft (3000 m) of the ice sheet in central Greenland. Annual dust layers indicate that the isotope record extends back about 100,000 years (Fig. 8.6).

From about 100,000 to 10,000 years ago, the last Ice Age was characterized by many abrupt periods of warming and cooling (Fig. 8.6). About 15,000 years ago, the climate suddenly warmed about 21°F (~12°C) in about a century (Fig. 8.7), causing catastrophic melting of the huge continental ice sheets covering vast areas in the world. This abrupt warming, known as the Bölling (~14,700 to ~14,200 years ago) (Fig. 8.8), was followed by rapid cooling at ~14,200 to ~14,000 years ago (the Older Dryas). Another period of abrupt warming (the Allerφd) ensued from ~14 000 to ~13,000 years ago (Fig. 8.8). About 13,000 years ago, temperatures again plunged sharply back to full glacial levels and remained for about 1300 years (the Younger Dryas). About ~11,700 years ago, the Younger

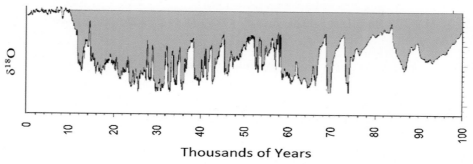

FIGURE 8.6   Oxygen isotope temperatures for the past 100,000 years from the GISP2 Greenland ice core. Blue = cold temperatures. *Plotted from data by Grootes, P.M., Stuiver, M., 1997. Oxygen 18/16 variability in Greenland snow and ice with 10⁻³- to 10⁵-year time resolution. Journal of Geophysical Research 102, 26455–26470.*

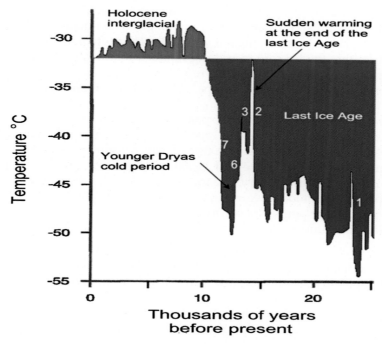

FIGURE 8.7   Greenland temperatures over the past 25,000 years recorded in the GISP2 ice core. Strong, abrupt warming is shown by nearly vertical rise of temperatures, strong cooling by nearly vertical drop of temperatures. *Modified from Cuffey, K.M., Clow, G.D., 1997. Temperature, accumulation, and ice sheet elevation in central Greenland through the last deglacial transition. Journal of Geophysical Research 102, 26383–26396.*

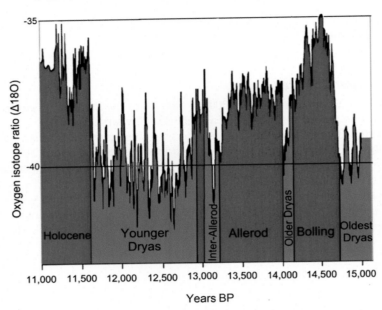

FIGURE 8.8   Large, abrupt fluctuations in temperature at the end of the last Ice Age 11,000 to 15,000 years ago.

Dryas ended suddenly when temperatures in Greenland soared ~21°F (~12°C) in less than a century, marking the beginning of the Holocene (Fig. 8.8) (Kobashi et al., 2008; Steffensen et al., 2008).

1.1.2 Magnitude and Rates of Climates Change

How do these past temperature oscillations compare with recent global warming (1977–1998) or with other warming periods over the past millennia? The answer to the question of magnitude and rates of climate change can be found in the $\delta^{18}O$ and borehole temperature data.

Temperature changes in the GISP2 core over the past 25,000 years are shown in Fig. 8.7, which is a portion of the original curve of Cuffey and Clow (1997). Places where the curve becomes nearly vertical signify times of very rapid temperature change. These are temperatures in Greenland, not global temperatures, but excellent correlation of the ice core temperatures with worldwide glacial fluctuations and correlation of modern Greenland temperatures with global temperatures confirm that the ice core record does indeed follow global temperature trends. For example, the portions of the curve from about 25,000 to 15,000 represent the last Ice Age, when huge ice sheets thousands of feet thick covered North America, northern Europe, and northern Russia and alpine glaciers readvanced far down-valley.

Comparison of warming and cooling periods in the past century to 100-year periods in the past 25,000 years show that the global warming experienced during the past century pales into insignificance when compared to the magnitude of profound climate reversals over the past 15,000 years. Fig. 8.9 shows comparisons of the largest magnitudes of warming/cooling events per century over the past 25,000 years. At least three warming events were 20–24 times the magnitude of warming over the past century, and four were 6–9 times the magnitude of warming over the past century. The magnitude of the only modern warming that might possibly have been caused by CO_2 (1978–2000) is insignificant compared to the earlier periods of warming.

Some of the more remarkable sudden climatic warming periods are listed next (Fig. 8.7).

- About 24,000 years ago, while the world was still in the grip of the last Ice Age and huge continental glaciers covered large areas, a sudden warming of about 20°F (7°C) occurred. Shortly thereafter, temperatures dropped abruptly about 11°F. Temperatures then remained cold for several thousand years but oscillated between about 5°F warmer and cooler.
- About 15,000 years ago, a sudden, intense, climatic warming of about 21°F (~12°C) caused dramatic melting of the large ice sheets that covered Canada and the northern United States, all of Scandinavia, and much of northern Europe and Russia.
- A few centuries later, temperatures again plummeted about 20°F (~11°C) and glaciers readvanced.
- About 14,000 years ago, global temperatures once again rose rapidly, about 8°F (~4.5°C), and glaciers receded.

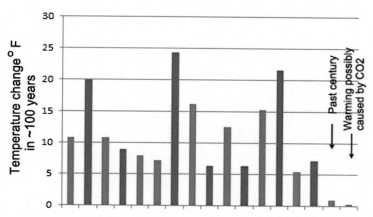

FIGURE 8.9 Magnitudes of the largest warming/cooling events over the past 25,000 years. Temperatures on the vertical axis represent rise or fall of temperatures in about a century. Each column represents the rise or fall of temperature shown on Fig. 8.7. Event number 1 is about 24,000 years ago and event number 15 is about 11,000 years old. The sudden warming about 15,000 years ago caused massive melting of these ice sheets at an unprecedented rate. The abrupt cooling that occurred from 12,700 to 11,500 years ago is known as the Younger Dryas cold period, which was responsible for readvance of the ice sheets and alpine glaciers. The end of the Younger Dryas cold period warmed by as much as 14°F (8°C) over about 40 years.

- About 13,300 years ago, global temperatures plunged again and glaciers readvanced.
- About 13,200 years ago, global temperatures increased rapidly, 9°F (~5°C), and glaciers receded.
- 12,700 years ago, global temperatures plunged sharply, 14°F (~8°C), and a 1300-year cold period, the Younger Dryas, began.
- After 1300 years of cold climate, global temperatures rose sharply, about 21°F (~12°C), 11,500 years ago, marking the end of the Younger Dryas cold period and the end of the Pleistocene Ice Age.

1.1.3 Holocene Temperatures

Fig. 8.10 shows $\delta^{18}O$ from the GISP2 ice core for the past 10,000 years (the Holocene) and Fig. 8.11 shows temperature reconstruction of the same core based on borehole temperatures. The isotope record begins at 1987 AD at the top of the core. Temperatures higher than those in 1987 are shown in red, lower temperatures in blue. The most striking thing about the curve is that temperatures for the first 8500 years of the 10,000-year record were higher than those in 1987 by 2–5°F (1–3°C). The last 1500 years, known as the Little Ice Age, were significantly cooler. Thawing out has occurred since the Little Ice Age, but temperatures are not yet back to where they had been for almost all of the Holocene.

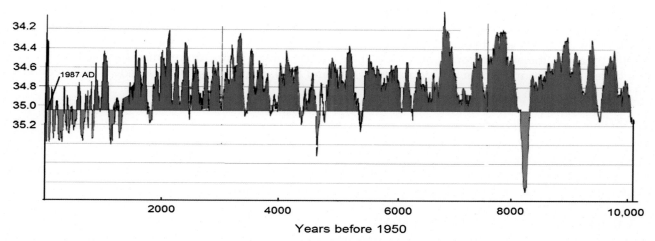

FIGURE 8.10 $\delta^{18}O$ from the GISP2 ice core for the past 10,000 years. Red areas represent temperatures warmer than those in 1987 (top of the core); blue areas were cooler. Almost all of the past 10,000 years were warmer than the past 1500 years. *Plotted from data by Grootes, P.M., Stuiver, M., 1997. Oxygen 18/16 variability in Greenland snow and ice with 10^{-3}- to 10^5-year time resolution. Journal of Geophysical Research 102, 26455–26470.*

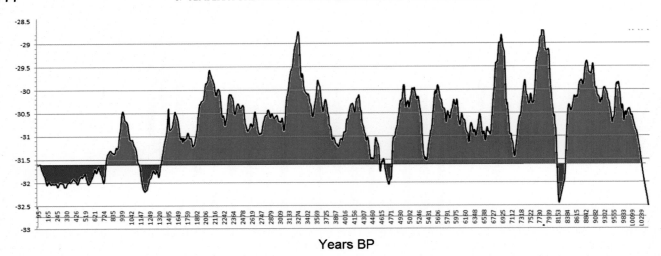

FIGURE 8.11 Temperature reconstruction of the GISP2 Greenland ice core. Red areas represent temperatures higher than recent temperatures. The past 1500 years have been significantly cooler than the preceding 8500 years. *Plotted from data in Alley, R.B., 2000. The Younger Dryas cold interval as viewed from central Greenland. Quaternary Science Reviews 19, 213–226.*

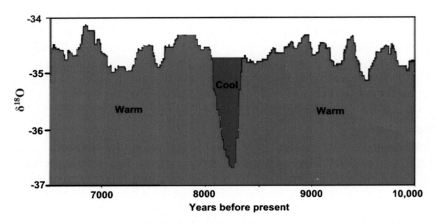

FIGURE 8.12 The 8200-year B.P. sudden climate change recorded in oxygen isotope ratios in the GISP2 ice core. *Plotted from data by Grootes, P.M., Stuiver, M., 1997. Oxygen 18/16 variability in Greenland snow and ice with 10^{-3}- to 10^5-year time resolution. Journal of Geophysical Research 102, 26455–26470.*

The early Holocene warm period was interrupted 8200 years ago by a sudden global cooling that lasted for a few centuries (Fig. 8.12). During this time, temperatures first plunged almost 7°F (3.75°C), and then warmed back up abruptly by about the same amount, all in less than two centuries. Temperatures then oscillated up and down half a dozen times for more than 3000 years (Figs. 8.10 and 8.11), culminating in the Minoan Warm Period about 3400 years ago.

1.1.4 Minoan Warm Period

The Minoan Warm Period was the warmest in the past 7000 years, reaching about 5°F (2.75°C) above recent temperatures in Greenland (Figs. 8.10, 8.11, and 8.13) about 3300 years ago. During this time, the Minoan culture flourished in the Mediterranean region. This warm period ended suddenly about 3100 years ago when temperatures plunged about 3.5°F (2°C) in Greenland (Fig. 8.13) into a cooler period.

1.1.5 Roman Warm Period (250 BC–400 AD)

Prior to the founding of the Roman Empire, Egyptian records show a cool climatic period from about 750 to 450 BC, and the Romans wrote that the Tiber River froze and snow remained on the ground for long periods (Singer and Avery, 2007). This cool period was followed by the Roman Warm Period, which was a time of warm climate from

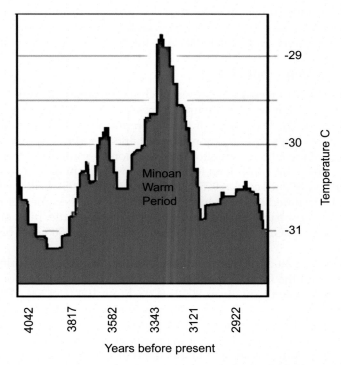

FIGURE 8.13   Paleotemperatures from the GISP2 Greenland ice core during the Minoan Warm Period.

about 250 BC to about 400 AD, when temperatures in Greenland rose to about 3.5°F (2°C) above modern temperatures. During this time, the Roman culture flourished and glaciers in the Alps retreated up-valley.

1.1.6 The Dark Ages (400–900 AD)

The Dark Ages coincided with a period of cooling from about 400 to ~900 AD. The Dark Ages were a time of cultural and economic deterioration, characterized by decline of knowledge, destruction of libraries, religious superstition, brutality, many wars, and generally hard times following the fall of the Roman Empire. Temperatures in Greenland fell ~5°F (2.75°C) from the highs of the Roman Warm Period (Fig. 8.14). A puzzling event apparently

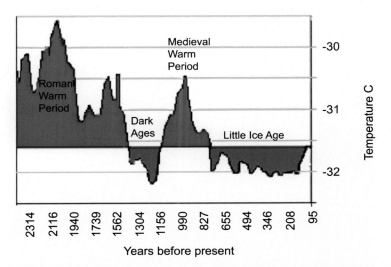

FIGURE 8.14   Paleotemperatures from the GISP2 Greenland ice core during the Roman Warm Period. *Plotted from data in Cuffey, K.M., Clow, G.D., 1997. Temperature, accumulation, and ice sheet elevation in central Greenland through the last deglacial transition. Journal of Geophysical Research 102, 26383–26396; Alley, R.B., 2000. The Younger Dryas cold interval as viewed from central Greenland. Quaternary Science Reviews 19, 213–226.*

occurred in 540 AD, when tree rings suggest greatly retarded growth, the sun appeared dimmed for more than a year, temperatures dropped in Ireland, Great Britain, Siberia, and North and South America, fruit did not ripen, and snow fell in the summer in southern regions. In 800 AD, the Black Sea froze, and in 829 AD the Nile River froze.

1.1.7 Medieval Warm Period (900–1300 AD)

The Medieval Warm Period (MWP) was a time of warm climate from about 900–1300 AD, when global temperatures were somewhat warmer than at present. Temperatures in the GISP2 ice core were about 2°F (1°C) warmer than modern temperatures (Fig. 8.14). The effects of the warm period were particularly evident in Europe, where grain crops flourished, alpine tree lines rose, many new cities arose, and the population more than doubled.

The Vikings took advantage of the climatic amelioration to colonize southern Greenland in 985 AD, when milder climates allowed favorable open-ocean conditions for navigation and fishing. This was close to the maximum Medieval warming recorded in the GISP2 ice core at 975 AD (Stuiver et al., 1995). Erik the Red explored Greenland from Iceland and gave it its name. He claimed land in southern Greenland and became a chieftain about 985 AD. The first Greenlanders brought grain seed, probably barley, oats, and rye, horses, cattle, pigs, sheep, and goats. The southern coastal area was forested at the time. Greenland settlements lasted about 500 years before cooling during the Little Ice Age ended the settlements.

About 620 farms have been excavated in Greenland. Longhouses, the central residences of farm dwellers, would house 10 to 20 people who worked the farm. Ten persons per farm would put the population in Greenland at more than 6000 people, it but could have been as many as 8000–9000. From 1000 to 1300 AD the settlements thrived under a climate favorable to farming, trade, and exploration. A cooling, steadily deteriorating climate began after 1300 AD and farming became impractical. Three churches, one large estate, and 95 farms have been excavated on the west coast of Greenland, mostly under permafrost. A bishop who travelled there about 1350 AD found that the settlement was completely abandoned. The Church abandoned Greenland in 1378 because ships could not get through the sea ice between Iceland and Greenland (Fig. 8.15).

During the Medieval Warm Period, wine grapes were grown as far north as England, where growing grapes is now not feasible and about 300 miles (500 km) north of present vineyards in France and Germany. Grapes are presently grown in Germany up to elevations of about 1800 ft (560 m), but from about 1100 to 1300 AD, vineyards extended up to about 2500 ft (780 m), implying that temperatures were warmer by about 2–2.5°F (1–1.4°C). Wheat and oats were grown around Trondheim, Norway, suggesting that the climate was about 2°F (1°C) warmer than present (Fagan, 2007).

1.1.8 The Little Ice Age (1300 AD to the 20th Century)

At the end of the Medieval Warm Period, ~1300 AD, temperatures in Greenland dropped ~7°F (~4°C) in ~20 years and the cold period that followed is known as the Little Ice Age. The colder climate that ensued for several centuries was devastating (see, eg, Grove, 2004; Singer and Avery, 2007; Fagan, 2000). Temperatures of the cold winters and cool, rainy summers were too low for growing of cereal crops, resulting in widespread famine and disease. Glaciers in Greenland advanced, alpine glaciers expanded worldwide, and pack-ice extended

FIGURE 8.15   Hvalsey Church, the place of the last recorded written record of the Norsemen in Greenland.

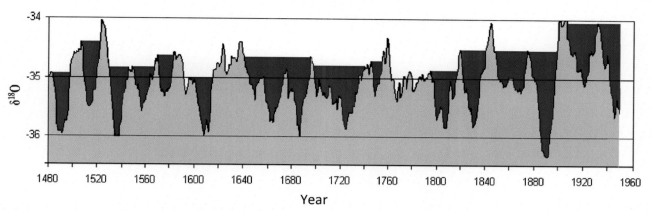

FIGURE 8.16   Oxygen isotope ratios from the GISP2 ice core showing alternating cool and warm periods during the Little Ice Age. Blue areas are cool periods.

southward in the North Atlantic. The population of Europe had become dependent on cereal grains as a food supply during the Medieval Warm Period, and when the colder climate, early snows, violent storms, and recurrent flooding swept Europe, massive crop failures occurred. Three years of torrential rains that began in 1315 led to the Great Famine of 1315−1317. The Thames River in London froze over, the growing season was significantly shortened, crops failed repeatedly, and wine production dropped.

Winters during the Little Ice Age were bitterly cold in many parts of the world. Advance of glaciers in the Swiss Alps encroached on farms and buried entire villages. The Thames River and canals and rivers of the Netherlands frequently froze over during the winter. New York Harbor froze in the winter of 1780 and people could walk from Manhattan to Staten Island. Sea ice surrounding Iceland extended for miles in every direction, closing many harbors. The population of Iceland decreased by half and the Viking colonies in Greenland died out in the early 1400s because they could no longer grow enough food there. In parts of China, warm weather crops that had been grown for centuries were abandoned. In North America, early European settlers experienced exceptionally severe winters.

Although the Little Ice Age was a time of overall cooler temperatures, it was by no means just a single event. Warm/cool temperatures fluctuated back and forth many times (Fig. 8.16). During cool episodes, glaciers advanced; and during warmer times, they retreated.

1.1.9 Wolf Minimum Cool Period (1280−1350 AD)

The earliest cool period of Little Ice Age occurred during the Wolf Solar Minimum. The change from the warm Medieval Warm Period to the cold of the Little Ice Age was abrupt and devastating. Repeated crop failures led to the Great Famine from about 1310 to 1322 AD. The GISP2 oxygen isotope record of the Wolf cool period is shown in Fig. 8.17.

1.1.10 Spörer Minimum Cool Period (1460−1550 AD)

The Spörer cool period occurred during the Spörer Solar Minimum, about 1440−1550 AD. The oxygen isotope record of the period is shown in Fig. 8.18.

1.1.11 Maunder Minimum Cool Period (1645−1710 AD)

The best-known cold period of the Little Ice Age was the Maunder, from 1645−1710 AD (Maunder, 1894, 1922; Eddy, 1976, 1977). The cooler temperatures of this time are evident in the GISP2 oxygen isotope record (Fig. 8.19), the Central England Temperature records (CET), global glacier advance, and extreme hardships in the historic record of Europe.

1.1.12 Dalton Minimum Cool Period (1790−1820 AD)

The Dalton cool period was a time of deep climatic cooling that invoked great hardship on the population of Europe and elsewhere. The cool temperatures of the Dalton are evident in the oxygen isotope record of the GISP2 ice core (Fig. 8.20) and confirmed by the historic record of the Central England Temperature data.

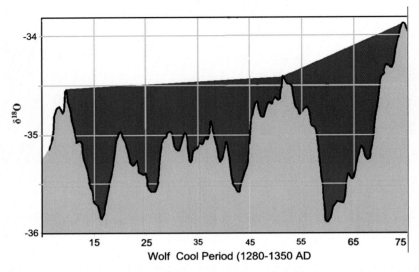

FIGURE 8.17   Oxygen isotope record of the Wolf Minimum cool period (blue) in the GISP2 ice core. *Plotted from data by Grootes, P.M., Stuiver, M., 1997. Oxygen 18/16 variability in Greenland snow and ice with 10^{-3}- to 10^5-year time resolution. Journal of Geophysical Research 102, 26455—26470.*

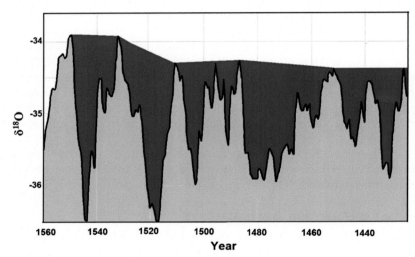

FIGURE 8.18   GISP2 oxygen isotope record of the Spörer cool period (blue). *Plotted from data by Grootes, P.M., Stuiver, M., 1997. Oxygen 18/16 variability in Greenland snow and ice with 10^{-3}- to 10^5-year time resolution. Journal of Geophysical Research 102, 26455—26470.*

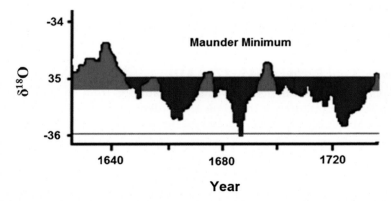

FIGURE 8.19   GISP2 oxygen isotope record of the Maunder cool period (blue). *Plotted from data by Grootes, P.M., Stuiver, M., 1997. Oxygen 18/16 variability in Greenland snow and ice with 10^{-3}- to 10^5-year time resolution. Journal of Geophysical Research 102, 26455—26470.*

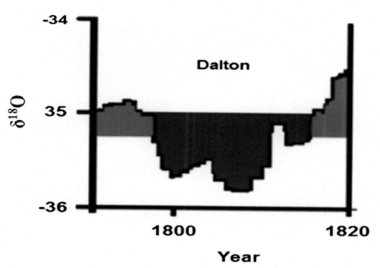

FIGURE 8.20   Oxygen isotope temperature record of the GISP2 ice core showing the Dalton cooling. *Plotted from data by Grootes, P.M., Stuiver, M., 1997. Oxygen 18/16 variability in Greenland snow and ice with 10^{-3}- to 10^5-year time resolution. Journal of Geophysical Research 102, 26455–26470.*

1.1.13 1880–1915 Cool Period

The climatic cooling from 1880 to ~1915 is apparent in the oxygen isotope data from the GISP2 ice core (Figs. 8.21 and 8.22), the CET, and in historic records (Fig. 8.23). During this cool period, alpine glaciers advanced down-valley nearly to their early Little Ice Age terminal positions.

1.1.14 1945–1977 Cool Period

Chylek et al. (2004) analyzed temperature histories of coastal stations in southern and central Greenland having almost uninterrupted temperature records between 1950 and 2000 (Fig. 8.24) and found that coastal Greenland's peak temperatures occurred between 1930 and 1940, after which subsequent decrease in temperature was so substantial and sustained that current coastal temperatures "are about 1°C below their 1940 values." At the summit of the Greenland Ice Sheet, the summer average temperature has decreased at the rate of 2.2°C per decade since the beginning of the measurements in 1987. Chylek et al. found that during the Greenland warming of the 1920s (Fig. 8.2), "average annual temperature rose between 2 and 4°C (and by as much as 6°C in the winter) in less

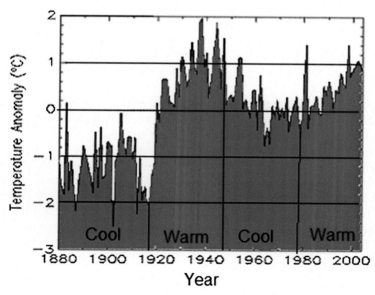

FIGURE 8.21   Temperature fluctuations in Greenland from 1880 to 2004.

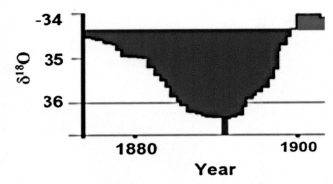

FIGURE 8.22   Oxygen isotope temperature record of the GISP2 ice core, showing the 1880–1915 cooling. *Plotted from data by Grootes, P.M., Stuiver, M., 1997. Oxygen 18/16 variability in Greenland snow and ice with 10^{-3}- to 10^{5}-year time resolution. Journal of Geophysical Research 102, 26455–26470.*

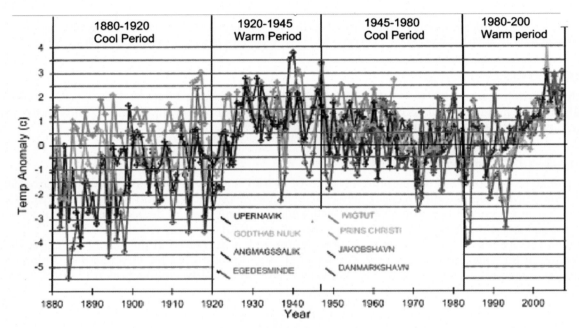

FIGURE 8.23   Temperatures since 1880 at eight Greenland stations. Temperatures were cool from 1880 to about 1920, and then warmed from 1920 to about 1945. Temperatures were cooler from ~1945 to ~1980, and then warmed again from 1980 to ~2004 to levels close to, but not exceeding, temperatures in the 1930s. *Modified from Jones et al. data set.*

than ten years." They called this the "great Greenland warming of the 1920s," and concluded that "since there was no significant increase in the atmospheric greenhouse gas concentration demonstrates that a large and rapid temperature increase can occur over Greenland, and perhaps in other regions of the Arctic, due to internal climate variability without a significant anthropogenic influence."

Two weather stations, Godthab Nuu and Angmagssalik, on opposite coasts of Greenland, have the longest records, dating back more than a century. Both show similar annual temperature patterns—strong warming in the 1920 and 1930s followed by cooling from 1950 to 1980 and warming from 1980 to 2005. With the exception of 2003 at one station, recent temperatures are only at mid-20th century levels. The significance of these recent temperature records is that they show that temperatures in the past several decades have not exceeded those of the 1930s.

1.1.15 Significance of Greenland Climate Changes

Oxygen isotope measurements of the GISP2 ice core show that about 8500 of the past 10,000 years were significantly warmer than recent decades. The most recent 1500 years, including the Little Ice Age, were cooler. If CO$_2$

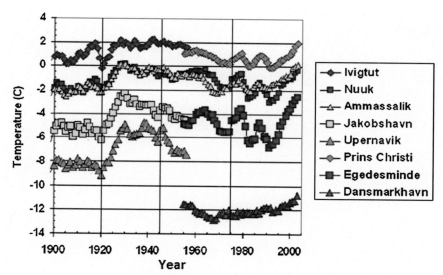

FIGURE 8.24   Annual five-year average temperatures from 1900 to 2005 for eight stations in Greenland. Temperatures during 1995—2005 were similar to 1920—1930 but did not exceed them. *From Chylek, P., Dubey, M.K., Lesins, G., 2006. Greenland warming of 1920—1930 and 1995—2005. Geophysical Research Letters 33.*

is indeed the cause of global warming, then global temperatures should mirror the rise in CO_2. For the past 1000 years, atmospheric CO_2 levels remained fairly constant at about 280 ppm (parts per million). Atmospheric CO_2 concentrations began to rise during the industrial revolution early in the 20th century but did not exceed about 300 ppm. The climatic warming that occurred between about 1915 and 1945 was not accompanied by significant rise in CO_2. In 1945, CO_2 emission began to rise sharply and by 1980 atmospheric CO_2 had risen to just under 340 ppm. During this time, however, global temperatures fell about 0.9°F (0.5°C) in the Northern Hemisphere and about 0.4°F (0.2°C) globally. Global temperatures suddenly reversed during the Great Climate Shift of 1977, when the Pacific Ocean switched from its cool mode to its warm mode with no change in the rate of CO_2 increase. The 1977—1998 warm cycle ended in 1999 and a new cool cycle began. If CO_2 is the cause of global warming, why did temperatures rise for 30 years (1915—1945) with no significant increase in CO_2? Why did temperatures fall for 30 years (1945—1977) while CO_2 was sharply accelerating? Logic dictates that this anomalous cooling cycle during accelerating CO_2 levels must mean that rising CO_2 is *not* the *cause* of global warming. Temperature patterns since the Little Ice Age (~1300—1860 AD) show a very similar pattern; 25—30 year periods of alternating warm and cool temperatures during overall warming from the Little Ice Age low. These temperature fluctuations took place well before any significant effect of anthropogenic atmospheric CO_2.

Temperature changes recorded in the GISP2 ice core from the Greenland Ice Sheet show that the magnitude of global warming experienced during the past century is insignificant compared to the magnitude of the profound natural climate reversals over the past 25,000 years, which preceded any significant rise of atmospheric CO_2. If so many much more intense periods of warming occurred naturally in the past without increase in CO_2, why should the mere coincidence of a small period of low magnitude warming this century be blamed on CO_2?

2. THE ARCTIC

The Arctic loosely encompasses the area within the Arctic Circle (66½°N) and adjacent areas. Most of the region is occupied by the Arctic Ocean, but also includes parts of northern Siberia, Norway, Alaska, Canada, and Greenland. Because there is no land at the North Pole, there are no glaciers there, only thin, floating sea ice. Thus, assertions by the news media and some climate scientists of "accelerating melting of the Arctic Ice Cap" are complete nonsense.

The Arctic Ocean lies in a large basin, the deepest parts of which lie in a deep trough along the Mid-Atlantic Ridge (Fig. 8.25). The North Pole lies at the western edge of the trough where depths exceed 16,000 ft (5000 m). Floating sea ice a few meters thick covers the entire Arctic Ocean (Fig. 8.26) in winter but partially melts during summer. Land adjacent to the Arctic Ocean includes areas of widespread permafrost 1000—2000 ft (300—600 m) deep.

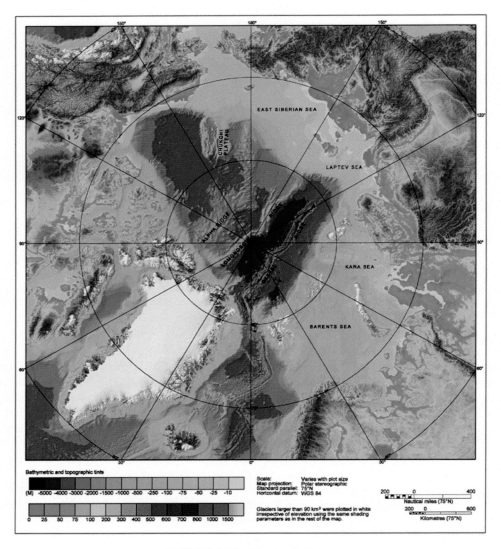

FIGURE 8.25   The Arctic Ocean.

FIGURE 8.26   Arctic sea ice. *NASA photo.*

IV. POLAR ICE

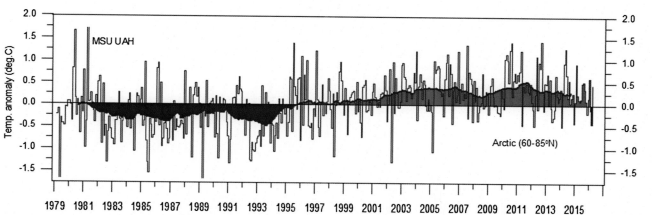

FIGURE 8.27   Satellite temperatures of the Arctic (60°−85°N) from 1979 to 2016. Note the cooling trend from 2011 to 2016. *Adapted from Spencer, UAH.*

2.1 Arctic Temperature

Because there is virtually no land in the Arctic Ocean, no permanent weather stations exist there, and the only temperature data available are from around the periphery and from satellites. Arctic temperatures have fluctuated between warm and cold many times in the past centuries and millennia, paralleling climatic oscillations in Greenland, Iceland, Europe, and the world. Fig. 8.27 is the UAH satellite temperature record for the troposphere in the Arctic between 60° and 85°N latitude from 1979 to 2016. It shows moderate cooling from 1979 to 1995, warming from 1995 to 2011, and cooling since 2011.

Figs. 8.28−8.30 are surface temperatures from weather stations in northern Scandinavia, Iceland, and Russia from 1880 to about 2010. The temperatures are not identical to the satellite temperatures (Fig. 8.27) because satellites measure troposphere temperatures and weather stations measure surface temperatures. The temperature patterns in Figs. 8.28−8.30 show temperature oscillations, cool from 1880 to about 1915−1920, strong warming from about 1915 to 1920 to about 1945, cooling from 1945 to the 1970s, warming from the 1970s to 2005, and cooling from 2005 to 2010 and later. Temperatures were warmer in the 1930s (without elevated CO_2) than in 2010.

Individual weather stations around the Arctic show remarkably consistent temperature patterns similar to those in Figs. 8.28−8.30. Fig. 8.31 shows temperatures for Ostrov Dikson, Russia (73.5°N, 80.4°E) from 1918 to 2016. Note the temperatures in the early 1940s were higher than in the past decade.

Fig. 8.32 shows temperatures at Reykjavik, Iceland from about1900−2016. It indicates a nearly identical oscillating warm/cold patterns as other Arctic stations. However, GISS (NASA) has corrupted these data by subtracting 2°F (1.2°C) from temperatures in the 1930s and 1940s but nothing after 1980 (Fig. 8.33). There is no justification for this major altering of the measured data.

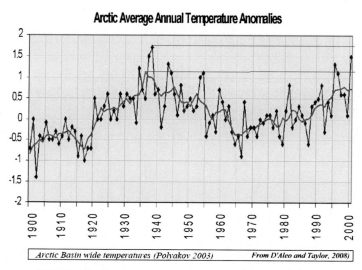

FIGURE 8.28   Arctic annual average surface temperatures.

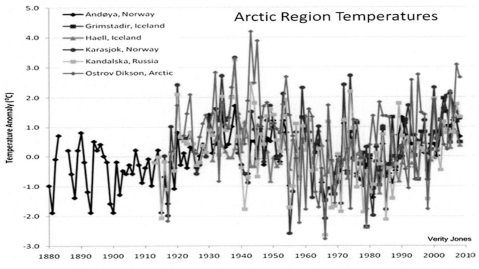

FIGURE 8.29   Arctic surface temperatures from 1880 to 2010.

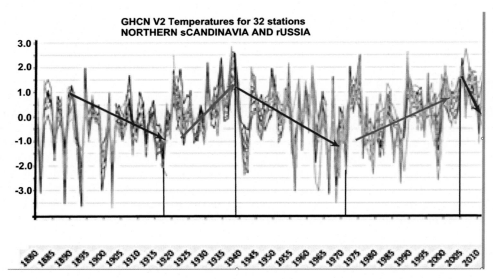

FIGURE 8.30   Arctic surface temperatures from 32 stations in northern Scandinavia and Russia.

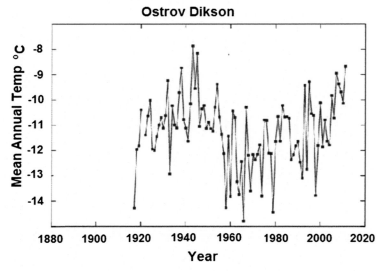

FIGURE 8.31   Temperatures at Ostrov Dikson, Russia from 1918 to 2016 showing warming from 1918 to the early 1940s, cooling from the early 1940s to the 1970s, and warming from the 1970s to 2016. Note that temperatures in the early 1940s were higher than in the past decade.

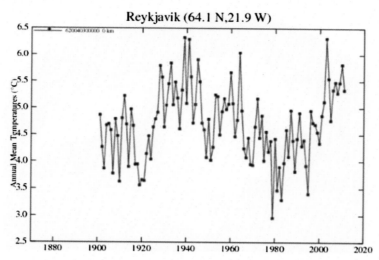

FIGURE 8.32   Temperatures in Reykjavik, Iceland from 1900 to 2016, showing the same warming/cooling oscillation as other Arctic stations.

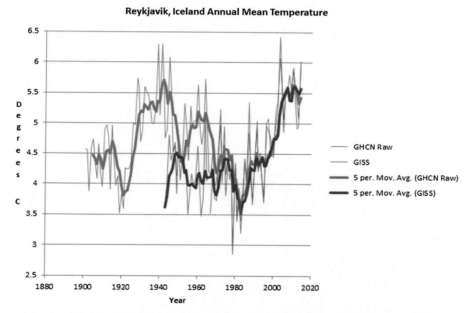

FIGURE 8.33   Temperatures in Reykjavik, Iceland showing corruption of data by GISS (NASA). GISS subtracted 2°F (1.2°C) from the measured temperatures of the 1930s and 1940s but subtracted nothing from temperatures after 1980, giving the false impression that recent warming was greater than earlier.

What these temperature data show is that Arctic temperatures have fluctuated between warm and cold every 25–30 years, just as global temperatures have, all within natural variability limits. The warming from about 1915 to 1945 without elevated atmospheric CO_2 was greater than the warming from about 1980 to 2000. Arctic temperatures in the 1930 and 1940s were higher than the past decade.

2.2 Arctic Sea Ice

No glaciers are present in the Arctic Ocean, but every day the news media and some climate scientists refer to "accelerated melting of the Arctic Ice Cap" even though there is no such thing. Only a few meters of floating ice covers the entire Arctic Ocean (Fig. 8.34).

Much attention has been focused on ice in the Arctic. Numerous assertions by government agencies, climate scientists, news media, politicians, and activists have predicted that the Arctic would be completely ice-free by

FIGURE 8.34   Arctic sea ice.

now. The U.S. National Snow and Ice Center predicted that all Arctic ice would be gone by 2012, and NASA predicted that the Arctic would be totally free of ice by 2013. Studies by the U.S. Dept. of Energy and the U.S. Navy predicted that the Arctic would completely free of ice by 2016. In 2009, Secretary of State John Kerry wrote "the Arctic will be ice-free in the summer of 2013, not in 2050, but four years from now." Al Gore said in 2009, "The North Pole ice cap is 40% gone already and could be completely and totally gone in the winter months in the next 5—10 years." Not only did none of this happen, but Arctic sea ice increased 58% from 2012, and the amount of 5-year old ice increased by 147%.

In winter, the entire Arctic Ocean is covered with floating ice a few meters thick (Fig. 8.35). Some of the ice melts in the summer, so the extent of ice cover varies annually from about 12 to 14 million km^2 in winter to

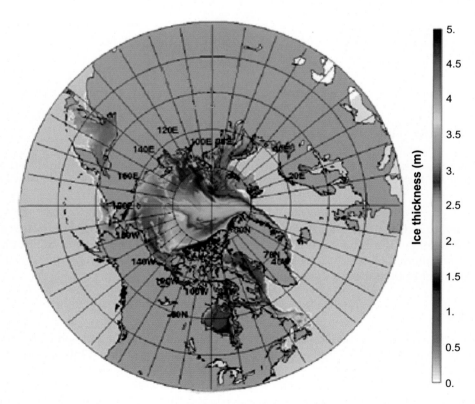

FIGURE 8.35   Arctic sea ice thickness, 2016. Most of the old, thicker ice is concentrated near the coast of North America.

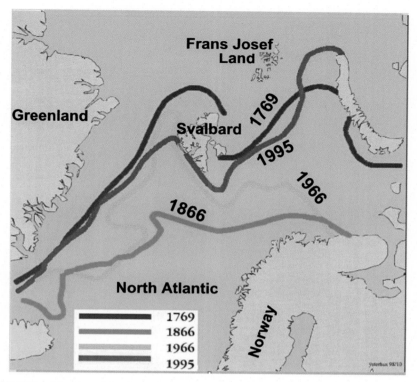

FIGURE 8.36   Margins of Arctic sea ice since 1769. *Modified from map by Norwegian Polar Institute.*

about 7 to 9 km^2 in the summer. The extent of ice also varies with climatic changes and has followed the periods of warming and cooling discussed above.

The positions of some of the margins of the Arctic sea ice in the North Atlantic since 1769 have been reconstructed by the Norwegian Polar Institute (Fig. 8.36). In 1769, the sea ice margin was north of Svalbard. In 1866, the sea ice margin was much farther south, touching the northern Scandinavian coast and extending southward to the coast of Iceland. It then retreated northward and by 1995 lay near the 1769 margin (Fig. 8.36).

Measurements of Arctic sea ice from satellite images began in 1979, just at the time of the "Great Climate Shift of 1977−1978," when temperatures abruptly switched from cool to warm as a result of the change in PDO (Pacific Decadal Oscillation) from cool to warm in one year. Thus, the ice record began at maximum Arctic ice extent and the beginning of the 1978−2000 warm period and declined accordingly (Fig. 8.37).

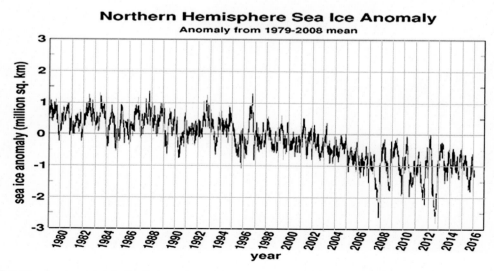

FIGURE 8.37   Arctic sea ice declined since 1979. *Arctic Climate Research, University of Illinois; Adapted from NOAA data.*

October Arctic Sea Ice Volume

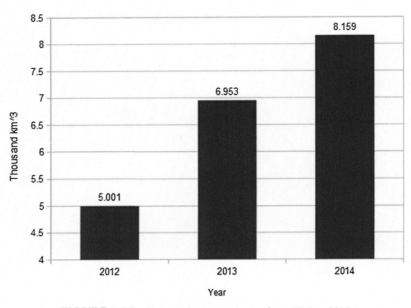

FIGURE 8.38   Increase in Arctic sea ice from 2012 to 2014.

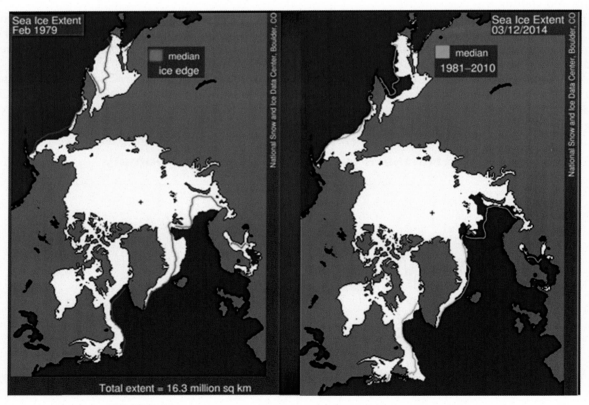

FIGURE 8.39   Comparison of the extent of Arctic sea ice in 1979 and 2014.

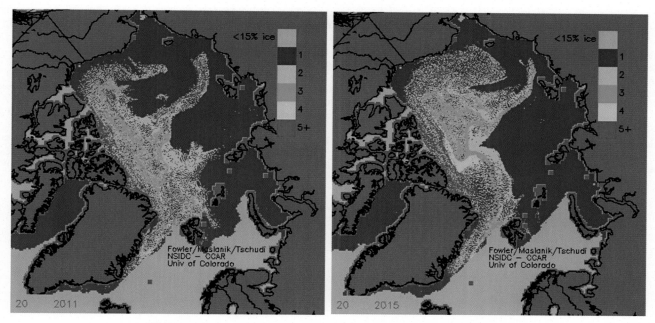

FIGURE 8.40 Comparison of Arctic thick, old, sea ice in 2011 and 2015.

In 2007, Arctic sea ice decreased to a record low level since satellite observations began in 1979. However, according to a NASA study, a large part of the decline from 2005 to 2007 was blowing the ice out of the Arctic basin by wind. Wind patterns compressed old, thick, sea ice and pushed it into the Transpolar Drift Stream, where it flowed out of the Arctic basin along East Greenland, leading to the 2007 record low total Arctic sea ice.

In general, Arctic sea ice diminished between 1979 and 2010, reaching a low in 2007, and then rebounding to higher levels by 2014 (Fig. 8.38). Fig. 8.39 shows a comparison of the extent of sea ice at the beginning of satellite measurement in 1979 and in 2014. Some decline in extent occurred along the margins in the North Atlantic and the Russian coast, but the extent of the main Arctic Ocean ice remains pretty much the same.

The amount of old, thick, sea ice has also varied, generally decreasing until about 2011, and then increasing sharply (147%) by 2015 (Fig. 8.40).

The Arctic temperature and sea ice data clearly show that what is happening in the Arctic is not at all unusual. Arctic temperature and sea ice extent have varied as global temperatures have fluctuated between warm and cool in an ongoing 60-year cycle over the past 500 years. Arctic temperatures were slightly higher in the 1930s and 1940s (without elevated CO_2 levels) than during the modern warm period (1978—2000). Correlation of Arctic temperature with the PDO and AMO are excellent, but no correlation exists between modern warming and CO_2.

References

Alley, R.B., 2000. The Younger Dryas cold interval as viewed from central Greenland. Quaternary Science Reviews 19, 213—226.

Chylek, P., Box, J.E., Lesins, G., 2004. Global warming and the Greenland ice sheet. Climatic Change 63, 201—221.

Chylek, P., Dubey, M.K., Lesins, G., 2006. Greenland warming of 1920—1930 and 1995—2005. Geophysical Research Letters 33.

Cuffey, K.M., Clow, G.D., 1997. Temperature, accumulation, and ice sheet elevation in central Greenland through the last deglacial transition. Journal of Geophysical Research 102, 26383—26396.

Easterbrook, D.J., 2011. Geologic evidence of recurring climate cycles and their implications for the cause of global climate changes: the past is the key to the future. In: Evidence-Based Climate Science. Elsevier Inc, pp. 3—51.

Eddy, J.A., 1976. The Maunder minimum. Science 192, 1189—1202.

Eddy, J.A., 1977. Climate and the changing sun. Climatic Change 1, 173—190.

Fagan, B., 2000. The Little Ice Age. Basic Books, NY, 246 p.

Fagan, B., 2007. The Great Warming: Climate Change and the Rise and Fall of Civilizations. Bloomsbury Press, 283 p.

Grootes, P.M., Stuiver, M., 1997. Oxygen 18/16 variability in Greenland snow and ice with 10^{-3}- to 10^5-year time resolution. Journal of Geophysical Research 102, 26455—26470.

Grove, J.M., 2004. Little Ice Ages: Ancient and Modern. Routledge, London, UK, 718 p.

Kobashi, T., Severinghaus, J.P., Barnola, J.M., 2008. $4 \pm 1.5°C$ abrupt warming 11,270 years ago identified from trapped air in Greenland ice. Earth and Planetary Science Letters 268, 397—407.

Kobashi, T., Kawamura, K., Severinghaus, J.P., Barnola, J.-M., Nakaegawa, T., Vinther, B.M., Johnsen, S.J., Box, J.E., 2011. High variability of Greenland surface temperature over the past 4000 years estimated from trapped air in an ice core. Geophysical Research Letters 38. http:// dx.doi.org/10.1029/2011GL049444.

Maunder, E.W., 1894. A prolonged sunspot minimum. Knowledge 17, 173—176.

Maunder, E.W., 1922. The prolonged sunspot minimum, 1645—1715. Journal of the British Astronomical Society 32, 140 p.

Singer, S.F., Avery, D., 2007. Unstoppable Global Warming Every 1,500 Years. Rowman & Littlefield Publishers, Inc., 278 p.

Steffensen, J.P., Andersen, K.K., Bigler, M., Clausen, H.B., Dahl-Jensen, D., Goto-Azuma, K., Hansson, M.J., Sigfus, J., Jouzel, J., Masson-Delmotte, V., Popp, T., Rasmussen, S.O., Roethlisberger, R., Ruth, U., Stauffer, B., Siggaard-Andersen, M., Sveinbjornsdottir, A.E., Svensson, A., White, J.W.C., 2008. High-resolution Greenland ice core data show abrupt climate change happens in few years. Science 321, 680—684.

Stuiver, M., Brasiunas, T.F., 1991. Isotopic and solar records. In: Bradley, R.S. (Ed.), Global Changes of the Past. Boulder University, Corporation for Atmospheric Research, pp. 225—244.

Stuiver, M., Grootes, P.M., Brasiunas, T.F., 1995. The GISP2 $\delta^{18}O$ record of the past 16,500 years and the role of the sun, ocean, and volcanoes. Quaternary Research 44, 341—354.

PART V

CARBON DIOXIDE

CHAPTER

9

Greenhouse Gases

D.J. Easterbrook

Western Washington University, Bellingham, WA, United States

A greenhouse gas is a gas that absorbs and emits infrared radiation. The primary greenhouse gases in the atmosphere are water vapor, carbon dioxide, methane, nitrous oxide, and ozone. Without greenhouse gases, the average temperature of Earth's surface would be about 15°C (27°F) colder than the present average of 14°C (57°F).

Water vapor is by far the most important greenhouse gas, accounting for up to 90—95% of the greenhouse effect (Fig. 9.1).

The greenhouse effect works as follows: Solar energy warms the Earth's surface during the day. The surface radiates infrared radiation (IR) upward into the atmosphere, where it is absorbed by water, CO_2, and methane This absorption then drives convection and evaporation (latent heat) to restore the lapse rate toward adiabatic stability. IR photons escape to space at higher altitudes.

The lapse rate and convection stops at the tropopause. The height of the tropopause is determined by the height where the net radiation loss to space exceeds the radiation absorbed from all lower levels. The greenhouse effect stops there, and the atmosphere cools by radiation alone. Direct infrared radiation in the main CO_2 bands is absorbed well below 1 km above the Earth's surface. Increasing levels of CO_2 merely cause the absorption to move closer to the surface. Doubling the amount of CO_2 does not double the amount of global warming. Any increase is, at most, logarithmic.

1. ROLE OF WATER VAPOR

Water vapor accounts for by far the largest greenhouse effect (Fig. 9.1). The reason for this is because water vapor emits and absorbs infrared radiation at many more wavelengths than any of the other greenhouse gases (Fig. 9.2), and there is much more water vapor in the atmosphere than any of the other greenhouse gases. The atmospheric water vapor content is highly variable and not easy to measure as a single global number.

The effect of water vapor on temperature is especially important because of the United Nations Intergovernmental Panel on Climate Change (IPCC) claim that CO_2 can cause catastrophic global warming. Because CO_2 is

Evidence-Based Climate Science, Second Edition
http://dx.doi.org/10.1016/B978-0-12-804588-6.00009-4

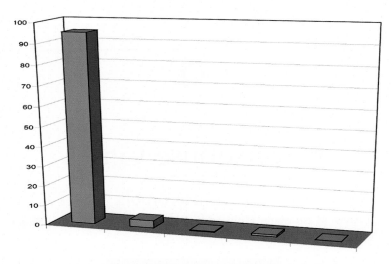

FIGURE 9.1   Greenhouse gas content of the atmosphere.

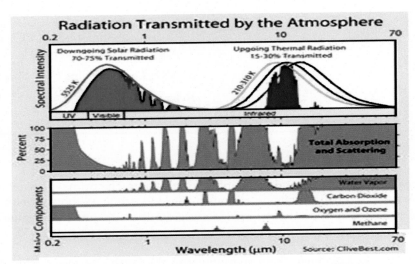

FIGURE 9.2   Wavelengths of greenhouse gases that emit and absorb infrared radiation.

not capable of causing significant global warming by itself, their contention is that increased CO_2 raises temperature slightly and that produces an increase in water vapor, which does have the capability of raising atmospheric temperature. If that is indeed the case, then as CO_2 rises, we should observe a concomitant increase in water vapor. However, Figs. 9.3 and 9.4 show that water vapor (relative humidity) between 10,000 and 30,000 feet declined from 1948 to 2014.

2. CARBON DIOXIDE

Atmospheric carbon dioxide (CO_2) is a nontoxic, colorless, odorless gas that constitutes a tiny portion of the Earth's atmosphere, making up only ∼0.040% of the atmosphere (Fig. 9.5). In every 100,000 molecules of air, 78,000 are nitrogen, 21,000 are oxygen, 2000−4000 are water vapor, and only 30 are carbon dioxide.

The carbon dioxide molecule is composed of one carbon atom covalently double bonded to two oxygen atoms (Fig. 9.6). The molecule has no electrical dipole, and consequently only two vibrational bands are observed in the IR spectrum.

Carbon dioxide is soluble in water, where it forms carbonic acid (H_2CO_3) according to the reversible reaction:

$$CO_2 + H_2O \rightleftharpoons H_2CO_3$$

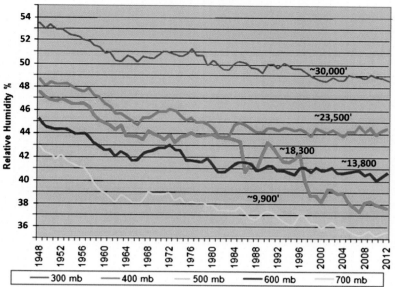

FIGURE 9.3   Decline of atmospheric water vapor between 10,000 and 30,000 feet from 1948 to 2012. *From data at Friendsofscience.org.*

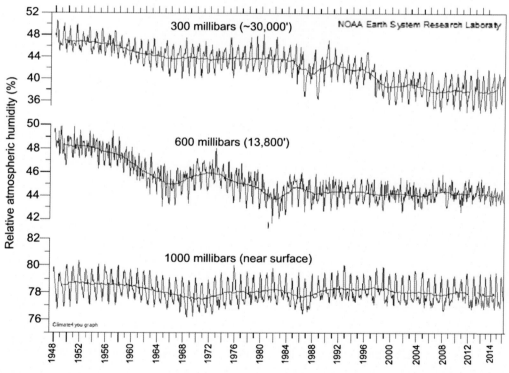

FIGURE 9.4   Decline in atmospheric water vapor (relative humidity) from 1948 to 2014. These data refute the claim that increasing CO_2 causes increased atmosphere water vapor, which, in turn, causes global warming. (From NOAA.)

Most carbon dioxide is not converted into carbonic acid, but remains as CO_2 molecules that do not affect the pH of the water. Seawater contains about 75 times as much CO_2 as fresh water and about 50 times as much CO_2 as air at 25°C. The solubility of CO_2 in water varies with the temperature of the water. Cold water can hold more CO_2 than warm water. As water temperature increases, the solubility of CO_2 decreases, so CO_2 is given off into the

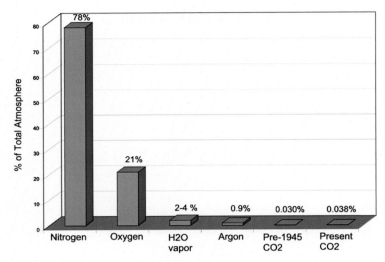

FIGURE 9.5   Composition of the atmosphere. CO_2 makes up only 0.04% of the atmosphere.

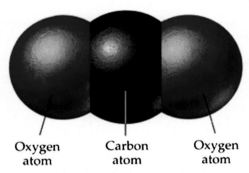

FIGURE 9.6   The carbon dioxide molecule.

atmosphere to establish a new equilibrium between the air and water. The high solubility and chemical reactivity of CO_2 permit ready exchange of CO_2 between the atmosphere and oceans. The amount of CO_2 in the atmosphere is determined largely by the temperature of the oceans, although emission from human sources has added significant amounts since 1950.

When global temperatures rise, as during interglacial periods, the amount of atmospheric CO_2 rises, and when temperatures decline, as during Ice Ages, atmospheric CO_2 declines. Measurements of CO_2 from air trapped in polar ice cores over tens of thousands of years show that atmospheric CO_2 concentrations typically vary from about 0.026–0.0285%, averaging about 0.028%. Higher CO_2 levels during the past interglacial periods do not indicate that CO_2 is the cause of the warmer interglacials because the CO_2 increase lagged Antarctic warming by 600–800 ± 200 years (Fischer et al., 1999; Caillon et al., 2003).

Water vapor accounts for up to 95% of greenhouse gases, with CO_2, methane, and a few other gases making up the remaining 5%. The greenhouse effect from CO_2 is only about 3.6%. Most of the greenhouse warming effect takes place early (Fig. 9.7). After that, the effect decreases exponentially (Fig. 9.6), so the rise in atmospheric CO_2 from 0.030% to 0.038% from 1950 to 2016 could have caused warming of only about 0.01°C. The total change in CO_2 of the atmosphere amounted to an addition of only one molecule of CO_2 per 10,000 molecules of air.

Atmospheric CO_2 rose slowly from the late 1800s to 1945. Emissions began to soar abruptly in 1945 after World War II (Fig. 9.8), but global temperatures cooled for 30 years instead of rising, as would be the case if CO_2 causes warming.

How does the present level of atmospheric CO_2 (0.04%) compare with long → term levels? Figs. 9.9 and 9.10 show some recent examples, and Fig. 9.11 shows CO_2 levels for the past 250 million years. From the mid-Jurassic Period

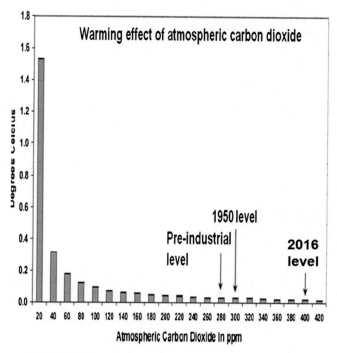

FIGURE 9.7   Warming effect of CO₂ *From D. Archibald.*

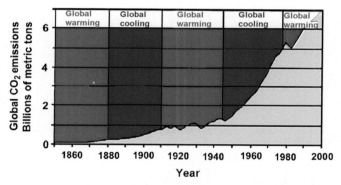

FIGURE 9.8   Global CO₂ emissions show no correlation with global warming.

into the early Cretaceous, atmospheric CO_2 was 0.2% to 0.24%, five to eight times the present level. CO_2 levels dropped steadily from the early Cretaceous to the mid-Tertiary (Fig. 9.11). For 200 million years prior to the mid-Tertiary, CO_2 levels were about two to eight times present levels.

At the abrupt 1977 "Great Climate Shift," when the global climate shifted from cooling to warming, no significant change occurred in the rate of increase of CO_2 (Fig. 9.12), suggesting that CO_2 had nothing to do with the shifting of the climate.

CO_2, which makes up only 0.040% of the atmosphere and constitutes only 3.6% of the greenhouse effect, has increased only 0.008% since emissions began to soar after 1945. How can such a tiny increment of CO_2 cause the 10°F increase in temperature predicted by CO_2 advocates? The obvious answer is that it can't. Computer climate modelers build into their models a high water vapor component, which they claim is due to increased atmospheric water vapor caused by very small warming from CO_2, and since water vapor makes up 95% of the greenhouse effect, they claim the result will be warming. The problem is that atmospheric water vapor has actually declined since 1948 (Figs. 9.3 and 9.4), not increased as demanded by climate models.

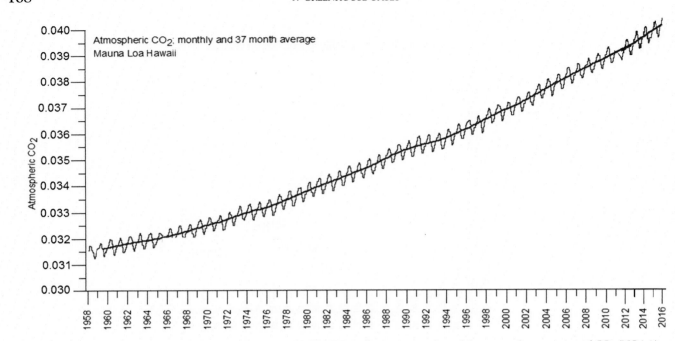

FIGURE 9.9   Rise of CO_2 since 1958 at Mauna Loa, Hawaii. Note what a tiny portion of the atmosphere consists of CO_2 (NOAA).

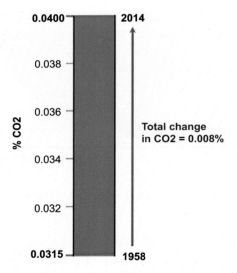

FIGURE 9.10   The total change in CO_2 content of the atmosphere from 1958 to 2014 was only 0.008%, not enough to cause any significant global warming.

3. GLOBAL WARMING AND CO_2 DURING THE PAST CENTURY

Atmospheric temperature measurements, glacier fluctuations, and oxygen isotope data from Greenland ice cores all record a cool period from about 1880 to about 1915. During this period, global temperatures were about 0.9°C (1.6°F) cooler than at present. From 1880 to 1890, temperatures dropped 0.35°C (0.6°F) in only 10 years. From 1890 to 1900, temperatures rose 0.25°C (0.45°F) in 10 years, after which temperatures dipped slightly (0.15°C (0.3°F) until about 1915.

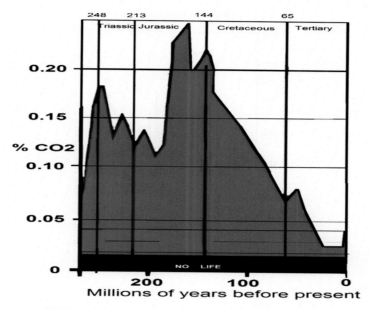

FIGURE 9.11   CO₂ levels for the past 250 million years.

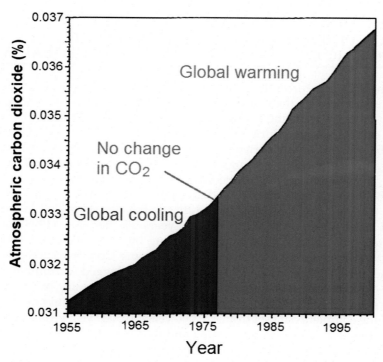

FIGURE 9.12   Lack of any change in atmospheric CO₂ at the switch from the 1945–1977 cool period to the 1978–2000 warm period.

During the 1915 to 1945 warm period, temperatures rose without significant increase in CO₂, showing that global warming occurs without any possibility of CO₂ as a cause because it occurred before CO₂ had risen significantly. CO₂ began to rise sharply after the end of World War II (1945) and continued for 30 years. But instead of causing global warming, as would be the case if CO₂ caused atmospheric warming, global cooling occurred for 30 years (1945–1977) during soaring CO₂. In 1977, the northeastern Pacific switched from its cool mode (where it had been

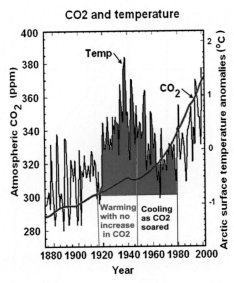

FIGURE 9.13   Lack of correlation of between CO_2 and global temperature.

since ~1945) to its warm mode, and global warming occurred from 1978 to about 2000. CO_2 continued to rise as it had since 1958, so the warm period corresponded to increased CO_2 as a matter of coincidence (Fig. 9.13).

4. GEOLOGIC EVIDENCE THAT GLOBAL WARMING CAUSES INCREASED ATMOSPHERIC CO_2—CO_2 DOES NOT CAUSE GLOBAL WARMING

For several decades, the IPCC has forcefully asserted that increased atmospheric causes global warming that will result in catastrophic consequences for the world. We can test this contention by looking at the timing of increased CO_2 and global warming during alternating Ice Ages and interglaciations. At the end of each Ice Age over the past 420,000 years, the global climate warmed during the following interglaciation and CO_2 rose. All we need to do is to see which came first, global warming or increased CO_2. If CO_2 caused the global warming, then the rise in CO_2 must precede global warming. If it lags global warming, it cannot possibly be the cause of the warming.

Measurements of CO_2 in air bubbles in ice of the Vostock core in Antarctica have been published by Petit et al. (1999), Fischer et al. (1999), Monnin et al. (2001), Mudelsee (2001), Caillon et al. (2003). Petit at al. (1999) measured CO_2 for 420,000 years of the Vostock ice core and found that as the climate cooled into an Ice Age, the decrease in atmospheric CO_2 lagged temperature by several thousand years. Fischer et al. (1999) found that in going from an Ice Age into a warm interglacial, rise in CO_2 lagged warming by 600 ± 400 years. Monnin et al. (2001) showed that rise in CO_2 lagged warming by 800 ± 600 years in the Dome Concordia ice core in Antarctica. Mudelsee (2001) found that over the full 420,000 years of the Vostock core, CO_2 lagged warming by 1300 ± 1000 years. Caillon et al. (2003) analyzed the Vostock core data and found that CO_2 lagged warming by 800 ± 200 years. All five studies of the Antarctic ice cores showed that CO_2 always lagged warming and thus could not be the cause of the warming.

Joanne Nova graphed the complete data set from 420,000 years to 5000 years from the original sources (Fig. 9.14). From these graphs, CO_2 obviously always lags warming and thus cannot be the cause of any of the warm interglacials of the past 420,000 years. The inescapable conclusion from these data is that CO_2 is not the cause of global warming. Global warming causes atmospheric CO_2 to rise.

5. CO_2 LAGS WARMING OVER SHORT TIME SPANS

Humlum et al. (2011, 2013) used data on atmospheric CO_2 and global temperatures for the period of January 1980 to December 2011 to investigate leads/lags between them. They found that changes in CO_2 always lag changes in temperature by 9.5—10 months and lag sea surface temperature by 11—12 months (Fig. 9.15).

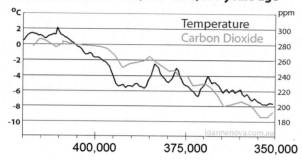

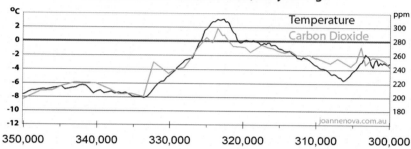

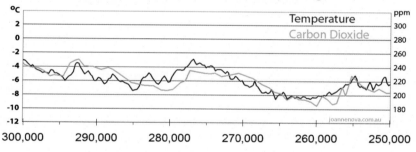

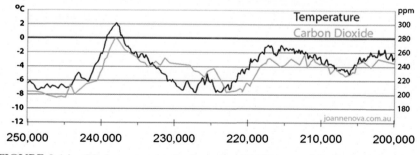

FIGURE 9.14   CO$_2$ lags warming for the entire 420,000 years of the Antarctic ice cores.

1. The overall global temperature change sequence of events appears to be from (1) the ocean surface to (2) the land surface to (3) the lower troposphere.
2. Changes in global atmospheric CO$_2$ lag about 11—12 months behind changes in global sea surface temperature.
3. Changes in global atmospheric CO$_2$ lag 9.5—10 months behind changes in global air surface temperature.
4. Changes in global atmospheric CO$_2$ lag about 9 months behind changes in global lower troposphere temperature (Humlum et al., 2011, 2013).

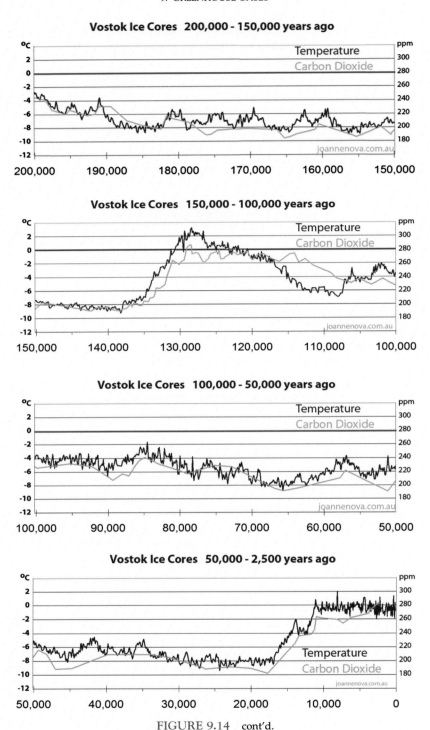

FIGURE 9.14   cont'd.

In general, Humlum et al. (2011, 2013) found that the CO_2 lag in surface temperature changes and lower troposphere temperature changes suggest a temperature sequence of events from the surface to the lower troposphere. Because cause must always precede effect, their observations demonstrate that modern changes in temperatures are not induced by changes in atmospheric CO_2, but rather, the opposite—changes in temperature drive changes in atmospheric CO_2. Changes in atmospheric CO_2 do not correlate with changes in human emissions.

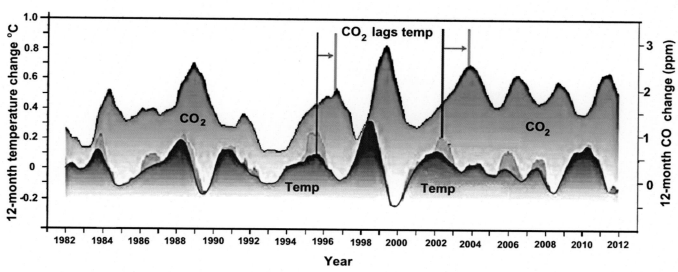

FIGURE 9.15 CO_2 lags warming over short time spans as well as long ones. The blue curve is temperature and the green curve is CO_2. Increase in CO_2 follows a rise in temperature, showing that warming causes a rise in CO_2 rather than CO_2 causing warming. *From Humlum, O., Solheim, J.-E., Stordahl, K., 2011. Identifying natural contributions to late Holocene climate change. Global and Planetary Change 79, 145—156.*

References

Caillon, N., Severinghaus, J.P., Jouzel, J., Barnola, J.-M., Kang, J., Lipenkov, V.Y., 2003. Timing of atmospheric CO_2 and Antarctic temperature changes across termination III. Science 299 (5613), 1728—1731.

Fischer, H., Wahlen, M., Smith, J., Mastroianni, D., Deck, B., 1999. Ice core records of atmospheric CO_2 around the last three glacial terminations. Science 283, 1712—1714.

Humlum, O., Solheim, J.-E., Stordahl, K., 2011. Identifying natural contributions to late Holocene climate change. Global and Planetary Change 79, 145—156.

Humlum, O., Stordahl, K., Solheim, J., 2013. The phase relation between atmospheric carbon dioxide and global temperature. Global and Planetary Change 100, 51—69.

Monnin, E., Indermühle, A., Dällenbach, A., Flückiger, J., Stauffer, B., Stocker, T.F., Raynaud, D., Barnola, J.-M., 2001. Atmospheric CO_2 concentrations over the last glacial termination. Science 291, 112—114.

Mudelsee, M., 2001. The phase relations among atmospheric CO_2 content, temperature and global ice volume over the past 420 ka. Quaternary Science Reviews 20, 583—589.

Petit, J.R., Jouzel, J., Raynaud, D., Barkov, N.I., Barnola, J.-M., Basile, I., Bender, M., Chappellaz, J., Davis, M., Delaygue, G., Delmotte, M., Kotlyakov, V.M., Legrand, M., Lipenkov, Y., Lorius, C., Pepin, L., Ritz, C., Saltzman, E., Stievenard, M., 1999. Climate and atmospheric history of the past 420,000 years from the Vostok ice core, Antarctica. Nature 399, 429—436.

CHAPTER

10

Is CO₂ Mitigation Cost Effective?

Christopher Monckton of Benchley

Science and Public Policy Institute, Washington, DC, United States

1. INTRODUCTION

Hitherto, economists have chiefly addressed the cost-effectiveness of climate mitigation globally. Here a much-simplified method, based on results of the Intergovernmental Panel on Climate Change (IPCC) that are taken as normative *ad argumentum*, is intended to enable even nonspecialist policy makers to rapidly estimate not only how much global warming any proposed CO₂ eduction policy may be expected to abate but also, on the assumption that the unit mitigation cost of all mitigation strategies worldwide is equivalent to that of the policy, its global abatement cost (which is the cost of abating a given quantum of projected future warming by measures of equivalent cost-effectiveness) and its unit mitigation cost in US dollars per Kelvin of warming abated. Brief case studies compare the costs of competing CO₂ reduction policies with one another and with published estimates of the welfare loss arising from unmitigated climate change.

As benchmarks, Stern (2006), adopting an intertemporal discount rate not exceeding 1.4%, estimates that the cost of abating the 3K 21st-century global warming the IPCC expects will be 0—3% of GDP (mean 1.5%) and that 1% of 21st-century global GDP would suffice to abate 5K global warming to 2100, against a global inaction cost of 5—20%, while Garnaut (2008), projecting 5.1K unmitigated global warming to 2100 and using a 1.35—2.65% discount rate, puts Australia's 21st-century mitigation and inaction costs at 3.2—4% and 6%, respectively. The economic literature

Evidence-Based Climate Science, Second Edition
http://dx.doi.org/10.1016/B978-0-12-804588-6.00010-0

puts the inaction cost at 1—4% of global 21st-century GDP and regards an intertemporal discount rate of 5% as normative (eg, Nordhaus, 2008; Murphy, 2008).

Warming beyond 2100 is not considered here. Since equilibrium temperature will not be reached for 1000—3000 years (Solomon et al., 2009), it is centennial-scale transient warming that is policy-relevant today—subsequent warming will occur too slowly over the millennia to do unavoidable harm. Costs external to the policy and benefits external to mitigation of CO_2 forcing are beyond the focus of this chapter, but the method may readily be adapted to encompass them. Warming of 3K and 3% uniform GDP growth in the 21st century are assumed; little error arises from assuming uniformity, and other rates may be chosen. A 5% discount rate is assumed on account of centennial-scale uncertainties, but lesser rates are also considered.

2. PROJECTED 21ST-CENTURY CO₂-DRIVEN WARMING

Stern (2006) concluded that, though previous projections had indicated 2—3K anthropogenic warming by 2100, causing a permanent global output loss estimated at 0—3% (mean 1.5%), more recent evidence suggested 5—6K warming by the end of this century, and possibly 10—11K warming by 2200.

However, the IPCC (2007) presented estimates of radiative forcing and warming this century under six emission scenarios (p. 18), to each of which it accorded equal weight. Taking their mean, a central estimate ΔT_{C21} of 21st-century warming is 2.8K (Table 10.1), of which 0.6K is already committed, so that the implicit central estimate of mean warming from 2000 to 2100 consequent upon all greenhouse gas emissions since 2000 is 2.2K, of which 70%, or 1.5K, is CO_2-driven.

Table 10.1 shows that, for each scenario, the IPCC's estimate of the bicentennial-scale transient—sensitivity parameter λ_{tra} is 0.5 K/(W/m²). IPCC (2001, p. 354, citing Ramanathan et al., 1985) took 0.5 K/(W/m²) as a typical climate-sensitivity parameter. Garnaut (2008) talks of keeping greenhouse gas rises to 450 ppmv CO_2 equivalent above the 280 ppmv prevalent in 1750, so as to hold 21st-century global warming since then to 2K, implying $\lambda_{tra} = 0.4$ K/(W/m²). This lesser rate, more suited to subcentennial-scale appraisals than the bicentennial scale 0.5 K/(W/m²), will be adopted here, though other values may readily be substituted.

To reflect the IPCC's wide error intervals, where λ_c is a central bicentennial-scale climate-sensitivity estimate, λ will fall on the 1 σ interval $[0.8\lambda_c, 1.2\lambda_c]$ (from the ±0.69K 1σ error-bar in IPCC, 2007, p. 798, Box 10.2), or on the >66%-probability interval $[0.6\lambda_c, 1.4\lambda_c]$ (derived IPCC, 2007, p. 12). So, where $\lambda_c = 0.5$ K/(W/m²), λ will fall on [0.4, 0.6] to 1 σ, and on [0.3, 0.7] with probability >0.66.

Recall that the IPCC (IPCC, 2001, p. 358, Table 6.2), following Myhre et al. (1998), takes the CO_2 forcing in W/m² as 5.35 times the logarithm of a given proportionate change C_b/C_a in CO_2 concentration, where C_a is the unperturbed value. Note that there was no statistically significant warming from 1997 to 2012 (RSS, 2012; UAH, 2012). Then projected CO_2 concentration C_{2100} in 2100, as the six-scenario mean, is 700 ppmv against 391 ppmv in 2011 (Conway and Tans, 2011), implying CO_2-driven warming from 2011 to 2100 of $\lambda_{tra}[5.35 \ln(C_{2100}/C_{2011})] = 0.4 [5.35 \ln(700/391)] = 1.25$K, similar to the 1.5K established earlier from IPCC (2007, Table SPM.3).

TABLE 10.1 Projected 21st-Century Anthropogenic Warming ΔT_{C21} (IPCC, 2007, p. 13, Table SPM.3) and Warming ΔT_{tra} and Total Radiative Forcings ΔF_{tra} From All Greenhouse Gases for 1900—2100 on All Emissions Scenarios, and CO_2 Concentration C_{2100} in 2100 (IPCC, 2007, p. 803, Fig. 10.26); and, Derived From These, the 200-Year Transient—Sensitivity Parameter $\lambda_{tra} = \Delta T_{tra}/\Delta F_{tra}$; the CO_2 Radiative Forcing $\Delta F_{tra,CO_2} = 5.35 \ln(C_{2100}/C_{1900})$ From 1900 to 2100, Taking C_{1900} as 300 ppmv; and the Ratio $q = \Delta F_{tra,CO_2}/\Delta F_{tra}$ of CO_2 Forcing to Total Greenhouse Gas Forcing

Scenario	ΔT_{C21} (K)	ΔT_{tra} (K)	ΔF_{tra} (W/m²)	C_{2100} (ppmv)	λ_{tra} [K/(W/m²)]	$\Delta F_{tra,CO_2}$ (W/m²)	q
AB	2.8	3.0	6.2	700	0.5	4.5	0.7
A1F1	4.0	4.5	9.1	960	0.5	6.2	0.7
A1T	2.4	2.5	5.1	570	0.5	3.4	0.7
A2	3.4	3.8	8.0	840	0.5	5.5	0.7
B1	1.8	2.0	4.1	520	0.5	2.9	0.7
B2	2.4	2.7	5.6	610	0.5	3.8	0.7
Mean	2.8	3.1	6.3	700	0.5	4.4	0.7

TABLE 10.2   Projected Anthropogenic Warming, 2000–2100

Projected anthropogenic warming, 2000–2100	Sources	ΔT_{C21} (K)
High-end projection	Stern (2006)	10–11
Central projection	Stern (2006)	5–6
Low-end projection	Stern (2006)	2–3
Mean projection: All anthropogenic warming	IPCC (2007)	2.8
… of which, warming not already committed	IPCC (2007)	2.2
… of which, fraction caused by CO_2 emissions	IPCC (2007)	1.5
Mean projection: anthropogenic global warming, by calculation	(vs. text supra)	1.25
Observed warming rate/century, 1950–2011	HadCRUt3	1.2
Mean projection: anthropogenic global warming (from CO_2 only)	$0.7\,\Delta T$	0.8
Mean projection: anthropogenic global warming (From CO_2 if half of warming was man made)	$0.7\,\Delta T/2$	0.4

Observed warming since 1950 has occurred at a rate equivalent to 1.2K/century (Brohan et al., 2006). Of this, 70%, or 0.8K/century, is attributable to CO_2; and, since the IPCC (2007) finds that up to half of the warming since 1950 might be natural, the centennial rate of CO_2-driven warming could have been as low as 0.4K/century. The IPCC's implicit 1.5K/century for the 21st century may accordingly be best seen as an upper bound rather than as a central estimate.

Table 10.2 summarizes official projections and observations of 21st-century warming:

3. METHOD

In this deliberately very simple method, only two case-specific inputs are required: C_y, the projected business-as-usual CO_2 concentration in the target final year y of the policy, and p, the proportion of projected global business-as-usual CO_2 emissions until year y that the policy is intended to abate.

Eq. (10.1) determines C_{pol}, the CO_2 concentration in parts per million by volume (somewhat below C_y) in year y that may be achievable by following a given policy to mitigate the radiative forcing from atmospheric CO_2 enrichment from 2010 (when $C_{2010} = 390$ ppmv) until year y. Eq. (10.1) also determines ΔT_{nix}, the quantum (in K) of transient global warming that the policy will abate if pursued until year y. Eq. (10.1) may be tuned to represent any forcing (see Table 10.4), but only warming attributable to the CO_2 forcing is demonstrated here.

$$\Delta T_{nix} = \lambda_{tra} \Delta F_{nix}$$

$$= 0.4 \left[5.35 \ \ln \left(\frac{C_y}{C_{pol}} \right) \right] \tag{10.1}$$

$$= 2.14 \ \ln \left(\frac{C_y}{C_y - p(C_y - 390)} \right).$$

Unit mitigation cost is here defined as the cost of abating 1K CO_2-driven global warming on the assumption that all measures to mitigate all CO_2-driven warming to year y are as cost effective as the policy under consideration. On the same assumption, the policy's *global abatement cost* is defined as the total cost from 2010 to year y (as a global cash cost, or a per capita cost, or a percentage of real global GDP) of abating all anthropogenic warming that the IPCC projects will occur by year y (ie, over the policy term) without mitigation.

Eq. (10.2) gives the unit mitigation cost M in US dollars per Kelvin of global warming abated. The lesser the value of M, the more cost effective is the policy, enabling policy makers to rapidly reliably compare the estimated mitigation cost-effectiveness of competing mitigation proposals.

$$M = \frac{x}{\Delta T_{nix}}. \tag{10.2}$$

TABLE 10.3   Business-as-Usual CO_2 Concentrations, 2010–2100, as the Mean of the Central Projections for All Emissions Scenarios (IPCC, 2007, p. 803, Fig. 10.26)

y	2010	2020	2030	2040	2050	2060	2070	2080	2090	2100
C_y	390	410	440	480	510	550	590	630	660	700

Global abatement cost: Where $w = 7 \times 10^9$ is world population, q is the fraction of total anthropogenic forcing attributable to CO_2, and $\Delta T_y = 2.14 \ln(C_y/C_{2010})/q$ is the projected anthropogenic global warming to year y, Eqs. (10.3)–(10.5) give the policy's global abatement cost over the term to year y in cash; per head of global population; and as a percentage of real global 21st-century GDP r over the term:

$$\text{Cash } G = M\Delta T_y \tag{10.3}$$

$$\text{Per capita } H = G/w \tag{10.4}$$

$$\text{As } \% \text{ global GDP } J = 100G/r \tag{10.5}$$

Derivation of Eq. (10.1),
Where λ is a climate-sensitivity parameter in $K/(W/m^2)$, consequent global warming in K may be expressed generally by Eq. (10.6):

$$\Delta T = \lambda \ \Delta F = \lambda[5.35 \ \ln(C_b/C_a)] \tag{10.6}$$

As a check-sum, taking $\lambda_{tra} = 0.4 \ K/(W/m^2)$ for 1900–2100, at CO_2 doubling Eq. (10.6) gives $0.4[5.35 \ln(2)] \approx$ 1.5K, on the model-derived transient-climate-response interval [1, 3]K (IPCC, 2007).

Where p, on [0, 1], is the fraction of future global emissions that a given CO_2 reduction policy is projected to abate by a target calendar year y, and C_y is the IPCC's projected unmitigated CO_2 concentration in year y, Eq. (10.7) gives C_{pol}, the somewhat lesser concentration in ppmv that may be expected to obtain in year y if the policy is followed:

$$C_{pol} = C_y - p(C_y - 390). \tag{10.7}$$

Table 10.3 gives central estimates of projected decadal values of C_y for 2010–2100.

Eq. (10.8), of similar form to Eq. (10.6), determines how much warming ΔT_{nix}, in Kelvin, a specific policy intended to cut CO_2 emissions will abate in the 21st century:

$$\Delta T_{nix} = \lambda_{tra}[5.35 \ \ln(C_y/C_{2010}) - 5.35 \ \ln(C_{pol}/C_{2010})$$

$$= \lambda_{tra}[5.35 \ \ln(C_y/C_{pol})]. \tag{10.8}$$

The second expressions of Eq. (10.8) and Eq. (10.1) are equivalent.

4. OTHER GREENHOUSE GASES

The method sketched here may be adapted to determine the unit mitigation cost and global abatement costs of greenhouse gases other than CO_2.

Table 10.4 summarizes some climate-relevant radiative forcing functions.

5. THE INTERTEMPORAL DISCOUNT RATE

By how much should future costs and benefits be discounted to net present value to take account of the uncertainties inherent in any long-term investment appraisal, such as that of a given policy's effect in reducing global warming?

Stern (2006) adopts a discount rate not exceeding 1.4% (it may in practice have been as low as 0.1%; Stern Review team, personal communication, 2006), well below HM Treasury's standard 3.5% "Green Book" rate, which is in turn somewhat below the 5% rate typical in the literature (eg, Nordhaus, 2008; Murphy, 2008).

Stern justifies his rate as follows: "The most straightforward and defensible interpretation (as argued in the Review) of [the utility discount factor] δ is the probability of existence of the world. In the Review, we took as our base

TABLE 10.4   Forcing Functions for Methane (CH$_4$), Nitrous Oxide (N$_2$O), the Chlorofluorocarbons CFC$_{11}$, CFC$_{12}$, Sulfur Hexafluoride (SF$_6$), and Sulfur dioxide (SO$_2$). Global Concentrations for SO$_2$ Are Unknown Because Its Concentration Is Highly Variable Both Spatially and Temporally

Trace gas	Radiative forcing ΔF g_y = business as usual; g_{pol} = lesser conc. after policy	Concentration in 2010
CH$_4$	$0.036(\mu_y^{0.5} - \mu_{pol}^{0.5}) + f(\mu_{pol}, v_{pol})^a - f(\mu_y, v_{pol})^a$	1816 ppbv
N$_2$O	$0.12(v_y^{0.5} - v_{pol}^{0.5}) + f(\mu_{pol}, v_{pol})^a - f(\mu_{pol}, v_y)^a$	324 ppbv
CFC$_{11}$	$0.00025(\beta_y - \beta_{pol})$	238 pptv
CFC$_{12}$	$0.00033(\gamma_y - \gamma_{pol})$	532 pptv
SF$_6$	$5.2^{-4}(\varphi_y - \varphi_{pol})$	7.31 pptv
SO$_2$	$-[0.03 \, \psi_y/\psi_{pol} + 0.08 \ln(1 + \psi_y/34.4)/(1 + \psi_{pol}/34.4)]$	Unknown

^{a}f$(\sigma,\tau) = 0.47 \ln[1 + 2.01 \times 10^{-5}(\sigma\tau)^{0.75} + 5.31 \times 10^{-15} \sigma(\sigma\tau)^{1.52}]$.

Reproduced from IPCC, 2007. In: Solomon, S., Qin, D., Manning, M., Chen, Z., Marquis, M., Avery, K.B., Tignor, M., Miller, H.L. (Eds.), Climate Change 2007: The Physical Science Basis. Contribution of Working Group I to the Fourth Assessment Report of the Intergovernmental Panel on Climate Change. Cambridge University Press, Cambridge, United Kingdom, and New York, NY, USA.

case $\delta = 0.1\%$/year, which gives roughly a one-in-ten chance of the planet not seeing out this century. … [Per-capita consumption growth] g is on average ∼1.3% in a world without climate change, giving an average consumption or social discount rate across the entire period of 1.4% (being lower where the impacts of climate change depress consumption growth)" (Dietz et al., 2007).

HM Treasury has moved in Stern's direction by adopting two reduced "climate change" discount rates that are initially commercial and are reduced after year 30 and again after year 75 to allow for "very-long-term, substantial, and for practical purposes irreversible wealth transfers between generations" (Grice, 2011; Lowe, 2008).

Over a century, these variable rates are equivalent to uniform rates of 3.22% and 2.75% respectively—closer to the Treasury's standard 3.5% "Green Book" discount rate than to Stern's 1.4% or the 1.35–2.65% in Garnaut (2008).

Klaus (2011) recommends a market approach: "To make a rational choice means to pay attention to inter-temporal relationships and to look at the opportunity costs. It is evident that … assuming a very low (near-zero) discount rate … neglect[s] the issue of time and of alternative opportunities. Using a low discount rate in global-warming models means harming current generations vis-à-vis future generations. Undermining current economic development also harms future generations."

Klaus continues: "Economists representing very different schools of thought, from Nordhaus (2008) to Murphy (2008), tell us convincingly that the discount rate—indispensable for any inter-temporal calculations—should be around the market rate of 5%, and that it should be close to the real rate of return on capital, because only that rate represents the opportunity cost of climate mitigation."

Accordingly, a 5% discount rate will be adopted in the illustrative case studies that follow, though other rates will be considered later.

Since warming is not occurring at the rate the IPCC predicts, the extreme warming rates of 5–6K or even 10–11K in Stern (2006) are not modeled here, though the method allows for them if required.

It will be assumed that the IPCC's central estimate of 2.8K anthropogenic warming will occur by 2100, and that the welfare loss arising from climate inaction will be 1.5% of GDP, on Stern's [0, 3]%-of-GDP interval.

6. WELFARE LOSS FROM INACTION

Table 10.5 adjusts the projected 21st-century 1.5, 3, 5, and 20% global costs of inaction given in Stern (2006), in line with the published discount rates shown:

Adjustments between a given inaction cost Z_s mentioned by Stern on the basis of a 1.4% discount rate d_s and the equivalent inaction cost Z_m on the basis of a 5% market rate d_m may be conveniently made using Eq. (10.9), where t is the term of the policy in years; a is each successive year from 1 to t; g is an assumed uniform annual percentage growth rate (here, 3%); d_m is the preferred discount rate (here, the 5% minimum market rate); d_s is the discount rate at which the inaction cost was determined (here, Stern's 1.4%); $|n|$ is the absolute value of n; and sgn(n) is the signum function.

TABLE 10.5 Estimates of the Welfare Loss Over 100 Years Owing to Climate Inaction at the Discount Rates Shown Are Determined by Multiplying Stern's 1.5%-, 3%-, 5%- and 20%-of-GDP Estimated 21st-Century Inaction Costs by the Ratio of 21st-Century GDP Discounted at Stern's 1.4% Rate to 21st-Century GDP Discounted at the Mean Rates Shown. Annual GDP Growth in the 21st Century Is Assumed Uniform at 3% Throughout the Table

Stern (2006) inaction costs adjusted for discount rate	Discount rate	Year 0–30	Year 31–75	Year 76–100	Cost: % of GDP	Cost: % of GDP	Cost: % of GDP	Cost: % of GDP
Units	%	%	%	%	%	%	%	%
Stern (2006)	1.40				1.50	3.0	5.0	20
Garnaut (2011) high	2.65				0.72	1.5	2.4	9.7
HM Tsy low (Lowe, 2008)	2.75	3.00	2.57	2.14	0.69	1.4	2.3	9.2
HM Tsy high (Grice, 2011)	3.22	3.50	3.00	2.50	0.54	1.1	1.8	7.3
HM Tsy Green Book rate	3.50				0.48	1.0	1.6	6.4
Market rate (s, 2011)	5.00				0.26	0.5	0.9	3.5

$$Z_m = Z_s \frac{\sum_{a=1}^{t}\left(1 + \frac{|g-d_m|}{100}\right)^{a \ \ \mathrm{sgn}(g-d_m)}}{\sum_{a=1}^{t}\left(1 + \frac{|g-d_s|}{100}\right)^{a \ \ \mathrm{sgn}(g-d_s)}} \tag{10.9}$$

Some published estimates of the welfare loss from failing to take action to mitigate global warming by reducing future emissions of CO_2 follow. The reviewed literature, summarized by Tol (2009a,b), suggests that the global inaction cost will be 1–5% of GDP: 1.0K warming will cost 2.5% of GDP (Tol, 2002); 2.5K warming will cost 0.9% (Nordhaus, 2006), 1.4% (Fankhauser, 1995), 1.5% (Nordhaus and Boyer, 2000), 1.7% (Nordhaus and Yang, 1996), 1.9% (Tol, 1995), 2.5% (Plamberk and Hope, 1996), or 0.0–0.1% of GDP (from market impacts only: Mendelsohn et al., 2000); 3K warming will cost 1.3–4.8% of GDP (Nordhaus, 1994a,b).

7. THE COST–BENEFIT RATIO

Since this much-simplified model excludes all costs and benefits external to those of CO_2 mitigation, the ratio of the global abatement cost J [Eq. (10.5)] of climate action, expressed as a percentage of GDP over the term of the policy, to the market GDP cost Z_m of the climate-related damage that is projected to arise from inaction is, for present purposes, the policy's cost-benefit ratio.

Nordhaus (2012) argues that the absolute difference between cost and benefit, rather than the ratio, should determine CO_2 mitigation policies:

> Suppose we were thinking about two policies. Policy A has a small investment in abatement of CO_2 emissions. It costs relatively little (say $1 billion) but has substantial benefits (say $10 billion), for a net benefit of $9 billion. Now compare this with a very effective and larger investment, Policy B. This second investment costs more (say $10 billion) but has substantial benefits (say $50 billion), for a net benefit of $40 billion. B is preferable because it has higher net benefits ($40 billion for B as compared with $9 for A), but A has a higher benefit-cost ratio (a ratio of 10 for A as compared with 5 for B). This example shows why we should, in designing the most effective policies, look at benefits minus costs, not benefits divided by costs.

However, the true choice is not between a small, but cost-effective, investment and a larger though less cost-effective investment that will yield a greater absolute benefit. A rational choice surely depends upon appraising the cost of acting to prevent global warming by way of a given CO_2 mitigation policy compared with the cost of the climate-related damage that might arise from inaction over the term of the policy.

In the latter context, the estimated cost of inaction is fixed, but various policy options for action to mitigate CO_2 emissions are available, wherefore the cost-benefit ratio is of no less relevance than the absolute cost or benefit of action against inaction.

8. ILLUSTRATIVE CASE STUDIES

In the brief illustrative case studies that follow, uniform real GDP growth of 3%/year from $60 tr/year in 2010 (World Bank, 2011) is assumed in all cases, with a further 2% cost escalator for the Australian emissions-trading scheme.

Since the 5% discount rate prevalent in the literature rather than Stern's 1.4% is adopted here, Stern's estimated welfare loss of 1.5–20% of GDP arising in the absence of any mitigation falls to 0.26–3.5% of GDP (Table 10.5). Since there is no sign of warming above the IPCC's central estimate, the least inaction cost, 1.5% of GDP discounted over the term, is the basis for comparison with the discounted costs of action to establish the cost-benefit ratio in each case study.

8.1 Case Study 1: US Carbon-Trading Bill

At $180 bn/year for 40 years, total $7.2 tr, discounted to $5 tr at p.v., the climate bill (HR 2454, 2009, s. 311) would have abated 83% of US CO_2 emissions by 2050. The United States emits 17% of global CO_2 (derived from Olivier and Peters, 2010, Table A1). Thus $p = 0.1411$. From Table 10.1, business-as-usual CO_2 concentration in 2050 would be 510 ppmv, falling to 493.1 ppmv [from Eq. (10.7)] via the bill. From Eq. (10.1), forcing abated is 0.2 W/m^2 and warming abated is 0.07K; from Eq. (10.2), unit mitigation cost is $69 tr/K; from Eq. (10.3), the global abatement cost of all projected warming to 2050 is \$56 tr, or [from Eq. (10.4)], \$8000 per capita of global population, or [from Eq. (10.5)], 3.4% of global GDP to 2050. The mitigation cost exceeds fivefold the benefit in preventing climate damage.

8.2 Case Study 2: UK Climate Change Act

At an officially estimated cost of \$1.2 tr by 2050, discounted to \$835 bn, the Climate Change Act (2008, s. 1(1)), aims to cut 80% of UK emissions, which are 1.5% of world emissions (derived from Olivier and Peters, 2010, Table A1). Thus $p = 0.012$. Business-as-usual CO$_2$ concentration of 510 ppmv in 2050 would fall to 508.6 ppmv via the Climate Change Act. Forcing abated is 0.015 W/m^2; warming abated is 0.006K; unit mitigation cost is \$138 tr/K; and global abatement cost to 2020 is \$113 tr, or \$16,000/head, or 6.8% of global GDP to 2050. Cost exceeds benefit ninefold.

8.3 Case Study 3: European Union Carbon Trading

European Union (EU) carbon trading costs \$92 bn/year (World Bank, 2009), here multiplied by 2.5 (implicit in Lomborg, 2007) to allow for nontrading mitigation measures. Total cost is \$2 tr at p.v. by 2020. The EU aims to halt 20% of its emissions, which are 13% of global emissions (from Boden and Marland, 2010; Boden et al., 2010). Thus $p = 0.026$. Business-as-usual CO$_2$ concentration of 410 ppmv in 2020 would fall to 409.5 ppmv via the policy. Forcing abated is 0.007 W/m^2; warming abated is 0.003K; unit mitigation cost is \$763 tr/K; and the global abatement cost of \$117 tr is \$17,000 per capita, or 21.5% of GDP to 2020. Mitigation costs 17.5 times the cost of climate-related damage in the absence of mitigation.

8.4 Case Study 4: California Cap and Trade

Under the Cap and Trade Act (AB 32 of 2006), which took full effect in August 2012, some \$182 billion per year (Varshney and Tootelian, 2009) will be spent for a decade on cap and trade and related measures. The report has been criticized for overstating costs: accordingly, one-quarter of this value will be taken over a 10-year term, giving a discounted cost of \$410 billion, to abate 25% of current emissions, which represent 8% of US emissions, which represent 18.7% of global emissions (derived from Olivier and Peters, 2010, Table A1). Thus $p = 0.0033$. CO$_2$ concentration would fall from a business-as-usual 410 to 409.93 ppmv by 2020. Forcing abated is 0.001 W/m^2; warming abated is 0.00034K; unit mitigation cost is \$1200 tr/K; global abatement cost is \$183 tr, or \$26,000/head, or 34% of global GDP to 2020. Action costs 28 times inaction.

8.5 Case Study 5: Thanet Wind Array

Subsidy to the world's largest wind farm, off the English coast, guaranteed at \$100 mn annually for its 20-year life, is \$1.6 bn at p.v. Rated output of the 100 turbines is 300 MW, but wind farms yield only 24% of rated capacity (Young, 2011, p. 1), so total output, at 72 MW, is 1/600 of mean 43.2 GW UK electricity demand (Department for Energy and Climate Change, 2011). Electricity is 33% of UK CO$_2$ emissions, which are 1.5% of global emissions, so $p = 8.333 \times 10^{-6}$. Business-as-usual CO$_2$ concentration in 2030 would be 440 ppmv, falling to 439.9996 ppmv as a result of the subsidy. Forcing abated is 0.000005 W/m^2; warming abated is 0.000002K; unit mitigation cost is \$800 tr/K; and the global abatement cost of almost \$300 tr is \$42,000/head, or 30% of GDP to 2030. Action costs 29 times inaction.

8.6 Case Study 6: Australia Cuts Emissions 5% in 10 Years

Carbon trading in Australia, as enacted by the Clean Energy legislation (Parliament of the Commonwealth of Australia, 2011), costs \$10.1 bn/year, plus \$1.6 bn/year for administration (Wong, 2010, p. 5), plus \$1.2 bn/year for renewables and other costs, a total of \$13 bn/year, rising at 5%/year, or \$130 bn by 2020 at n.p.v., to abate 5% of current emissions, which represent 1.2% of world emissions (derived from Boden and Marland, 2010; Boden et al., 2010). Thus $p = 0.0006$. CO$_2$ concentration would fall from a business-as-usual 412 to 411.987 ppmv after

10 years. Forcing abated is 0.0002 W/m^2; warming abated is 0.00006K; unit mitigation cost is \$2000 tr/K; global abatement cost of projected warming to 2020 is \$300 trillion, or \$45,000/head, or 59% of global GDP to 2020. Action costs 48 times inaction.

8.7 Case Study 7: Oldbury Primary School Wind Turbine

On March 31, 2010, Sandwell Council, England, answered a freedom-of-information request, disclosing that it had spent \$9694 (£5875) on buying and installing a small wind turbine like one at a primary school in Oldbury, which had in a year generated 209 KWh (McCauley, 2011) — enough to power a single 100 W reading lamp for <3 months. Assuming no maintenance costs, and discounting revenues of \$0.18 (11p)/KWh for 20 years to p.v. of \$623, net project cost is \$9070. $p = 209$ KWh/365 days/24 h/43.2 GW $\times$ 0.33 $\times$ 0.015 = 2.76 $\times 10^{-12}$. CO_2 concentration of 440 ppmv will fall to 439.9999999999 ppmv. Forcing abated is 0.000000000002 W/m^2; warming abated is 0.000000000001K; is \$13,500 tr/K; and the global abatement cost, at close to \$5000 tr, is \$700,000/head, or 500% of global GDP to 2030. Action costs almost 500 times inaction.

8.8 Case Study 8: London Bicycle-Hire Scheme

In 2010 the mayor of London set up what he called a "Rolls Royce" scheme at US\$130 mn for 5000 bicycles (>\$26,000 per bicycle). Transport represents 15.2% of UK emissions (from Office for National Statistics, 2010, Table C). Cycling represents 3.1 bn of the 316.3 bn vehicle miles traveled on UK roads annually (Department for Transport, 2011). There are 23 mn bicycles in use in Britain (Cyclists' Touring Club, 2011). Global emissions will be cut by 1.5% of 15.2% of 3.1/316.3 times 5000/23 mn. Thus $p = 4.886 \times 10^{-9}$. If the lifetime of bicycles and docking stations is 20 years, business-as-usual CO_2 concentration of 440 ppmv will fall to 439.9999998 ppmv through the scheme. Forcing abated is 0.000000003 W/m^2; warming abated is 0.000000001K; unit mitigation cost exceeds \$110,000 tr/K; and the global abatement cost of \$40,000 tr is \$5.8 mn per capita, or 4000% of global GDP to 2030. Action costs almost 4000 times inaction.

9. RESULTS

Government estimates of abatement cost (cases 1 and 2) are of the same order as those in Stern (2006), Garnaut (2008), and the reviewed literature. However, the costs of specific measures prove considerably higher than all such estimates, which have proven optimistic. Gesture policies (cases 7 and 8) are particularly cost-ineffective. However, this analysis is strictly confined to comparing the costs of taking climate action now with those of climate-related damage that might arise if no action were taken, excluding all other costs and benefits. In particular, benefits from investment in alternative or renewable energy are excluded, since they are likely to be comfortably exceeded by the opportunity losses arising from the diversion of substantial resources from the productive sector in the form of mitigation costs. Opportunity losses are also excluded from the accounting.

This analysis is not a complete study of all the costs and benefits of attempted climate mitigation. Its focus is on enabling policy makers to understand the relationship between the IPCC's implicit central estimates of subcentennial-scale transient climate sensitivity and of the forcing and warming likely to be forestalled by a given mitigation policy, allowing a first approximation of how much (or how little) global warming that policy may forestall. The new climatological equations derived here can readily be adapted for detailed cost-benefit analyses and comparisons beyond the scope of this chapter.

Table 10.6 summarizes the results of the case studies.

In Table 10.7, the effect of various intertemporal discount rates is illustrated by comparing the mitigation and inaction costs of the Australian government's carbon dioxide tax policy (case study 5) after applying Stern's and Garnaut's rates, as well as the Treasury's standard 3.5% flat rate and the 5% minimum market rate. Over longer periods than a decade, differing discount rates have a greater impact.

Even the use of Stern's minimalist discount rate shows that the global abatement cost of the carbon tax—ie, the cost of abating all global warming over the decade to 2020 if all measures to mitigate global warming from all anthropogenic causes had unit mitigation costs equal to the Australian government's proposal—will greatly exceed even Stern's maximum 20% cost of climate-related damage arising from worldwide inaction.

TABLE 10.6   Summary of Case Study Results, Assuming a 5% Intertemporal Discount Rate, Uniform 3% Annual GDP Growth, and a 1.5%-of-GDP Inaction Cost

Case study [#]	Warming abated (K by year y)	Unit mitigation cost ($ tr/K)	Abate. cost (%GDP)	Action/inaction ratio
[1] US cap and trade	0.07K by 2050	69	3.4	5×
[2] UK Climate Act	0.006K by 2050	138	6.8	9×
[3] EU carbon trading	0.003K by 2020	763	22	18×
[4] California AB 32	0.0+ K by 2020	1155	33	26×
[5] Thanet Wind Array	0.0+ K by 2030	803	30	29×
[6] Australia 5% cut	0.0+ K by 2020	2082	59	48×
[7] School windmill	0.0+ K by 2030	13,500	504	488×
[8] London cycle hire	0.0+ K by 2030	109,000	4090	3956×

TABLE 10.7   The Policy Cost x of Case Study 6; the Unit Mitigation Cost M; the Per Capita Global Abatement Cost J; the Cash Global Abatement Cost H in Cash and as a Percentage of GDP; the Upper and Lower Bounds of the Global Welfare Loss Interval I Arising From Inaction, Expressed as Percentages of GDP; and the Climate Action/Inaction Ratio of H (Expressed as a Percentage of GDP) to the Bounds of I

Stern	Garnaut 2		Green Book market
Discount rate 1.4%	2.65%	3.5%	5.0%
Case study 5: cost x $159 bn	$148 bn	$141 bn	$130 bn
Unit mitigation cost $2.8 qd/K	$2.6 qd/K	$2.5 qd/K	$2.2 qd/K
Global abatement cost $378 tr	$352 tr	$336 tr	$310 tr
Global abatement cost per capita $54,000	$50,000	$48,000	$44,000
Global abatement cost as % GDP 58%	58%	58%	58%
Global inaction cost 1.5—20%	1.4—18.6%	1.3—17.8%	1.2—16.4%
Action/inaction ratio 2.9—38	3.1—41	3.2—42	3.5—48

bn, billion; *mn*, million; *qd*, quadrillion; *tr*, trillion.

10. DISCUSSION

For the sake of simplicity and accessibility, the focus of the method is deliberately narrow. Potential benefits external to CO$_2$ mitigation, changes in global warming potentials, variability in the global-GDP growth rate, or relatively higher mitigation costs in regions with lower emission intensities are ignored, for little error arises. GDP growth rates and climate-inaction costs are assumed as uniform, though in practice little climate-related damage would arise unless global temperature rose by at least 2°C above today's temperatures.

Given the small quanta of warming abated by CO$_2$ eduction policies, as well as the breadth of the intervals of published estimates of inaction and mitigation costs, the greater complexity of adopting nonuniform GDP growth rates and climate-inaction costs may in any event be otiose.

The case studies suggest official projections may be optimistic against the unit mitigation costs of specific policies. Based on the US and UK governments' estimates, the global abatement cost of their policies would be five and nine times the cost of inaction, respectively; however, the global abatement costs of the EU's carbon trading scheme and the taxpayer subsidy to the world's largest wind farm would be 18 and 29 times inaction, respectively, with smaller schemes proving considerably less cost effective still. In general, smaller projects seem less cost effective than larger projects, but projects of any scale are not cost-effective.

A substantial reduction in global CO_2 emissions, maintained over centuries, might offset some of the warming caused by the preexisting increase in atmospheric CO_2 concentration from 278 ppmv in 1750 to 390 ppmv in 2010.

After a sufficiently long period of global emissions reduction ($y \gg 2100$), it may become justifiable to reduce the value 390 in the denominator of Eq. (10.1) stepwise toward the preindustrial CO_2 concentration 278 ppmv, increasing ΔT_{nix} and consequently improving cost-effectiveness by reducing M. However, within the 21st century even the immediate and total elimination of CO_2 emissions would only abate ~ 1.5K global warming.

For numerous reasons, Eq. (10.1) and the case studies tend to overstate the warming that any CO_2-reduction policy may abate, and also to overstate unit mitigation costs. The IPCC takes CO_2's mean atmospheric residence time as 50,200 years; if so, little mitigation will occur within the 21st century. It is here assumed that any policy-driven reduction in CO_2 concentration occurs at once, when it would be likely to occur stepwise to year y, halving the warming otherwise abated by that year and doubling the unit mitigation cost.

In some cases, it is assumed that the policy will meet the emissions-reduction target on its own, ignoring the often heavy cost of all other mitigation measures intended to contribute to achievement of the target. In most case studies' capital costs only are counted and running costs excluded. Capital costs external but essential to a project, such as provision of spinning-reserve generation for wind turbines on windless days, are excluded. Emissions from project construction and installation, such as concrete bases for wind turbines, are ignored, as are costs and CO_2 emissions arising from necessary external operating expenditures such as spinning-reserve for wind turbines.

If the IPCC's central projections exaggerate the warming that may arise from a given increase in atmospheric CO_2 concentration, the warming abated may be less than shown. Though emissions are rising in accordance with the IPCC's A2 emissions scenario, concentration growth has been near linear for a decade, so that outturn by 2100 may be considerably below the IPCC's mean estimate of 700 ppmv.

The climate-sensitivity parameter λ_{tra} used in the case studies is centennial scale; accordingly, over the shorter periods covered by the studies a somewhat lesser coefficient (allowing for the fact that longer-term temperature feedbacks may not yet have acted) and consequently less warming abated would reduce unit mitigation costs.

Finally, all opportunity losses from diverting resources to global warming mitigation are ignored.

11. CONCLUSIONS

The case studies indicate that governments' initial abatement cost estimates have proven optimistic. It is unlikely that any policy to abate global warming by taxing, trading, regulating, reducing, or replacing greenhouse gas emissions will prove cost-effective solely on grounds of the welfare benefit foreseeable from global warming mitigation alone.

High abatement costs, and the negligible returns in warming abated, imply that focused adaptation to the consequences of such future warming as may occur may be considerably more cost effective than any attempted mitigation. Mitigation policies inexpensive enough to be affordable are likely to prove ineffective, while policies costly enough to be effective will be unaffordable. Since the opportunity cost of mitigation is heavy, the question arises whether mitigation should be attempted at all.

Acknowledgments

This chapter is based on lectures to the 43rd (2010) and 45th (2012) Seminars on Planetary Emergencies of the World Federation of Scientists at Erice, Sicily, to the Prague School of Economics in May 2011, and to the Third Los Alamos Conference on Global and Regional Climate Change, Santa Fe, in November 2011. I am most grateful for comments from Dr. Petr Chylek of the Los Alamos National Laboratory; Professor Tim Congdon, formerly of the Monetary Policy Committee of the Bank of England; Dr. Christopher Essex, Professor and Departmental Chair of Applied Mathematics in the University of Western Ontario; Dr. David Evans, formerly of the Australian government's Carbon Accounting Office; Dr. Laurence Gould, Professor of Physics at the University of Hartford; Dr. Vaclav Klaus, President of the Czech Republic; and Dr. Fred Singer, Professor Emeritus of Environmental Sciences at the University of Virginia.

References

AB 32, 2006. Global Warming Solutions Act. California State Legislature.
Boden, Marland, 2010. Global CO_2 Emissions from Fossil-Fuel Burning, Cement Manufacture, and Gas Flaring, 1751–2007. Carbon Dioxide Information and Analysis Center, Oak Ridge, Tennessee, USA.
Boden, et al., 2010. Ranking of the World's Countries by 2007 Total CO_2 Emissions from Fossilfuel Burning, Cement Production, and Gas Flaring. Carbon Dioxide Information and Analysis Center, Oak Ridge, Tennessee, USA.

Brohan, P., Kennedy, J.J., Harris, I., Tett, S.F.B., Jones, P.D., 2006. Uncertainty estimates in regional and global observed temperature changes: a new dataset from 1850. Journal of Geophysical Research 111, D12106. http://dx.doi.org/10.1029/2005JD006548.

Climate Change Act, 2008. http://www.legislation.gov.uk/ukpga/2008/27/section/1.

Conway, T., Tans, P., 2011. Recent Trends in Globally-Averaged CO_2 Concentration. NOAA/ESRL. http://www.esrl.noaa.gov/gmd/ccgg/trends/global.html#global.

Cyclists' Touring Club, 2011. Cyclists' Touring Club Facts and Figures. http://www.ctc.org.uk/resources/Campaigns/CTC-Facts+figs_rpt.pdf.

Department for Energy and Climate Change, 2011. UK Energy Statistics: Electricity. www.decc.gov.uk/en/content/cms/statistics/source/electricity/electricity.aspx.

Department for Transport, 2011. Table TRA0101, Road Traffic by Vehicle Type, Great Britain, 1950—2009. http://www.dft.gov.uk/pgr/statistics/datatablespublications/roads/traffic/#tables.

Dietz, S., Hope, C., Stern, N., Zenghelis, D., 2007. Reflections on the Stern review (1) a robust case for strong action to reduce the risks of climate change. World Economics 8 (1), 121—168.

Fankhauser, 1995. Valuing Climate Change — The Economics of the Greenhouse, first ed. EarthScan, London, UK.

Garnaut, 2008. The Garnaut Climate Change Review: Final Report. Cambridge University Press, Port Melbourne, Australia, 680 pp. ISBN: 9780521744447.

Garnaut, R., 2011. Australia in the Global Response to Climate Change: Garnaut Climate Change Review Update 2011. Cambridge University Press, Melbourne.

Grice, 2011. The Green Book: Appraisal and Evaluation in Central Government: Treasury Guidance. The Stationery Office, London, UK. http://www.hm-treasury.gov.uk/d/green_book_complete.pdf.

HR 2454, 2009. American Clean Energy & Security Bill. In: 111th Congress, Washington, DC.

IPCC, 2001. In: Houghton, J.T., Ding, Y., Griggs, D.J., Noguer, M., van der Linden, P.J., Dai, X., Maskell, K., Johnson, C.A. (Eds.), Climate Change 2001: The Scientific Basis: Contribution of Working Group I to the Third Assessment Report of the Intergovernmental Panel on Climate Change. Cambridge University Press, Cambridge, United Kingdom, and New York, NY, USA.

IPCC, 2007. In: Solomon, S., Qin, D., Manning, M., Chen, Z., Marquis, M., Avery, K.B., Tignor, M., Miller, H.L. (Eds.), Climate Change 2007: The Physical Science Basis. Contribution of Working Group I to the Fourth Assessment Report of the Intergovernmental Panel on Climate Change. Cambridge University Press, Cambridge, United Kingdom, and New York, NY, USA.

Klaus, May 10, 2011. The Global Warming Doctrine Is Not a Science. Conference on The Science and Economics of Climate Change. Downing College, Cambridge, UK.

Lomborg, March 21, 2007. Perspective on Climate Change, Testimony Before the Committee on Science and Technology of the US House of Representatives.

Lowe, 2008. Intergenerational Wealth Transfers and Social Discounting: Supplementary Green Book Guidance. HM Treasury, London. ISBN:978-1-84532-419-3.

McCauley, March 31, 2011. Letter of Response to a Request Under the Freedom of Information Act Received From Peter Day, Sandwell Borough Council, Oldbury, England. www.whatdotheyknow.com/request/wind_turbine#incoming-163689.

Mendelsohn, et al., 2000. Country-specific market impacts of climate change. Climatic Change 45 (3—4), 553—569.

Murphy, 2008. Some simple economics of climate changes. In: Paper Presented to the MPS General Meeting, Tokyo, September 8.

Myhre, et al., 1998. New estimates of radiative forcing due to well mixed greenhouse gases. Geophysical Research Letters 25 (14), 2715—2718. http://dx.doi.org/10.1029/98GL01908.

Nordhaus, W.D., 1994a. Expert opinion on climate change. American Scientist 82 (1), 45—51.

Nordhaus, W.D., 1994b. Managing the Global Commons: The Economics of Climate Change. MIT Press, Cambridge, Mass., and London, England.

Nordhaus, W.D., 2006. Geography and macroeconomics: new data and new findings. Proceedings of the National Academy of Sciences of the United States of America 103 (10), 3510—3517.

Nordhaus, W.D., 2008. A Question of Balance: Weighing the Options on Global Warming Policies. Yale University Press.

Nordhaus, W.D., March 22, 2012. Why the Global Warming Skeptics Are Wrong. New York Review of Books.

Nordhaus, Boyer, 2000. Warming the World: Economic Models of Global Warming. MIT Press, Cambridge, Mass., and London, England.

Nordhaus, Yang, 1996. RICE: a regional dynamic general-equilibrium model of optimal climate-change policy. American Economic Review 86 (4), 741—765.

Office for National Statistics, 2010. Statistical Bulletin, Greenhouse Gas Emissions Intensity Falls in 2008. www.statistics.gov.uk/pdfdir/ea0610.pdf.

Olivier and Peters, 2010. Mondiale emissies koolstofdioxide door gebruik fossiele brandstoffen en cementproductie, 1990—2009. PBL Netherlands Environmental Assessment Agency, Den Haag, Netherlands.

Parliament of the Commonwealth of Australia, 2011. Exposure Draft of the Clean Energy Bill. http://www.climatechange.gov.au/government/submissions/clean-energy-legislativepackage/~/media/publications/clean-energy-legislation/exposure-draft-clean-energy-bill-2011pdf.pdf.

Plamberk, Hope, 1996. PAGE95 — an updated valuation of the impacts of global warming. Energy Policy 24 (9), 783—793.

Ramanathan, et al., 1985. Trace gas trends and their potential role in climate change. Journal of Geophysical Research 90, 5547—5566.

RSS, 2012. Monthly MSU Global Mean Land and Sea Lower-Troposphere Temperature Anomalies. http://www.remss.com/data/msu/monthly_time_series/RSS_Monthly_MSU_AMSU_Channel_TLT_Anomalies_Land_and_Ocean_v03_3.txt.

Solomon, et al., 2009. Irreversible climate change due to carbon dioxide emissions. Proceedings of the National Academy of Sciences of the United States of America 106 (6), 74—79. http://dx.doi.org/10.1073/pnas:0812721106.

Stern, N., 2006. The Economics of Climate Change: The Stern Review. Cambridge University Press, Cambridge, United Kingdom, and New York, NY, USA.

Tol, 1995. The damage costs of climate change: toward more comprehensive calculations. Environmental and Resource Economics 5 (4), 353—374.

Tol, 2002. Benchmark and dynamic estimates of the damage costs of climate change. Environmental and Resource Economics 21 (1), 47—73, and 21 (2), 135—160.

Tol, 2009a. The economic effects of climate change. Journal of Economic Perspectives 23 (2), 29—51.

Tol, 2009b. An Analysis of Mitigation as a Response to Climate Change. Copenhagen Consensus Center, Copenhagen Business School, Frederiksberg, Denmark.

UAH, 2012. Monthly Global Mean Land and Sea Lower-Troposphere Temperature Anomalies. http://vortex.nsstc.uah.edu/data/msu/t2lt/tltglhmam_5.4.

Varshney, S.B., Tootelian, D.H., 2009. Cost of AB 32 on California Small Businesses. California Small Business Round Table.

Wong, 2010. Portfolio Budget Statements 2010—11: Budget-Related Paper No. 1.4. Climate Change and Energy Efficiency Portfolio. Commonwealth of Australia, Canberra, Australia.

World Bank, 2009. State and Trends of the Carbon Market. Washington, DC. http://siteresources.worldbank.org/INTCARBONFINANCE/Resources/State___Trends_of_the_Carbon_Market_2009-FINAL_26_May09.pdf.

World Bank, 2011. Gross Domestic Product 2009, World Development Indicators. http://siteresources.worldbank.org/DATASTATISTICS/Resources/GDP.pdf.

Young, S., 2011. Analysis of UK Wind Power Generation, November 2008 to December 2010. John Muir Trust, Edinburgh, 25 pp. http://www.jmt.org/assets/pdf/wind-report.pdf.

OCEANS

11

Relationship of Multidecadal Global Temperatures to Multidecadal Oceanic Oscillations

J.S. D'Aleo[1], D.J. Easterbrook[2]

[1]American Meteorological Society, Hudson, NH, United States;
[2]Western Washington University, Bellingham, WA, United States

1. INTRODUCTION

The sun and ocean undergo regular changes on regular and predictable time frames. Temperatures likewise have exhibited changes that are cyclical. Sir Gilbert Walker was generally recognized as the first to find large-scale oscillations in atmospheric variables. As early as 1908, while on a mission to explain why the Indian monsoon sometimes failed, he assembled global surface data and did a thorough correlation analysis.

On careful interpretation of statistical data, Walker and Bliss (1932) were able to identify three pressure oscillations:

1. A flip-flop on a big scale between the Pacific Ocean and the Indian Ocean, which he called the Southern Oscillation (SO).
2. A second oscillation, on a much smaller scale, between the Azores and Iceland, which he named the North Atlantic Oscillation.
3. A third oscillation between areas of high and low pressure in the North Pacific, which Walker called the North Pacific Oscillation.

Evidence-Based Climate Science, Second Edition
http://dx.doi.org/10.1016/B978-0-12-804588-6.00011-2

Walker further asserted that the SO is the predominant oscillation, which had a tendency to persist for at least one to two seasons. He went so far in 1924 as to suggest the SOI had global weather impacts and might be useful in predicting the world's weather. He was ridiculed by the scientific community at the time for these statements. Not until four decades later was the Southern Oscillation recognized as a coupled atmosphere pressure and ocean temperature phenomenon (Bjerknes, 1969), and more than two additional decades passed before it was shown to have statistically significant global impacts and could be used to predict global weather/climate, at times many seasons in advance. Walker was clearly a man ahead of his time.

Global temperatures, ocean-based teleconnections, and solar variances interrelate with each other. A team of mathematicians (Tsonis et al., 2003, 2007), led by Dr. Anastasios Tsonis, developed a model suggesting that known cycles of the Earth's oceans—the Pacific Decadal Oscillation, the North Atlantic Oscillation, El Niño (Southern Oscillation), and the North Pacific Oscillation—all tend to synchronize with each other. The theory is based on a branch of mathematics known as Sychronized Chaos. The model predicts the degree of coupling to increase over time, causing the solution to "bifurcate," or split. Then, the synchronization vanishes. The result is a climate shift. Eventually the cycles begin to synchronize again, causing a repeating pattern of warming and cooling, along with sudden changes in the frequency and strength of El Niño events. They show how this has explained the major shifts that have occurred, including those in 1913, 1942, and 1978. These may be in the process of synchronizing once again, with the likely impact on climate very different from what has been observed over the last several decades.

2. THE SOUTHERN OSCILLATION INDEX

The Southern Oscillation Index (SOI) is the oldest measure of large-scale fluctuations in air pressure occurring between the western and eastern tropical Pacific (ie, the state of the Southern Oscillation) during El Niño and La Niña episodes (Walker and Bliss, 1932). Traditionally, this index has been calculated based on the differences in air pressure anomaly between Tahiti and Darwin, Australia. In general, smoothed time series of the SOI correspond very well with changes in ocean temperatures across the eastern tropical Pacific. The negative phase of the SOI represents below-normal air pressure at Tahiti and above-normal air pressure at Darwin. Prolonged periods of negative SOI values coincide with abnormally warm ocean waters across the eastern tropical Pacific typical of El Niño episodes. Prolonged periods of positive SOI values coincide with abnormally cold ocean waters across the eastern tropical Pacific typical of La Niña episodes.

As an atmospheric observation-based measure, SOI is subject not only to underlying ocean temperature anomalies in the Pacific but also intraseasonal oscillations, like the Madden-Julian Oscillation (MJO). The SOI often shows month-to-month swings, even if the ocean temperatures remain steady due to these atmospheric waves. This is especially true in weaker El Niño or La Niña events, as well as La Nadas (neutral ENSO) conditions. Indeed, even week-to-week changes can be significant. For that reason, other measures are often preferred.

2.1 Niño 3.4 Anomalies

On February 23, 2005, NOAA announced that the NOAA National Weather Service, the Meteorological Service of Canada, and the National Meteorological Service of Mexico reached a consensus on an index and definitions for El Niño and La Niña events (also referred to as the El Niño Southern Oscillation or ENSO). Canada, Mexico, and the United States all experience impacts from El Niño and La Niña.

The index was called the ONI, and is defined as a three-month average of sea surface temperature departures from normal for a critical region of the equatorial Pacific (Niño 3.4 region; 120−170W, 5N−5S). This region of the tropical Pacific contains what scientists call the "equatorial cold tongue," a band of cool water that extends along the equator from the coast of South America to the central Pacific Ocean. North America's operational definitions for El Niño and La Niña, based on the index, are:

El Niño: A phenomenon in the equatorial Pacific Ocean characterized by a positive sea surface temperature departure from normal (for the 1971−2000 base period) in the Niño 3.4 region greater than or equal in magnitude to 0.5°C (0.9°F), averaged over three consecutive months.
La Niña: A phenomenon in the equatorial Pacific Ocean characterized by a negative sea surface temperature departure from normal (for the 1971−2000 base period) in the Niño 3.4 region greater than or equal in magnitude to 0.5°C (0.9°F), averaged over three consecutive months.

3. MULTIVARIATE ENSO INDEX

Wolter (1987) attempted to combine oceanic and atmospheric variables to track and compare ENSO events. He developed the Multivariate ENSO Index (MEI) using the six main observed variables over the tropical Pacific. These six variables are sea-level pressure (P), zonal (U) and meridional (V) components of the surface wind, sea surface temperature (S), surface air temperature (A), and total cloudiness fraction of the sky (C).

The MEI is calculated as the first unrotated principal component (PC) of all six observed fields combined. This is accomplished by normalizing the total variance of each field first, and then performing the extraction of the first PC on the covariance matrix of the combined fields (http://www.cdc.noaa.gov/people/klaus.wolter/MEI/mei.htmlWolter and Timlin, 1993).

In order to keep the MEI comparable, all seasonal values are standardized with respect to each season and to the 1950−1993 reference period. Negative values of the MEI represent the cold ENSO phase (La Niña), while positive MEI values represent the warm ENSO phase (El Niño). Fig. 11.2 is a plot of the three indices since 2000 (Wolter and Timlin, 1993).

On, Fig.11.2 Niño 34 is well correlated with the MEI. The SOI is much more variable month to month than the MEI and Niño 34. The MEI and Niño are more reliable determinants of the true state of ENSO, especially in weaker ENSO events.

4. THE PACIFIC DECADAL OSCILLATION

The first hint of a Pacific basin-wide cycle was the recognition of a major regime change in the Pacific in 1977 that became known as the Great Pacific Climate Shift (Fig. 11.1). Later, this shift was shown to be part of a cyclical regime change with decadal-like ENSO variability (Zhang et al., 1996, 1997; Mantua et al., 1997) and given the name Pacific Decadal Oscillation (PDO) by fisheries scientist Steven Hare (1996) while researching connections between Alaska salmon production cycles and Pacific climate.

Mantua et al. (1997) found the PDO is a long-lived El Niño-like pattern of Pacific climate variability. While the two climate oscillations have similar spatial climate fingerprints, they have very different behavior in time. Two main characteristics distinguish PDO from ENSO: (1) twentieth-century PDO "events" persisted for 20 to 30 years, while

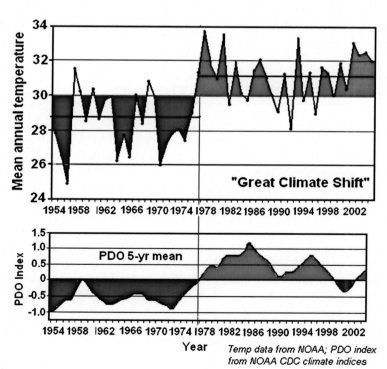

FIGURE 11.1 Correlation of the great Pacific Climate Shift (GPCS) and the Pacific Decadal Oscillation.

FIGURE 11.2   Atmospheric CO_2 showed no change across the Great Pacific Shift so could not have been the cause of it.

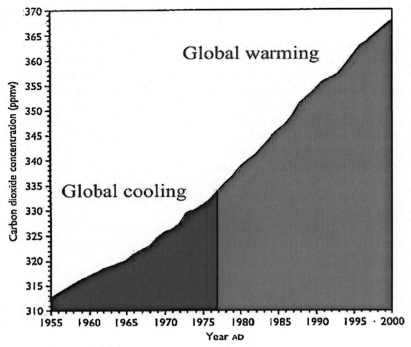

typical ENSO events persisted for 6 to 18 months; (2) the climatic fingerprints of the PDO are most visible in the North Pacific/North American sector, while secondary signatures exist in the tropics—the opposite is true for ENSO. Note in Fig. 11.2 how CO_2 showed no change during this PDO shift, suggesting it was unlikely to have played a role.

A study by Gershunov and Barnett (1998) showed that the PDO has a modulating effect on the climate patterns resulting from ENSO. The climate signal of El Niño is likely to be stronger when the PDO is highly positive; conversely, the climate signal of La Niña will be stronger when the PDO is highly negative. This does not mean that the PDO physically controls ENSO, but rather that the resulting climate patterns interact with each other. The annual PDO and ENSO (Mulitvariate ENSO Index) track well since 1950.

5. FREQUENCY AND STRENGTH OF ENSO AND THE PDO

Warm PDOs are characterized by more frequent and stronger El Niños than La Niñas. Cold PDOs have the opposite tendency. Fig. 11.4 shows how well one ENSO measure, Wolter's MEI, correlates with the PDO. McLean et al. (2009) showed that the mean monthly global temperature (GTTA) using the University of Alabama Huntsville MSU temperatures corresponds in general terms with the another ENSO measure, the Southern Oscillation Index (SOI) of seven months earlier. The SOI is a rough indicator of general atmospheric circulation and thus global climate change.

Temperatures also follow suit (Fig. 11.5). El Niños and the warm-mode PDOs have similar land-based temperature patterns, as do cold-mode PDOs and La Niñas.

There is a strong similarity between PDO and ENSO ocean basin patterns. Land temperatures also are very similar between the PDO warm modes and El Niños and the PDO cold modes and La Niñas. Not surprisingly, El Niños occur more frequently during the PDO warm phase and La Niñas during the PDO cold phase. It maybe that ocean circulation shifts drive it for decades favoring El Niños, which leads to a PDO warm phase or La Niñas and a PDO cold phase (the proverbial chicken and egg), but the 60-year cyclical nature of this cycle is well established (Fig. 11.6).

About 1947, the PDO switched from its warm mode to its cool mode, and global climate cooled from then until 1977, despite the sudden soaring of CO_2 emissions. In 1977, the PDO switched back from its cool mode to its warm mode, initiating what is regarded as "global warming" from 1977 to 1998 (Fig. 11.7).

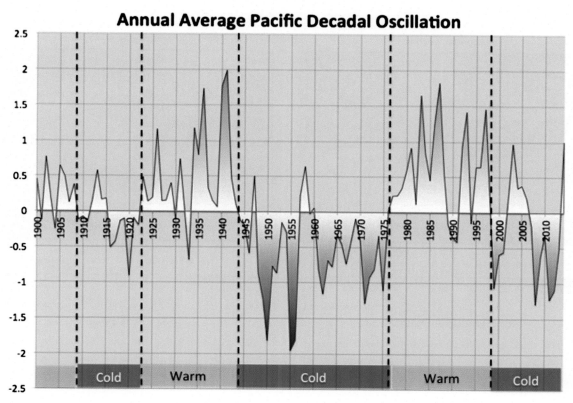

FIGURE 11.3   Annual average PDO 1900–2009. Note the multidecadal cycle.

During the past century, global climates have consisted of two cool periods (1880–1915 and 1945–1977) and two warm periods (1915–1945 and 1977–1998). In 1977, the PDO switched abruptly from its cool mode, where it had been since about 1945, into its warm mode and global climate shifted from cool to warm (Miller et al., 1994). This rapid switch from cool to warm has become known as the Great Pacific Climatic Shift (Fig. 11.1). Atmospheric CO_2 showed no unusual changes across this sudden climate shift and was clearly not responsible for it. Similarly,

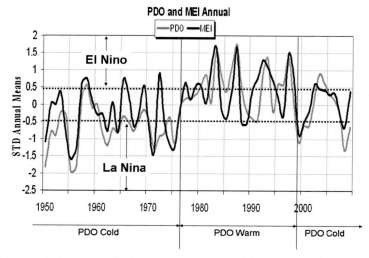

FIGURE 11.4   Annual average PDO and MEI from 1950 to 2007 clearly correlate well. Note how the ENSO events amplify or diminish the favored PDO state.

Temperature Correlations

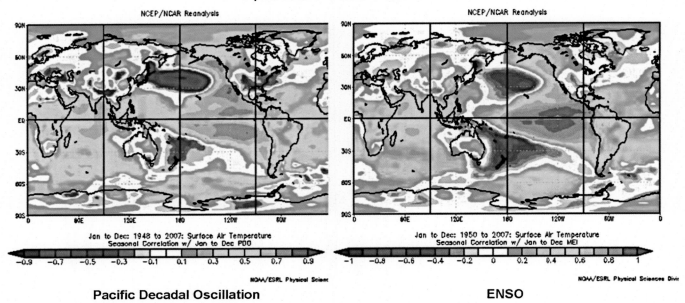

Pacific Decadal Oscillation   **ENSO**

FIGURE 11.5   PDO and ENSO global annual temperature correlations compared. A positive PDO and El Niño have an almost identical pattern. Reverse the colors for the negative PDO which correlates with La Niña.

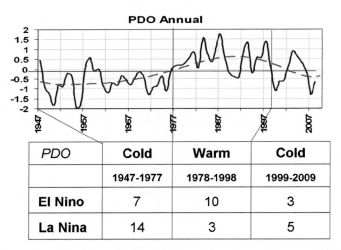

PDO	Cold	Warm	Cold
	1947-1977	1978-1998	1999-2009
El Nino	7	10	3
La Nina	14	3	5

FIGURE 11.6   During the PDO cold phases, La Niña winters dominate (14 to 7 in the 1947 to 1977 cold phase) while in the warm phase from 1977 to 1998, the El Niño winters had a decided frequency advantage of 10 to 3 (ENSO state based on MEI).

the global warming from ~1915 to ~1945 could not have been caused by increased atmospheric CO_2 because that time preceded the rapid rise of CO_2, and when CO_2 began to increase rapidly after 1945, 30 years of global cooling occurred (1945–1977). The two warm and two cool PDO cycles during the past century (Fig. 11.3) have periods of about 25–30 years.

The PDO flipped back to the cold mode in 1999. The change can be seen with this sea surface temperature difference image of the decade after the GPCS minus the decade before the GPCS (Fig. 11.8).

Verdon and Franks (2006) reconstructed the positive and negative phases of the PDO back to AD 1662 based on tree ring chronologies from Alaska, the Pacific Northwest, and subtropical North America as well as coral fossil from Rarotonga located in the South Pacific. They found evidence for this cyclical behavior over the whole period (Fig. 11.9).

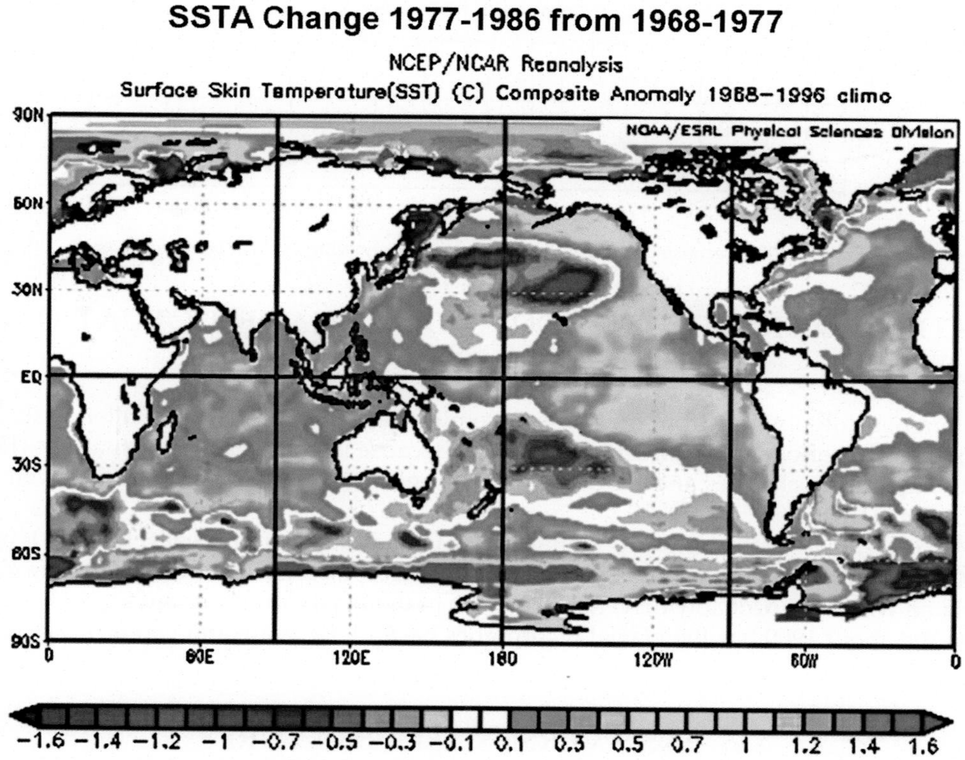

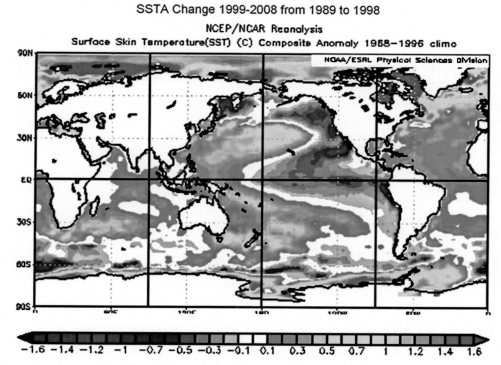

FIGURE 11.7 Difference in average sea surface temperatures between the decade prior to the GPCS and the decade after the GPCS. Yellow and green colors indicate warming of the NE Pacific off the coast of North America relative to what it had been from 1968 to 1977. Note the cooling in the west central North Pacific.

FIGURE 11.8 Sea surface temperature difference image of the decade after the GPCS minus the decade before the GPCS. Note the strong cooling in the eastern Pacific and the warming of the west central North Pacific.

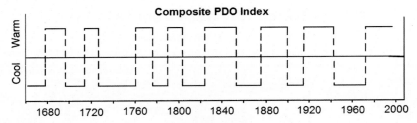

FIGURE 11.9   Verdon and Franks (2006) reconstructed the PDO back to 1662 showing cyclical behavior over the whole period.

6. CORRELATION OF THE PDO AND GLACIAL FLUCTUATIONS IN THE PACIFIC NORTHWEST

The ages of moraines down-valley from the present Deming glacier on Mt. Baker (Fuller, 1980; Fuller et al., 1983) match the ages of the cool periods in the Greenland ice core. Because historic glacier fluctuations (Harper, 1993) coincide with global temperature changes and PDO, these earlier glacier fluctuations could also well be due to oscillations of the PDO (Fig. 11.10).

Glaciers on Mt. Baker, Washington show a regular pattern of advance and retreat (Fig. 11.11), which matches the Pacific Decadal Oscillation (PDO) in the NE Pacific Ocean (Easterbrook and Kovanen, 2000). The glacier fluctuations are clearly correlated with, and probably driven by, changes in the PDO. An important aspect of this is that the PDO record extends to the about 1900 but the glacial record goes back many years and can be used as a proxy for older climate changes.

7. ENSO VERSUS TEMPERATURES

Douglass and Christy (2008) compared the Niño 34 region anomalies to the tropical UAH lower troposphere and showed a good agreement, with some departures during periods of strong volcanism. During these volcanic events, high levels of stratospheric sulfate aerosols block incoming solar radiation and produce multiyear cooling of the atmosphere and oceans. A similar comparison of UAH global lower tropospheric data with the MEI Index also shows good agreement, with some departure during periods of major volcanism in the early 1980 and 1990s. Alaskan temperatures clearly show discontinuities associated with changes in the PDO.

8. THE ATLANTIC MULTIDECADAL OSCILLATION (AMO)

Like the Pacific, the Atlantic exhibits multidecadal tendencies and a characteristic tripole structure (Figs. 11.12 and 11.13). For a period that averages about 30 years, the Atlantic tends to be in what is called the warm phase, with

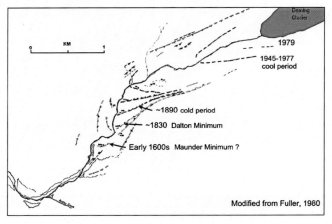

FIGURE 11.10   Ice marginal deposits (moraines) showing fluctuations of the Deming glacier, Mt. Baker, Washington, corresponding to climatic warming and cooling in Greenland ice cores.

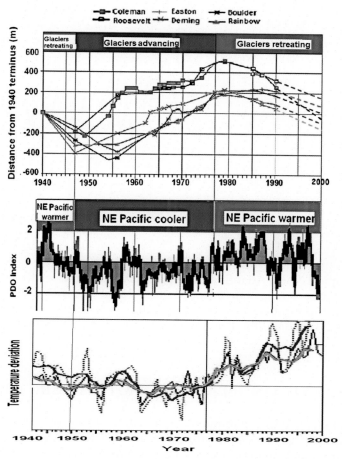

FIGURE 11.11   Correlation of glacial fluctuations, global temperature changes, and the PDO.

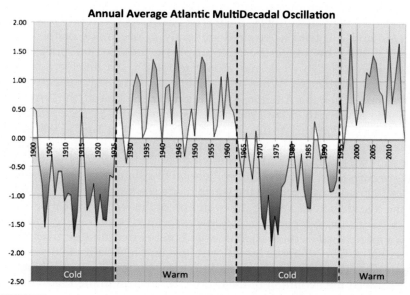

FIGURE 11.12   AMO annual mean (STD) showing a very distinct 60–70 year cycle.

FIGURE 11.13   Annual average AMO and NAO compared. Note the inverse relationship with a slight lag of the NAO to the AMO.

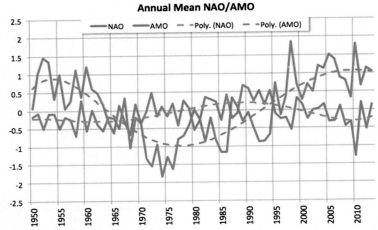

warm temperatures in the tropical North Atlantic and far North Atlantic and relatively cool temperatures in the central (west central) area. Then the ocean flips into the opposite (cold) phase, with cold temperatures in the tropics and far North Atlantic and a warm central ocean. The AMO (Atlantic sea surface temperatures standardized) is the average anomaly standardized from 0 to 70N. The AMO has a period of 60+ years maximum to maximum and minimum to minimum.

9. THE NORTH ATLANTIC OSCILLATION, ARCTIC OSCILLATION, AND THE AMO

The North Atlantic Oscillation (NAO), first found by Walker in the 1920s, is the north—south flip-flop of pressures in the eastern and central North Atlantic (Walker and Bliss, 1932). The difference of normalized MSLP anomalies between Lisbon, Portugal, and Stykkisholmur, Iceland has become the widest used NAO index and extends back in time to 1864 (Hurell, 1995), and to 1821 if Reykjavik is used instead of Stykkisholmur and Gibraltar instead of Lisbon (Jones et al., 1997). Hanna et al. (2006) showed how these cycles in the Atlantic sector play a key role in temperature variations in Greenland and Iceland. Kerr (2000) identified the NAO and AMO (Fig. 11.13) as key climate pacemakers for large-scale climate variations over the centuries.

The Arctic Oscillation (AO), also known as the Northern Annular Mode Index (NAM), is defined as the amplitude of the pattern defined by the leading empirical orthogonal function of winter monthly mean NH MSLP anomalies poleward of 20°N (Thompson and Wallace, 1998, 2000). The NAM/AO is closely related to the NAO.

Like the PDO, the NAO and AO tend to be predominantly in one mode or the other for decades at a time; although, since like the SOI, they are a measure of atmospheric pressure and subject to transient features, they tend to vary much more from week to week and month to month. An inverse relationship exists between the AMO and NAO/AO decadal tendencies. When the Atlantic is cold (AMO negative), the AO and NAO tend more often to the positive state, when the Atlantic is warm, on the other hand, the NAO/AO tend to be more often negative. The AMO trip-pole of warmth in the 1960s was associated with a predominantly negative NAO and AO, while the cold phase was associated with a distinctly positive NAO and AO in the 1980s and early 1990s (Fig. 11.14). The relationship is a little more robust for the cold (negative AMO) phase than for the warm (positive) AMO. There tends to be considerable intraseasonal variability of these indices that relate to other factors (stratospheric warming and cooling events that are correlated with the Quasi-Biennial Oscillation, or QBO, for example).

A flip of the AMO after 1995 from cold back to the warm mode is shown in Fig. 11.15, which shows the change the decade after 1995 from the decade before. We see the evolution of the warm horseshoe pattern of the +AMO mode.

Fig. 11.16 shows the pressure and temperature patterns associated with the AO and NAO phases in the winter. The positive AO leads to a more zonal (west to east) flow and milder maritime air over the continents in winter. The negative AO, on the other hand, favors a more meridional flow and high latitude blocking high pressure, directing Arctic or even Siberian air southeast into the eastern and central United States and Siberian air west into Europe.

Eden and Jung (2001) showed the evolution of the tripole from warm in the 1960s to cold by the early 1980s. We showed in Fig. 11.17 how the warm tripole correlated with predominantly negative NAO and the cold tripole with

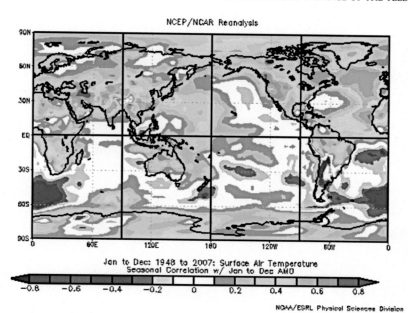

FIGURE 11.14 Correlation of the AMO with annual surface temperatures. Positive AMO correlates positively with temperatures (warmth) in most of the Northern Hemisphere. A negative AMO would correlate with cooler temperatures.

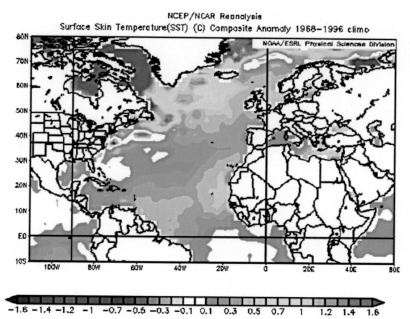

FIGURE 11.15 Difference in sea surface temperatures 1996–2004 compared to 1986–1995. It shows the evolution to the warm AMO.

positive NAO. The warm tripole showed itself strongly in the 2009–2010 winter, which led to the cold winter in Eurasia with frigid Siberian air moving west under polar blocking high pressure. The NAO and AO were the lowest for any winter since 1950 (Figs. 11.18 and 11.19).

10. SYNCHRONIZED DANCE OF THE TELECONNECTIONS

The record of natural climate change and the measured temperature record during the last 150 years gives no reason for alarm about dangerous warming caused by human CO_2 emissions. Predictions based on past warming and cooling cycles over the past 500 years accurately predicted the present cooling phase (Easterbrook, 2001,

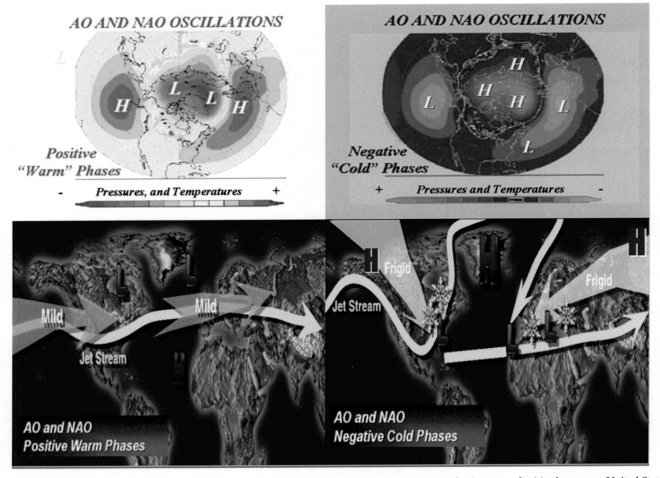

FIGURE 11.16   AO and NAO phases and the winter jet streams and weather (temperatures and wintry weather) in the eastern United States and Eurasia.

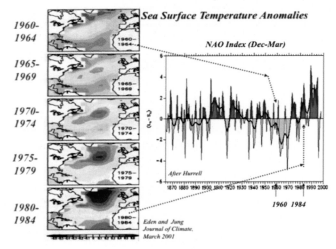

FIGURE 11.17   The positive AMO during the 1960s favored the persistent negative NAO/AO in the December to March cold season. When the AMO turned cold by the 1980s through 1994, the positive NAO was favored.

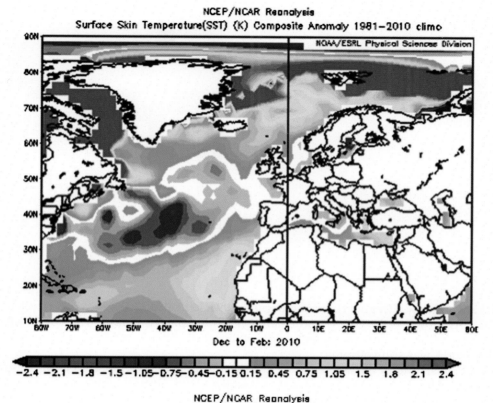

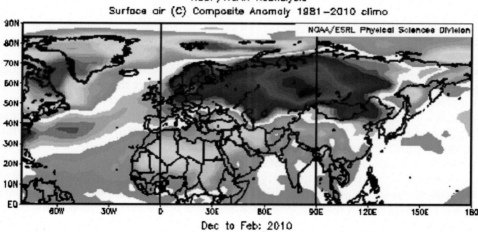

FIGURE 11.18   In the El Niño winter of 2009–2010, a strong positive AMO with warm water in the far North Atlantic (top) favored a strongly negative NAO/AO and blocking high pressure that fed frigid Siberian air west into Europe (bottom).

2005, 2006a,b, 2007, 2008a,b,c) and the establishment of cool Pacific sea surface temperatures confirms that the present cool phase will persist for several decades.

Latif and his colleagues at Leibniz Institute at Germany's Kiel University predicted the new cooling trend in a paper published in 2009, and warned of it again at an IPCC conference in Geneva in September 2009:

> A significant share of the warming we saw from 1980 to 2000 and at earlier periods in the 20th Century was due to these cycles—perhaps as much as 50 per cent. They have now gone into reverse, so winters like this one will become much more likely. Summers will also probably be cooler, and all this may well last two decades or longer. The extreme retreats that we have seen in glaciers and sea ice will come to a halt. For the time being, global warming has paused, and there may well be some cooling.

FIGURE 11.19   The AO for December to February in 2009—2010 was the most negative of any winter since 1950.

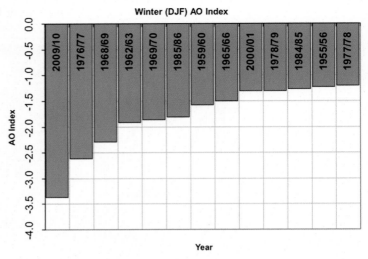

According to Latif and his colleagues (Latif and Barnett, 1994; Latif et al., 2009), this in turn relates to much longer-term shifts—what are known as the Pacific and Atlantic multidecadal oscillations (MDOs). For Europe, the crucial factor here is the temperature of the water in the middle of the North Atlantic, which is now several degrees below its average when the world was still warming.

Prof Anastasios Tsonis, head of the University of Wisconsin Atmospheric Sciences Group, has shown (2007) that these MDOs move together in a synchronized way across the globe, abruptly flipping the world's climate from a warm mode to a cold mode and back again in 20—30-year cycles.

They amount to massive rearrangements in the dominant patterns of the weather, and their shifts explain all the major changes in world temperatures during the 20th and 21st Centuries. We have such a change now and can therefore expect 20 or 30 years of cooler temperatures.

The period from 1915 to 1940 saw a strong warm mode, reflected in rising temperatures, but from 1940 until the late 1970s, the last MDO cold-mode era, the world cooled, despite the fact that carbon dioxide levels in the atmosphere continued to rise. Many of the consequences of the recent warm mode were also observed 90 years ago. For example, in 1922, the *Washington Post* reported that Greenland's glaciers were fast disappearing, while Arctic seals were "finding the water too hot." The *Post* interviewed Captain Martin Ingebrigsten, who had been sailing the eastern Arctic for 54 years: "He says that he first noted warmer conditions in 1918, and since that time it has gotten steadily warmer. Where formerly great masses of ice were found, there are now moraines, accumulations of earth and stones. At many points where glaciers formerly extended into the sea they have entirely disappeared. As a result, the shoals of fish that used to live in these waters had vanished, while the sea ice beyond the north coast of Spitsbergen in the Arctic Ocean had melted. Warm Gulf Stream water was still detectable within a few hundred miles of the Pole."

In contrast, 56 percent of the surface of the United States was covered by snow. "That hasn't happened for several decades," Tsonis pointed out. "It just isn't true to say this is a blip. We can expect colder winters for quite a while." He recalled that towards the end of the last cold mode, the world's media were preoccupied by fears of freezing. For example, in 1974, a *Time* magazine cover story predicted "Another Ice Age," saying "Man may be somewhat responsible—as a result of farming and fuel burning [which is] blocking more and more sunlight from reaching and heating the Earth."

Tsonis observed "Perhaps we will see talk of an ice age again by the early 2030s, just as the MDOs shift once more and temperatures begin to rise." Although the two indices (PDO and AMO) are derived in different ways, they both represent a pattern of sea surface temperatures, a tripole with warm in the high latitudes and tropics and colder in between, especially west, or vice versa. In both cases the warm modes were characterized by general global warmth and the cold modes with general broad climatic cooling, though each with regional variations. Normalizing and adding the two indices makes them more comparable. A positive AMO + PDO should correspond to an above-normal temperature, and the negative below normal. Indeed that is the case for U.S. temperatures (NCDC USHCN v2), as shown in Fig. 11.20.

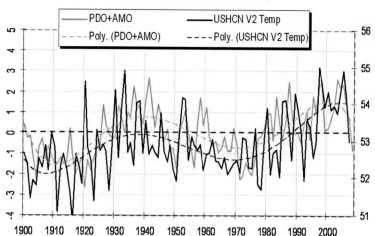

PDO+AMO vs USHCN V2 Annual Temp

FIGURE 11.20 NASA GISS version of NCDC USHCN version 2 versus PDO + AMO. The multidecadal cycles with periods of 60 years match the USHCN warming and cooling cycles. Annual temperatures end at 2007.

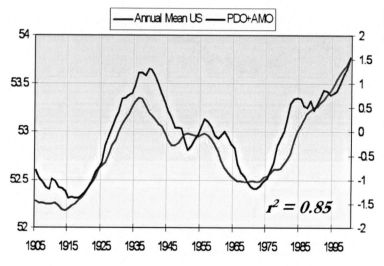

FIGURE 11.21 With 22-point smoothing, the correlation of U.S. temperatures and the ocean multidecadal oscillations is clear, with an r^2 of 0.85.

$$r^2 = 0.85$$

With an 11-year smoothing of the temperatures and PDO + AMO to remove any effect of the 11-year solar cycles, an even better correlation with an r^2 of 0.85 is obtained (Fig. 11.21).

Fig. 11.22 shows The AMO/PDO regression fit to USHCN version 2. The PDO/AMO works well in predicting temperatures (Fig. 11.23). The divergence in recent years is in part due to the adjustments made to version 2, most notably the removal of the Urban Heat Island adjustment.

11. USING WARM AND COLD POOLS IN OPERATIONAL SEASONAL FORECASTING

The PDO and AMO and ENSO have been successfully used to predict seasonal temperatures when combined with other teleconnections. The value of looking beyond just the tropics and ENSO has been shown by many including Namias (1959, 1978, 1981), Latif and Barnett (1994), Hurell (1995, 1996), Mantua et al. (1997), Kerr (2000), Marshall et al. (2001), Moore et al. (2002), Rodwell et al. (1999) Eden and Willebrand (2001), and Perlwitz et al. (2009). Huug van den Dool at CP with a Constructed Analog model matches years back to the middle 1950s with the current and recent months' sea surface temperature patterns in the oceans from 45N to 45S. We often find our analogs that include ocean decadal oscillations, solar and ENSO more often agree with the CPC CA model

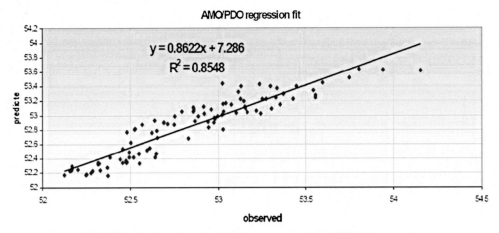

FIGURE 11.22   The AMO/PDO regression fit to USHCN version 2.

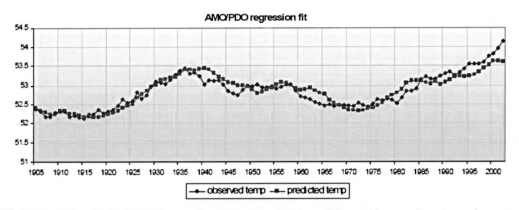

FIGURE 11.23   Using the PDO/AMO to predict temperatures works well here with some divergence after around 2000.

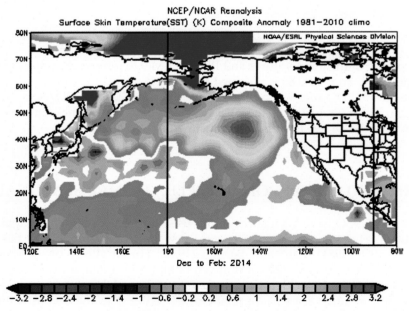

FIGURE 11.24   During 2013, warm water spread into the Gulf of Alaska, where it was anchored during the winter of 2013–2014.

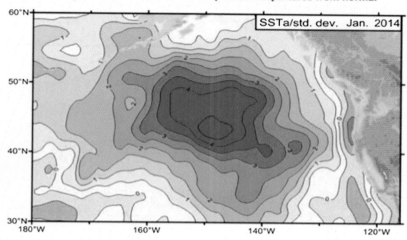

FIGURE 11.25 The water reached over four standard deviations positive in January. *CMOS January 2014 temperature departure (in standard deviations) from the 1981—2013 Reynolds NCEP dataset. http://www.oceannetworks.ca/sites/default/files/pdf/science/CMOS-Bulletin-April2014-Freeland.pdf.*

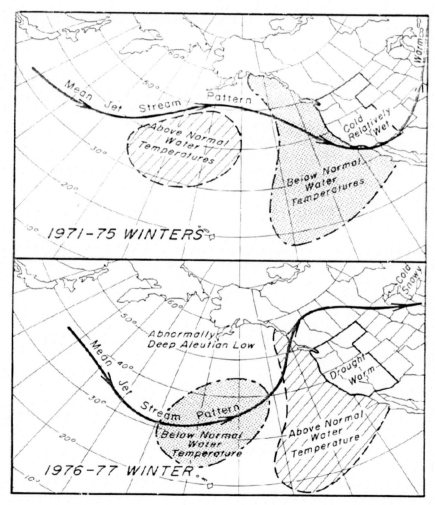

FIGURE 11.26 Jerome Namias (1977) used warm and cold water pools to explain the GPCS in 1976. Warm water replaced cold water off the west coast, favoring the dry western ridge and cold and snowy eastern trough.

FIGURE 11.27 Warm water has appeared in other winters, including the frigid winters of 1916–1917 and 1917–1918. This is seen in the Kaplan SST reconstruction for the winter of 1917–1918 (ESRL PSD reanalysis).

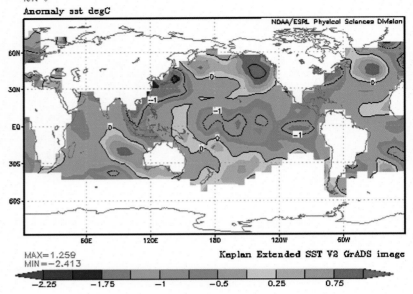

than the dynamic coupled ocean and atmospheric global models. We usually find the dynamic models converge on the analog composite forecasts as the seasons approach. We have shown skill at times several seasons into the future.

Our statistical approach looks at the ocean patterns when significant outside the 45N to 45S band. The value of that has been clearly shown the last few winters.

In the fall of 2013, it was clear a very strong warm pool was developing and showing stability in the Gulf of Alaska (Fig. 11.24). It strengthened to over four standard deviations by January (Fig. 11.25).

The idea of warm and cold pools in the middle and even higher latitudes originated decades earlier with Jerome Namias. He published his ideas (Namias, 1959, 1978, 1981) about how these pools anchored deviations in the jet streams and favored the position of troughs and ridges and anomalies of temperatures and precipitation. He used his analysis (Fig. 11.26) to show how the flip (GPCS) in 1976 moved the favored trough from the western United States to the eastern United States and kicked off a serious drought in California and the West. This suggested the warm evolution following the cold water off the West Coast in 2010–2012, which had favored a western cool trough leading to cold weather in the winter of 2013–2014 in the central region and East as in 1976–1977 and 1977–1978.

We also recalled that a warm pool had developed in 1916 and continued in the winters of 1916–1917 and 1917–1918, which like 1976–1977 and 1977–1978 were frigid and snowy in the Great Lakes and northeastern states (Figs. 11.27 and 11.28).

Our forecasts issued as early as July talked about the "historic" cold winter into the Great Lakes. Most dynamical model forecasts and National Center official forecast from NOAA in the United States and Environment Canada even in late November were calling for a warm winter near the Great Lakes and east. Our final November forecast and actual findings are shown in Fig. 11.29.

Fig. 11.30 shows the situation evolved in 2014 so that by the fall, the warm pool in the Gulf of Alaska drifted slightly east and grew to the south with a second warm pool off of Baja California (named the Baja Niño). Cold pools developed northwest of Hawaii and in the North Atlantic. Also an El Niño Modoki (an El Niño with the warm water settling near the dateline with colder water near South America) had developed. When we correlated each of these anomaly centers with prior years, we found very similar patterns (the synchronization dance of the teleconnections that Tsonis discussed).

Our forecast (Fig. 11.31) reflected the ocean drivers and again correctly forecast the hard winter in the northeast (the coldest winter in the entire record back to 1895 for the 10 northeast states) and DC and parts of Maritime Canada. For the second straight year, all-time snow records also occurred.

It should be noted that the climate models and official government forecasts again were warm, especially in the Northeast, where the coldest condition occurred.

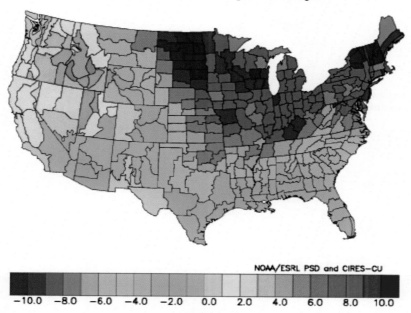

FIGURE 11.28   NOAA PSD reanalysis of temperature anomalies in the winter of 1917–1918.

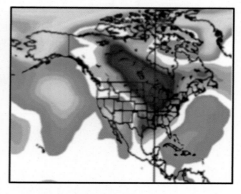

FIGURE 11.29   Weatherbell analytics winter 2013–2014 forecast (left) versus actual (right).

12. SHORT-TERM WARM/COOL CYCLES FROM THE GREENLAND ICE CORE

Variation of oxygen isotopes in ice from Greenland ice cores is a measure of temperature. Most atmospheric oxygen consists of ^{16}O but a small amount consists of ^{18}O, an isotope of oxygen that is somewhat heavier. When water vapor (H_2O) condenses from the atmosphere as snow, it contains a ratio of $^{16}O/^{18}O$ that reflects the temperature at that time. When snow falls on a glacier and is converted to ice, it retains an isotopic 'fingerprint' of the temperature conditions at the time of condensation. Measurement of the $^{16}O/^{18}O$ ratios in glacial ice hundreds or thousands of years old allows reconstruction of past temperature conditions (Stuiver and Grootes, 2000; Stuiver and Brasiunas, 1991, 1992; Grootes and Stuiver, 1997; Stuiver et al., 1995; Grootes et al., 1993). High-resolution ice core data show that abrupt climate changes occurred in only a few years (Steffensen et al., 2008). The GISP2 ice core data

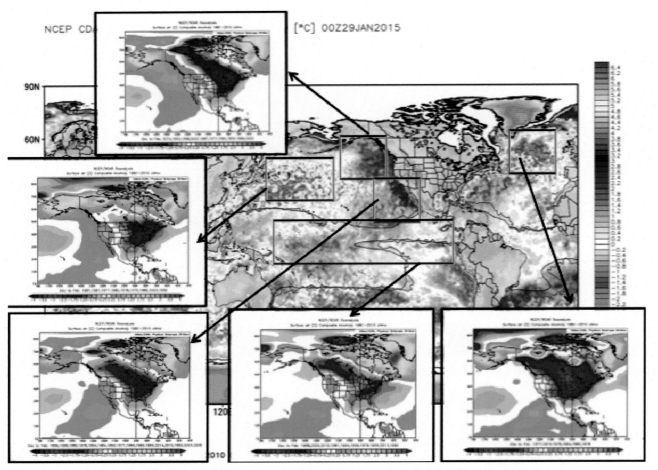

FIGURE 11.30   Significant warm and cold pools entering the winter of 2014—2015 each correlated with a cold eastern North America, which was used with other teleconnections to make another high-confidence winter forecast.

2014/15 Winter Anomalies

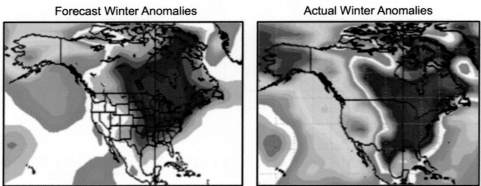

FIGURE 11.31   Weatherbell forecast November to March temperature anomalies versus the actual anomalies. The strong warm pools off both western Canada and Baja California made the West warmer but the eastern North American cold was correctly captured.

of Stuiver and Grootes (2000) can be used to reconstruct temperature fluctuations in Greenland over the past 500 years (Fig. 11.21). Fig. 11.32 shows a number of well-known climatic events. For example, the isotope record shows the Maunder Minimum, the Dalton Minimum, the 1880—1915 cool period, the 1915—~1945 warm period, and the ~1945—1977 cool period, as well as many other cool and warm periods. Temperatures fluctuated between

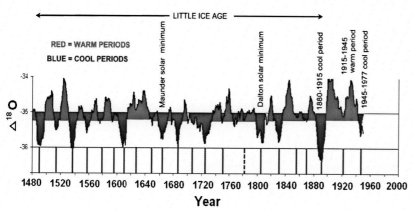

FIGURE 11.32   Cyclic warming and cooling trends in the past 500 years. *Plotted from GISP2 data, Stuiver, M., Grootes, P.M., 2000. GISP2 oxygen isotope ratios. Quaternary Research, 54 (3).*

warm and cool at least 22 times between 1480 AD and 1950 (Fig. 11.32). None of the warming periods could have possibly been caused by increased CO_2 because they all preceded rising CO_2.

Only one out of all of the global warming periods in the past 500 years occurred at the same time as rising CO_2 (1977–1998). About 96% of the warm periods in the past 500 years could not possibly have been caused by rise of CO_2. The inescapable conclusion of this is that CO_2 is not the cause of global warming. The Greenland ice core isotope record matches climatic fluctuations recorded in alpine glacier advances and retreats.

13. WHERE ARE WE HEADED DURING THE COMING CENTURY?

The cool phase of PDO is now entrenched. We have shown how the two ocean oscillations drive climate shifts. The PDO's effect is later amplified by the AMO. Each time this has occurred in the past century, global temperatures have remained cool for about 30 years.

No statistically significant global warming has taken place since 1998 (Fig. 11.33). Winter regional cooling has occurred during the past two decades in the United States and parts of Europe (Hanna and Cappelen, 2003). A very likely reason for global cooling over the past decade is the switch of the Pacific Ocean from its warm mode (where it has been from 1977 to 1998) to its cool mode in 1999. Each time this has occurred in the past century, global temperatures have remained cool for about 30 years. Thus, the current sea surface temperatures not only explain why we have had stasis or global cooling for the past 10 years, but also should assure that cooler temperatures will continue for several more decades. There will be brief bounces upwards with periodic El Niños, as we have seen in late 2009 and early 2010 (and may see in 2016), but they will give way to cooling as the favored La Niña states returns. With a net La Niña tendency, the net result should be cooling.

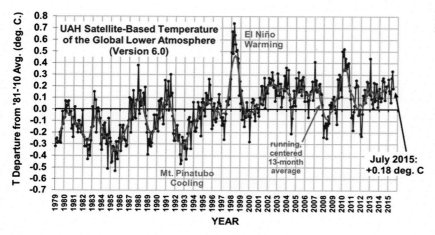

FIGURE 11.33   UAH globally averaged lower atmospheric temperatures.

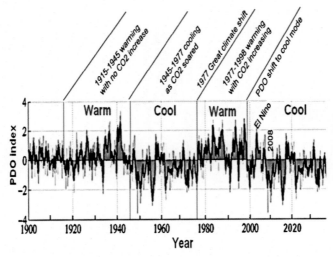

FIGURE 11.34   Using past behavior of the PDO to predict the future.

13.1 Predictions Based on Past Climate Patterns

The past is the key to understanding the future. Past warming and cooling cycles over the past 500 years were used by Easterbrook (2001, 2005, 2006a,b, 2007, 2008a,b,c) to accurately predict the cooling phase that is now happening. Establishment of cool Pacific sea surface temperatures since 1999 indicates that the cool phase will persist for the next several decades. We can look to past natural climatic cycles as a basis for predicting future climate changes. The climatic fluctuations over the past few 100 years suggest ∼30-year climatic cycles of global warming and cooling, on a general warming trend from the Little Ice Age cool period. If the trend continues as it has for the past several centuries, global temperatures for the coming century might look like those in Fig. 11.34. The left side of Fig. 11.34 is the warming/cooling history of the past century. The right side of the graph shows that we have entered a global cooling phase that fits the historic pattern very well. The switch to the PDO cool mode to its cool mode virtually assures cooling global climate for several decades.

Three possible projections are shown in Fig. 11.35: (1) moderate cooling (similar to the 1945 to 1977 cooling); (2) deeper cooling (similar to the 1945 to 1977 cooling); or (3) severe cooling (similar to the 1790 to 1830 cooling). Only time will tell which of these will be the case, but at the moment, the sun is behaving very similar to the Dalton Minimum (sunspot cycles 4/5), which was a very cold time. This is based on the similarity of sunspot cycle 23 to cycle 4 (which immediately preceded the Dalton Minimum).

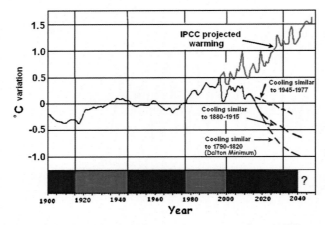

FIGURE 11.35   The projected climate for the century based on climatic patterns over the past 500 years and the switch of the PDO to its cool phase.

As the global climate and solar variation reveal themselves in a way not seen in the past 200 years, we will surely attain a much better understanding of what causes global warming and cooling. Time will tell. If the climate continues its ocean cycle cooling and the sun behaves in a manner not witnessed since 1800, we can be sure that climate changes are dominated by the sun and sea and that atmospheric CO_2 has a very small role in climate changes. If the same climatic patterns, cyclic warming and cooling, that occurred over the past 500 years continue, we can expect several decades of moderate to severe global cooling.

References

Bjerknes, J., 1969. Atmospheric teleconnections from the equatorial Pacific. Monthly Weather Review 97, 153−172.

Douglass, D.H., Christy, J.R., 2008. Limits on CO_2 climate forcing from recent temperature data of earth. Energy and Environment.

Easterbrook, D.J., 2001. The next 25 years: global warming or global cooling? Geologic and oceanographic evidence for cyclical climatic oscillations. Geological Society of America, Abstracts With Program 33, 253.

Easterbrook, D.J., 2005. Causes and effects of abrupt, global, climate changes and global warming. Geological Society of America, Abstracts With Program 37, 41.

Easterbrook, D.J., 2006a. Causes of abrupt global climate changes and global warming predictions for the coming century. Geological Society of America, Abstracts With Program 38, 77.

Easterbrook, D.J., 2006b. The cause of global warming and predictions for the coming century. Geological Society of America, Abstracts With Program 38, 235−236.

Easterbrook, D.J., 2007. Geologic evidence of recurring climate cycles and their implications for the cause of global warming and climate changes in the coming century. Geological Society of America Abstracts With Programs 39, 507.

Easterbrook, D.J., 2008a. Solar influence on recurring global, decadal, climate cycles recorded by glacial fluctuations, ice cores, sea surface temperatures, and historic measurements over the past millennium. In: Abstracts of American Geophysical Union Annual Meeting, San Francisco.

Easterbrook, D.J., 2008b. Implications of glacial fluctuations, PDO, NAO, and sun spot cycles for global climate in the coming decades. Geological Society of America, Abstracts With Programs 40, 428.

Easterbrook, D.J., 2008c. Correlation of climatic and solar variations over the past 500 years and predicting global climate changes from recurring climate cycles. In: Abstracts of 33rd International Geological Congress, Oslo, Norway.

Easterbrook, D.J., Kovanen, D.J., 2000. Cyclical oscillation of Mt. Baker glaciers in response to climatic changes and their correlation with periodic oceanographic changes in the northeast Pacific Ocean. Geological Society of America, Abstracts With Program 32, 17.

Eden, C., Jung, T., 2001. North Atlantic interdecadal variability: oceanic response to the North Atlantic oscillation (1865−1997). Journal of Climate 14 (5), 676−691.

Eden, C., Willebrand, J., 2001. Mechanism of interannual to decadal variability of the North Atlantic circulation. Journal of Climate 14 (10), 2266−2280.

Fuller, S.R., Easterbrook, D.J., Burke, R.M., 1983. Holocene glacial activity in five valleys on the flanks of Mt. Baker, Washington. Geological Society of America, Abstracts With Program 15, 43.

Fuller, S.R., 1980. Neoglaciation of Avalanche Gorge and the Middle Fork Nooksack River Valley, Mt. Baker, Washington (M.S. thesis). Western Washington University, Bellingham, Washington.

Gershunov, A., Barnett, T.P., 1998. Interdecadal modulation of ENSO teleconnections. Bulletin of the American Meteorological Society 79, 2715−2725.

Grootes, P.M., Stuiver, M., 1997. Oxygen 18/16 variability in Greenland snow and ice with 10^3 to 10^5-year time resolution. Journal of Geophysical Research 102, 26455−26470.

Grootes, P.M., Stuiver, M., White, J.W.C., Johnsen, S.J., Jouzel, J., 1993. Comparison of oxygen isotope records from the GISP2 and GRIP Greenland ice cores. Nature 366, 552−554.

Hanna, E., Cappelen, J., 2003. Recent cooling in coastal southern Greenland and relation with the North Atlantic Oscillation. Geophysical Research Letters 30.

Hanna, E., Jonsson, T., Olafsson, J., Valdimarsson, H., 2006. Icelandic coastal sea surface temperature records constructed: putting the pulse on air-sea-climate interactions in the Northern North Atlantic. Part I: comparison with HadISST1 open-ocean surface temperatures and preliminary analysis of long-term patterns and anomalies of SSTs around Iceland. Journal of Climate 19, 5652−5666.

Hare, S.R., 1996. Low Frequency Climate Variability and Salmon Production (Ph.D. dissertation). School of Fisheries, University of Washington, Seattle, WA.

Harper, J.T., 1993. Glacier fluctuations on Mount Baker, Washington, U.S.A., 1940−1990, and climatic variations. Arctic and Alpine Research 4, 332−339.

Hurrell, J.W., 1995. Decadal trends in the North Atlantic Oscillation, regional temperatures and precipitation. Science 269, 676−679.

Hurrell, J.W., 1996. Influence of variations in extratropical wintertime teleconnections on Northern Hemisphere temperature. Geophysical Research Letters 23, 665−668.

Jones, P.D., Johnson, T., Wheeler, D., 1997. Extension in the North Atlantic Oscillation using early instrumental pressure observations from Gibralter and southwest Iceland. Journal of Climatology 17, 1433−1450.

Kerr, R.A., 2000. A North Atlantic climate pacemaker for the centuries. Science 288, 1984−1986.

Latif, M., Barnett, T.P., 1994. Causes of decadal climate variability over the North Pacific and North America. Science 266, 634−637.

Latif, M., Park, W., Ding, H., Keenlyside, N., 2009. Internal and External North Atlantic sector variability in the Kiel climate model. Meteorologische Zeitschrift 18, 433−443.

Mantua, N.J., Hare, S.R., Zhang, Y., Wallace, J.M., Francis, R.C., 1997. A Pacific interdecadal climate oscillation with impacts on salmon production. Bulletin of the American Meteorological Society 78, 1069−1079.

Marshall, J., Johnson, H., Goodman, J., 2001. A study of the interaction of the North Atlantic Oscillation with ocean circulation. Journal of Climate 14, 1399–1421.

McLean, J.D., de Freitas, C.R., Carter, R.M., 2009. Influence of the southern oscillation on tropospheric temperature. Journal of Geophysical Research 114.

Miller, A.J., Cayan, D.R., Barnett, T.P., Graham, N.E., Oberhuber, J.M., 1994. The 1976–77 climate shift of the Pacific Ocean. Oceanography 7, 21–26.

Moore, G.W.K., Holdsworth, G., Alverson, K., 2002. Climate change in the North Pacific region over the past three centuries. Nature 420, 401–403.

Namias, J., 1959. Recent seasonal interactions between North Pacific waters and the overlying atmospheric circulation. Journal of Geophysical Research 64, 631–646.

Namias, J., 1977. Mean Jet Stream Pattern in Relation to Eastern Pacific Ocean Water Temperatures During a Relatively Wet, Cool Period in Western U.S.A. Compared to the Same Elements During a Warm Drought Period in the Western U.S.A. (from WRTA NO 77-8).

Namias, J., 1978. Multiple causes of the North American abnormal winter 1976–1977. Monthly Weather Review 106, 279–295.

Namias, J., Douglas, A.V., Cayan, A., 1981. Large-scale changes in North Pacific and North American weather patterns in recent decades. Monthly Weather Review 110, 1851–1862.

Perlwitz, J., Hoerling, M., Eischeid, J., Xu, T., Kumar, A., 2009. A strong bout of natural cooling in 2008. Geophysical Research Letters 36.

Rodwell, M., Rowell, D., Folland, C., 1999. Oceanic forcing of the wintertime North Atlantic oscillation and European climate. Nature 398, 320–323.

Steffensen, J.P., Andersen, K.K., Bigler, M., Clausen, H.B., Dahl-Jensen, D., Goto-Azuma, K., Hansson, M.J., Sigfus, J., Jouzel, J., Masson-Delmotte, V., Popp, T., Rasmussen, S.O., Roethlisberger, R., Ruth, U., Stauffer, B., Siggaard-Andersen, M., Sveinbjornsdottir, A.E., Svensson, A., White, J.W.C., 2008. High-resolution Greenland ice core data show abrupt climate change happens in few years. Science 321, 680–684.

Stuiver, M., Brasiunas, T.F., 1991. Isotopic and solar records. In: Bradley, R.S. (Ed.), Global Changes of the Past. Boulder University, Corporation for Atmospheric Research, pp. 225–244.

Stuiver, M., Brasiunas, T.F., 1992. Evidence of solar variations. In: Bradley, R.S., Jones, P.D. (Eds.), Climate Since A.D. 1500. Routledge, London, pp. 593–605.

Stuiver, M., Grootes, P.M., 2000. GISP2 oxygen isotope ratios. Quaternary Research 54 (3).

Stuiver, M., Grootes, P.M., Brasiunas, T.F., 1995. The GISP2 ^{18}O record of the past 16,500 years and the role of the sun, ocean, and volcanoes. Quaternary Research 44, 341–354.

Thompson, D.W.J., Wallace, J.M., 1998. The Arctic Oscillation signature in the wintertime geopotential height and temperature fields. Geophysical Research Letters 25, 1297–1300.

Thompson, D.W.J., Wallace, J.M., 2000. Annular modes in the extratropical circulation. Part 1: month-to-Month variability. Journal of Climate 13, 1000–1016.

Tsonis, A.A., Hunt, G., Elsner, G.B., 2003. On the relation between ENSO and global climate change. Meteorological Atmospheric Physics 1–14.

Tsonis, A.A., Swanson, K.L., Kravtsov, S., 2007. A new dynamical mechanism for major climate shifts. Geophysical Research Letters 34.

Verdon, D.C., Franks, S.W., 2006. Long-term behaviour of ENSO: interactions with the PDO over the past 400 years inferred from paleoclimate records. Geophysical Research Letters 33.

Walker, G., Bliss, 1932. World weather V. Memoirs Royal Meteorological Society 4, 53–84.

Wolter, K., 1987. The Southern Oscillation in surface circulation and climate over the tropical Atlantic, Eastern Pacific, and Indian Oceans as captured by cluster analysis. Journal of Climate and Applied Meteorology 26, 540–558.

Wolter, K., Timlin, M.S., 1993. Monitoring ENSO in COADS with a seasonally adjusted principal component index. In: Proceedings of the 17th Climate Diagnostics Workshop. Oklahoma Climatological Survey, CIMMS and the School of Meteorology, University of Oklahoma, Norman, OK, pp. 52–57.

Zhang, Y., Wallace, J.M., Battisti, D., 1997. ENSO-like interdecadal variability: 1990–1993. Journal of Climatology 10, 1004–1020.

Zhang, Y., Wallace, J.M., Iwasaka, N., 1996. Is climate variability over the North Pacific a linear response to ENSO? Journal of Climatology 9, 1468–1478.

12

Sea Level Changes as Observed in Nature

N.-A. Mörner

Paleogeophysics & Geodynamics, Saltsjöbaden, Sweden

1. INTRODUCTION

In the past, we observed records of past sea-level positions and tried to date them by different means, primarily by radiocarbon, with the aim of producing a curve of postglacial changes in sea level that could be used as a regional or global standard (eg, Fairbridge, 1961; Shepard, 1963; Jelgersma, 1966; Mörner, 1969; Tooley, 1974). Today, the observationally based sea-level research has, to a large extent, been replaced by model-based scenarios trying to establish graphs of recent and near-future sea level aimed at the preconceived idea that sea level must be in an accelerating mode due to the hypothesis of anthropogenic global warming (as launched and driven by the IPCC, 2001; 2007, 2013) (eg, Hoffmann, 2007; Rahmstorf, 2007; Church et al., 2007; Cazenave et al., 2009; Kemp et al., 2011; White et al., 2014; Grinsted et al., 2015; Little et al., 2015; Romm, 2015; Lorbacher et al., 2015). By doing so, this not only violates observational facts but also geoethical principles (Mörner, 2015a,b).

2. "POSTER SITES" FOR LOBBYISTS

Places like Tuvalu, the Maldives, Kiribati, and Bangladesh have been cited as areas where sea level is rapidly rising, soon to flood these sites. All such claims are scenario-based lobbying attempts, with little or no anchoring in real, observational facts.

Evidence-Based Climate Science, Second Edition
http://dx.doi.org/10.1016/B978-0-12-804588-6.00012-4

2.1 The Islands of Tuvalu

Michael Crichton (2004) wrote a novel, *State of Fear*, in which he explains how climate fanatics operate, and even planned that the people of Tuvalu should sue the U.S. government for polluting the atmosphere with CO_2 and ultimately causing sea level to rise enough to threaten flooding the islands of Tuvalu. He also revealed that the plans were cancelled at a late stage when it was realized that the sea-level-rise scenario was not strong enough: "The lawsuit is a [deleted] disaster." He continues: "And that Scandinavian guy, that sea level expert. He's becoming a pest. He's even attacking the IPCC for incompetence."

The only factual basis for a meaningful judgement of the sea-level situation in the Islands of Tuvalu is to consult their tide gauge records. The tide gauge on the island of Funafuti has been in operation since 1978. It shows a total absence of any long-term sea-level rise; just variability around a stable position, broken by three major lows at the ENSO years of 1983, 1992, and 1998, as illustrated in Fig. 12.1 (Mörner, 2007a,b, 2010a, 2011a). The same message applies for the new high-resolution SEAFRAME tide gauge station of Funafuti (Gray, 2010; Mörner, 2011a).

The situation is indeed weird: observational facts say "no rise in sea level," but still people continue to drive the rising sea-level illusion, and it doesn't become better—rather the opposite—when the message is broadcasted by Ban Ki Moon and the Secretary-General for the Pacific Islands Forum (Young, 2011). It upsets observational facts and ethical principles. What is even worse, it steals the limelight from real problems in the real world (Mörner, 2011a).

The Islands of Tuvalu are not threatened by any sea-level rise; only by normal extreme events such as tsunamis and "king tides" (Lin et al., 2014).

The Pacific island chain of Vanuatu was claimed to be doomed by flooding and, thus, people were in need of evacuation. The observational facts tell a different story: a general absence of a sea-level rise since 1993 (Mörner, 2007a; Gray, 2010).

2.2 The Maldives

For a long time, it was common among alarmists to claim that the Maldives were already on their way to becoming flooded due to a rapidly rising sea level (Titus, 1989; IPCC, 2001; Bryant, 2004; Hoffmann, 2007). The former President, Mohamed Nasheed, was a strong proponent of this flooding hypothesis, and he set up a number of events (or rather shows) to manifest it; eg, a cabinet meeting under water in 2009.

In 2000, we launched an International Sea Level Project in the Maldives (INQUA, 2000; Tooley, 2000; Mörner et al., 2004). It soon became obvious that sea level is not at all in a rapidly rising mode (Mörner et al., 2004; Mörner, 2007a,c). After several expeditions, lots of fieldwork by a group of international experts, and 57 radiocarbon dates, it was possible to present a quite detailed curve of sea-level changes over the last 6000 years (Mörner, 2007c, 2011b). The data records seven sea-level oscillations in the past 4000 years with amplitudes ranging between 1.2 and 0.5 m, thought to be driven by regional oceanographic—climatic factors, including the high/low latitude interchanges of water masses in response to changes in Earth's rate of rotation at grand solar maxima/minima of the last millennium (Mörner, 2015c, Fig. 2). The record of the last several centuries is of special interest. It records a +0.5 m level in the 17th century, a low sea level (below present zero) in the 18th century, a rise at 1790, about +0.2 m high level up to 1960—1970, and virtually stable level over the last 40—50 years (Mörner, 2011b, 2013, 2014a) as illustrated in Fig. 12.2 (top graph). This implied a total reevaluation of the actual sea-level situation in the Maldives, all based on hard observational facts: stratigraphy, morphology, and radiocarbon-dated corals in growing positions.

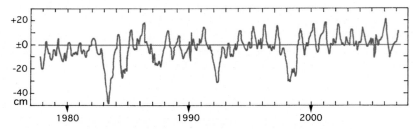

FIGURE 12.1   The tide gauge record since 1978 for Tuvalu. Since 1985 there are no signs of any sea-level rise (for 1978—1985, there is a small rise that might be due to site compaction). Three major ENSO events with significant drops in sea level are recorded in 1983, 1992, and 1998 (plus one in 2010). The new high-resolution record, starting in 1993, confirms the absence of any sea-level rise. *From Mörner, N.-A., 2010a. Some problems in the reconstruction of mean sea level and its changes with time. Quaternary International 221, 3—8, http://dx.doi.org/10.1016/j.quaint.2009.10.044 and Mörner, N.-A., 2010b. The Greatest Lie Ever Told, third ed. P&G-print, 20 pp.*

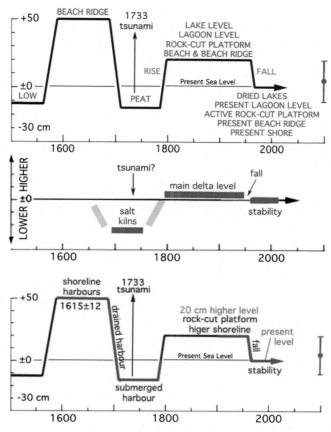

FIGURE 12.2   Observed, documented, and dated sea-level changes during the last 500 years in the Indian Ocean. *Top*: the Maldives (Mörner, 2007c). *Middle*: Bangladesh (Mörner, 2010c). *Bottom*: Goa, India (Mörner, 2013). The agreement is striking. All three curves show (1) a stability during the last 40–50 years; (2) a fall around 1960; (3) a 20-cm higher level 1790–1960; (4) a rise around 1790; (5) a distinct low level, not quantified, in the 18th century; and (6) a +50–60 cm high level in the 17th century. A tsunami event in the year 1733 is recorded both in the Maldives and Goa, and probably in Bangladesh, too (Mörner, 2015e).

A tree in a vulnerable shore position on the Island of Viligili had remained in this position from the early 1950s (Mörner, 2007a,c, 2011b; Murphy, 2007; Mortensen, 2004). The slightest rise in sea level and it would be gone. Therefore, its existence 60 years back in time was an indication that sea level, indeed, was not rising—a tree cannot lie (Mörner, 2011b). In 2003, "it was pulled down by a group of Australian scientists" according to testimony of local eyewitnesses working close by (Mörner, 2007a,c, 2011b). The story even became a theater play, "The Heretic" by Richard Bean (2011; played for a full house at The Royal Court Theatre in 2011, and in Melbourne in 2012).

In an "Open letter to President Mohamed Nasheed of the Maldives" (Mörner, 2009), I explained the situation, but never got any reply back. Still, it should be a very happy message: "no flooding." But the President got angry and "wanted his flooding"; ie, his argument for asking for economical support.

2.3 The Islands of Kiribati

When Tuvalu and the Maldives fell as "poster sites" for the AGW proponents, because sea level has not risen there, alarmists turned their attention to the Island of Kiribati (or Christmas Island), and again Ban Ki Moon and the Secretary-General for the Pacific Islands Forum spoke of the need for urgent action to save the people (eg, Young, 2011), but tide gauge records provide a clear message of an absence of sea-level rise since 1994 (Gray, 2010; Mörner, 2011a). Coastal erosion is sometimes serious in Kiribati (Chauvin and Hilaire, 2015), but this is not a sign of sea-level rise (Mörner, 2015d), just what is normal at exposed coastal segments.

At the 44th Pacific Islands Forum (PIF) held in Majuro in the Marshall Islands, the Secretary-General of PIF again stressed the serious situation of many Pacific islands in view of sea-level rise and island flooding. This sounded very weird because the tide gauge in Majuro remains virtually flat since 1993 (Mörner and Harris, 2013). At the meeting,

Connie Hedegaard, European Commissioner for Climate Action, stated: "It is my ambition to make the European Union and the Pacific region partners in advancing the global climate change agenda," implying that PIF gets economic support and the European Union gets something back: ie, cries of help in view of disastrous sea-level rise and flooding. The charade must go on.

2.4 Bangladesh

Bangladesh is notorious in the fear of coastal flooding and predictions of disasters with a massive crowd of sea-level refugees. According to Pachauri (Gunyon, 2011), it is "the world's most vulnerable country." In 2005, the United Nations Environment Programme predicted that "climate change would create 50 million climate refugees by 2010" (Watts, 2011)—and still today, we have not seen any true "climate refugee." Time reveals the truth.

The morphological and stratigraphic field evidence in the Sundarban Delta tell a quite different story, however: a low level in the 18th century; a main delta surface build-up at about 10–20 cm higher level in the 19th and early 20th centuries; a 10–20 cm fall in sea level around 1960; and stable sea-level conditions for the last 50 years (Mörner, 2010b,d), as illustrated in Fig. 12.2 (middle graph).

The fall in sea level recorded by morphology and stratigraphy in the Sunderban Delta is firmly backed up by tide gauge records in both in Mumbai and Visakhapatnam, where a rapid sea-level fall of 12 cm is recorded between 1955 and 1962 (Mörner, 2010c, Fig. 2). Both tide gauges recorded virtually stable conditions during the last 60 years.

The low level in the 18th century is partly documented by an intra-mud unconformity with sand deposition (Mörner, 2010c) and by submerged salt-producing kilns (Hanebuth et al., 2013). The stratigraphy of the submerged kilns is somewhat complicated and partly chaotic between the sites and also within the sites. The radiocarbon dates do not provide any order. The OSL dates are quite consistent, and should, in my opinion, be split up into a lower level of around −1.5 m with a mean date of 1698 ± 40 years and an upper level around −0.7 m with a mean age of 1727 ± 25 years. The age of the upper level is very close to the age of tsunami events recorded both in the Maldives and Goa (Mörner, 2013, 2015e), as noted in Fig. 12.2 (middle graph).

In conclusion, field evidence (Mörner, 2010c) provides a sea-level record very similar to those obtained in the Maldives and Goa (Fig. 12.2) with an absence of any sea-level rise over the last 60 years. Indeed, "the world's most vulnerable country" is not threatened by any eustatic sea-level rise, but rather internal population problems in the Dhaka area and general poverty. Its coasts are repeatedly cursed by destructive cyclones, however. Moreover, the lowlands are repeatedly flooded due to extreme river discharge of rainwater from the inlands. These problems are very severe, but they have nothing to do with coastal sea level.

2.5 Goa, India

In connection with Session 3, "The Illusive Sea Level Threat in the Indian Ocean" at the ICAMG7 international conference in Goa (October 2011), I was able to record an unusually clear documentation of the sea-level changes during the last 500 years. The coastal morphology showed: (1) an older level of about +2 m, (2) an upper, very clear sea level at +0.6 m level with notches in older deposits (Fig. 12.3A), (3) a very clear rock-cut platform at about +0.2 m, and (4) a present-day rock-cut platform, which has remained stable since the fall from the previous rock-cut platform in subrecent time (Fig. 12.3C and D). There can be no doubt that the 0.2–0.1 m sea-level fall represents the 1955–1962 fall recorded in the Mumbai tide gauge (Mörner, 2010c).

The stratigraphy records a main marine unit dated at 1615 ± 12 cal years BP and a thin and irregular marine layer dated at 1720 ± 42 cal years BP. The older age obviously provides an age of the +0.6 m level. The younger age falls very close to the 1733 tsunami event in the Maldives, and the appearance of this layer in the field gives the impression of probably being a tsunami layer.

Coastal archaeological data (provided by A.G. Gaur) include an old harbor construction, now a few 100 m inland and about +0.8 m above sea level, and a submarine level with harbor construction from around 1700–1750 AD right on the present beach and with steps going down to −1.5 m (Fig. 12.3B).

A painting in the Museum of Goa shows the Ancient City of Goa with two harbors; an upper abandoned harbor with five shipwrecks in it, and a lower, active, anchoring of ships along the riverbank. The upper level seems to depict the high sea-level situation of the 17th century and the lower level the low level of the 18th century.

The combined data provide a very good and solid reconstruction of the sea-level changes during the last 500 years (Mörner, 2013) as illustrated in Fig. 12.2 (bottom). The conclusion is twofold: (1) the sea-level curve of the last 500 years in Goa is almost identical to those of the Maldives and Bangladesh (Fig. 12.2), and (2) during the last 60 years sea level has remained stable (Special Correspondent, 2011).

FIGURE 12.3　Main sea-level stages during the last 500 years in Goa: (A) shore notch of the +50−60 cm high sea level from the 17th century, (B) partly submerged building of the low sea level in the 18th century, (C) dead cliff of the pre-1960 sea level and a double rock-cut platform; a +20-cm dead surface (brown) and an active surface (gray), and (D) the dead cliff of the pre-1960 sea level and the present shore of the post-1960 sea level.

2.6　The "Poster Sites" Decoded

In key site after key site in the Pacific and the Indian Ocean, previously claimed to be in the process of disastrous flooding due to rapid sea-level rise, it is here demonstrated that observational facts indicate something quite different: insignificant to zero sea-level rise over the last 20−50 years.

In this situation, measured and observed facts must, of course, overrule opinions and computer modeling. In addition, the broadcasting of views contradicted by facts represents a violation of geoethical principles (Mörner, 2015a).

Fig. 12.2 summarizes the records in the Indian Ocean with respect to the Maldives, Bangladesh, and Goa in SW India (Mörner, 2013, 2014a). Similar results were also obtained from the Island of Minicoy in the Laccadives (Mörner, 2011b), at Perth in East Australia (Mörner and Parker, 2013), Saint Paul Island (Testut et al., 2010), and in Qatar (Mörner, 2015d). The message is clear: the regional eustatic sea level in the Indian Ocean has remained unchanged over the last 40−50 years.

3.　TEST AREAS OF EUSTATIC CHANGES IN SEA LEVEL

The eustatic component of sea level was previously defined on the basis of available observational facts. For recent centuries, this usually meant analyses of available tide gauge records. Gutenberg (1941) arrived at a value of 1.1 mm/year, and Fairbridge and Krebs (1962) found 1.2 mm/year.

Mörner (1973) observed that there was a bend both in the uplifted Stockholm tide gauge and the subsiding Amsterdam tide gauge. Therefore, the isostatic and eustatic components of the relative sea-level changes of the two tide gauges could be isolated. For the period 1840–1950, the eustatic component was a mean rise of 1.1 mm/year. A mean global eustatic sea-level rise of 1.1 mm/year fits perfectly well with the deceleration in the Earth's rate of rotation over the same time period (Mörner, 1996, 2013).

The Permanent Service for Mean Sea Level, established in 1933, today has a database including 2133 tide gauges (PSMSL, 2015). NOAA uses a selection of 204 tide gauges scattered all over the globe (NOAA, 2012). The data have a Gaussian distribution (removing some outliers) with a mean value of +1.6 mm/year for 182 stations (Mörner, 2013, 2014a), later revised to a mean of +1.14 mm/year for 184 stations (Mörner, 2015f).

The PSMSL (2015) database includes 170 stations with a length of 60 years or more. The mean rate of sea-level rise of those stations is +0.25 ± 0.19 mm/year with no component of acceleration (Parker and Ollier, 2015; Parker, 2016).

It must be noted that a global network of tide gauges have a tendency to overestimate the sea-level rise because of local to site-specific subsidence due to sediment loading in river delta environments and harbor constructing loading at the stations (Mörner, 2004, 2010a, 2014a). In some areas and some sites, we are able to quantify the crustal component, allowing a better estimate of the true eustatic component. Consequently, they can be used as test areas or test sites of regional eustatic sea-level changes, described in the next sections.

3.1 The Kattegatt Sea

Sea-level data preserved along the northwest European coasts—uplifted around Fennoscandia and subsided along the North Sea coast—provide a unique database for the definition of the eustatic component. This was quite successfully done for past sea-level change in the Late Glacial and Holocene (Mörner, 1969, 1971, 1973, 1980). Having established the isostatic uplift (in Fennoscandia) and subsidence (in the North Sea coast) in great detail, we are also able to handle the delicate issue of present to future sea-level changes in a much better way than elsewhere in the world (Mörner, 2014b).

The Kattegatt Sea (an embayment between the North Atlantic and the Baltic) is probably the best test area in the world for regional eustasy (Mörner, 2014b). We have a closely determined and dated shoreline spectrum of the last 14,000 years (Mörner, 1969). A constant direction of tilting and isobase system cover an area of about 300 × 250 km. Laterally the area is bounded by faults and lineaments beyond which there are changes in the uplift geometry (Mörner, 1969). The zero isobase (ie, the boundary between uplift to the NE and subsidence to the SW) has remained stable in the Great Belt region for the last 8000 years (Mörner, 1973, 2014b), as illustrated in Fig. 12.4.

Knowing the crustal component of uplift (Fig. 12.4) and having the measured tide gauge values (red dots in Fig. 12.4), it is easy to determine the eustatic component (cf. Mörner, 2014a, Table 1): the value is 0.9 mm/year over that last 125 years.

The details are as follows: R − I = E; ie, Relative sea level − Isostasy = Eustasy (Mörner, 2014a,b, 2015f):

Nyborg	+1.01 ± 0.16	−0.1	+0.91
Korsör	+0.81 ± 0.18	±0.0	+0.81
Aarhus	+0.63 ± 0.11	+0.28	+0.91
Klagshamn	+0.64 ± 0.40	+0.3	+0.94
Varberg	−0.86 ± 0.16	+1.75	+0.89

Hansen (2015) and Hansen et al. (2015) made excellent analyses of available tide gauges in the Eastern North Sea, Skagerack, Kattegatt, and South Baltic and arrived at a eustatic component of 1.18 mm/year. This is close to the 1.1 mm/year of Mörner (1973) for the eustatic sea-level component in the Northeast European area for the period 1840–1950.

Stockholm has the second longest tide gauge record in the world. The rate of absolute glacial isostatic uplift is closely defined at 4.9 mm/year (Mörner, 1973). The present tide gage record gives a mean rate of relative sea-level changes of −3.8 mm/year. Consequently, the eustatic component can be set at 1.1 mm/year sea-level rise (Mörner, 2015f).

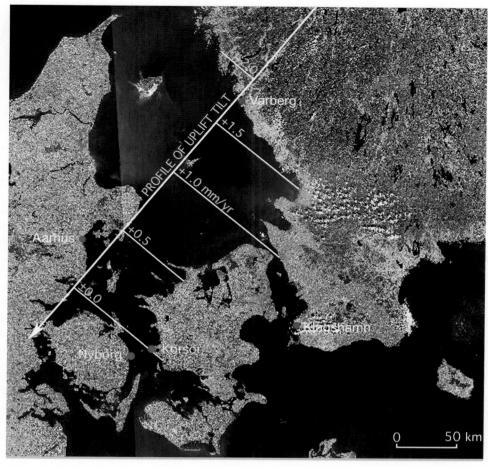

FIGURE 12.4   The uplift in the Kattegatt sea is known in great detail. The location of the zero isobase (or axis of tilting) has remained stable for 8000 years in the Great Belt area. Knowing the rate of absolute uplift and the relative sea level in the five tide gauge stations used (*red dots*), the eustatic factor can be determined with high precision at +0.9 mm/year (Mörner, 2014b).

3.2 The North Sea Coast

The southeast coast of the North Sea is dominated by long-term, postglacial, crustal subsidence. In Amsterdam, the subsidence is known to be of a rate of 0.4 mm/year (Mörner, 1973; Kooijmans, 1974), a value that should also apply for the Ijmuiden tide gauge. At Cuxhaven, the subsidence is estimated at 1.4 mm/year (Mörner, 2010d, 2013). Brest, on the other hand, seems to represent a more or less stable area over the last 10,000 years (Mörner, 1969, 1973).

With the crustal component reasonably well established, we can assess the eustatic components from the tide gauge values presented by PSMSL (2015) as follows:

Cuxhaven	$+2.53 \pm 0.13$	-1.4	$+1.14$
Amsterdam	$+1.5$	-0.4	$+1.1$
Ijmuide	$+1.47 \pm 0.30$	-0.4	$+1.05$
Brest	$+0.05 \pm 0.10$	~ 0.0	$+1.05$

Therefore, it seems fair to conclude that the eustatic component of the North Sea region has been on the order of 1.1 ± 0.5 mm/year for the last 100–150 years.

3.3 Venice

The long-term subsidence rate of this part of the Po delta is very well recorded over the last 300 years (Mörner, 2007a) at a mean rate of 2.3 mm/year (Fig. 12.5). The tide gauge record gives a relative sea-level rise of 2.4 mm/year for the past 140 years (PSMSL, 2015), implying a eustatic component of ±0.1 mm/year over the past 140 years, with a negative trend after 1970 (Mörner, 2007b, 2014a). Consequently, the regional eustatic component of the Mediterranean Sea seems to be on the order of ±0.0 mm/year over that last 150 years.

Fig. 12.5 (originally from Mörner, 2007a) illustrates the data set available. The early 18th- century sea-level data comes from extremely exact algae rims in a mass-production of pictures by the application of the camera obscura technique by Canoletto.

3.4 Connecticut

In salt marshes along the east coast of the United States, the foraminferal assemblages of dated sediment samples have been used to record the changes in local mean-high-water (Scott and Medioli, 1980). An interesting test of the methodology comes from work of the Dutch research group lead by Orson van de Plassche (van de Plassche, 1991, 2000; Varekamp et al., 1992; van de Plassche et al., 1998). The problems are highlighted in Fig. 6 of van de Plassche (2000), where six MHW-curves from four different marshes are compared. The differences are significant, sometimes even opposed. This implies large uncertainties with respect to quality and local differences.

In the Hammock River marsh (Connecticut), the Dutch group has undertaken an extensive analysis with respect to integrated litho-, bio-, and chrono-stratigraphy (Varekamp et al., 1992; van de Plassche et al., 1998; van de Plassche, 1991, 2000). They showed that the sea-level resolution was strongly dependent on the number and quality of radiocarbon dates. Few records are as well dated as the Hammock River marsh, and it can be used as a regional standard record (Mörner, 2010a, 2013, 2015g). I have previously stressed that "this opens for pending revisions of other records of sea level changes based only on a limited number of dates" (Mörner, 2010a), and I may now add: "and on single core analyses." The Hummock River marsh sea-level curve (van de Plassche, 2000) is shown in Fig. 12.6. It is considered to provide a fair representation of the regional changes in eustatic sea level in the last 1500 years (Mörner, 2010a, 2013, 2015g).

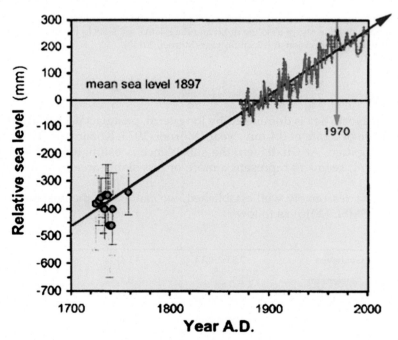

FIGURE 12.5   Venice is built of delta deposits and experiences a long-term subsidence (*blue line*) of −2.3 mm/year. Sea level (*pink line*) fluctuated around this line from 1870 to 1970, implying an absence of eustatic sea-level rise. After 1970, there is even a sea-level lowering recorded. The PSMSL (2015) mean tide gauge value is given as +2.4 mm/year, giving a eustatic component of +0.1 mm/year when calibrated for the 2.3 mm/year subsidence factor. *From Mörner, N.-A., 2007a. The Greatest Lie Ever Told, first ed. P&G-print, 20 pp., with the author's permission.*

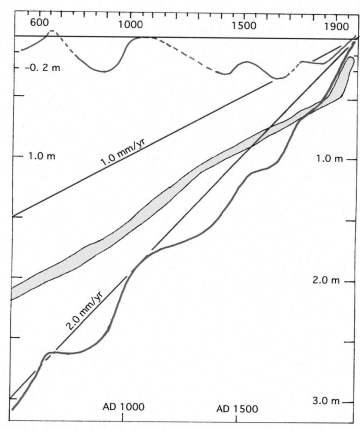

FIGURE 12.6   The sea-level curve of Hammock River (red; van de Plassche, 2000), proposed to be a regional eustatic standard curve of the last 1500 years (Mörner, 2010a, 2013, 2015d), and the same curve as deformed by a 2.0 mm/year subsidence (*purple curve*). The sea-level graph of Sandy Point (yellow; Kemp et al., 2011) is incompatible with the deformed standard curve, and must hence represent a poor and erroneous representation of actual sea level-changes.

In subsiding areas, this sea-level curve must be deformed according to the rules of Fig. 3 in Mörner (2013, from Fig. 3 of 1971). The lines of 1.0 mm/year and 2.0 mm/year subsidence are given in Fig. 12.6. Below the 2.0 mm/year subsidence line follows the Hummock River sea-level graph as deformed by the subsidence. The five sea-level oscillations can still be identified.

As the Hummock River sea-level curve is considered to be a reasonable regional eustatic standard (Mörner, 2015g), other sea-level reconstruction should follow it also in submerged situations. Hence, it can be used as some sort of test.

Kemp et al. (2011) have presented graphs established from single core records from North Carolina. In Fig. 12.6, their sea-level curve for Sandy Point has been added (yellow curve). There are very bad agreements between their curve and the deformed Hummock River curve: the upper 0.5 m agrees reasonably well with the 2.0 mm/year deformed Hummock River curve, but the lower 1.5 m has no relation at all, rather with a line of 1.5 mm/year subsidence. Therefore, we must assume that there are serious problems related to the reconstruction of sea-level changes at Sandy Point, which implies that it cannot be cited as evidence of a recent acceleration in eustatic sea-level rise. This single core record is deceptive, and simply overruled by the much better Hummock River curve.

3.5 Guyana—Surinam

An excellent sea-level record from French Guyana and Surinam shows the 18.6 tidal cycle for 2.5 full cycles (Gratiot et al., 2008; Mörner, 2010a). The cyclic behavior shows ups and downs around a stable zero level of ±0.0 mm/year (Mörner, 2010a, Fig. 3).

Again, we find a record suggesting stable sea-level conditions over the last 50 years and a value of regional eustasy of about ±0.0 mm/year. Satellite altimetry from the same area gives a rise of about 3.0 mm/year, which generated a comment (Mörner, 2014a): "there is a message in this difference, to say the least."

3.6 Fremantle, Australia

The Fremantle tide gauge at Perth gives an interesting record with variable trend segments and a probable subsidence on the order of 1.4 mm/year (Mörner and Parker, 2013, Fig. 2).

Calibrating for subsidence leaves little or no sea-level rise; virtually stable conditions over the last 60 years, and full stability over the last 15 years—in good agreement with the records from the Maldives, Goa, and Bangladesh (Fig. 12.2).

3.7 Qatar

In Qatar, there is a coastal segment with three stages of stable sea-level conditions (Mörner, 2014c, 2015d); the oldest at +0.3 m date from 5000 BP (Mörner, 2014c) and the youngest at ±0.0 m, being the presently active beach. The morphology indicates stable coastal sea-level conditions for decades to a century or more. Consequently, coastal morphology here contradicts claims of a present sea-level rise.

3.8 Summary of Test Site Records

The Northeast European coasts offer excellent means of separating the crustal component from the relative sea-level records at a number of tide gauges, and hence to define the regional eustatic component. The regional eustatic component is 0.9 for the Kattegatt, 1.1 for the North Sea, and 1.1 for Stockholm, which gives a very firmly defined mean value of 1.0 ± 0.1 mm/year for the last 125—150 years.

In Venice, Guyana—Surinam, and Qatar, the sea seems to have remained unchanged over the last century.

The Hammock River sea-level curve seems to be a reasonable curve of regional eustatic changes over the last 1500 years. It effectively rules out claims of hockey-stick-like sea-level accelerations in the 20th century (eg, Kemp et al., 2011).

4. SATELLITE ALTIMETRY BEFORE "CORRECTIONS"

The satellite altimetry records (NOAA, 2015a; UC, 2015) offer problems because of subjective adjustments (Mörner, 2010d, 2013, 2015f). The first records of 1993—2000 lacked a rising component (MEDIAS, 2000; Mörner, 2004). Later, rising trends were introduced in order to cope with loading models and GIA corrections (Mörner, 2010d, 2013).

A special analysis has been devoted to the reestablishment of the original values of the satellite altimetry records (Mörner, 2015f). When this is done, the NOAA (2015a) record changes from 2.9 ± 0.4 mm/year to a mean value of 0.45 mm/year for the period 1993—2015, and the UC (2015) from 3.3 ± 0.4 mm/year to a mean value of 0.65 mm/year for the period 1992—2015 (for graphs and calculations, see Mörner, 2015f).

Suddenly, there is a general agreement among all the different sources of information for the establishment of present regional-to-global sea-level changes (Mörner, 2015f) as illustrated in Fig. 12.7.

5. DISCUSSION

To begin with, we may ask: *Why would sea level change?* For the present-to-future period, ocean level may change for three main reasons:

1. If the ocean volume would increase due to glacial melting.
2. If the water column would expand due to heating.
3. If the water masses would move laterally (currents, rotation).

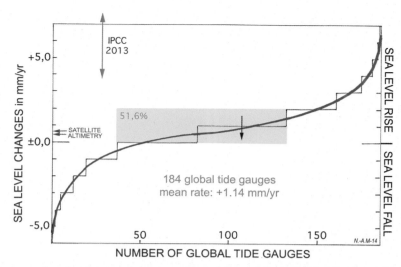

FIGURE 12.7   The new spectrum of sea-level changes after removal of erroneous "corrections" applied to the satellite altimetry records. *Yellow zone* gives the peak values of recorded tide gauge rates. *Blue arrow* indicates that several of those sites refer to subsiding sites overestimating the eustatic factor. Now the different records of sea-level changes (ie, tide gauges, coastal morphology, and satellite altimetry) give a congruent picture of a mean global sea-level rise within the zone ranging from ±0.0 to +1.0 mm/year. The IPCC estimates alone is now hanging "in the air" above all the other records. *From Mörner, N.-A., 2015f. Glacial isostasy: regional — not global. International Journal of Geosciences 6, 577—592.*

All of these factors can be related to rates and amplitudes (Mörner, 1996, 2011c).

1. During the deglacial period, with maximum climate stress and enormous masses of ice to melt, sea level rose at rates of about 10 mm/year (1 m/century). This value can be held as an ultimate frame for any possible ice-melting effect (Mörner, 2011c). Any realistic value for a present-to-future sea-level rise must be well below this value.
2. The thermal expansion in the open oceans can hardly exceed 10 cm (Mörner, 1996, 2011c, 2013). Towards the coasts, the effect rapidly decreases with depth and becomes zero at the shore (because there is no water to expand and there will be no flush inlands).
3. The lateral water mass movements seem to have dominated the local-to-regional sea-level changes over the last 6000 years or so (Mörner, 1995, 1996), and are well expressed in satellite altimetry maps (eg, NOAA, 2015b).

The group of sea-level experts within the INQUA commission on *Sea Level Changes and Coastal Evolution*, therefore, estimated the probable sea-level change by year 2100 at +10 cm ± 10 cm (INQUA, 2000), later updated to +5 cm ± 15 cm (Mörner, 2004).

Fig. 12.8 gives the observed changes in sea level over the last 300 years and the predictions and estimates for the present century up to 2100. There are oscillations up and down in the order of ±1.0 mm/year around the present sea-level value. The IPCC predictions for year 2100 lie far above the observed variations. The INQUA (2000) and Mörner (2004) values, on the other hand, lie close to the observed values over the past 300 years.

As illustrated in Fig. 12.7, we now have a quite congruent picture of real values of regional eustasy: ranging between ±0.0 and 1.0 mm/year. In Table 12.1, the actual values as given in the text are listed from the highest (which include a component of subsidence and hence must be considered too high) to the lowest (referring to key sites in the Indian Ocean, the Pacific, and areas like Venice, Guyana—Surinam, and Qatar).

5.1 Setting Frames and Likelihoods for Sea Level at 2100

The year 2100 has become some sort of magic boundary for IPCC's predictions of climate and sea level. The sea-level predictions are illustrated in Fig. 12.8. All the data here presented are in good agreement with the prediction of Mörner (2004) of +5 cm ± 15 cm by 2100.

One may also assess the sea-level changes by 2100 by applying frame values (cf. Mörner, 2011c) of the impossible, not likely, possible, and probable maximum and minimum values of sea-level changes, as done in Fig. 12.9.

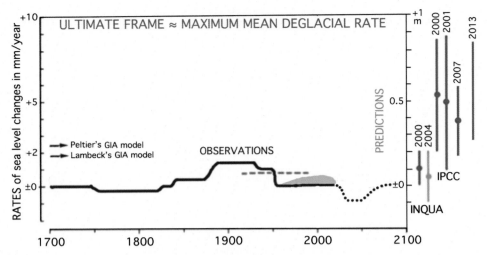

FIGURE 12.8   Combined data of observed sea-level changes (in mm/year) for the last 300 years, and predictions for year 2100. *Modified from Mörner, N.-A., 2004. Estimating future sea level changes from past records. Global and Planetary Change 40, 49—54.*

TABLE 12.1   Summary of Regional Eustatic Values

Rates observed (mm/year)	Region, site, or database	References
+1.14	Mean of 184 NOAA stations	Mörner (2015f)
+1.0 ± 0.1	Mean of NE European test sites	Mörner (2014a,b)
+0.65	NOOA satellite altimetry revised	Mörner (2015f)
+0.45	UC satellite altimetry revised	Mörner (2015f)
+0.25 ± 0.19	Mean of 170 > 60 years PSMSL sites	Parker and Ollier (2015)
~±0.0	Test sites of Venice, Guyana—Surinam, Qatar	Mörner (2014a)
~±0.0	Tuvalu, Vanuatu, Kiribati key sites	Mörner (2007a, 2014a)
~±0.0	The Maldives, Goa, Bangladesh	Mörner (2007a, 2013)

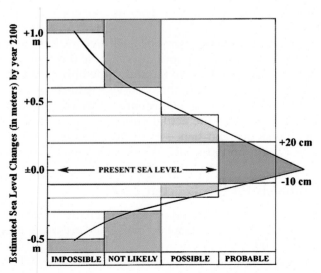

FIGURE 12.9   Frames and likelihoods of sea-level changes up to the year 2100 with the most probable values ranging between −10 cm and +20 cm. *From Mörner, N.-A., 2014a. Sea level changes in the 19—20th and 21st centuries. Coordinates X (10), 15—21.*

6. CONCLUSIONS

Figs. 12.7 and 12.8, and Table 12.1, summarize the situation with respect to "sea-level changes as observed in nature" and the establishment of meaningful eustatic sea-level values. This value is now firmly set at ±0.0 to +1.0 mm/year sea-level rise. The frames and likelihood diagram of Fig. 12.9 gives a sea-level rise by 2100 ranging between +20 cm and −10 cm (ie, +5 cm ± 15 cm, just as proposed by Mörner, 2004). In this situation, the conclusions are obvious and straightforward:

1. Regional and global sea levels are not in a rapidly rising mode. This is obvious and clear if we analyze available observational facts in test areas as well as key sites (or old "poster sites" for proponents of the IPCC scenarios).
2. The regional-to-global component of eustatic changes in sea-level varies between +1.0 mm/year and ±0.0 mm/year. In key sites of the Indian Ocean and the Pacific, eustastic sea level has remained stable for the last 40−50 years (or more).
3. The best estimates of possible sea-level changes by 2100 is somewhere between +20 cm and −10 cm or +5 cm ± 15 cm.
4. This implies that there is no longer any realistic sea-level threat, and no true danger of disastrous coastal flooding.
5. Horror scenarios with a sea-level rise of 1 m or more by 2100 violate physical laws, observational facts, and geoethical principles, and must now be discarded.

References

Bean, R., 2011. The Heretic. Oberon Modern Plays, 115 pp.
Bryant, N., July 28, 2004. Maldives: Paradise Soon to Be Lost. BBC News. http://news.bbc.co.uk/2/hi/south_asia/3930765.stm.
Cazenave, A., Dominh, K., Guinehut, S., Berthier, E., Llovel, W., Rammien, G., Ablain, M., Larnicol, G., 2009. Sea level budget over 2003−2008: a reevaluation from GRACE space gravimetry, satellite altimetry and Argo. Global and Planetary Change 65, 83−88.
Chauvin, R., Hilaire, E., 2015. Climate Changes in the Marshall Islands and Kiribati, Before and After − Interactive. http://www.theguardian.com/environment/ng-interactive/2015/mar/11/climate-change-in-the-marshall-islands-and-kiribati-before-and-after-interactive.
Church, J.A., White, N.J., Hunter, J.R., 2007. Sea-level rise at tropical Pacific and Indian Ocean islands. Global and Planetary Change 53, 155−168.
Crichton, M., 2004. State of Fear. HarperCollins Publishers, New York.
Fairbridge, R.W., Krebs, O.A., 1962. Sea level and the southern oscillation. Geophysical Journal 6 (4), 532−545.
Fairbridge, R.W., 1961. Eustatic changes in sea level. Physics and Chemistry of the Earth 4, 99−185.
Gratiot, N., Anthony, E.J., Gardel, A., Gaucherel, C., Proisy, C., Wells, J.T., 2008. Significant contribution of the 18.6 year tidal cycle to regional coastal changes. Nature Geoscience 1, 169−172. http://dx.doi.org/10.1038/ngeo127.
Gray, V.R., 2010. South Pacific Sea Level: A Reassessment, pp. 1−24. SPPI Original Paper.
Grinsted, A., Jerejeva, S., Riva, R.E.M., Dahl-Jensen, D., 2015. Sea level rise projections for northern Europe under RCP8.5. Climate Research 64, 15−23.
Gunyon, B., April 1, 2011. Pachauri Opens Door to New Climate Adaptation Science. Yahoo News. http://news.yahoo.com/pachauri-opens-door-climate-adaptation-science-20110401-091700-868.html.
Gutenberg, B., 1941. Changes in sea level, postglacial uplift and mobility of the earth's interior. Geological Society of America Bulletin 52, 721−772.
Hanebuth, T.J.J., Kudrass, H.R., Linstädter, J., Islam, B., Zander, A.M., 2013. Rapid coastal subsidence in the central Ganges-Brahmaputra delta (Bangladesh) since the 17th century deduced from submerged salt-producing kilns. Geology 41, 987−990. http://dx.doi.org/10.1130/G34646.1.
Hansen, J.M., Aagaard, T., Kuijpers, A., 2015. Sea-level forcing by synchronization of 56- and 74-year oscillations with Moon's nodal tide on the northwest European shelf (Eastern North sea to central Baltic sea). Journal of Coastal Research 31, 1041−1056.
Hansen, J.M., 2015. Sea-level effects of NOA and AMO: synchronization and amplitude locking by the Lunar Nodal Oscillation in the North Sea and Baltic embayment. In: Mörner, N.-A. (Ed.), Planetary Influence on the Sun and the Earth, and a Modern Book-burning. Nova Sci. Publ., pp. 51−70 (Chapter 5).
Hoffmann, J., 2007. The Maldives and Rising Sea Levels. ICE Case Studies, p. 206. http://www1.american.edu/ted/ice/maldives.htm.
IGCP, 2001. Third Assessment Report. The Intergovernmental Panel of Climate Change.
IGCP, 2007. Fourth Assessment Report. The Intergovernmental Panel of Climate Change.
IGCP, 2013. Fifth Assessment Report. The Intergovernmental Panel of Climate Change.
INQUA, 2000. The Commission on "Sea Level Changes and Coastal Evolution". www.pog.se/sea. www.pog.nu (2005).
Jelgerma, S., April 1966. Sea-level changes during the last 10,000 years. In: Proc. Int. Symp. World Climate from 8,000 to 0 RC., R. Meteorol. Soc., London. Imp. Coll., London, pp. 54−71.
Kemp, A.C., Horton, B.P., Donnelly, J.P., Mann, M.E., Vermeer, M., Rahmstorf, S., 2011. Climate related sea-level variations over the past two millennia. Proceedings of the National Academy of Sciences 108 (27), 11017−11022.
Kooijmans, L.P.L., 1974. The Reine/Meuse Delta: Four Studies on its Prehistoric Occupation and Holocene Geology. E.J. Brill, Leiden, pp. 1−421.
Lin, C.-C., Ho, C.-R., Cheng, Y.-H., 2014. Interpreting and analysing king tide in Tuvalu. Natural Hazards and Earth System Science 14, 209−217.
Little, C.M., Horton, R.B., Kopp, R.E., Oppenheimer, M., Vecchi, G.A., Villarini, G., 2015. Joint projections of US East Coast sea level and storm surge. Nature Climate Change 5, 1114−1120. http://dx.doi.org/10.1038/NCLIMATE2801 (Advance Online Publication).

Lorbacher, K., Nauels, A., Meinshausen, M., 2015. Complementing thermosteric sea level rise estimates. Geoscientific Model Development 8, 2723–2734. http://dx.doi.org/10.5194/gmd-8-2723-2015.

MEDIAS, 2000. Satellite-based altimetry reveals physical ocean. Medias Newsletter 12, 9–17.

Mörner, N.-A., Harris, T., September 14, 2013. U.S. Being Snookered into Another Kyoto? PJ Media. http://pjmedia.com/blog/u-s-being-snookered-into-another-kyoto/?singlepage=true.

Mörner, N.-A., Parker, A., 2013. Present-to-future sea level changes: the Australian case. Environmental Science: An Indian Journal 8 (2), 43–51.

Mörner, N.-A., Tooley, M., Possnert, G., 2004. New perspectives for the future of the Maldives. Global and Planetary Change 40, 177–182.

Mörner, N.-A., 1969. The Late Quaternary history of the Kattegat Sea and the Swedish west coast; deglaciation, shorelevel displacement, chronology, isostasy and eustasy. Sveriges Geologiska Undersoekning C-640 1–487.

Mörner, N.-A., 1971. Eustatic changes during the last 20,000 years and a method of separating the isostatic and eustatic factors in an uplifted area. Palaeogeography, Palaeoclimatology, Palaeoecology 9, 153–181.

Mörner, N.-A., 1973. Eustatic changes during the last 300 years. Palaeogeography, Palaeoclimatology, Palaeoecology 13, 1–14.

Mörner, N.-A., 1980. The northwest European "sea-level laboratory" and regional Holocene eustasy. Palaeogeography, Palaeoclimatology, Palaeoecology 29, 281–300.

Mörner, N.-A., 1995. Earth rotation, ocean circulation and paleoclimate. GeoJournal 37 (4), 419–430.

Mörner, N.-A., 1996. Sea level variability. Zeitschrift für Geomorphologie N.F., Supplement-Bd 102, 223–232.

Mörner, N.-A., 2004. Estimating future sea level changes from past records. Global and Planetary Change 40, 49–54.

Mörner, N.-A., 2007a. The Greatest Lie Ever Told, first ed. P&G-print. 20 pp.

Mörner, N.-A., 2007b. The Sun rules the climate. There's no danger of global sea level rise. In: 21st Century Science & Technology, Fall 2009, pp. 30–34.

Mörner, N.-A., 2007c. Sea level changes and tsunamis, environmental stress and migration overseas. The case of the Maldives and Sri Lanka. Internationales Asienforum 38, 353–374.

Mörner, N.-A., 2009. Open letter to the President of the Maldives. New Concepts in Global Tectonics Newsletter, No. 53 80–83.

Mörner, N.-A., 2010a. Some problems in the reconstruction of mean sea level and its changes with time. Quaternary International 221, 3–8. http://dx.doi.org/10.1016/j.quaint.2009.10.044.

Mörner, N.-A., 2010b. The Greatest Lie Ever Told, third ed. P&G-print. 20 pp.

Mörner, N.-A., 2010c. Sea level changes in Bangladesh. New observational facts. Energy & Environment 21 (3), 235–249.

Mörner, N.-A., 2010d. There is no alarming sea level rise!. In: 21st Century Science & Technology, Winter 2010/2011, pp. 12–22.

Mörner, N.-A., 2011a. A Sad and Twisted Story or Stealing the Limelight from Real Problems in the Real World. http://icecap.us/images/uploads/A_sad_and_twisted_story.doc.pdf.

Mörner, N.-A., 2011b. The Maldives: a measure of sea level changes and sea level ethics. In: Easterbrook, D.J. (Ed.), Evidence-Based Climate Science. Elsevier, pp. 197–209 (Chapter 7).

Mörner, N.-A., 2011c. Setting the frames of expected future sea level changes by exploring past geological sea level records. In: Easterbrook, D.J. (Ed.), Evidence-Based Climate Science. Elsevier, pp. 185–196 (Chapter 6).

Mörner, N.-A., 2013. Sea level changes: past records and future expectations. Energy & Environment 24 (3–4), 509–536.

Mörner, N.-A., 2014a. Sea level changes in the 19-20th and 21st centuries. Coordinates X (10), 15–21.

Mörner, N.-A., 2014b. Deriving the eustatic sea level component in the Kattegatt Sea. Global Perspectives on Geography 2, 16–21.

Mörner, N.-A., 2014c. The flooding of Ur in Mesopotamia in new perspectives. Archaeological Discovery 3, 26–31.

Mörner, N.-A., 2015a. Geoethics: the principle of ethics in natural sciences. In: Proceedings of the Conference on Geoethics, Prague 2015. http://dx.doi.org/10.13140/RG.2.1.1953.4168.

Mörner, N.-A., 2015b. Climate fundamentalism. In: Mörner, N.-A. (Ed.), Planetary Influence on the Sun and the Earth, and a Modern Book-Burning. Nova Sci. Publ., pp. 167–174 (Chapter 15).

Mörner, N.-A., 2015c. Multiple planetary influences on the Earth. In: Mörner, N.-A. (Ed.), Planetary Influence on the Sun and the Earth, and a Modern Book-Burning. Nova Sci. Publ., pp. 39–50 (Chapter 4).

Mörner, N.-A., 2015d. Coastal erosion and coastal stability. In: Barens, D. (Ed.), Coastal and Beach Erosion: Processes, Adaptation Strategies and Environmental Impacts. Nova Sci Publ., pp. 69–82 (Chapter 4).

Mörner, N.-A., 2015e. The AD 1733 tsunami in the Indian Ocean. In: 4th ITFS Symposium, Thailand, March 2015. Programme and Abstracts, p. 41.

Mörner, N.-A., 2015f. Glacial isostasy: regional — not global. International Journal of Geosciences 6, 577–592.

Mörner, N.-A., 2015g. Sea level changes as recorded in nature itself. International Journal of Engineering Research and Applications 1 (5), 124–129.

Mortensen, L., 2004. Doomsday Called Off. TV-documentary. See Also: "Maldives Will Avoid Extinction". Danish TV, Copenhagen. http://climateclips.com/archives/117.

Murphy, G., 2007. Claim that sea level is rising is a total fraud (interview). In: 21st Century Science & Technology, Fall 2007, pp. 33–37.

NOAA, 2012. Tides & Currents, Sea Level Trends. http://tidesandcurrents.noaa.gov/sltrends/sltrends.shtml.

NOAA, 2015a. Laboratory for Satellite Altimetry/sea Level Rise: Sea Level Trend Map. http://www.star.nesdis.noaa.gov/sod/lsa/SeaLevelRise/LSA_SLR_maps.php.

NOAA, 2015b. Laboratory for Satellite Altimetry/Sea Level Rise: Sea Level Data and Plots. http://www.star.nesdis.noaa.gov/sod/lsa/SeaLevelRise/LSA_SLR_timeseries.php.

Parker, A., Ollier, C.D., 2015. Sea level rise for India since the start of tide gauge records. Arabian Journal of Geosciences 8 (9), 6483–6495.

Parker, A., 2016. Sea level rises: arguing the nanometre to defocus from the missed meter. Environmental Science: An Indian Journal 12 (1), 22–29.

van de Plassche, O., van der Borg, K., de Jong, A.F.M., 1998. Sea level–climate correlation during the past 1400 year. Geology 26, 319–322.

van de Plassche, O., 1991. Late Holocene sea-level fluctuations on the shore of Connecticut inferred from transgressive and regressive overlap boundaries in salt march deposits. Journal of Coastal Research 11, 159–180.

van de Plassche, O., 2000. North Atlantic climate-ocean variations and sea level in Long Island Sound, Connecticut, since 500 cal year A.D. Quaternary Research 53, 89–97.

PSMSL, 2015. Table of Relative Mean Sea Level Secular Trends. http://www.psmsl.org/products/trends/trends.txt.

Rahmstorf, S., 2007. A semi-empirical approach to projecting future sea level rise. Science 315, 368–370.

around the world. It is not objective science but rather a sensationalist prediction that has little basis in fact or logic. Required here is an appeal for critical thinking among the populace in order to distinguish between the factual and the predictive and between sober language and apocalyptic predictions.

2. THE HISTORICAL RECORD OF CO_2 AND TEMPERATURE IN THE ATMOSPHERE

All the CO_2 in the atmosphere came from inside the Earth. During the early life of the planet, the Earth was much hotter, and there was much more volcanic activity than there is today. The heat of the core caused carbon and oxygen to combine to form CO_2, which became a significant part of the Earth's early atmosphere, perhaps the second most abundant component after nitrogen, until photosynthesis evolved. Most of the CO_2 in the oceans comes from the atmosphere, although some is injected directly from ocean vents.

It is widely accepted that the concentration of CO_2 was higher in the Earth's atmosphere before modern-day life forms evolved during the Cambrian Period, which began 544 million years ago. It was also at that time that a number of marine species evolved the ability to control calcification, an example of the more-general term "biomineralization" (Weiner and Dove, 2003). This allowed these species to build hard shells of calcium carbonate ($CaCO_3$) around their soft bodies, thus providing a type of armor plating. Early shellfish such as clams arose more than 500 million-years ago, when atmospheric CO_2 was 10—15 times higher than it is today (Virtual Fossil Museum, 2015). Clearly, the pH of the oceans did not cause the extinction of corals or shellfish or they would not be here today. Why, then, are we told that even at today's much lower level, CO_2 is already causing damage to calcifying species?

The most common argument is along the lines of "today's species of corals and shellfish are not adapted to the level of CO_2 that ancient species were familiar with. Acidification is happening so quickly that species will not be able to adapt to higher levels of CO_2." This is nonsensical in that from a biochemical perspective there is no reason to believe these species have lost their ability to calcify at the higher CO_2 levels that existed for millions of years in the past. The ancestors of every species alive today survived through millennia during which conditions sometimes changed very rapidly, such as when an asteroid caused the extinction of dinosaurs and many other species 65 million years ago. While many more species became extinct than are alive today, it must be said that those species that came through these times have proven the most resilient through time and change.

As far as is known, there was only one other period in the Earth's history when CO_2 was nearly as low as it has been during the past 2.5 million years of the Pleistocene Ice Age. During the late Carboniferous Period and into the Permian and Triassic Periods, CO_2 was drawn down from about 4000 ppm to about 400 ppm, probably owing to the advent of vast areas of forest that pulled CO_2 out of the atmosphere and incorporated it into wood and thus into coal (see Fig. 13.1). We know from Antarctic ice cores that CO_2 was drawn down to as low as 180 ppm during the Pleistocene, only 30 ppm above the threshold for the survival of plants, at the peak of glacial advances (see Fig. 13.2). These periods of low atmospheric CO_2, as is the case at present, are the exception to the much longer periods when CO_2 was more than 1000 ppm, and often much higher.

For this reason alone, the possibility that present and future atmospheric CO_2 levels will cause significant harm to calcifying marine life should be questioned. However, a number of other factors bring the ocean acidification hypothesis into question.

3. THE ADAPTATION OF SPECIES TO CHANGING ENVIRONMENTAL CONDITIONS

People have a tendency to assume that it takes thousands or millions of years for species to adapt to changes in the environment. This is not the case. Even species with relatively long breeding periods can adapt relatively quickly when challenged by rapidly changing environmental conditions (Boeye et al., 2013). In fact, it is rapidly changing environmental conditions that foster rapid evolutionary change and adaptation. Stephen Jay Gould explains this well in his classic *Wonderful Life*, which focuses on the Cambrian Explosion and the evolution of vast numbers of species beginning 544 million years ago (Gould, 1989).

Most of the invertebrates that have developed the ability to produce calcium carbonate armor are capable of relatively rapid adaptation to changes in their environment due to two distinct factors. First, they reproduce at least annually and sometimes more frequently. This means their progeny are tested on an annual basis for suitability to a changing environment. Second, these species produce thousands to millions of offspring every time they

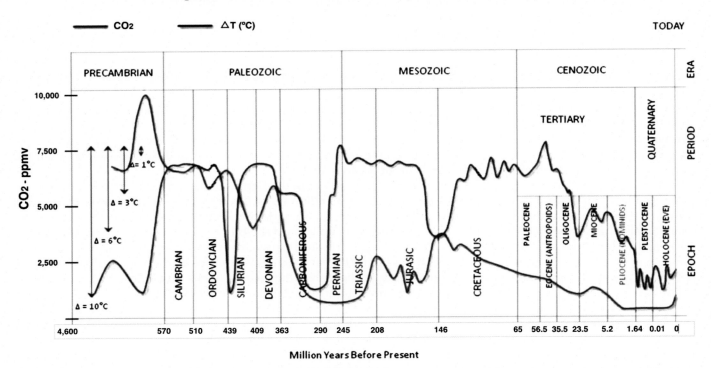

FIGURE 13.1   Reconstruction of CO_2 and temperature over the Earth's history. Note that these two parameters are not highly correlated over the past 600 million years. This does not indicate a lockstep cause-and-effect relationship. (Interpretation by Nahle, 2009; Scotese, 2002; Ruddiman, 2001; Pagani et al., 2005).

reproduce. This greatly increases the chance that genetic mutations that are better suited to the changes in environmental conditions will occur in some offspring.

A number of studies have demonstrated that change in an organism's genetic make-up, or genotype, is not the only factor that allows species to adapt to changing environmental conditions. Many marine species inhabit coastal waters for some or all of their lives, where they are exposed to much wider ranges of pH, CO_2, O_2, temperature, and salinity than occur in the open ocean. Two distinct physiological mechanisms exist whereby adaptation to environmental change can occur much more rapidly than by changes in the genotype through genetic evolution.

The first of these is phenotypic plasticity, which is the ability of one genotype to produce more than one phenotype when exposed to different environments (Price et al., 2003). In other words, a specific genotype can express itself differently due to an ability to respond in different ways to variations in environmental factors. This helps to explain how individuals of the same species with nearly identical genotypes can successfully inhabit very different

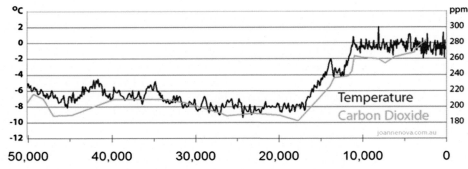

FIGURE 13.2   Cryostratigraphic reconstructions of air temperature and CO_2 concentration anomalies from Vostok station, Antarctica, 50,000 to 2500 years BP. CO_2 concentration fell to a little above 180 ppmv 18 ka BP (JoNova, 2013).

environments. Examples of this in humans are the ability to acclimatize to different temperature regimes and different altitudes. There is no change in the genotype, but there are changes in physiology.

The second and more fascinating factor is transgenerational plasticity, which is the ability of parents to pass their adaptations to their offspring (Jablonka and Raz, 2009). One recent study pointed out that "contemporary coastal organisms already experience a wide range of pH and CO_2 conditions, most of which are not predicted to occur in the open ocean for hundreds of years—if ever" (Murray et al., 2014). The authors used what they called "a novel experimental approach that combined bi-weekly sampling of a wild, spawning fish population (Atlantic silverside *Menidia menidia*) with standardized offspring CO_2 exposure experiments and parallel pH monitoring of a coastal ecosystem." The parents and offspring were exposed to CO_2 levels of 1200 ppm and 2300 ppm compared with today's ambient level of 400 ppm. The scientists report that "early in the season (April), high CO_2 levels significantly…reduced fish survival by 54% (2012) and 33% (2013) and reduced 1—10 day post-hatch growth by 17% relative to ambient conditions." However, they found that "offspring from parents collected later in the season became increasingly CO_2-tolerant until, by mid-May, offspring survival was equally high at all CO_2 levels." This indicates that a coastal species of fish is capable of adapting to high levels of CO_2 in a very short time. It also indicates that this same species would not even notice the relatively slow rate at which CO_2 is increasing in the atmosphere today.

The changes that have occurred to the Earth's climate over the past 300 years since the peak of the Little Ice Age around 1700 are in no way unusual or unique in history. During the past 3000 years, a blink in geological time, there has been a succession of warm periods and cool periods. There is no record of species extinction due to climatic change during these periods.

4. THE BUFFERING CAPACITY OF SEAWATER

Over the millennia, the oceans have received minerals dissolved in rainwater from the land. Most of these are in the form of ions such as chloride, sodium, sulfate, magnesium, potassium, and calcium. Underwater hydrothermal vents and submarine volcanoes also contribute to the salt content. These elements have come to make up about 3.5% of seawater by mass, thus giving seawater some unique properties compared with fresh water. It is widely believed by oceanographers that the salt content of the sea has been constant for hundreds of millions, even billions, of years, as mineralization on the sea floor balances new salts entering the sea (Holland et al., 2006; Ocean Health).

The salt content of seawater provides it with a powerful buffering capacity, the ability to resist change in pH when an acidic or basic compound is added to the water. For example, 1 μmol of hydrochloric acid added to 1 kg of distilled water at pH 7.0 (neutral) causes the pH to drop to nearly 6.0. If the same amount of hydrochloric acid is added to seawater at pH 7, the resulting pH is 6.997, a change of only 0.003 of a pH unit. Thus, seawater has approximately 330 times the buffering capacity of freshwater. In addition to the buffering capacity, there is another factor, the Revelle factor, named after Roger Revelle, former director of the Scripps Institute of Oceanography. The Revelle factor determines that if atmospheric CO_2 is doubled, the dissolved CO_2 in the ocean will only rise by 10% (Zeebe and Wolf-Gladrow, 2008).

It is widely stated in the literature that the pH of the oceans was 8.2 before industrialization (1750) and that owing to human CO_2 emissions it has since dropped to 8.1 (Zeebe, 2012; National Oceanic and Atmospheric Administration, 2014). No one measured the pH of ocean water in 1750. The concept of pH was not conceived of until 1909, and an accurate pH meter was not available until 1924. The assertion that more than 250 years ago the ocean pH was 8.2 is an estimate rather than an actual measurement. Measuring pH accurately in the field to 0.1 of a pH unit is not a simple procedure even today. In addition, for two reasons, there is no global-scale monitoring of the pH of the oceans: First, genuine oceanographers know the overwhelming buffering capacity of seawater, so they do not expect the global acid—base balance to change, and second, there is no automated instrument available for measuring pH.

The predictions of change in ocean pH owing to CO_2 in the future are based on the same assumptions that resulted in the estimate of pH 8.2 in 1750 when we have no measurement of the pH of the oceans at that time. By simply extrapolating from the claim, generated by a computer model, that pH has dropped from 8.2 to 8.1 during the past 265 years, the models calculate that pH will drop by 0.3 of a pH by 2100.

Many scientists have simply repeated the claim that the ocean's pH has dropped by 0.1 during the past 265 years, as if it is an established fact. They should be challenged to provide observational data from 1750 that supports their inference. Observations from three eminent oceanographers, including Harald Sverdrup, former director of the Scripps Institute of Oceanography, in a book that covers all aspects of ocean physics, chemistry, and biology, bring into question these scientists' assertion. The book was written before the subject of climate change and carbon dioxide became politicized.

Sea water is a very favorable medium for the development of photosynthetic organisms. It not only contains an abundant supply of CO_2, but removal or addition of considerable amounts results in no marked changes of the partial pressure of CO_2 and the pH of the solution, both of which are properties of importance in the biological environment.

If a small quantity of a strong acid or base is added to pure water, there are tremendous changes in the numbers of H^+ and OH ions present, but the changes are small if the acid or base is added to a solution containing a weak acid and its salts or a weak base and its salts. This repression of the change in pH is known as buffer action, and such solutions are called buffer solutions. Seawater contains carbonic and boric acids and their salts and is, therefore, a buffer solution. Let us consider only the carbonate system. Carbonate and bicarbonate salts of strong bases, such as occur in seawater, tend to hydrolyze, and there are always both H^+ and OH ions in the solution. If an acid is added, carbonate is converted to bicarbonate and the bicarbonate to carbonic acid, but as the latter is a weak acid (only slightly dissociated), relatively few additional hydrogen ions are set free. Similarly, if a strong base is added, the amount of carbonate increases, but the OH ions formed in the hydrolysis of the carbonate increase only slightly. The buffering effect is greatest when the hydrogen ion concentration is equal to the dissociation constant of the weak acid or based that is, when the concentration of the acid is equal to that of its salt (Sverdrup et al., 1942).

In addition, a study has been published in which the pH of the oceans was reconstructed from 1908 to 1988, based on the boron isotopic composition of a long-lived massive coral from Flinders Reef in the western Coral Sea of the southwestern Pacific (Pelejero et al., 2005). The report concluded that there was no notable trend toward lower isotopic values over the 300-year period investigated. This indicates that there has been no change in ocean pH over that period at this site. This study, in which actual measurements of a reliable proxy were made, is much more credible than an estimate based on assumptions in computer models that have not been verified.

In many ways, the assertions made about the degree of pH change caused by a given level of atmospheric CO_2 are analogous to the claims made about the degree of atmospheric temperature rise that might be caused by a given level of atmospheric CO_2. This is termed "sensitivity," and the literature becomes very confusing when the subject is researched. Perhaps the assumptions used to estimate future ocean pH are as questionable as those used to estimate the increase in temperature from increases in atmospheric CO_2 (see Fig. 13.3).

The most serious problem with the assertion that pH has dropped from 8.2 to 8.1 since 1750 is that there is no consistent universal pH in the world's oceans. The pH of the oceans varies far more than 0.1 on a daily, monthly, annual, and geographic basis. In the offshore oceans, pH typically varies geographically from 7.5 to 8.4, or 0.9 of a pH unit. A study in offshore California shows that pH can vary by 1.43 of a pH unit on a monthly basis (Hofmann et al., 2011). This is nearly five times the change in pH that computer models forecast during the next 85 years to 2100. In coastal areas that are influenced by runoff from the land, pH can be as low as 6.0 and as high as 9.0.

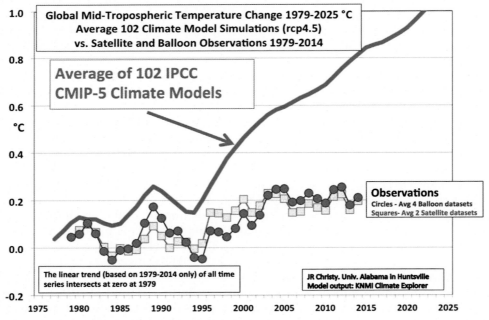

FIGURE 13.3   Average of 102 IPCC CMIP-5 climate model predictions compared with realworld observations from four balloon and two satellite data sets (U.S. House Committee on Natural Resources, 2015).

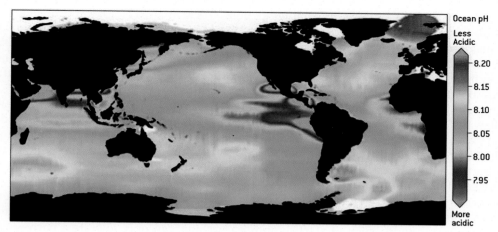

FIGURE 13.4   World map depicting the pH of the oceans, including the large area of lower pH seawater off the west coast of South America. To be correct, the scale of ocean pH on the right should read "More Basic" at he top and "Less Basic") at the bottom (Doney, 2006).

The Humboldt Current, a large area of ocean upwelling off the coasts of Chile and Peru, has among the lowest pH values found naturally in the oceans (see Fig. 13.4). The pH of this seawater is 7.7–7.8 (Egger, 2011). If the ocean average pH is now 8.1, the water in the Humboldt Current is already at a lower pH than is predicted by 2100. Upwelling waters tend to be lower in pH than other areas of the ocean for two reasons. First, the water has been at a depth where the remains of sea creatures fall down and decompose into nutrients, tending to drive pH down. Second, the water that is upwelling to form the Humboldt Current is water that downwelled (sank) around Antarctica, and being cold, it had a high solubility for CO_2 at the ocean—atmosphere interface. Ocean water that sinks at the poles eventually comes to the surface where it is warmed, thus outgassing some of the CO_2 that was absorbed in the Antarctic and the Arctic (Blanco et al., 2002).

Despite its low pH, the upwelling waters of the Humboldt Current produce 20% of the world's wild fish catch, which consists largely of anchovies, sardines, and mackerel. The basis for the food chain includes large blooms of coccolithophores, a calcifying phytoplankton that produces symmetrical calcium carbonate plates to protect itself from predators. The White Cliffs of Dover are composed of the shells of coccolithophores. To quote from one of the more thoroughly researched papers on the subject, "These biome-specific pH signatures disclose current levels of exposure to both high and low dissolved CO_2, often demonstrating that resident organisms are already experiencing pH regimes that are not predicted until 2100." The authors remark, "The effect of Ocean Acidification (OA) on marine biota is quasi-predictable at best" (Hofmann et al., 2011). It is refreshing to read an opinion that is not so certain about predicting the future of an ecosystem as complex as the world's oceans.

Scientists working at oceanographic institutes in the United Kingdom and Germany published a paper in 2015 that explored the possibility that the asteroid that struck the Earth 65 million years ago caused ocean acidification.

Along with the extinction of terrestrial and marine dinosaurs, 100% of ammonites and 90% of coccolithophores, both calcifying species, became extinct. The study considered the possibility that 6500 Gt (billion tons) of carbon as CO_2 were produced by the vaporization of carbonaceous rock and wildfires because of the impact. The authors concluded, "Our results suggest that acidification was most probably not the cause of the extinctions" (Tyrrell, 2015).

Six-thousand five-hundred Gt is the equivalent of 650 years of CO_2 emissions at the current global rate of about 10 Gt carbon as CO_2 per year. Given that atmospheric CO_2 concentration was about 1000 ppm at the time of the impact, the addition of 6500 Gt of carbon as CO_2 would have raised the concentration to approximately 4170, which is about 10 times higher than in 2015 and about five times higher than it may be in 2100.

5. THE ABILITY OF CALCIFYING SPECIES TO CONTROL THE BIOCHEMISTRY AT THE SITE OF CALCIFICATION

All organisms are able to control the chemistry of their internal organs and biochemical processes. The term "homeostasis" means that an organism can maintain a desirable state of chemistry, temperature, etc. within itself under a range of external conditions (Wood, 2011). This is especially necessary in a marine environment, because the salinity of the ocean is not compatible with the metabolic processes that take place in an organism. The general term for an important part of homeostasis is "osmoregulation." There are two biological strategies for accomplishing it. The osmoregulators, which include most fishes, maintain their internal salinity at a different level from their environment. This requires energy to counteract the natural osmotic pressure that tends to equalize an organism's internal salinity with the salinity of the water it inhabits. The osmoconformers, which include most invertebrate species, maintain their salt content at the same osmotic pressure as their environment, but they alter the make-up of the salts inside themselves compared with their surroundings (Saladin, 2016).

The osmoregulators are best illustrated by the examples of freshwater fish, saltwater fish, and fish that are able to live in both fresh water and salt water. Freshwater fish must be able to retain salts in their bodies, and so are therefore able to repel and expel fresh water and to recover salts from their kidneys before excretion. Saltwater fish are able to retain water while excreting salts through their gills. Fish such as salmon and eels, which spend part of their lives in freshwater and part in saltwater, are able to transform their bodily functions as they move from one environment to the other (American Museum of Natural History).

The osmoconformers save energy by maintaining a salt concentration that is the same as their environment, but like the osmoregulators, they change the make-up of the salt mixture to allow critical biological functions to occur internally. Some osmoconformers, such as starfish and sea urchins, can only tolerate a narrow range of external salinity, while others, such as mussels and clams, can isolate themselves from the environment by closing their shells and can tolerate a wide range of external salinity (McClary, 2014).

Osmoregulation is a good example of how species are able to adapt to environments that would otherwise be hostile to life. Controlled calcification is another biological function that depends on species' ability to alter and control their internal chemistry.

The ocean acidification narrative is based almost entirely on the chemistry of seawater and the chemistry of calcium and carbon dioxide. It is true that the shell of a dead organism will gradually dissolve in water with a lowered pH (National Oceanic and Atmospheric Administration); however, it cannot be inferred directly from this that the shell, or carapace, or coral structure of a species will dissolve under similar pH while the organism is alive. Even if some dissolution is occurring, as long as the organism builds calcium carbonate faster than it dissolves, the shell will grow. If this were not the case, it would be impossible for the duck mussel, *Anodonta anatina*, to survive in a laboratory experiment at pH 3.0 for 10 days without significant shell loss (Mäkelä and Oikari, 1992). This is an extreme example, as it is outside natural conditions. It is, however, well established that calcification growth in freshwater species of mussels and clams occurs at pH 6.0, well into the range of genuine acidity. The Louisiana pearlshell, *Margaritifera hembeli*, is actually restricted to waters with a pH of 6.0−6.9. In other words, it requires acidic water to survive (Haag, 2012). This does not mean that all marine species that calcify will tolerate pH 6.0, only that there are organisms that can calcify at much lower pH than is found in ocean waters today or that are projected even under extreme scenarios.

The coccolithophores account for about 50% of all calcium carbonate production in the open oceans. A laboratory study found that "the coccolithophore species *Emiliania huxleyi* are significantly increased by high CO_2 partial pressures" and that "over the past 220 years there has been a 40% increase in average coccolith mass" and that "in a scenario where the PCO_2 in the world's oceans increases to 750 ppmv, coccolithophores will double their rate of calcification and photosynthesis" (Iglesias-Rodriguez et al., 2008).

This is good news for the ocean's primary production and fisheries production up the food chain. It demonstrates that higher levels of CO_2 will not only increase productivity in plants, both terrestrial and aquatic, but will also boost the productivity of one—if not the most important—of the calcifying species in the oceans.

The reason that calcifying marine organisms can calcify under a wider range of pH values than one would expect from a simple chemical calculation is that they can control their internal chemistry at the site of calcification. The proponents of dangerous ocean acidification are not considering this. If the internal biology of organisms were strictly determined by the chemical environment around them, it is unlikely there would be any life on Earth.

As mentioned earlier, it was at the beginning of the Cambrian Period approximately 540 million years ago that marine species of invertebrates evolved the ability to control the crystallization of calcium carbonate as an armor plating to protect themselves from predators. It is hypothesized that this ability stemmed from a long-standing previous ability to prevent spontaneous calcium carbonate crystallization to protect essential metabolic processes. Surprisingly, the common denominator in the anticalcification—calcification history is mucus, often referred to as "slime" (Marin et al., 1996).

The abstract from the paper cited earlier sums up this hypothesis well:

> The sudden appearance of calcified skeletons among many different invertebrate taxa at the PrecambrianCambrian transition may have required minor reorganization of pre-existing secretory functions.

In particular, features of the skeletal organic matrix responsible for regulating crystal growth by inhibition may be derived from mucus epithelial excretions. The latter would have prevented spontaneous calcium carbonate overcrusting of soft tissues exposed to the highly supersaturated Late Proterozoic ocean…, a putative function for which we propose the term "*anticalcification*." We tested this hypothesis by comparing the serological properties of skeletal water-soluble matrices and mucus excretions of three invertebrates—the scleractinian coral *Galaxea fascicularis* and the bivalve molluscs *Mytilus edulis* and *Mercenaria mercenaria*. Cross-reactivities recorded between muci and skeletal water-soluble matrices suggest that these different secretory products have a high degree of homology. Furthermore, freshly extracted muci of *Mytilus* were found to inhibit calcium carbonate precipitation in solution (Martin et al., 1996).

The authors found that the muci produced by a coral, a mussel, and a clam were chemically very similar, indicating inheritance from a common ancestor earlier in the Precambrian. The mucus produced by invertebrates has a number of known functions. It assists with mobility, acts as a barrier to disease and predators, helps with feeding, acts as a homing device, and prevents desiccation (Denny, 1989). The authors postulate that the mucus is also central in the calcification process. This explains how the chemistry at the site of calcification can be isolated from the chemistry of the seawater. Calcification can occur in and under the mucus layer where the organism can control the chemistry.

The creation of a shell requires certain biochemical processes. The periostracum, a leathery layer on the outside of the shell, folds around the lip of the shell to form an enclosed pocket called the extrapallial space. Within this space at the lip of the shell, the calcification process occurs, resulting in growth of the shell. The concentration of calcium ions is intensified by ion pumps within the extrapallial space, allowing crystallization to occur. The mucus within the space contains hormones that direct the pattern of calcium carbonate crystal deposit to result in a smooth layer of new shell. This is a classic case of an organism controlling the biochemistry within itself, despite fluctuations in the external environment that would not permit such sophisticated functions (Encyclopedia Brittanica, 2015).

The references cited make it clear that species that calcify have a high degree of sophistication in controlling the calcification process. The clear implication is that calcification can be successfully achieved despite a varying range of environmental conditions that would interfere with or end the process if it were not controlled. This does not appear to have been considered by the authors who propose that ocean acidification will exterminate a large proportion of calcifying species within a few decades.

Much of the concern about ocean acidification in the literature focuses on carbonate chemistry. When the pH of seawater lowers, the bicarbonate ion (HCO$_3$) becomes more abundant while the carbonate ion (CO$_3$) becomes less abundant. This is predicted to make it more difficult for calcifying species to obtain the CO$_3$ required for calcification. It does not appear to be considered that the calcifying species may be capable of converting HCO$_3$ to CO$_3$.

There are very few references to journal articles after 1996 that investigate the biochemical processes involved in calcification. The paper cited earlier by Marin et al. (1996) is the most thorough investigation and discussion of the subject found. Yet there are hundreds, if not thousands, of articles that predict dire consequences from ocean acidification during this century. Some basic science on the nature of calcification may help in reaching a sound conclusion on the environmental impact of higher levels pf CO$_2$ in the future.

A recent study published in the Proceedings of the National Academy of Sciences highlights how resilient coral reefs are to changes in ocean pH. A five-year study of the Bermuda coral reef shows that during spurts in growth and calcification, the seawater around the reef undergoes a rapid reduction in pH (Yeakel et al., 2015). This reduction in pH is clearly not causing a negative reaction from the reef, as it is associated with rapid growth. The study found that the reason the pH dropped during growth spurts is due to the CO$_2$ emitted by the reef due to increased respiration. It was determined that the growth spurts were the result of offshore blooms of phytoplankton drifting in to the reef and providing an abundant food supply for the reef polyps. The conclusion from the study is that coral growth can increase even though the growth itself results in a reduction in pH in the surrounding seawater. A summary of the study in New Scientist concluded, "These corals didn't seem to mind the fluctuations in local acidity that they created, which were much bigger than those we expect to see from climate change. This may mean that corals are well equipped to deal with the lower pH levels" (Slezak, 2015). It follows from the earlier discussion that this is likely due to the fact that the coral polyps can control their own internal pH despite the decrease in pH in their environment.

6. A WARMER OCEAN MAY EMIT CO$_2$ BACK INTO THE ATMOSPHERE

While today's atmosphere contains about 850 Gt of carbon as CO$_2$, the oceans contain 38,000 Gt of carbon, nearly 45 times as much as the atmosphere. The ocean either absorbs or emits CO$_2$ at the ocean–atmosphere interface, depending on the CO$_2$ concentrations in the atmosphere and the sea below, and the salinity and temperature of

FIGURE 13.5   All peer-reviewed experimental results for pH decrease of 0.0–0.3 from present value (Tans, 2009). (Prediction of range of actual expected pH change in gray.) Five parameters are included: calcification, metabolism, growth, fertility, and survival. Note that the overall trend is positive for all studied up to 0.30 units of pH reduction.

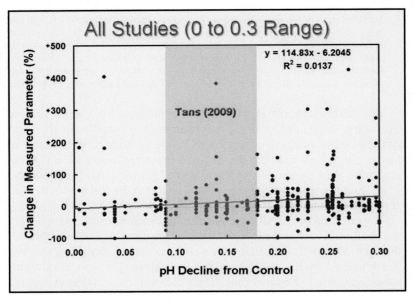

the sea. At the poles, where seawater is coldest and densest and has the highest solubility for CO_2, seawater sinks into the abyss, taking CO_2 down with it. In regions of deep seawater upwelling such as off the coasts of Peru, California, West Africa, and the northern India Ocean, seawater rich in CO_2 fertilizes plankton blooms that feed great fisheries. The phytoplankton near the surface consumes some of the CO_2, and some is outgassed to the atmosphere.

As mentioned above, we do not have the ability to determine how much CO_2 is absorbed by the oceans, how much is outgassed back into the atmosphere, or the net effect of these phenomena (Cho, 2014). The Intergovernmental Panel on Climate Change implicitly admits this lack of knowledge when it sets the estimate of the CO_2 feedback on the exceptionally wide interval of 25–225 ppmv K−1. What we do know is that if the oceans warm as the proponents of human-caused global warming say they will, the oceans will tend to release CO_2 into the atmosphere because warm seawater at 30°C can dissolve only about half as much CO_2 as cold seawater at 4°C. This will be balanced against the tendency of increased atmospheric CO_2 to result in more absorption of CO_2 by the oceans. It does not appear as though anyone has done the calculation of the net effect of these two competing factors under varying circumstances.

7. SUMMARY OF EXPERIMENTAL RESULTS ON EFFECT OF REDUCED PH ON CALCIFYING SPECIES

In his thorough and inclusive analysis of peer-reviewed experimental results on the effect of reduced pH on five factors (calcification, metabolism, growth, fertility, and survival) among marine calcifying species, Craig Idso of the CO_2 Science Website provides a surprising insight. Beginning with 1103 results from a wide range of studies, the results are narrowed down to those within a 0.0 to 0.3 reduction in pH units (CO_2 Science, 2015). A review of these many studies, all of which use direct observation of measured parameters, indicates that the overall predicted effect of increased CO_2 on marine species would be positive rather than negative (see Fig. 13.5). This further reinforces the fact that CO_2 is essential for life, that CO_2 is at an historical low concentration during this Pleistocene Ice Age and that more CO_2 rather than less would be generally beneficial to life on Earth.

8. CONCLUSIONS

There is no solid evidence that ocean acidification is the dire threat to marine species that many researchers have claimed. The entire premise is based upon an assumption of what the average pH of the oceans was 265 years ago, when it was not possible to measure pH at all, never mind over all the world's oceans. Laboratory experiments in which pH was kept within a range that may feasibly occur during this century show a slight positive effect on five critical factors: calcification, metabolism, growth, fertility, and survival.

Of most importance is the fact that those raising the alarm about ocean acidification do not take into account the ability of living species to adapt to a range of environmental conditions. This is one of the fundamental characteristics of life itself.

References

American Museum of Natural History. Surviving in Salt Water. http://www.amnh.org/exhibitions/past-exhibitions/water-h2o-life/life-in-water/surviving-insalt-water.

Bates, N.R., Best, M.H.P., Neely, K., Garley, R., Dickson, A.G., Johnson, R.J., 2012. Detecting anthropogenic carbon dioxide uptake and ocean acidification in the North Atlantic Ocean. Biogeosciences 9 (7), 2509–2522. http://dx.doi.org/10.5194/bg-9-2509-2012.

Blanco, J., Daneri, G., Escribano, R., Guzmán, L., Morales, C., Osses, J., Pizarro, G., Quiñones, R., Rosales, S.A., Serra, R., December 2002. Integrated Overview of the Oceanography and Environmental Variability of the Humboldt Current System." Chile, IFOP-IMARPE-ONUDI. http://humboldt.iwlearn.org/es/informacion-y-publicacion/Modulo1_PyVariabil_AmbientalVol1.pdf.

Boeye, J., Travis, J.M.J., Stoks, R., Bonte, D., 2013. More rapid climate change promotes evolutionary rescue through selection for increased dispersal distance. Evolutionary Applications 6 (2), 353–364. http://dx.doi.org/10.1111/eva.12004.

Caldeira, K., Wickett, M.E., 2003. Oceanography: anthropogenic carbon and ocean pH. Nature 425 (6956), 365.

Cho, R., July 30, 2014. Solving the Mysteries of Carbon Dioxide. State of the Planet, Earth Institute, Columbia University. http://blogs.ei.columbia.edu/2014/07/30/solving-the-mysteries-of-carbon-dioxide/.

CO_2 Science, 2015. Ocean Acidification Database. See also http://www.co2science.org/subject/o/subject_o.php. http://www.co2science.org/data/acidification/results.php.

Denny, M.W., 1989. Invertebrate mucous secretions: functional alternatives to vertebrate paradigms. Symposia of the Society for Experimental Biology 43, 337–366.

Doney, S.C., March 2006. The Dangers of Ocean Acidification. Scientific American. http://www.precaution.org/lib/06/ocean_acidification_from_c02_060301.pdf.

Donohue, R.J., Roderick, M.L., McVicar, T.R., Farquhar, G.D., 2013. Impact of CO_2 fertilization on maximum foliage cover across the globe's warm, arid environments. Geophysical Research Letters 40 (12), 3031–3035. http://dx.doi.org/10.1002/grl.50563.

Egger, M., August 2011. Ocean Acidification in the Humboldt Current System (Master thesis). Swiss Federal Institute of Technology.

Encyclopedia Brittanica, 2015. Periostracum. http://www.britannica.com/science/periostracum.

Gould, S.J., 1989. Wonderful Life: The Burgess Shale and the Nature of History. W.W. Norton and Company.

Haag, W.R., 2012. North American Freshwater Mussels: Natural History, Ecology, and Conservation. Cambridge University Press, Cambridge.

Hofmann, G.E., Smith, J.E., Johnson, K.S., Send, U., Levin, L.A., Micheli, F., Paytan, A., Price, N.N., Peterson, B., Takeshita, Y., Matson, P.G., Crook, E.D., Kroeker, K.J., Gambi, M.C., Rivest, E.B., Frieder, C.A., Yu, P.C., Martz, T.R., 2011. High-frequency dynamics of ocean pH: a multi-ecosystem comparison. PLoS One 6 (12), e28983. http://dx.doi.org/10.1371/journal.pone.0028983.

Holland, H.D., 2006. The geological history of seawater. In: Elderfeld, H. (Ed.), The Oceans and Marine Geochemistry: Treatise on Geochemistry, vol. 6. Elsevier.

Iglesias-Rodriguez, D., Halloran, P.R., Rickaby, R.E.M., Hall, I.R., Colmenero-Hidalgo, E., Gittins, J.R., Green, D.R.H., Tyrrell, T., Gibbs, S.J., von Dassow, P., Rehm, E., Armbrust, E.V., Boessenkool, K.P., April 2008. Phytoplankton calcification in a high-CO_2 world. Science 320 (5874), 336–340. http://www.sciencemag.org/content/320/5874/336.full#F1.

Jablonka, E., Raz, G., 2009. Transgenerational epigenetic inheritance: prevalence mechanisms, and implications for the study of heredity and evolution. Quarterly Review of Biology 84 (2), 131–176.

JoNova, 2013. The 800 Year Lag in CO_2 After Temperature — Graphed. http://joannenova.com.au/global-warming-2/ice-core-graph/.

Mäkelä, T.P., Oikari, A.O.J., December 1992. The effect of low water pH on the ionic balance in freshwater mussel Anadonta anatina L. Annales Zoologici Fennici 29, 169–175. http://www.sekj.org/PDF/anzf29/anz29-169-175.pdf.

Marin, F., Smith, M., Isa, Y., Muyzer, G., Westbroek, P., 1996. Skeletal matrices, muci, and the origin of invertebrate calcification. Proceedings of the National Academy of Sciences of the United States of America 93 (4), 1554–1559.

McClary, M., June 20, 2014. Osmoconformer. Encyclopedia of the Earth. http://www.eoearth.org/view/article/155074/.

McKitrick, R., 2014. HAC-robust measurement of the duration of a trendless subsample in a global climate time series. Open Journal of Statistics 4 (7), 527–535. http://dx.doi.org/10.4236/ojs.2014.47050.

Murray, C.S., Malvezzi, A., Gobler, C.J., Baumann, H., 2014. Offspring sensitivity to ocean acidification changes seasonally in a coastal marine fish. Marine Ecology Progress Series 504, 1–11.

Nahle, N., July 2009. Cycles of Global Climate Change. Biology Cabinet. http://www.biocab.org/carbon_dioxide_geological_timescale.html.

National Oceanic and Atmospheric Administration. What is Ocean Acidification? PMEL Carbon Program. http://www.pmel.noaa.gov/co2/story/What+is+Ocean+Acidification%3F.

National Oceanic and Atmospheric Administration, May 2014. Ocean Acidification in the Pacific Northwest. http://www.noaa.gov/factsheets/OA18PNWFacts14V4.pdf.

Natural Resources Defense Council, August 2009. Ocean Acidification: The Other CO_2 Problem. https://www.nrdc.org/oceans/acidification/files/NRDCOceanAcidFSWeb.pdf.

Ocean Health, Chemistry of Seawater. http://oceanplasma.org/documents/chemistry.html.

Pagani, M., et al., 2005. Marked decline in atmospheric carbon dioxide concentrations during the paleocene. Science 309 (5734), 600–603.

Pelejero, C., Calvo, E., McCulloch, M.T., Marshall, J.F., Gagan, M.K., Lough, J.M., Opdyke, B.N., 2005. Preindustrial to modern interdecadal variability in coral reef pH. Science 309, 2204–2207.

Pelejero, C., Calvo, E., Hoegh-Guldberg, O., 2010. Paleo-perspectives on ocean acidification. Trends in Ecology and Evolution 25 (6), 332–344.

Peylin, P., Law, R.M., Gurney, K.R., Chevallier, F., Jacobson, A.R., Maki, T., Niwa, Y., Patra, P.K., Peters, W., Rayner, P.J., Rödenbeck, C., van der Laan-Luijkx, I.T., Zhang, X., 2013. Global atmospheric carbon budget: results from an ensemble of atmospheric CO_2 inversions. Biogeosciences, 10 (10), 6699–6720. http://dx.doi.org/10.5194/bg-10-6699-2013 quoting Wanninkhof, R., et al., 2013. Global ocean carbon uptake: magnitude, variability and trends. Biogeosciences 10, 1983–2000. http://dx.doi.org/10.5194/bg10-1983-2013.

Pickrell, J., July 15, 2004. Oceans Found to Absorb Half of All Man-Made Carbon Dioxide. National Geographic News. http://news.nationalgeographic.com/news/2004/07/0715_040715_oceancarbon.html.

Pretzsch, H., Biber, P., Schütze, G., Uhl, E., Rötzer, T., 2014. Forest stand growth dynamics in Central Europe have accelerated since 1870. Nature Communications 5 (4967). http://dx.doi.org/10.1038/ncomms5967.

Price, T.D., Qvarnström, A., Irwin, D.E., 2003. The role of phenotypic plasticity in driving genetic evolution. Proceedings of the Royal Society B: Biological Sciences 270 (1523), 1433–1440. http://dx.doi.org/10.1098/rspb.2003.2372.

Ruddiman, W.F., 2001. Earth's Climate: Past and Future. W. H. Freeman, New York, NY.

Saladin, K.S., 2016. Osmoregulation. Biology Encyclopedia Forum. http://www.biologyreference.com/Oc-Ph/Osmoregulation.html.

Scotese, C.R., 2002. Analysis of the Temperature Oscillations in Geological Eras.

Slezak, M., November 9, 2015. Growing Corals Turn Water More Acidic Without Suffering Damage. New Scientist. https://www.newscientist.com/article/dn28468-growing-corals-turn-water-more-acidic-without-suffering-damage/.

Sverdrup, H.U., Johnson, M.W., Fleming, R.H., 1942. The Oceans, Their Physics, Chemistry, and General Biology. Prentice-Hall, New York. http://ark.cdlib.org/ark:/13030/kt167nb66r/.

Tans, P., 2009. An accounting of the observed increase in oceanic and atmospheric CO_2 and an outlook for the future. Oceanography 22, 26–35.

Tyrrell, T., Merico, A., McKay, D.I.A., 2015. Severity of ocean acidification following the end-Cretaceous asteroid impact. Proceedings of the National Academy of Sciences of the United States of America 112 (21), 6556–6561.

U.S. House Committee on Natural Resources, May 13, 2015. CEQ Draft Guidance for GHG Emissions and the Effects of Climate Change, Testimony of Prof. John R. Christy. University of Alabama in Huntsville. http://docs.house.gov/meetings/II/II00/20150513/103524/HHRG-114-II00-Wstate-ChristyJ-20150513. pdf.

Virtual Fossil Museum, 2015. The Cambrian Period (544-505 mya). http://www.fossilmuseum.net/Paleobiology/Paleozoic_paleobiology.htm.

Weiner, S., Dove, P.M., 2003. An overview of biomineralization processes and the problem of the vital effect. Reviews in Mineralogy and Geochemistry 54 (1), 1–29.

Wood, J.M., 2011. Bacterial osmoregulation: a paradigm for the study of cellular homeostasis. Annual Review of Microbiology 65, 215–238. http://dx.doi.org/10.1146/annurev-micro-090110-102815.

Yeakel, K.L., Anderson, A.J., Bates, N.R., Noyes, T.J., Collins, A., Garley, R., 2015. Shifts in coral reef biogeochemistry and resulting acidification linked to offshore productivity. Proceedings of the National Academy of Sciences of the United States of America. http://www.pnas.org/content/early/2015/11/04/1507021112.abstract.

Zeebe, R.E., Wolf-Gladrow, D.A., 2008. Carbon dioxide, dissolved (ocean). In: Gornitz, V. (Ed.), Encyclopedia of Paleoclimatology and Ancient Environments, Earth Science Series. Kluwer Academic Publishers.

Zeebe, R.E., 2012. History of seawater carbonate chemistry atmospheric CO_2, and ocean acidification. Annual Review of Earth and Planetary Science 40, 141–165. http://dx.doi.org/10.1146/annurev-earth-042711-105521.

PART VII

SOLAR INFLUENCES ON CLIMATE

Cause of Global Climate Changes: Correlation of Global Temperature, Sunspots, Solar Irradiance, Cosmic Rays, and Radiocarbon and Berylium Production Rates

D.J. Easterbrook

Western Washington University, Bellingham, WA, United States

1. SOLAR VARIATION—GRAND MINIMA

At the end of the Medieval Warm Period, ~1300 AD, temperatures dropped dramatically and the cold period that followed is known as the Little Ice Age. The periods of colder climate that ensued for five centuries were devastating. The population of Europe had become dependent on cereal grains as a food supply during the Medieval Warm Period, and with the colder climate, early snows, violent storms, and recurrent flooding that swept Europe, massive crop failures occurred, resulting in widespread famine and disease (Fagan, 2000; Grove, 2004). Glaciers in Greenland and elsewhere began advancing and pack ice extended southward in the North Atlantic, blocking ports and affecting fishing. Three years of torrential rains that began in 1315 led to the Great Famine of 1315—1317.

Evidence-Based Climate Science, Second Edition
http://dx.doi.org/10.1016/B978-0-12-804588-6.00014-8

The Little Ice Age was not a time of continuous cold climate, but rather repeated periods of cooling and warming, each of which occurred during times of solar minima, characterized by low sunspot numbers, low total solar irradiance (TSI), decreased solar magnetism, increased cosmic ray intensity, and increased production of radiocarbon and beryllium in the upper atmosphere.

Centuries of observations of the sun have shown that sunspots, solar irradiance, and solar magnetism vary over time, and these phenomena correlate very well with global climate changes on Earth. A number of solar Grand Minima, periods of reduced solar output, have been recognized (Fig. 14.1).

1.1 Wolf Minimum (1290—1320 AD)

The Wolf Minimum was a period of low sunspot numbers (SSNs) and TSI between about 1300 and 1320 AD. It occurred during the cold period that marked the end of the Medieval Warm Period (MWP) and the beginning of the Little Ice Age (LIA) about 1300 AD.

The change from the warmth of the MWP to the cold of the LIA was abrupt and devastating, leading to the Great Famine from 1310 to 1322. The winter of 1309—1310 AD was exceptionally cold. The Thames River froze over and poor people were especially affected. The year 1315 AD was especially bad. Jean Desnouelles wrote at the time, "Exceedingly great rains descended from the heavens and they made huge and deep mud-pools on the land. Throughout nearly all of May, June, and August, the rains did not stop." Corn, oats, and hay crops were beaten to the ground, August and September were cold, and floods swept away entire villages. Crop harvests in 1315 AD were a disaster, affecting an enormous area in northern Europe. In places, up to half of farmlands were eroded away, cold, wet weather prevented grain harvests, and fall plantings failed, triggering famines.

In 1316 AD, spring rain continued, again impeding the sowing of grain crops, and harvests failed once again. Diseases increased, newborn and old people died of starvation, and multitudes scavenged anything edible. Whole communities disappeared and many farms were abandoned. The year 1316 was the worst for cereal crops in the entire Middle Ages. Cattle couldn't be fed, hay wouldn't dry and couldn't be moved so it just rotted. Thousands of cattle froze during the bitterly cold winter of 1317—1318 and many others starved. The cold immobilized shipping. Rain in 1317—1318 continued through the summer and people suffered for another seven years. The coincidence of sudden cooling of the climate from the warm Medieval Warm Period to the harsh cold climate of the Little Ice Age during the Wolf Minimum was not just a coincidence, as shown by at least five later, similar instances.

1.2 Sporer Minimum (1410—1540)

The Sporer Minimum occurred from about 1410 to 1540 (Fig. 14.1). Like the Wolf Minimum, the Sporer coincided with a cold period (Fig. 14.2).

1.3 Maunder Minimum

The Maunder Minimum is the most famous cold period of the Little Ice Age. Temperatures plummeted in Europe (Figs. 14.3—14.7), the growing season became shorter by more than a month, the number of snowy days increased

Selected solar activity events

Event	Approx dates	
Medieval maximum	1100	1250
Wolf minimum	1280	1350
Spörer Minimum	1460	1550
Maunder Minimum	1645	1715
Dalton Minimum	1790	1820
1880-1915 Minimum	1880	1915
1945-1977 Minimum	1945	1977

FIGURE 14.1　Solar minima.

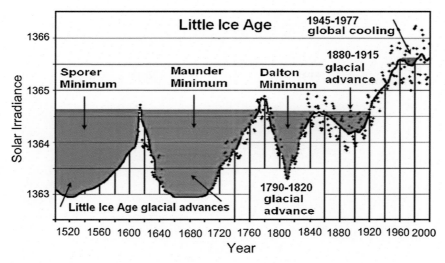

FIGURE 14.2   Relationship of solar minima, solar irradiance, and glacier advances. Blue areas were cool periods. Cool climates prevailed in all six solar minima since 1300 AD.

FIGURE 14.3   1663 painting by Jan Grifier of the frozen Thames River in London during the Maunder Minimum.

from a few to 20—30, the ground froze to several feet, alpine glaciers advanced all over the world, glaciers in the Swiss Alps encroached on farms and buried villages, tree-lines in the Alps dropped, sea ports were blocked by sea ice that surrounded Iceland and Holland for about 20 miles, wine grape harvests diminished, and cereal grain harvests failed, leading to mass famines (Fagan, 2007). The Thames River and canals and rivers of the Netherlands froze over during the winter (Fig. 14.3). The population of Iceland decreased by about half. In parts of China, warm-weather crops that had been grown for centuries were abandoned. In North America, early European settlers experienced exceptionally severe winters.

1.3.1 Sunspots

Sunspots are temporary dark spots on the surface of the sun (Fig. 14.8), where concentrations of magnetic field flux inhibit convection, reducing surface temperature. They occur in pairs and may last from a few days to a few months before eventually disappearing. Sunspots expand and contract as they move across the surface of the sun, ranging from 16 km (10 mi) to 160,000 km (100,000 mi) in diameter. Sunspots usually appear in groups. Sunspot activity cycles about every 11 years. The point of highest sunspot activity during a cycle is known as solar maximum, and the point of lowest activity as solar minimum.

FIGURE 14.4   Glaciers in the Alps advanced during the Little Ice Age.

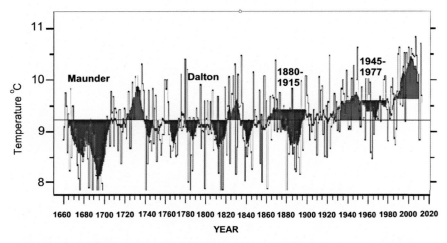

FIGURE 14.5   Central England temperatures (CET) recorded continuously since 1658. Blue areas are reoccurring cool periods; red areas are warm periods. All times of solar minima were coincident with cool periods in central England.

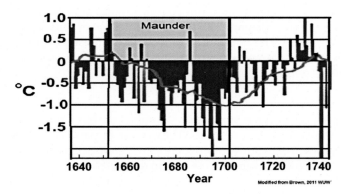

FIGURE 14.6   CET during the Maunder Minimum.

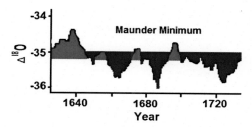

FIGURE 14.7   Oxygen isotope record, GISP2 Greenland ice core showing the Maunder Minimum. Blue area is cool, red is warm. The isotope record shows the same cooling as the CET.

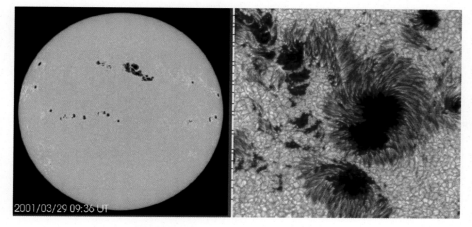

FIGURE 14.8   Sunspots. *NASA images.*

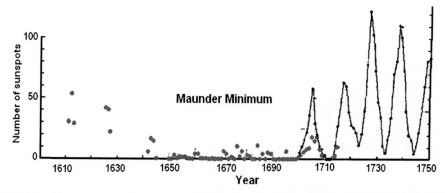

FIGURE 14.9   Sunspots during the Maunder Minimum. From 1645 to 1700, many years had no sunspots.

When Galileo perfected the telescope in 1609, scientists could see sunspots for the first time. They were of such interest that records were kept of the number of sunspots observed, and although not perfectly accurate due to cloudy days, lost records, and so on, the records show a remarkable pattern for more than a century (Fig. 14.9) (Maunder, 1894, 1922; Eddy, 1976, 1977; Soon, 2005; Hoyt and Schatten, 1997, 1998; Lean et al., 1995, 2002). From 1600 to 1715 AD, very few sunspots were seen, and from 1645 to 1715, many years had no sunspots at all, despite the fact that many scientists with telescopes were actively looking for them. The longest known minimum (about 50 years) of virtually no sunspots occurred during the Maunder Minimum. After 1715 AD, the number of observed sunspots increased sharply from near zero to 50—100 (Fig. 14.9) and the global climate warmed.

1.3.2 Total Solar Irradiance

TSI is the solar radiative power per unit area incident on the Earth's upper atmosphere. It is usually measured in watts per square meter (W/m^2). It has varied historically, reaching maximums during periods of high sunspot numbers and minimums during of low sunspot numbers (Fig. 14.10). TSI drops to lowest values during Solar Minimums and during sunspots lows.

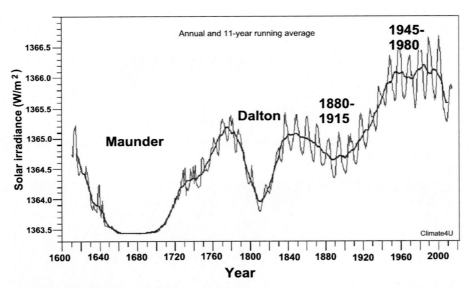

FIGURE 14.10   Total solar irradiance from 1600 to 2014 AD. *Modified from Lean, J.L., Beer, J., Bradley, R., 1995, Reconstruction of solar irradiance since 1610: implications for climatic change. Geophysics Research Letters 22, 3195–3198.*

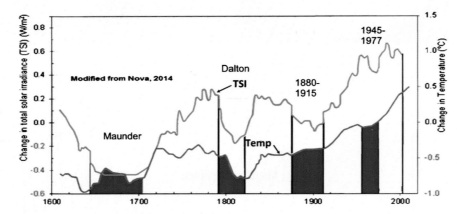

FIGURE 14.11   Correlation of total solar irradiance (TSI) and temperature. *Modified from Nova, 2014.*

1.3.3 *Temperature, Sunspots, and Total Solar Irradiance*

Temperatures during the Maunder closely correlate with SSNs and TSI (Fig. 14.11). When SSNs and TSI are low, temperatures are also low, and when SSNs and TSI are high, temperatures are high.

1.4 Dalton Minimum

The Dalton Minimum was a period of low SSNs and TSI from about 1790 to 1820. Like the Maunder, it was a time of intense cooling and great hardship, although not as bad as the Maunder. Widespread famines due to crop failures spread across Europe. Several notable events occurred during the Dalton, including the French Revolution and Napoleon's defeat in Russia because of a bitterly cold winter.

The cool temperatures of the Dalton show up both in the CET and GISP2 Greenland ice core (Figs. 14.12 and 14.13).

1.4.1 *Sunspots*

Sunspots declined sharply during the Dalton Minimum (Fig. 14.14).

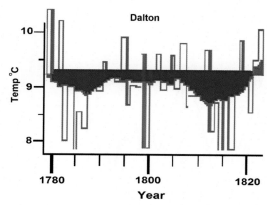

FIGURE 14.12   CET during the Dalton Minimum.

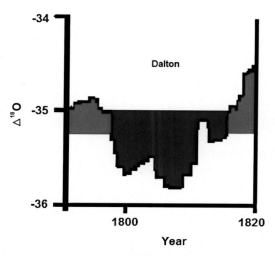

FIGURE 14.13   Oxygen isotope record from the GISP2 Greenland ice core. *Data from Grootes and Stuiver.*

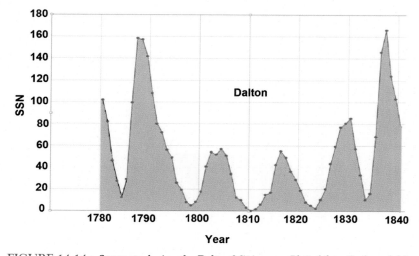

FIGURE 14.14   Sunspots during the Dalton Minimum. *Plotted from Svalgaard data.*

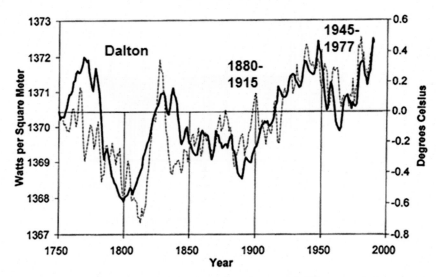

FIGURE 14.15   Solar irradiance and global temperature from 1750 to 1990. During this 250-year period, temperature and total solar irradiance curves follow a remarkably similar pattern. *Modified from Hoyt, D.V., Schatten, K.H., 1997, The Role of the Sun in Climate Change: Oxford University, p. 279.*

1.4.2 Total Solar Irradiance

TSI also declined sharply during the Dalton Minimum (Fig. 14.10).

1.4.3 Temperature and Total Solar Irradiance

During the Dalton Minimum, temperatures dropped, closely following TSI (Fig. 14.15).

1.5 1880−1915 Minimum

Temperatures dropped sharply beginning about 1880, and the 1850−1880 warm period came to a close and the climate cooled (Figs. 14.16−14.18). Alpine glaciers advanced down valley to terminal positions not far from their maximums during the early part of the LIA.

1.5.1 Sunspots

The number of sunspots declined during the 1880−1915 cool period (Fig. 14.19).

1.5.2 Total Solar Irradiance and Temperature

TSI declined during the 1880−1915 cool period (Fig. 14.10). Temperature closely followed TSI (Fig. 14.20).

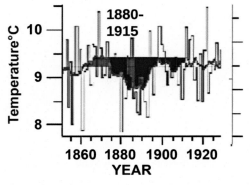

FIGURE 14.16   CET during the 1880−1915 cool period.

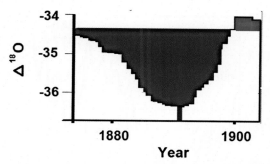

FIGURE 14.17   Oxygen isotope record from the GISP2 Greenland ice core. Blue area is cool. *Data from Grootes, P.M., Stuiver, M., 1997. Oxygen 18/16 variability in Greenland snow and ice with 10^3 to 10^5-year time resolution: Journal of Geophysical Research 102, 26455–26470.*

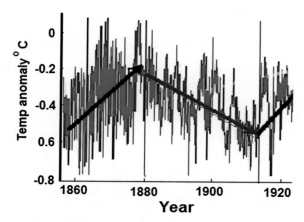

FIGURE 14.18   Global temperature, 1880 to 1915 cool period. *Modified from HADRUT3.*

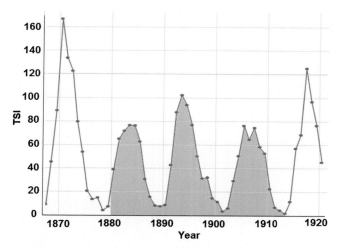

FIGURE 14.19   Sunspots during the 1880–1915 cool period.

1.6 1945–1977 Minimum

Following the early 20th-century warm period (1915–1945), the climate cooled for 30 years (Figs. 14.21 and 14.22).

1.6.1 Sunspots

The number of sunspots declined during the 1945–1977 cool period (Fig. 14.23).

1.6.2 Total Solar Irradiance and Temperature

Decline in TSI coincided with 30 years of global cooling temperatures from 1945 to 1977 (Fig. 14.20), even as CO_2 emissions soared.

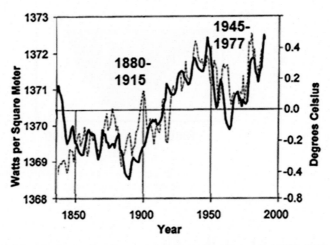

FIGURE 14.20   Total solar irradiance and temperature during the 1880–1915 cool period.

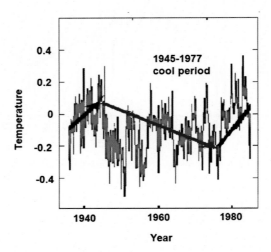

FIGURE 14.21   Temperature during the 1945–1977 cool period. *HADCRUT3.*

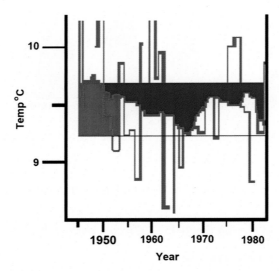

FIGURE 14.22   CET during the 1945–1977 cool period. (Blue = cool, red = warm.)

VII. SOLAR INFLUENCES ON CLIMATE

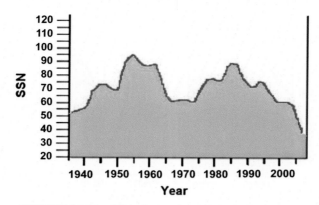

FIGURE 14.23 SSNs during the 1945–1977 cool period.

2. RADIOCARBON ($^{14}C_6$) PRODUCTION RATES

Radiocarbon ($^{14}C_6$) is produced in the upper atmosphere by collision of neutrons with nitrogen atoms ($^{14}N_7$), which has the effect of knocking a proton out the nucleus of the nitrogen atoms, thus decreasing the atomic number by one to form $^{14}C_6$:

$$^{14}N_7 + n = {}^{14}C_6 + {}^{1}H_1$$

Radiocarbon ($^{14}C_6$) differs from $^{12}C_6$, the most abundant isotope of carbon, in that its mass is greater (14 compared to 12) and its nucleus is radioactive, emitting a beta particle and an electron anti-neutrino. As a result, one of the neutrons in the $^{14}C_6$ decays to a proton, thus increasing the atomic number by 1 to form $^{14}N_7$. The half–life of the radioactive decay is 5730 years:

$$^{14}C_6 = {}^{14}N_7 + \beta-$$

Radiocarbon readily combines with oxygen to form carbon dioxide (CO_2), which mixes with other CO_2 in the atmosphere to produce a constant ratio of radioactive ^{14}C to stable ^{12}C in carbon dioxide.

Nitrogen is so abundant in the atmosphere that the amount of radiocarbon produced by this reaction is nearly equal to the number of neutrons generated by cosmic radiation. Thus the production rate of radiocarbon is a measure of cosmic radiation in the upper atmosphere. An important aspect of these reactions is that since the rate of production of radiocarbon depends on the incidence of cosmic radiation, if we can determine the production rate of ^{14}C, we can use radiocarbon as a measure of past cosmic radiation. This can be done with old tree rings, whose calendar age can be determined by counting the number of rings and the radiocarbon age can be measured by ^{14}C dating of the same ring. The amount of deviance of the ^{14}C date from the calendar age is a measure of the production rate of radiocarbon, known as $\delta^{14}C$.

Papers by Stuiver and others have shown the correlation between radiocarbon production and sunspots (Stuiver, 1961, 1994; Stuiver and Brasiunas, 1991; Stuiver and Quay, 1980; Stuiver et al., 1991, 1995). Fig. 14.24 shows changes in radiocarbon production rates ($\delta^{14}C$) since 1600 AD. $\delta^{14}C$ was higher in both the Maunder and Dalton Solar Minimums.

Figs. 14.25–14.27 show the correlation of radiocarbon production rates and temperature. Note how closely temperatures follow radiocarbon production rates and that the Wolf, Sporer, Maunder, and Dalton Solar Minimums are all characterized by high radiocarbon production rates, ie, higher incoming cosmic radiation.

3. BERYLIUM-10 (^{10}BE$_4$) PRODUCTION RATES

$^{10}Be_4$ is a radioactive isotope of the most common beryllium, $^{9}Be_4$, formed by cosmic ray spallation of oxygen in the atmosphere (Figs. 14.28–14.30). It decays by beta decay to boron-10 with a half-life of 1.39×10^6 years. Like radiocarbon, because ^{10}Be is produced in the atmosphere by cosmic radiation, it can also be used to measure the incidence of cosmic radiation. ^{10}Be is soluble in atmospheric precipitation and accumulates in glacial ice, where it is preserved and can be measured and dated by counting annual ice layers. Good correlation between ^{14}C and ^{10}Be fluxes

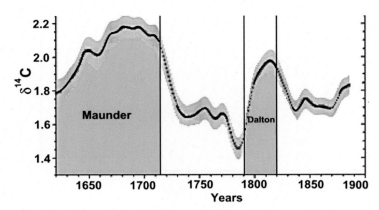

FIGURE 14.24 $\delta^{14}C$ changes since 1600 AD. Note the high values during the Maunder and Dalton Solar Minimums.

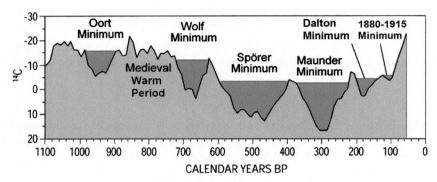

FIGURE 14.25   Correlation of ^{14}C with Oort, Wolf, Sporer, Maunder, Dalton, and 1880–1915 Solar Minimums. Each solar minimum was a period of high ^{14}C production and each corresponded to a cold climate.

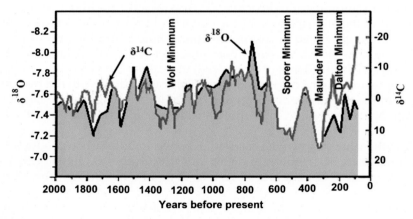

FIGURE 14.26   Correlation of temperature ($\delta^{18}O$) and radiocarbon production rates ($\delta^{14}C$). Temperature closely follows radiocarbon production rates ($\delta^{14}C$).

(Fig. 14.31) indicates that both are a result of changes in cosmic radiation, since their transport processes to their place of accumulation are so different—^{14}C is measured from tree rings and ^{10}Be is measured from glacial ice cores. The relationship of ^{10}Be to cosmic radiation is confirmed by the correlation of ^{10}Be and solar magnetic flux (Fig. 14.32).

3.1 Berylium-10 (^{10}Be) and Sunspots

Both ^{10}Be and SSNs are directly related to solar magnetism, so it is not surprising that ^{10}Be correlates well with sunspots (Fig. 14.33).

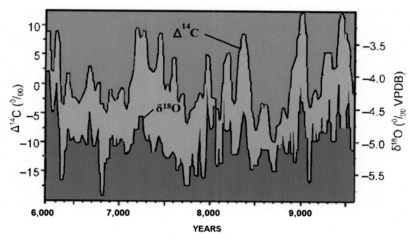

FIGURE 14.27   Close correlation of radiocarbon production (δ^{14}C) and temperature (δ^{18}O) from a stalagmite in Oman (Matter et al., 2001).

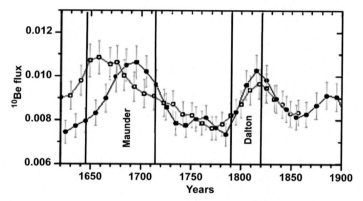

FIGURE 14.28   Depositional flux of ^{10}Be in ice cores in Greenland (red) and Antarctica (blue). Note the high ^{10}Be values during the Maunder and Dalton Solar Minimums. *Modified from Usoskin, I.G., et al., 2015. The Maunder minimum (1645–1715) was indeed a grand minimum: a reassessment of multiple datasets. Astronomy Astrophysics 581, 1–19.*

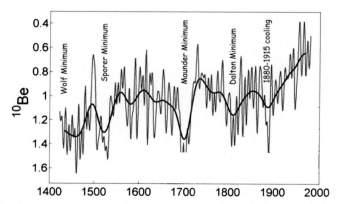

FIGURE 14.29   Fluctuation of ^{10}Be as a measure of cosmic ray incidence. Note that ^{10}Be production rates were high for the Wolf, Sporer, Maunder, Dalton, and 1880–1915 solar minima.

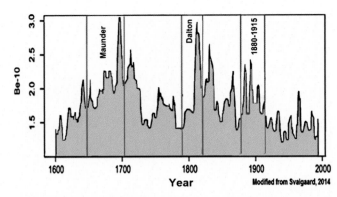

FIGURE 14.30 ^{10}Be production was higher during the Maunder, Dalton, and 1880–1915 Solar Minimums, indicating that cosmic ray incidence was higher then. *Modified from Svalgaard.*

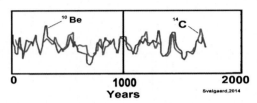

FIGURE 14.31 ^{10}Be and ^{14}C fluxes over the past 2000 years show good correlation, despite having very different paths of accumulation. *Modified from Svalgaard.*

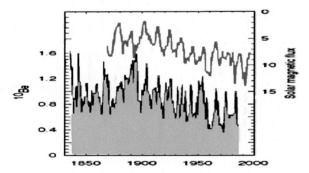

FIGURE 14.32   Variation of ^{10}Be concentration in Greenland ice cores and the solar magnetic field derived from geomagnetic measurements. *Modified from Kirkby, J., 2008. Cosmic rays and climate. Surveys in Geophysics 28, 5.*

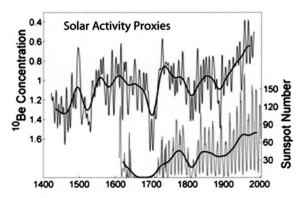

FIGURE 14.33   Correlation of ^{10}Be concentration from the Dye-3 Greenland ice core and SSNs. (Beer et al., 1994.)

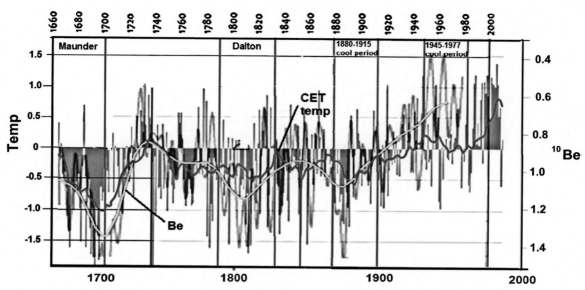

FIGURE 14.34   Correlation of [10]Be and temperature. Note that the Maunder, Dalton, 1880–1915, and 1945–1977 cold periods were all characterized by high rates of [10]Be production, indicating increased incidence of cosmic rays.

3.2 Berylium-10 ([10]Be) Production and Temperature

Fig. 14.34 shows that high production rates of [10]Be occurred during the Maunder, Dalton, 1880–1915, and 1945–1977 cold periods, indicating higher incidence of cosmic rays during the cold episodes.

4. COSMIC RAY INCIDENCE AND CLIMATE

Studies of the effects of cosmic radiation on the atmosphere date back to the 1912 development of the Wilson cloud chamber by C.T.R. Wilson, who showed that cosmic radiation would create trails of condensation in water-saturated air. The role of cosmic rays in climate change was suggested by Ney (1959) and by Dickinson (1975), and more recently by Svensmark (2006), Svensmark and Calder (2007), Svensmark and Friis-Christensen (1997), Svensmark et al. (2007), Usoskin et al. (2004a,b), Kirkby (2008), Marsh and Svensmark (2000), and others.

Cosmic rays consist of two types of high-energy radiation: (1) galactic cosmic rays (GCRs), high−energy particles originating outside the solar system; and (2) high-energy particles (mostly protons) emitted by the sun during solar events. They may produce showers of secondary particles that penetrate and impact the Earth's atmosphere. About 99% of primary cosmic rays entering the Earth's atmosphere are nuclei of atoms; about 90% are simple protons and 9% alpha particles. About 1% consists of electrons.

High-energy protons passing through the atmosphere cause ionization and produce nuclei for condensation of water droplets. Condensation tends to occur readily in the atmosphere because it is often supersaturated with water vapor. Clouds reflect incoming solar irradiance, which results in atmospheric cooling. Clouds account for about 28 Wm^{-2} of global cooling, so even small changes in cloud cover can have a significant effect on climate. Low-altitude, layered clouds covering large areas are most effective in reflecting incoming solar radiation and make the greatest contribution to atmospheric cooling. Cosmic-ray-produced clouds may provide the key to understanding global climate. Increased cosmic ray flux creates clouds, which increase albedo and results in global cooling (Fig. 14.35). This mechanism explains the observed synchronicity of global climate changes, abrupt climate reversals, and climate changes on all time scales. Thus, cloud-generating cosmic rays provide a satisfactory explanation for both long-term and short-term climate changes (Easterbrook, 2001, 2005, 2006a,b, 2007, 2008a,b,c, 2009, 2011a,b, 2014).

As discussed above, cosmic rays produce radiocarbon and berylium−10 isotopes in the upper atmosphere in amounts proportional to the incidence of incoming radiation. This is reflected in the coincidence of ^{14}C and [10]Be production rates (Fig. 14.36).

Fig. 14.37 summarizes the relationships between low solar magnetic field, sunspots, galactic cosmic rays, cloud formation, albedo, and global cooling.

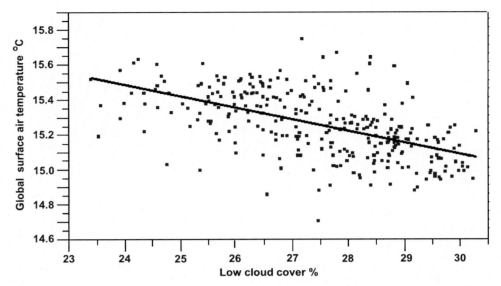

FIGURE 14.35   Global atmospheric cooling by increasing cloud cover. As cloud cover increases, temperatures decline.

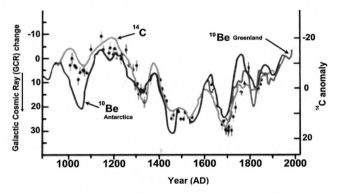

FIGURE 14.36   Correlation of ^{14}C and [10]Be production rates for the past 1000 years. *Modified from Kirkby, J., 2008. Cosmic rays and climate. Surveys in Geophysics 28, 5.*

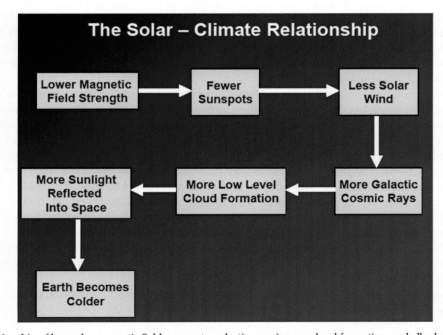

FIGURE 14.37   Relationship of low solar magnetic field, sunspots, galactic cosmic rays, cloud formation, and albedo in causing global cooling.

5. CONCLUSIONS

Excellent correlations can be made between global temperature change, sunspots, TSI, ^{14}C and [10]Be production in the upper atmosphere, cosmic ray incidence, and albedo from cloud generation. Global cooling coincided with changes in sunspot activity, TSI, solar flux, cosmic ray incidence, and rates of production of ^{14}C and [10]Be in the upper atmosphere during the Oort, Wolf, Maunder, Dalton, 1880—1915, and 1945—1977 Solar Minimums. Increased ^{14}C and [10]Be production during times of increased cosmic radiation serve as a proxy for solar activity.

Ionization in the atmosphere caused by cosmic rays causes increased cloudiness that reflects incoming sunlight and cools the Earth. The amount of cosmic radiation is greatly affected by the sun's magnetic field, so during times of weak solar magnetic field, more cosmic radiation reaches the Earth, creating more cloudiness and cooling the atmosphere.

This mechanism accounts for the global synchronicity of climate changes, abrupt climate reversals, and climate changes on all time scales. Thus, cloud-generating cosmic rays provide a satisfactory explanation for both long-term and short-term climate changes.

References

Beer, J., Joos, F., Lukasczyk, C., Mende, W., Rodriguez, J., Sikegenthaler, U., Stellmacher, R., 1994. [10]Be as an indicator of solar variability and climate. In: Nesme-Ribes, E. (Ed.), The Solar Engine and Its Influence on Terrestrial Atmosphere and Climate. Springer—Verlag, Berlin, pp. 221—233.

Dickinson, R.E., 1975. Solar variability and the lower atmosphere. Bulletin of American Meteorological Society 56, 1240.

Easterbrook, D.J., 2001. The next 25 years: global warming or global cooling? Geologic and oceanographic evidence for cyclical climatic oscillations. Geological Society of America 33, 253. Abstracts with Program.

Easterbrook, D.J., 2005. Causes and effects of abrupt, global, climate changes and global warming. Geological Society of America 37, 41. Abstracts with Program.

Easterbrook, D.J., 2006a. Causes of abrupt global climate changes and global warming predictions for the coming century. Geological Society of America 38, 77. Abstracts with Program.

Easterbrook, D.J., 2006b. The cause of global warming and predictions for the coming century. Geological Society of America 38, 235—236. Abstracts with Program.

Easterbrook, D.J., 2007. Geologic evidence of recurring climate cycles and their implications for the cause of global warming and climate changes in the coming century. Geological Society of America 39, 507. Abstracts with Programs.

Easterbrook, D.J., 2008a. Solar influence on recurring global, decadal, climate cycles recorded by glacial fluctuations, ice cores, sea surface temperatures, and historic measurements over the past millennium. In: Abstracts of American Geophysical Union Annual Meeting, San Francisco.

Easterbrook, D.J., 2008b. Implications of glacial fluctuations, PDO, NAO, and sun spot cycles for global climate in the coming decades. Geological Society of America 40, 428. Abstracts with Programs.

Easterbrook, D.J., 2008c. Correlation of climatic and solar variations over the past 500 years and predicting global climate changes from recurring climate cycles. In: Abstracts of 33rd International Geological Congress, Oslo, Norway.

Easterbrook, D.J., 2009. The role of the oceans and the sun in late Pleistocene and historical glacial and climatic fluctuations. Geological Society of America 41, 33. Abstracts with Programs.

Easterbrook, D.J., 2011a. Geologic evidence of recurring climate cycles and their implications for the cause of global climate changes: the past is the key to the future. In: Evidence-based Climate Science. Elsevier Inc, pp. 3—51.

Easterbrook, D.J., 2011b. Climatic implications of the impending grand solar minimum and cool Pacific Decadal Oscillation: the past is the key to the future—what we can learn from recurring past climate cycles recorded by glacial fluctuations, ice cores, sea surface temperatures, and historic measurements. Geologic Society of America 43, 34. Abstracts with Programs.

Easterbrook, D.J., 2014. Synchroneity of multiple Younger Dryas and Allerod moraines in the Fraser Lowland with late Pleistocene moraines in North America, Europe, New Zealand, and South America. Geological Society of America 46, 349. Abstracts with Programs.

Eddy, J.A., 1976. The Maunder Minimum. Science 192, 1189—1202.

Eddy, J.A., 1977. Climate and the changing sun. Climatic Change 1, 173—190.

Fagan, B., 2000. The Little Ice Age. Basic Books, N.Y., p. 246.

Fagan, B., 2007. The Great Warming: Climate Change and the Rise and Fall of Civilizations. Bloomsbury Press, p. 283.

Grootes, P.M., Stuiver, M., 1997. Oxygen 18/16 variability in Greenland snow and ice with 10^3 to 10^5-year time resolution. Journal of Geophysical Research 102, 26455—26470.

Grove, J.M., 2004. Little Ice Ages: Ancient and Modern. Routledge, London, p. 718.

Hoyt, D.V., Schatten, K.H., 1997. The Role of the Sun in Climate Change. Oxford University, p. 279.

Hoyt, D.V., Schatten, K.H., 1998. Group sunspot numbers: a new solar activity reconstruction. Solar Physics 179, 189—219.

Kirkby, J., 2008. Cosmic rays and climate. Surveys in Geophysics 28, 5.

Lean, J.L., Beer, J., Bradley, R., 1995. Reconstruction of solar irradiance since 1610: implications for climatic change. Geophysics Research Letters 22, 3195—3198.

Lean, J.L., Wang, Y.M., Sheeley, N.R., 2002. The effect of increasing solar activity on the Sun's total and open magnetic flux during multiple cycles: implications for solar forcing of climate. Geophysics Research Letters 29, 2224.

Marsh, N.D., Svensmark, H., 2000. Low cloud properties influenced by cosmic rays. Physics Revue Letters 85, 5004—5007.

Matter, A., Neff, U., Fleitmann, D., Burns, S., Mangini, A., 2001. 350,000 years of climate variability recorded in speleothems from Oman. GeoArabia (Manama) 6, 315–316.

Maunder, E.W., 1894. A prolonged sunspot minimum. Knowledge 17, 173–176.

Maunder, E.W., 1922. The prolonged sunspot minimum, 1645–1715. Journal of the British Astronomical Society 32, 140.

Ney, E.P., 1959. Cosmic radiation and the weather. Nature 183, 451–452.

Soon, W., 2005. Variable solar irradiance as a plausible agent for multidecadal variations in the Arctic–wide surface air temperature record of the past 130 years. Geophysical Research Letters 32, L16712.

Stuiver, M., 1961. Variations in radio carbon concentration and sunspot activity. Geophysical Research 66, 273–276.

Stuiver, M., 1994. Atmospheric ^{14}C as a proxy of solar and climatic change. In: Nesme-Ribes (Ed.), The Solar Engine and Its Influence on Terrestrial Atmosphere and Climate. Springer–Verlag, Berlin, pp. 203–220.

Stuiver, M., Brasiunas, T.F., 1991. Isotopic and solar records. In: Bradley, R.S. (Ed.), Global Changes of the Past. Boulder University, Corporation for Atmospheric Research, pp. 225–244.

Stuiver, M., Quay, P.D., 1980. Changes in atmospheric carbon-14 attributed to a variable Sun. Science 207, 11–19.

Stuiver, M., Grootes, P.M., Brasiunas, T.F., 1995. The GISP2 d^{18}O record of the past 16,500 years and the role of the sun, ocean, and volcanoes. Quaternary Research 44, 341–354.

Stuiver, M., Braziunas, T.F., Becker, B., Kromer, B., 1991. Climatic, solar, oceanic, and geomagnetic influences on late-glacial and Holocene atmospheric ^{14}C/^{12}C change. Quaternary Research 35, 1–24.

Svensmark, H., 2006. Imprint of galactic dynamics on Earth's climate. Astronomische Nachrichten 327, 866–870.

Svensmark, H., Calder, N., 2007. The Chilling Stars: A New Theory of Climate Change. Icon Books, Allen and Unwin Pty Ltd, p. 246.

Svensmark, H., Friis-Christensen, E., 1997. Variation of cosmic ray flux and global cloud cover–a missing link in solar–climate relationships. Journal of Atmospheric and Solar–Terrestrial Physics 59, 1125–1132.

Svensmark, H., Pedersen, J.O., Marsh, N.D., Enghoff, M.B., Uggerhøj, U.I., 2007. Experimental evidence for the role of ions in particle nucleation under atmospheric conditions. Proceedings of the Royal Society 463, 385–396.

Usoskin, I.G., Marsh, N.D., Kovaltsov, G.A., Mursula, K., Gladysheva, O.G., 2004a. Latitudinal dependence of low cloud amount on cosmic ray induced ionization. Geophysis Research Letters 31, L16109. http://dx.doi.org/10.1029/2004GL019507.

Usoskin, I.G., Mursula, K., Solanki, S.K., Schussler, M., Alanko, K., 2004b. Reconstruction of solar activity for the last millenium using ^{10}Be data. Astronomy and Astrophysics 413, 745–751.

Usoskin, I.G., et al., 2015. The Maunder minimum (1645–1715) was indeed a grand minimum: a reassessment of multiple datasets. Astronomy Astrophysics 581, 1–19.

CHAPTER

15

Solar Changes and the Climate

J.S. D'Aleo

American Meteorological Society, Hudson, NH, United States

1. INTRODUCTION

The Intergovernmental Panel on Climate Change (IPCC) AR4 report discussed some of the research on variance of solar irradiance and uncertainties related to indirect solar influences, ultraviolet and solar wind/geomagnetic activity. They admit that ultraviolet radiation, by warming through ozone chemistry, and geomagnetic activity, through the reduction of cosmic rays, could have an effect on low clouds, but in the end chose to ignore the indirect effect. They stated:

> Since TAR, new studies have confirmed and advanced the plausibility of indirect effects involving the modification of the stratosphere by solar UV irradiance variations (and possibly by solar-induced variations in the overlying mesosphere and lower thermosphere), with subsequent dynamical and radiative coupling to the troposphere. Whether solar wind fluctuations (Boberg and Lundstedt, 2002) or solar-induced heliospheric modulation of galactic cosmic rays (Marsh and Svensmark, 2000) also contribute indirect forcings remains ambiguous.
>
> *AR4 2.7.1.3*

In the end, the AR4 chose to ignore the considerable peer-reviewed literature on total solar forcing in favor of Wang et al. (2005), who used an untested flux transport model with variable meridional flow hypothesis that reduced net, long-term variance of direct solar irradiance since the Little Ice Age by a factor of seven. This may ultimately prove to be AR4's version of the AR3's "hockey stick" debacle.

Though AR5 gave more attention to other possible solar amplifiers, they defended the model assumption that greenhouse gases were the principle driver.

Evidence-Based Climate Science, Second Edition
http://dx.doi.org/10.1016/B978-0-12-804588-6.00015-X

SC 23 showed an activity decline not previously seen in the satellite era (McComas et al., 2008; Smith and Balogh, 2008; Russell et al., 2010). Most current estimations suggest that the forthcoming solar cycles will have lower TSI than those for the past 30 years (Abreu et al., 2008; Lockwood et al., 2009; Rigozo et al., 2001; Russell et al., 2010). Also there are indications that the mean magnetic field in sunspots may be diminishing on decadal level. A linear expansion of the current trend may indicate that of the order of half the sunspot activity may disappear by about 2015 (Penn and Livingston, 2006). These studies only suggest that the Sun may have left the 20th century grand maximum and not that it is entering another grand minimum. However, much more evidence is needed and at present there is very low confidence concerning future solar forcing estimates. Nevertheless, even if there is such decrease in the solar activity, there is a high confidence that the TSI Radiative Forcing variations will be much smaller in magnitude than the projected increased forcing due to GHG.

2. THE EARTH–SUN CONNECTION

The Sun is the ultimate source of climate-related energy on Earth. Its rays heat the planet and drive the churning motions of its atmosphere. The amount of energy that the Sun emits varies over an 11-year cycle, which also governs the extent and strength of sunspots on the Sun's surface and radiation storms that can wreak havoc with satellites and electric grids, but that cycle changes the total amount of energy reaching Earth by only about 0.1 percent. This has presented a conundrum for meteorologists in explaining how such a small variation could drive major changes in weather patterns on Earth.

Although the Sun's brightness/irradiance changes only slightly with solar cycles, the indirect effects of enhanced solar activity, including warming of the atmosphere in low and mid-latitudes by ozone reactions due to increased ultraviolet radiation; in higher latitudes by geomagnetic activity; and generally by increased solar radiative forcing due to fewer clouds caused by cosmic ray reduction, may greatly magnify the total solar effect on temperatures. The following is an assessment of the ways the Sun may influence weather and climate on short and long time scales.

2.1 The Sun Plays Direct and Indirect Roles in Climate

The Sun changes its activity on time scales that vary from 27 days to 11, 22, 80, 106, 212 years, and more. A more active Sun is brighter due to the dominance of faculae over cooler sunspots, resulting in increased solar irradiance. The amount of change of solar irradiance, based on satellite measurements since 1978 during the course of an 11-year cycle, is only 0.1% (Willson and Hudson, 1988), causing many to conclude that the solar effect is negligible. Cycle 23 has declined 0.15%. Over long cycles since the Maunder Minimum, irradiance changes are estimated to be as high as 0.4% (Hoyt and Schatten, 1997; Lean, 2000; Lockwood and Stamper, 1999; Fligge and Solanki, 2000).

However, this does not take into account the Sun's eruptional activity (flares, solar wind bursts from coronal mass ejections, and solar wind bursts from coronal holes), which may have a much greater effect. This takes on more importance since Lockwood et al. (1999) showed how the total magnetic flux leaving the Sun has increased by a factor of 2.3 since 1901. This eruptional activity may enhance warming through ultraviolet-induced ozone chemical reactions in the high atmosphere or ionization in higher latitudes during solar-induced geomagnetic storms. In addition, Svensmark (2007), Palle Bago and Butler (2000), and Tinsley and Yu (2002) have documented possible effects of the solar cycle on cosmic rays and through them, the amount of low cloudiness. These other indirect factors suggest that solar variance may be a much more important driver for climate change than currently assumed. Because it is more easily measured, tracking eruptional activity from solar irradiance, solar irradiance measurements have been used as a surrogate or proxy for the total solar effect.

2.2 Correlations with Total Solar Irradiance

Studies vary on the importance of direct solar irradiance, especially in recent decades. Lockwood and Stamper (1999) estimated that changes in solar luminosity can account for 52% of the change in temperatures from 1910 to 1960, but just 31% of the change from 1970 to 1999. Scafetta and West (2006a,b,c) concluded that total solar irradiance (TSI) accounted for up to 50% of the warming since 1900 and 25–35% since 1980. They noted that recent departures from earlier trends may result from spurious nonclimatic contamination of surface observations, such as heat-island and land-use effects (Pielke et al., 2002; Kalnay and Cai, 2003). Their analysis was done using global databases that may suffer from station dropout and improper adjustment for missing data, which increased in the 1990s. In 2007, in a follow-up paper, they noted the Sun could account for as much as 69% of the changes since 1900.

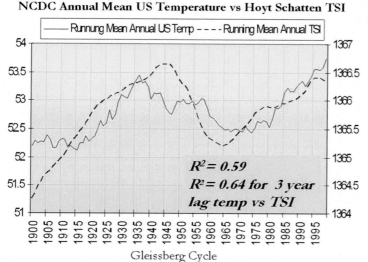

NCDC Annual Mean US Temperature vs Hoyt Schatten TSI

$R^2 = 0.59$
$R^2 = 0.64$ for 3 year lag temp vs TSI

Gleissberg Cycle

FIGURE 15.1   USHCN annual mean temperature (11-year running mean) correlated with Hoyt-Schatten Total Solar Irradiance (also 11-year running mean).

The original USHCN database, though regional in nature, would have been a better station database to use for analysis of change as it is more stable, has less missing data, and has a better scheme for adjusting for missing data, changes in siting, and urbanization. An independent analysis was conducted using the USHCN data and TSI data obtained from Hoyt and Schatten. The annual TSI composite record was constructed by Hoyt and Schatten (1997) and updated in 2005 utilizing all five historical proxies of solar irradiance, including sunspot cycle amplitude, sunspot cycle length, solar equatorial rotation rate, fraction of penumbral spots, and decay rate of the 11-year sunspot cycle.

Fig. 15.1 shows the 11-year running mean solar irradiance versus a similar 11-year running mean of NCDC USHCN v1 annual mean U.S. temperatures. It confirms a strong correlation (*r*-squared of 0.59). The correlation increased to an *r*-squared value of 0.654 if you introduce a lag of 3 years for the mean USHCN data to the mean TSI. This is close to the 5-year lag suggested by Wigley and used by Scafetta and West. The highest correlation occurred with a 3-year lag.

In recent years, satellite missions designed to measure changes in solar irradiance have encountered problems. As Judith Lean has noted, the problem is that no single sensor collected data over the entire time period from 1979, "forcing a splicing from different instruments, each with their own accuracy and reliability issues, only some of which we are able to account for." Fröhlich and Lean (1998) suggested no increase in solar irradiance in the 1980s and 1990s.

Richard Willson, principal investigator of NASA's ACRIM experiments, found specific errors in the data set used by Lean and Fröhlich to bridge the gap between the ACRIM satellites (Willson, 1997; Willson and Mordvinov, 2003). When the more accurate data set was used, a trend of 0.05% per decade was seen, which could account for warming since 1979 (Fig. 15.2).

Soon (2005) showed clear connections between solar changes and the Earth's climate. Soon showed that Arctic temperatures, with no urbanization contamination, correlated with solar irradiance far better than with the greenhouse gases over the last century (Fig. 15.3). The 10-year running mean of TSI versus Arctic air temperature shows a strong correlation (*r*-squared of 0.79), compared to a correlation versus greenhouse gases of just 0.22.

3. WARMING DUE TO ULTRAVIOLET EFFECTS THROUGH OZONE CHEMISTRY

Although solar irradiance varies slightly over the 11-year cycle, radiation at longer UV wavelengths is known to increase by several percent (6—8% or more), with still larger changes (factor of two or more) at extremely short UV and X-ray wavelengths (Baldwin and Dunkerton, 2005).

Energetic solar flares increase the UV radiation by 16%. Ozone in the stratosphere absorbs this excess energy, and heat has been shown to propagate downward, affecting general circulation in the troposphere. Shindell et al. (1999)

FIGURE 15.2   Richard Willson (ACRIMSAT) composite TSI showing trend of +0.05% per decade from successive solar minima (2003).

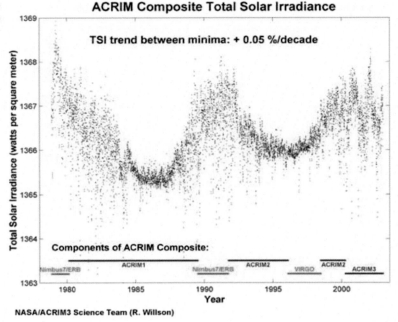

used a climate model that included ozone chemistry to reproduce this warming during high-flux (high UV) years. Labitzke and Van Loon (1988) and Labitzke in numerous papers have shown that high flux, which correlates very well with UV, produces stratosphere warming in low and middle latitudes in winter, with subsequent dynamic and radiative coupling to the troposphere. The winter of 2001–2002, when Cycle 23 had a very strong high flux, second maxima (Fig. 15.4) provided verification of the conclusions of Shindell and Labitzke and Van Loon.

The warming that took place with the high flux from September 2001 to April 2002 caused the northern winter polar vortex to shrink (Fig. 15.5) and the southern summer vortex to break into two centers for the first time ever observed. This disrupted flow patterns and may have contributed to the brief summer breakup of the Larsen ice sheet (Figs. 15.6–15.8).

HUNASA reported UH on the use of the Shindell et al. (1999) Ozone Chemistry Climate Model to explain the Maunder Minimum (Little Ice Age) (Fig. 15.9).

Their model showed that when the Sun was quiet in 1680, the climate was much colder than when it became active again 100 years later. "During this period, very few sunspots appeared on the surface of the Sun, and the overall brightness of the Sun decreased slightly. Already in the midst of a colder-than-average period called the Little Ice Age, Europe and North America went into a deep freeze: alpine glaciers extended over valley farmland; sea ice crept south from the Arctic; and the famous canals in the Netherlands froze regularly—an event that is rare today."

Lockwood et al. (2010) verified that solar activity does seem to have a direct correlation with climate by influencing North Atlantic blocking (NAO), as Shindell et al. (1999) have shown. The scope of the study was limited to areas of reliable, continuous temperature record going back to the Little Ice Age. They noted further:

> Solar activity during the current sunspot minimum has fallen to levels unknown since the start of the 20th century. The Maunder minimum (about 1650–1700) was a prolonged episode of low solar activity, which coincided with more severe winters in the United Kingdom and continental Europe. Motivated by recent relatively cold winters in the UK, we investigate the possible connection with solar activity. We identify regionally anomalous cold winters by detrending the Central England temperature (CET) record using reconstructions of the northern hemisphere mean temperature.

> We show that cold winter excursions from the hemispheric trend occur more commonly in the UK during low solar activity, consistent with the solar influence on the occurrence of persistent blocking events in the eastern Atlantic. We stress that this is a regional and seasonal effect relating to European winters and not a global effect. Average solar activity has declined rapidly since 1985 and cosmogenic isotopes suggest an 8% chance of a return to Maunder minimum conditions within the next 50 years (Lockwood, 2010): the results presented here indicate that, despite hemispheric warming, the UK and Europe could experience more cold winters than during recent decades.

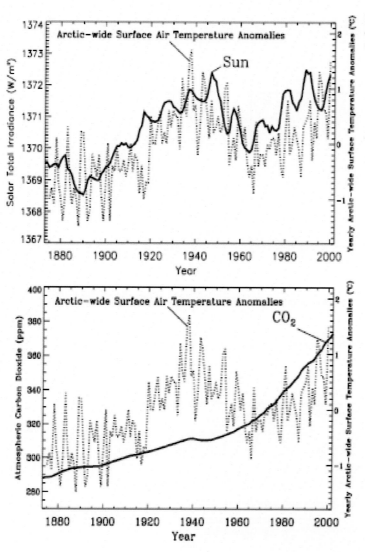

FIGURE 15.3 Arctic Basin wide air temperatures (Polyokov) correlated with Hoyt Schatten TSI and with annual average CO_2 (Soon, 2005).

We note the NAO has ties to the Arctic Oscillation (AO), and together they have great influence on the hemisphere. The 2009–2010 winter, with a record negative Arctic oscillation and persistent negative NAO (Fig. 15.10), was the coldest in the UK and the southeastern United States since 1977–1978, coldest in Scotland since 1962–1963, and coldest ever recorded in parts of Siberia. The coldest weather since 1971–1972 was reported in parts of North China.

The IPCC AR5 expanded on the AR4 discussion of the potential role UV in cyclical changes of global temperatures, including the global warming from 1979 to 1998, the modeling of which they admitted was not yet "robust."

> As UV heating of the stratosphere over a SC has the potential to influence the troposphere indirectly, through dynamic coupling, and therefore climate (Haigh, 1996; Gray et al., 2010), the UV may have a more significant impact on climate than changes in TSI alone would suggest. Although this indicates that metrics based only on TSI are not appropriate, UV measurements present several controversial issues and modelling is not yet robust.
>
> *AR5 8.4.1.4.2*

3.1 Tropical Effects

Elsner et al. (2010) found the probability of three or more hurricanes hitting the United States goes up drastically during low points of the 11-year sunspot cycle. Years with few sunspots and above-normal ocean temperatures

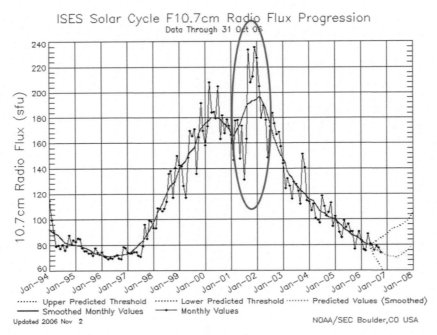

FIGURE 15.4   NOAA SEC solar flux (10.7 cm) during Cycle 23. Note the second solar max with extremely high flux from September 2001 to April 2002.

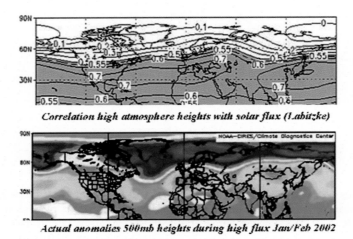

FIGURE 15.5   Labitzke correlated stratospheric heights with solar flux and actual height anomalies in the mid-troposphere during the high-flux mode of the second solar max in early 2002.

spawn a less stable atmosphere and consequently more hurricanes. Years with more sunspots and above-normal ocean temperatures yield a more stable atmosphere and thus fewer hurricanes (Fig. 15.11). Elsner et al. found that radiation can vary more than 10 percent in parts of the ultraviolet range between the high and low of the sunspot cycle. The more sunspots, the more ultraviolet radiation, which heats the atmosphere below.

The Sun's yearly average radiance during its 11-year cycle only changes about 0.1%, according to NASA's Earth Observatory. But the warming in the ozone layer can be much more profound because ozone absorbs ultraviolet radiation. Elsner et al. found that between the high and low of the sunspot cycle, radiation can vary more than 10% in parts of the ultraviolet range. With more sunspots, and therefore ultraviolet radiation, the warmer ozone layer heats the atmosphere below. They showed evidence that increased UV light from solar activity can influence a hurricane's power even on a daily basis.

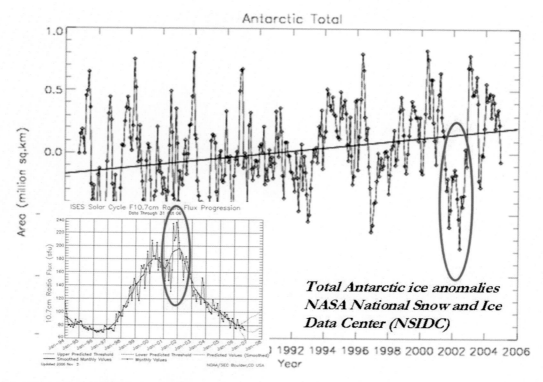

FIGURE 15.6   NASA NSIDC satellite-derived total Antarctic ice extent anomalies from 1979 to 2005. Note the drop-off with the Larsen ice sheet break-up in the summer of 2002 corresponding to major atmospheric changes during the high-flux second solar max.

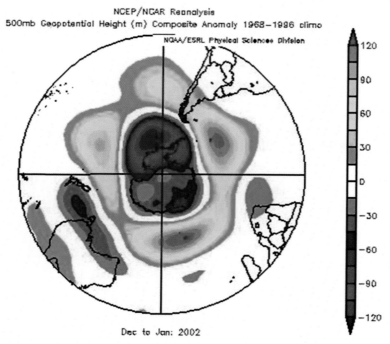

FIGURE 15.7   December 2001 to January 2002 500-mb height anomalies for Southern Hemisphere. Note the ring of warming with the high-flux induced UV ozone chemistry as a ring surrounding a shrunken polar vortex as seen in the Northern Hemisphere in Fig. 15.2. Note how the vortex actually became a dipole with weakness in center. The changing winds and currents very likely contributed to the ice break of the Larsen ice sheet.

FIGURE 15.8 Larsen ice sheet break-up late summer 2000 following strong solar flux break-up of southern polar vortex.

FIGURE 15.9 Shindell ozone chemistry model forecast of the difference between the quiet solar period of the Maunder Minimum and the active late 18th century.

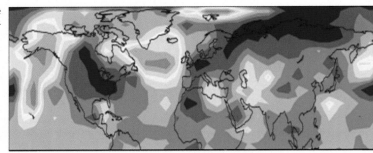

FIGURE 15.10 The North Atlantic Oscillation (NAO) tracks well with the solar activity, most specifically with the Ap geomagnetic index that tends to lag the solar sunspot cycle. A very quiet Sun is characterized by persistent periods of a negative NAO and usually the AO, most notably in winter.

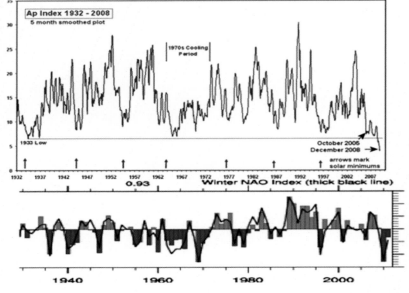

Hurricanes and the sunspot theory

Increased solar activity such as sunspots can warm upper layers of Earth's atmosphere, making the atmosphere more stable and decreasing hurricanes. Sunspot activity varies on an 11-year cycle. Researchers at Florida State University theorize that hurricane activity may increase as sunspots decrease. **Here's how:**

FIGURE 15.11 Hodges and Elsner found that the probability of three or more hurricanes hitting the United States goes up drastically during low points of the 11-year sunspot cycle. Years with few sunspots and above-normal ocean temperatures spawn a less stable atmosphere and consequently more hurricanes. Years with more sunspots and above-normal ocean temperatures yield a more stable atmosphere and thus fewer hurricanes.

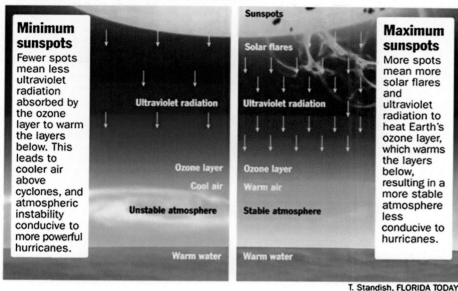

Minimum sunspots
Fewer spots mean less ultraviolet radiation absorbed by the ozone layer to warm the layers below. This leads to cooler air above cyclones, and atmospheric instability conducive to more powerful hurricanes.

Ultraviolet radiation

Ozone layer

Cool air

Unstable atmosphere

Warm water

Sunspots

Solar flares

Ultraviolet radiation

Ozone layer

Warm air

Stable atmosphere

Warm water

Maximum sunspots
More spots mean more solar flares and ultraviolet radiation to heat Earth's ozone layer, which warms the layers below, resulting in a more stable atmosphere less conducive to hurricanes.

T. Standish, FLORIDA TODAY

3.2 Other Effects

Meehl et al. (2009) found that ozone in the stratosphere and sea-surface temperatures in the Pacific Ocean respond during solar maximum in a way that amplifies the Sun's influence on some aspects of air movement. This can intensify winds and rainfall, change sea surface temperatures and cloud cover over tropical and subtropical regions, and influence global weather. An international team of scientists led by the National Center for Atmospheric Research (NCAR) used more than a century of weather observations and three powerful computer models to tackle this question. The study found that the Sun impacts two seemingly unrelated regions, water in the tropical Pacific Ocean and air in the stratosphere from about 6 miles (10 km) above Earth's surface to about 31 miles (50 km).

"The sun, the stratosphere, and the oceans are connected in ways that can influence events such as winter rainfall in North America," said lead author of the study, Gerald Meehl of NCAR. Understanding the role of the solar cycle can provide added insight as scientists work toward predicting regional weather patterns for the coming decades.

3.3 Geomagnetic Storms and High-Latitude Warming

When major solar eruptive activity (ie, coronal mass ejections, flares) takes place and charged particles encounter the earth, ionization in the high atmosphere leads to increased auroras. This ionization leads to warming of the high atmosphere, which, like ultraviolet warming of the stratosphere, works its way down into the middle troposphere with time.

Fig. 15.12 is an example of upper-level conditions two weeks after a major geomagnetic storm. Note the ring of warmth (higher than normal mid-tropospheric heights) surrounding the magnetic pole.

3.4 Solar Winds, Cosmic Rays, and Clouds

A key aspect of the Sun's effect on climate is the indirect effect on the of galactic cosmic ray flux (GCR) on the atmosphere. GCR is an ionizing radiation that supports low cloud formation. As the Sun's output increases, the solar wind shields the atmosphere from GCR flux. Consequently, the increased solar irradiance is accompanied by reduced low cloud cover, amplifying the climatic effect. Likewise, when solar output declines, increased GCR

FIGURE 15.12   Anomaly at 500 mb two weeks after a major geomagnetic storm in 2005. Warmth seen in approximate location and shape of auroral ring.

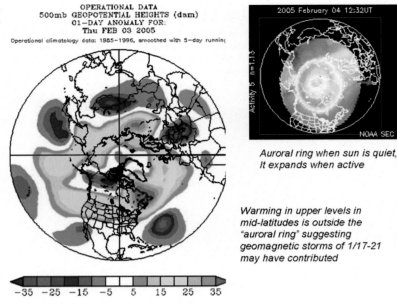

Auroral ring when sun is quiet, It expands when active

Warming in upper levels in mid-latitudes is outside the "auroral ring" suggesting geomagnetic storms of 1/17-21 may have contributed

flux enters the atmosphere, increasing low cloudiness and adding to the cooling effect associated with the diminished solar energy.

The conjectured mechanism connecting GCR flux to cloud formation received experimental confirmation in the recent laboratory experiments of Svensmark (2007) and Svensmark and Friis-Christensen (1997) in which he demonstrated exactly how cosmic rays could make water droplets.

Palle and Butler (2002) showed how low clouds changed with the 11-year cycle in inverse relation to the solar activity (Fig. 15.13). Changes of 1—2% in low cloudiness could have a significant effect on temperatures through changes in albedo.

FIGURE 15.13 Cosmic ray neutrons are inversely proportional to solar activity and directly proportional to low cloudiness. *From Palle Bago, E., Butler, C.J., 2001. Sunshine records from Ireland: cloud factors and possible links to solar activity and cosmic rays. International Journal of Climatology 21, 709—729; Palle, E., Butler, C.J., 2002. The proposed connection between clouds and cosmic rays: cloud behavior during the past 50—120 years. Journal Atmospheric Solar Terrestrial Physics 64, 327—337.*

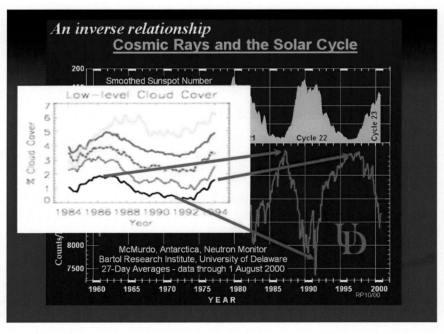

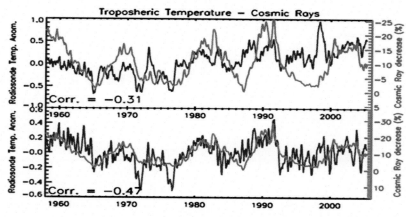

FIGURE 15.14   Tropospheric cosmic rays versus radiosonde temperature anomalies raw and bottom filtered and detrended (Svensmark and Friis-Christensen, 1997).

Svensmark and Friis-Christensen (1997) correlated tropospheric temperature with cosmic rays. Fig. 15.14 consists of two graphs. The first graph compares tropospheric temperature (blue) to cosmic rays (red). The second graph removes El Niño, volcanoes, and a **linear warming trend** of 0.14°C per decade.

The purpose of CERN's CLOUD experiment was "to study and quantify the cosmic ray-cloud mechanism in a controlled laboratory experiment" and answer the question of whether or not—and to what extent—climate is influenced by solar/cosmic ray variability. Jasper Kirkby of CERN found compelling evidence that indeed there could be such a connection (Figs. 15.15 and 15.16).

Cloud observations

- Original GCR-cloud correlation made by Svensmark & Friis-Christensen, 1997
- Many studies since then supporting or disputing solar/GCR - cloud correlation
- Not independent - most use the same ISCCP satellite cloud dataset
- No firm conclusion yet - requires more data - but, if there is an effect, it is likely to be restricted to certain regions of globe and at certain altitudes & conditions
- Eg. correlation (>90% sig.) of low cloud amount and solar UV/GCR, 1984-2004:

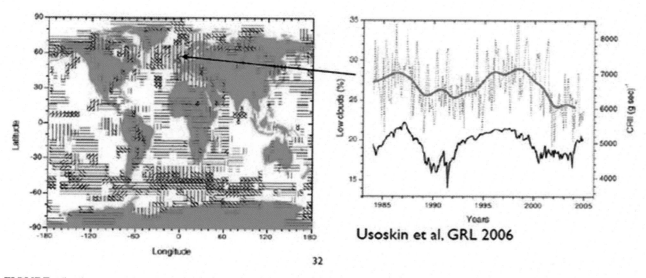

Usoskin et al. GRL 2006

FIGURE 15.15   Jasper Kirkby of CERN as an introduction to CERN's CLOUD experiment summarized the state of understanding.

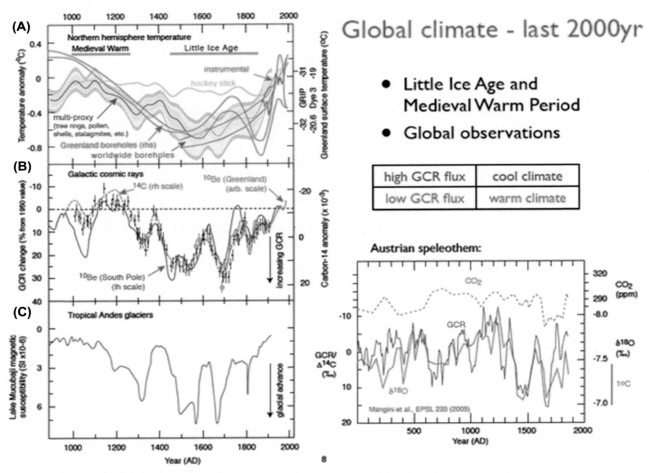

FIGURE 15.16   Excellent correlations of galactic cosmic rays and temperatures for the Northern Hemisphere, Greenland, tropical Andes, and Austria (Jasper Kirkby of CERN).

The influence of cosmic rays also appears on long geologic time scales. Using a combination of increased radiative forcing through cosmic ray reduction and the estimated changes in total solar luminosity (irradiance) over the last century, Shaviv (2005) estimated that the Sun could be responsible for up to 77% of temperature changes over the 20th century (Figs. 15.17 and 15.18).

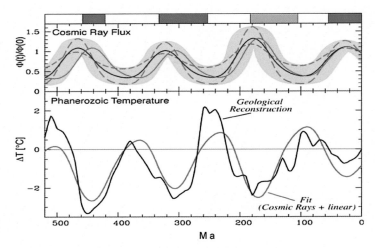

FIGURE 15.17   Cosmic ray flux plus irradiance versus geological reconstruction of temperatures (Shaviv, 2005).

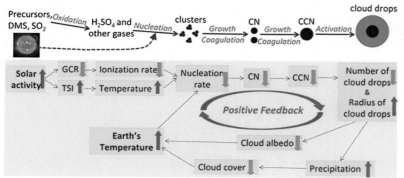

FIGURE 15.18 Schematic illustration of a mechanism amplifying the effect of solar variability through the influence of GCR ionization and temperature on particle formation, and a positive nucleation feedback (enhanced solar activity → more TSI and less GCR → reduced nucleation and aerosol abundance → less aerosol cooling → increased temperature).

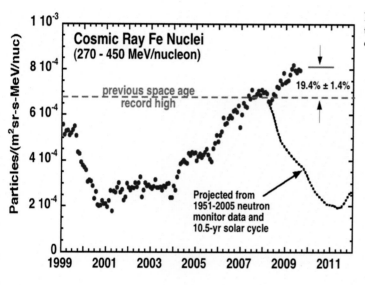

FIGURE 15.19 NASA cosmic ray monitoring data showing that the number of particles was approximately 19.4% higher than any other time since 1951.

The IPCC AR5 report evaluated the cosmic ray effect, but concluded with "high confidence" that the solar-modulated cosmic ray flux did not affect temperatures in a climatologically significant way.

The Effects of Cosmic Rays on Clouds Changing cloud amount or properties modify the Earth's albedo and therefore affect climate. It has been hypothesized that cosmic ray flux create atmospheric ions which facilitates aerosol nucleation and new particle formation with a further impact on cloud formation (Dickinson, 1975; Kirkby, 2007). High solar activity means a stronger heliospheric magnetic field and thus a more efficient screen against cosmic rays. Under the hypothesis underlined above, the reduced cosmic ray flux would promote fewer clouds amplifying the warming effect expected from high solar activity. There is evidence from laboratory, field and modelling studies that ionization from cosmic ray flux may enhance aerosol nucleation in the free troposphere (Merikanto et al., 2009; Mirme et al., 2010; Kirkby, 2011). However, there is high confidence (medium evidence and high agreement) that the cosmic ray— ionization mechanism is too weak to influence global concentrations of cloud condensation nuclei or their change over the last century or during a SC in a climatically significant way (Harrison and Ambaum, 2010; Erlykin and Wolfendale, 2011; Snow-Kropla et al., 2011). **AR5 8.4.1.5**

Yu and Luo (2014) found that the effect of solar-cycle variations on cloud condensation nuclei was up to one order of magnitude higher than reported in several previous studies. According to NASA, cosmic rays recently reached a space-age high (Fig. 15.19).

3.5 Solar Cycles 23 and 24

Cycle 23 has been a cycle that we have not seen in two centuries. Irradiance dropped 50% more than in recent cycles, and the solar wind was at the lowest levels of the satellite age. More than 800 days were without sunspots, well more than double that of recent cycles. Cycle 23 lasted three years longer than Cycle 22, the longest since Cycle 6 during the Dalton Solar Minimum that peaked in 1810 (Figs. 15.20—15.24).

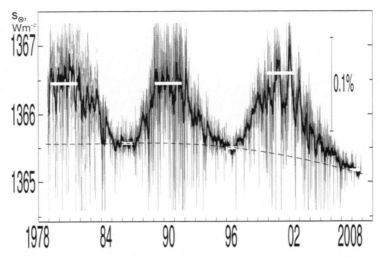

FIGURE 15.20   Irradiance in Cycle 23 dropped about 50% more than prior minima (with a change of near 0.15% from maxima) (NASA).

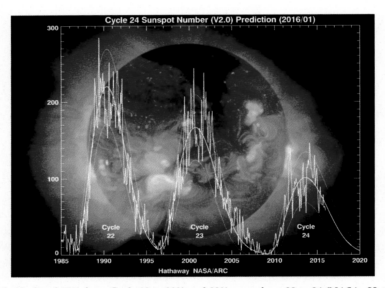

FIGURE 15.21   Sunspot peaks declined 25% from Cycle 22 to 23% and 33% more from 23 to 24 (NASA—Hathaway).

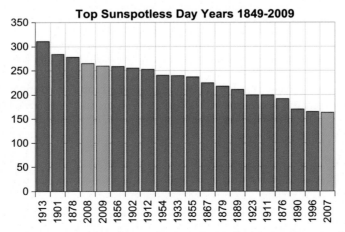

FIGURE 15.22   In the last solar minimum, 3 years (2007, 2008, and 2009) ranked in the top 20 years of days without sunspot since 1850.

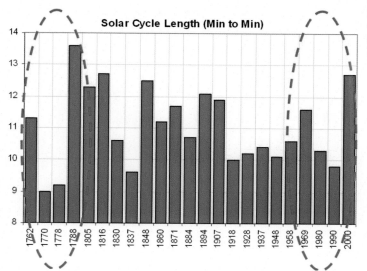

FIGURE 15.23 The length of Cycle 23 (minimum to minimum) was 12.6 years, the greatest since cycles in the Dalton Minimum. The maximum to maximum length from Cycle 23 to 24 was close to 14 years.

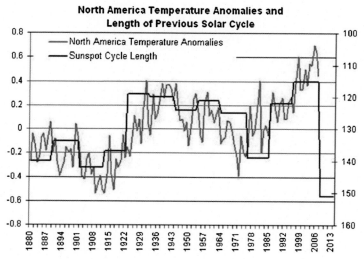

FIGURE 15.24 Historically, North American temperatures have correlated well with solar cycle length. Note the rapid increase in length for Cycle 23, implying an upcoming cooling (Friis-Christensen and Lassen, 1991).

4. THE "PAUSE" AND CLIMATE PROJECTIONS

Global temperatures, as measured by satellite, weather balloons, and ground station data have seen a cessation of the warming, observed from 1979 to 1998, for more than 18 years (Fig. 15.25). Annual temperatures have flat-lined and, in many locations, winter temperatures have declined for more than 20 years. Many explanations have been offered for why this has occurred despite the 11% increase in CO_2 levels. They include clouds, Arctic changes, ocean cycles, and the Sun. These factors are used to predict climate on a multi-seasonal basis, whereas water vapor changes and cloud changes are incorrectly modeled. Changes in the sun over time have matched global temperatures very well, far better than CO_2.

Clilverd et al. (2006) successfully forecast the decline in Cycles 23 and 24 using a statistical model of past cycles. Their regression-based statistical model projected a minimum rivaling at least the Dalton Solar Minimum, a quiet Sun and cold period from ∼1790 to ∼1820 (Figs. 15.26–15.28).

A Maunder-like minimum may have commenced in 2014 and may become most severe around 2055 (Abdussamatov, 2012, 2013). Abdussamatov (2012) quantified a declining trend in TSI and predicts further declining TSI. Abdussamatov (2013) modeled a Maunder-type solar minimum around 2043, with deep cooling around 2060.

FIGURE 15.25   Satellites and weather balloons and most surface station data sets have shown no warming for more than 18 years, even as CO_2 has increased. This corresponds to the start of a decline in solar activity.

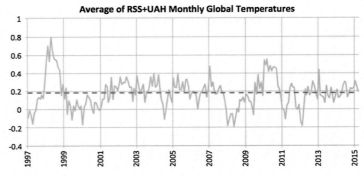

FIGURE 15.26   Clilverd et al. (2006) did a regression analysis of various solar cycles and built a model that showed success in predicting past cycles. The model suggests that a Dalton-like minimum should be starting.

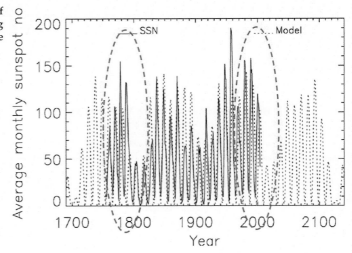

FIGURE 15.27   Long-term solar sunspot activity shows cyclical behavior with a relative minimum in the early 1900s, a deeper one in the early 1800s (Dalton Solar Minimum), and a very deep long one in the 1600s (Maunder Solar Minimum). Note the grand "modern maximum" in the 20th century, which we have been descending from the last two cycles.

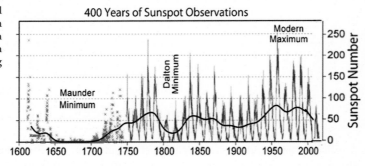

4.1 Secular Cycles: Combined Natural Factors

Temperature trends coincide with ocean and solar TSI cyclical trends. Fig. 15.29 overlays standardized ocean temperature indices (PDO + AMO) and Hoyt Schatten/Willson TSI and USHCN version 2 temperatures. A 60-year cycle is clearly shown, including the observed warming trend. The similarity with the ocean multidecadal cycle phases also suggest the Sun plays a role in their oscillatory behavior. Scafetta (2010) presents evidence for this 60-year cyclical behavior.

Dr. Don Easterbrook has used various options of a 60-year repeat of the mid-20th century solar/ocean-induced cooling, Dalton Minimum, and a Maunder Minimum scenarios to present the empirical forecast range of options given in Fig. 15.30.

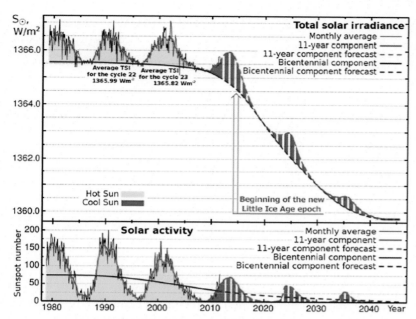

FIGURE 15.28   Solar irradiance and sunspots with projections to 2050 suggest a Dalton/Maunder like minimum commencing in 2014 and becoming most severe around 2055 (Abdussamatov, 2012, 2013).

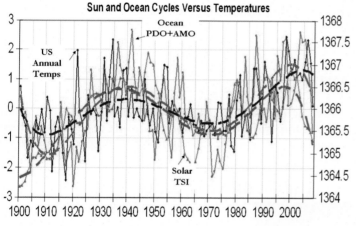

FIGURE 15.29   Solar TSI (Hoyt/Schatten TSI calibrated to Willson AMCRIMSAT TSI) and PDO + AMO (STD) versus the USHCN annual plots with polynomial smoothing.

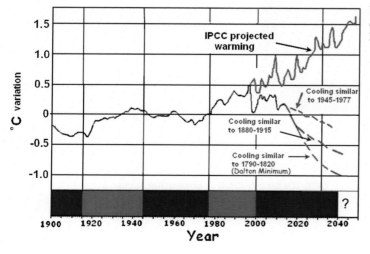

FIGURE 15.30   Projected future temperatures from the IPCC, to ocean/solar 60-year cycle cooling, to a Dalton Minimum, to a Maunder Minimum (Easterbrook and D'Aleo, 2010).

FIGURE 15.31 Sea-level change rate (mm/year) versus reconstructed solar flux (watts per square meter) since 1920 (Shaviv).

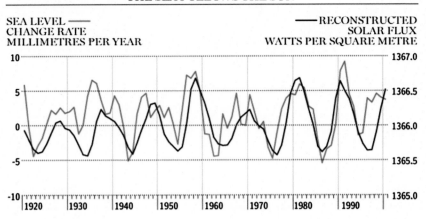

4.2 Ocean Changes Support Solar Influences

Shaviv (2008) showed additional empirical evidence that the Sun's influence on climate is much larger than expected from variations in TSI—the only solar forcing that is considered by the IPCC. The full forcing, which is large, can be quantified by studying sea level as it is linked to heat going into the oceans and therefore radiative forcing through thermal expansion.

This can be seen in Fig. 15.31, where tide-gauge-based sea level change rate, averaging close to 2 mm a year, is seen to vary in sync with the solar cycle. The amount of heat inferred from this large correlation corresponds to at least six times the forcing of the irradiance alone. However, this empirical evidence and its implications are ignored in models considered by the IPCC.

5. SUMMARY

Although the Sun's brightness or irradiance changes only slightly with solar cycles, the indirect effects of solar activity reduction may greatly magnify the total solar effect on temperatures, including warming of the atmosphere (1) in low and mid-latitudes by ozone reactions due to increased ultraviolet radiation, (2) in higher latitudes by

FIGURE 15.32 The total solar effect, including amplifiers versus IPCC brightness (irradiance)-based forcing.

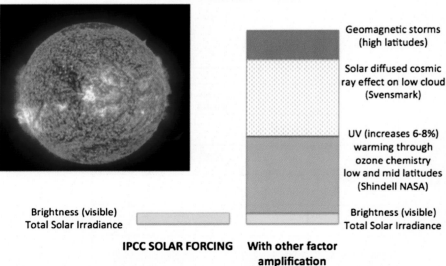

geomagnetic activity, and (3) generally by increased radiative forcing due to fewer clouds caused by cosmic rays. The total solar forcing is likely greatly underestimated in IPCC reports while the greenhouse effect is greatly exaggerated (Fig. 15.32).

A much more detailed literature review of these likely important solar amplifiers was provided in a summary in the Climate Change Reconsidered II: Physical Science Chapter 3 (2013), written by Willie Soon and Sebastian Luning.

References

Abdussamatov, H.I., 2012. Bicentennial decrease of the total solar irradiance leads to unbalanced thermal budget of the Earth and the Little Ice Age. Applied Physics Research 4, 178—184. http://dx.doi.org/10.5539/apr.v4n1p187. ISSN:1916-9639.

Abdussamatov, H.I., 2013. Grand minimum of the total solar irradiance leads to the Little Ice Age. Journal of Geology and Geoscience 2, 113. http://dx.doi.org/10.4172/2329-6755.1000113.

Abreu, J.A., Beer, J., Steinhilber, F., Tobias, S.M., Weiss, N.O., 2008. For how long will the current grand maximum of solar activity persist? Geophysical Research Letters 35, L20109.

Baldwin, M.P., Dunkerton, T.J., 2005. The solar cycle and stratospheric—tropospheric dynamical coupling. Journal of Atmospheric Science 9, 1—5.

Boberg, F., Lundstedt, H., 2002. Solar wind variations related to fluctuations of the North Atlantic Oscillation. Geophysical Research Letters 29 (15), 13-1—13-4.

Clilverd, M.A., Clarke, E., Ulich, T., Rishbeth, H., Jarvis, M.J., 2006. Predicting solar cycle 24 and beyond. Space Weather 4, 1—7.

Dickinson, R., 1975. Solar variability and lower atmosphere. American Meteorological Society Bulletin 56, 1240—1248.

Easterbrook, D.J., D'Aleo, J.S., 2010. Multidecadal tendencies in Enso and global temperatures related to multidecadal oscillations. Energy & Environment 21, 436—460.

Elsner, J.B., Jagger, T.H., Hodges, R.E., 2010. Daily tropical cyclone intensity response to solar ultraviolet radiation. Geophysical Research Letters 37.

Erlykin, A., Wolfendale, A., 2011. Cosmic ray effects on cloud cover and their relevance to climate change. J. Atmos. Solar Terrestrial Physics 73, 1681—1686.

Fligge, M., Solanki, S.K., 2000. The solar spectral irradiance since 1700. Geophysical Research Letters 27, 2157—2160.

Fröhlich, C., Lean, J., 1998. The sun's total irradiance: cycles, trends, and related climate change uncertainties since 1976. Geophysical Research Letters 25, 4377—4380.

Friis-Christensen, E., Lassen, K., 1991. Length of the solar cycle: an indicator of solar activity closely associated with climate. Science 254, 698—700.

Gray, L., et al., 2010. Solar influences on climate. Reviews of Geophysics 48, 1—53.

Haigh, J.D., 1996. The impact of solar variability on climate. Science 272, 981—984.

Harrison, R., Ambaum, M., 2010. Observing Forbush decreases in cloud at Shetland. Journal of Atmospheric Solar Terrestrial Physics 72, 1408—1414.

Hoyt, D.V., Schatten, K.H., 1997. The Role of the Sun in Climate Change. Oxford University Press, New York, 288p.

Kalnay, E., Cai, M., 2003. Impact of urbanization and land-use change on climate. Nature 423, 528—531.

Kirkby, J., 2007. Cosmic rays and climate. Surveys in Geophysics 28, 333—375.

Kirkby, J., 2011. Role of sulphuric acid, ammonia and galactic cosmic rays in atmospheric aerosol nucleation. Nature 476, 429—433.

Labitzke, K., 2001. The global signal of the 11-year sunspot cycle in the stratosphere: differences between solar maxima and minima. Meteorology Zeitschrift 10, 83—90.

Labitzke, K., Van Loon, H., 1988. Association between the 11-year solar cycle, the QBO and the atmosphere. Part I: The troposphere and stratosphere in the northern hemisphere in winter. Journal Atmospheric Terrestrial Physics 50, 197—206.

Lean, J., 2000. Evolution of the sun's spectral irradiance since the Maunder Minimum. Geophysical Research Letters 27, 2425—2428. http://dx.doi.org/10.1029/2000GL000043.

Lockwood, M., Rouillard, A.P., Finch, I.D., 2009. The rise and fall of open solar flux during the current grand solar maximum. Applied Journal 700, 937—944. http://dx.doi.org/10.1088/0004-637X/700/2/937.

Lockwood, M., Harrison, R.G., Woollings, T., Solanki, S.K., 2010. Are cold winters in Europe associated with low solar activity? Environmental Research Letters 5 (2).

Lockwood, M., 2010. Solar change and climate: an update in the light of the current exceptional solar minimum. Proceedings of the Royal Society, A 466, 303—329.

Lockwood, M., Stamper, R., 1999. Long-term drift of the coronal source magnetic flux and the total solar irradiance. Geophysical Research Letters 26, 2461—2464. http://dx.doi.org/10.1029/1999GL900485.

Lockwood, M., Stamper, R., Wild, M.N., 1999. A doubling of the Sun's coronal magnetic field during the past 100 years. Nature 399 (6735), 437. http://dx.doi.org/10.1038/20867.

Marsh, N.D., Svensmark, H., 2000. Low cloud properties influenced by cosmic rays. Physical Review Letters 85, 5004—5007.

McComas, D., Ebert, R., Elliott, H., Goldstein, B., Gosling, J., Schwadron, N., Skoug, R., 2008. Weaker solar wind from the polar coronal holes and the whole Sun. Geophysics Research Letters 35, 1—5.

Meehl, G.A., Arblaster, J.M., Matthes, K., Sassi, F., van Loon, H., 2009. Amplifying the Pacific climate system response to a small 11 year solar cycle forcing. Science 325, 1114—1118.

Merikanto, J., Spracklen, D., Mann, G., Pickering, S., Carslaw, K., 2009. Impact of nucleation on global CCN. Atmospheric Chemistry and Physics 9, 8601—8616.

Mirme, S., Mirme, A., Minikin, A., Petzold, A., Horrak, U., Kerminen, V.-M., Kulmala, M., 2010. Atmospheric sub-3 nm particles at high altitudes. Atmospheric Chemistry and Physics 10, 437—451.

Palle, E., Butler, C.J., 2002. The proposed connection between clouds and cosmic rays: cloud behavior during the past 50—120 years. Journal Atmospheric Solar Terrestrial Physics 64, 327—337.

Palle Bago, E., Butler, C.J., 2000. The influence of cosmic rays on terrestrial clouds and global warming. Astronomical Geophysics 41, 4.18–4.22.

Penn, M., Livingston, W., 2006. Temporal changes in sunspot umbral magnetic fields and temperatures. The Astrophysical Journal 649, L45–L48.

Pielke Sr., R.A., Marland, G., Betts, R.A., Chase, N., Eastman, J.L., Niles, J.O., Niyogi, D.D.S., Running, S.W., 2002. The influence of land-use change and landscape dynamics on the climate system: relevance to climate-change policy beyond the radiative effect of greenhouse gases. Philosophical Transactions of the Royal Society of London A360, 1705–1719.

Rigozo, N.R., Echer, E., Vieira, L.E.A., Nordemann, D.J.R., 2001. Reconstruction of Wolf sunspot numbers on the basis of spectral characteristics and estimates of associated radio flux and solar wind parameters for the last millennium. Solar Physics 203, 179–191.

Russell, C., Luhmann, J., Jian, L., 2010. How unprecedented a solar minimum? Reviews of Geophysics 48, 3287.

Scafetta, N., West, B.J., 2006a. Phenomenological solar contribution to the 1900–2000 global surface warming. Geophysical Research Letters 33, L05708. http://dx.doi.org/10.1029/2005GL025539.

Scafetta, N., West, B.J., 2006b. Reply to comment by J. L. Lean on "Estimated solar contribution to the global surface warming using the ACRIM TSI satellite composite". Geophysical Research Letters 33, L15702. http://dx.doi.org/10.1029/2006GL025668.

Scafetta, N., West, B.J., 2006c. Phenomenological solar signature in 400 years of reconstructed Northern Hemisphere temperature. Geophysical Research Letters 33, L17718. http://dx.doi.org/10.1029/2006GL027142.

Scafetta, N., 2010. Empirical evidence for a celestial origin of the climate oscillations and its implications. Journal of Atmospheric and Solar-Terrestrial Physics 72, 951–970. http://dx.doi.org/10.1016/j.jastp.2010.04.015.

Shaviv, N., 2008. Using the oceans as a calorimeter to quantify the solar radiative forcing. Journal of Geophysical Research 113 (A11), 1240–1248.

Shaviv, N.J., 2005. On climate response to changes in cosmic ray flux and radiative budget. Journal of Geophysical Research 110, A08105.

Shindell, D.T., Rind, D., Balachandran, N., Lean, J., Lonergan, P., 1999. Solar cycle variability, ozone, and climate. Science 284, 305–308.

Smith, E., Balogh, A., 2008. Decrease in heliospheric magnetic flux in this solar minimum: recent Ulysses magnetic field observations. Geophysical Research Letters 35, L22103.

Snow-Kropla, E., Pierce, J., Westervelt, D., Trivitayanurak, W., 2011. Cosmic rays, aerosol formation and cloud-condensation nuclei: sensitivities to model uncertainties. Atmospheric Chemistry and Physics 11, 4001–4013.

Soon, W.H., 2005. Variable solar irradiance as a plausible agent for multidecadal variations in the Arctic-wide surface air temperature record of the past 130 years. Geophysical Research Letters 32. http://dx.doi.org/10.1029/2005GL023429.

Svensmark, H., Friis-Christensen, E., 1997. Variation of cosmic ray flux and global cloud cover- a missing link in solar -climate relationships. Journal of Atmospheric and Solar-Terrestrial Physics 59, 1125–1132.

Svensmark, H., 2007. Cosmoclimatology: a new theory emerges. Astronomy & Geophysics 48, 1.18–1.24. http://dx.doi.org/10.1111/j.1468-4004.2007.48118.x.

Tinsley, B.A., Yu, F., 2002. Atmospheric ionization and clouds as links between solar activity and climate. In: American Geophysical Union Monograph, Solar Variability and Its Effects on the Earth's Atmosphere and Climate System.

Wang, Y.-M., Lean, J.L., Sheeley Jr, N.R., 2005. Modeling the Sun's magnetic field and irradiance since 1713. The Astrophysical Journal 625, 522–538. http://dx.doi.org/10.1086/429689.

Willson, R., 1997. Total solar irradiance trend during solar cycles 21 and 22. Science 277, 1963–1965.

Willson, R.C., Mordvinov, A.V., 2003. Secular total solar irradiance trend during solar cycles 21–23. Geophysical Research Letters 30, 1199.

Willson, R.C., Hudson, H.S., 1988. Solar luminosity variations in solar cycle 21. Nature U332U, 810–812.

Yu, F., Luo, G., 2014. Effects of solar variation on particle formation and cloud condensation nuclei. Environmental Research Letters 9 (4), 045004.

16

The Sun's Role in Climate

S. Lüning[1], F. Vahrenholt[2]

[1]Independent Researcher, Lisbon, Portugal; [2]German Wildlife Foundation, Hamburg, Germany

1. INTRODUCTION

The Sun is by far the most important source of energy for Earth's oceans, atmosphere, land, and biosphere. As solar activity is known to vary over time, an important field of activity in modern climate sciences is the quantification of possible climatic effects that such solar activity changes might have. The discussion is currently overshadowed by an enigma: According to reports of the Intergovernmental Panel on Climate Change (IPCC), solar activity changes are interpreted to have extremely small climatic significance. The radiative forcing (RF) attributed by the fifth assessment report (AR5) of the IPCC (2013a) to changes in solar irradiance of $0.05 \, W/m^2$ is dwarfed by the values assumed for anthropogenic greenhouse gases such as CO_2 ($1.68 \, W/m^2$) or CH_4 ($0.97 \, W/m^2$) (Fig. 16.1). At the same time, a very large number of scientific studies find a clear and strong relationship between historical climate change and solar activity. As it is unlikely that the solar–climate link might have ceased, a significant involvement of the sun also in present-day climate change appears plausible.

The IPCC must be aware of this challenge but unfortunately does not address the obvious issue in its dedicated Frequently Asked Question paragraph entitled "Is the Sun a Major Driver of Recent Changes in Climate?" which is

Evidence-Based Climate Science, Second Edition
http://dx.doi.org/10.1016/B978-0-12-804588-6.00016-1

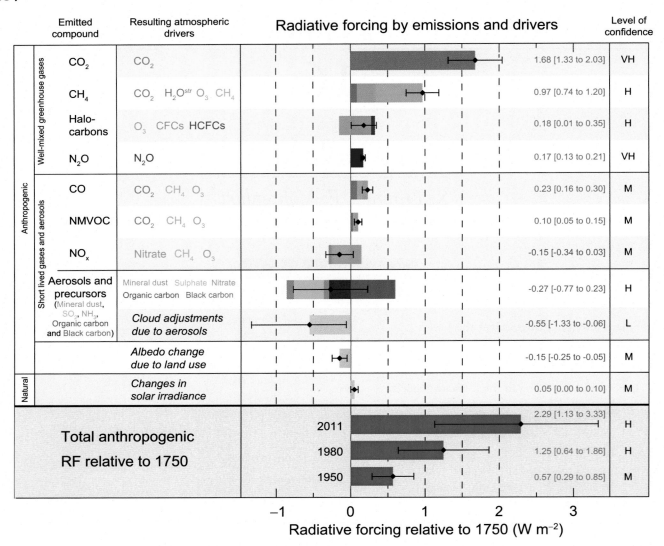

FIGURE 16.1   Radiative forcing according to the fifth IPCC assessment report. *From Summary for Policy Makers, IPCC, 2013b. Summary for Policymakers, In: Stocker, T.F., Qin, D., Plattner, G.-K., Tignor, M., Allen, S.K., Boschung, J., Nauels, A., Xia, Y., Bex, V., Midgley, P.M., (Eds.), Climate Change 2013: The Physical Science Basis. Contribution of Working Group I to the Fifth Assessment Report of the Intergovernmental Panel on Climate Change, Cambridge University Press, Cambridge, United Kingdom and New York, NY, US.*

included in the solar chapter of the AR5 report (WG1, FAQ 5.1, p. 392–393) (IPCC, 2013a). The IPCC's conclusion in the FAQ matches the low RF:

> The mechanisms that amplify the regional effects of the relatively small fluctuations of TSI [Total Solar Irradiance] in the roughly 11-year solar cycle involve dynamical interactions between the upper and the lower atmosphere, or between the ocean sea surface temperature and atmosphere, and have little effect on global mean temperatures.

In this study, we investigate if the very low RF value presently assumed by the IPCC for solar activity changes is compatible with paleoclimatologic reconstructions for the preindustrial Holocene. In particular, we will focus on natural millennial-scale cycles, which often were of a similar magnitude as the climate change observed over the past 150 years.

2. SOLAR ACTIVITY CHANGES OVER THE LAST 10,000 YEARS

Solar activity changes have long been known to be controlled by a series of cycles. The reconstruction of solar activity beyond the optical observational period is based on cosmogenic nuclides (^{14}C, ^{10}Be, ^{36}Cl), which are

TABLE 16.1 Solar Activity Cycles

Cycle Name	Average Period in Years	Fluctuation Range in Years
Schwabe	11	9–14
Hale	22	18–26
Gleissberg	87	60–120
Suess/de Vries	210	180–220
Eddy	1000	900–1100
Hallstatt	2300	2200–2400

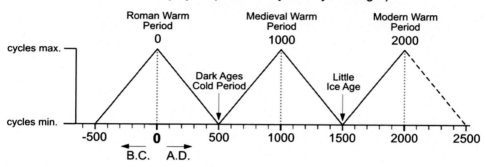

FIGURE 16.2   Historical maxima and minima of the 1000-year Eddy solar activity cycle (schematic sketch). We currently find ourselves at the plateau area of the maximum, and thus for the coming centuries we do not anticipate a further increase in solar activity, contrasting with the steep climb of the 19th and 20th centuries. Sometime during the course of the coming 100 years, the long-term decline will begin and will lead from the current Eddy Maximum to the Eddy Minimum, which is expected to occur around the year 2500.

generated by cosmic rays (Abreu et al., 2010). The stronger the cosmic rays, the higher the concentrations of the cosmogenic nuclides, which means the sun was less active because the solar magnetic field shielded the Earth from the galactic particle showers. Solar activity cycles are quasi-periodical. Most prominent are the 11- and 22-year cycles, which are termed Schwabe and Hale, respectively. Due to great inertia of the climate system, it is however mostly the longer periodic cycles of 90 years (Gleissberg), 200 years (Suess/de Vries), 1000 years (Eddy), and 2300 years (Hallstatt) that are worth studying as potential climate drivers (Table 16.1). The cycles have occurred in the past and continue to exist in the present and future.

The Eddy and Hallstatt Cycles are the basic building blocks of solar millennial-scale cycles. The Hallstatt Cycle has been described by various authors, eg, Suess (1980), Sonett and Finney (1990), Damon and Sonett (1991), Vasiliev and Dergachev (2002), Dreschhoff (2008), and Steinhilber et al. (2010). The last minimum of the 2300-year-long Hallstatt Cycle occurred in the phase between 1300 and 1800 during the Little Ice Age (LIA). Based on extrapolation of the Hallstatt Cycle into the future, the next Hallstatt solar irradiation minimum may be expected to occur about 1500 years from today (Steinhilber et al., 2010).

Closer inspection of the Hallstatt Cycle reveals a characteristic double peak whose crests are about 1000 years apart (Steinhilber et al., 2010). Therefore, the Hallstatt Cycle can be divided into two single oscillations, each being about 1000 years long (Ma, 2007). This "semi-Hallstatt Cycle" may correspond to the Eddy Cycle (Abreu et al., 2010), which is expected to reach its next minimum in about 500 years, whereby we currently find ourselves at a plateau area of a solar irradiation high point (Fig. 16.2).

3. MEDIEVAL WARM PERIOD

Reconstructions of total solar irradiance (TSI) indicate a 400-year-long period of elevated solar activity centered around 1000 AD (Steinhilber et al., 2009) (Fig. 16.3). TSI declined rapidly around 1350 AD, and remained low most of

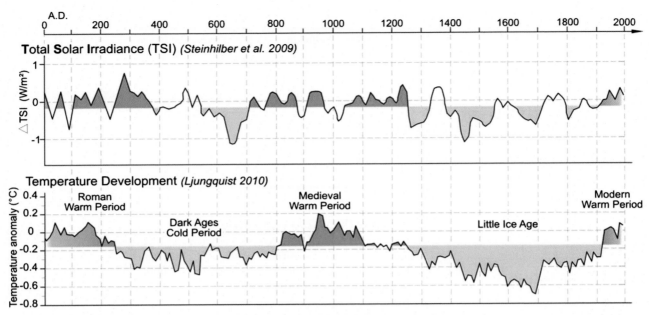

FIGURE 16.3   Long-term synchronicity of solar activity (Steinhilber et al., 2009) and the temperature development (extra-tropical Northern Hemisphere) during the last 2000 years (Ljungqvist, 2010).

the time until it increased again around 1850 AD. Solar activity reached some of the highest intensities of the entire Holocene in the second half of the 20th century (Solanki et al., 2004). Considering these well-documented variations, why the IPCC attempts to downplay these trends is unclear. In the AR5 Chapter 5, the IPCC surprisingly claims:

> Total solar irradiance (TSI, Chapter 8) is a measure of the total energy received from the sun at the top of the atmosphere. It varies over a wide range of time scales, from billions of years to just a few days, though variations have been relatively small over the past 140 years IPCC (2013a); FAQ 5.1, p. 392.

The long-lasting period of elevated solar activity interestingly coincides with a warm phase from 950 to 1250 AD, which has been well documented from various parts of the world. Temperatures during this Medieval Warm Period (MWP) or Medieval Climate Anomaly (MCA) were generally at a similar level to today. Ljungqvist et al. (2012) and Christiansen and Ljungqvist (2012) demonstrated that the MWP occurred on all continents of the Northern Hemisphere and that the warming of the 20th century hence must be considered to be still within the range of natural variability over the last 12 centuries. The subsequent period of low TSI notably corresponds to the LIA, a well-known cold period.

Unfortunately, current climate models cannot reproduce the MWP warming (Fig. 16.4), as the IPCC openly acknowledges in its AR5 climate report (IPCC, 2013a, Chapter 5.3.5):

> The reconstructed temperature differences between MCA and LIA…indicate higher medieval temperatures over the NH continents…The reconstructed MCA warming is higher than in the simulations, even for stronger TSI changes and individual simulations…The enhanced gradients are not reproduced by model simulations…and are not robust when considering the reconstruction uncertainties and the limited proxy records in these tropical ocean regions…This precludes an assessment of the role of external forcing and/or internal variability in these reconstructed patterns.

4. IS SOMETHING MISSING IN THE CLIMATE MODELS?

The MWP/MCA warming therefore still remains unexplained. Could climate models do better if they were to consider a much stronger solar influence on climate, ie, higher values for solar radiative forcing? Currently only the plain changes of the TSI are considered. A growing number of studies, however, propose solar amplifier mechanisms that involve solar radiation changes in the ultraviolet (UV) band, stratosphere–troposphere coupling, stratospheric ozone, and cosmic ray–cloud interaction. These processes are still poorly understood and therefore have not yet been built into the models. Nevertheless, the strong solar–climate link found in many

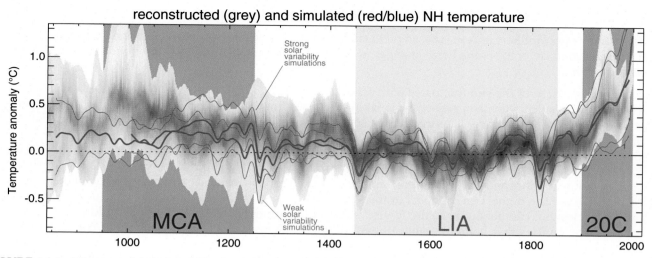

FIGURE 16.4   Climate models (red and blue lines) fail to reproduce documented high temperatures of the Medieval Warm Period (gray shaded area). *From Fig. 5.8 of IPCC AR5 WG1, IPCC, 2013a. Climate Change 2013: The Physical Science Basis. Contribution of Working Group I to the Fifth Assessment Report of the Intergovernmental Panel on Climate Change, Cambridge University Press, Cambridge, United Kingdom and New York, NY, USA, 1535 p.*

palaeoclimatologic reconstructions indicates that the current models are incomplete and should be treated with caution. Successful hindcast capability is a crucial requirement before models qualify to be used for future climate simulations.

4.1 Solar UV Changes and Stratosphere—Troposphere Coupling

UV fluctuates within the 11-year-solar cycle with a magnitude that is far greater than that of total irradiation, namely a few percentage points compared to the 0.1% change for total solar irradiance (Bard and Frank, 2006; Gray et al., 2010; Haigh et al., 2010; Meehl et al., 2009). Furthermore, Ermolli et al. (2013) found that variations in UV radiation have been previously underestimated in climate models by a factor of 4—6. The elevated UV radiation during the solar activity maximum spurs the formation of ozone at altitudes of 50—15 km. A large number of oxygen molecules (O_2) get converted into ozone (O_3) through the added UV energy input. A higher ozone concentration in turn catches more UV radiation and converts the energy into warmth, which then leads to warming of the ozone layer and the stratosphere. Satellite measurements over the last few years have already documented corresponding changes in ozone concentration and temperature in the stratosphere, and even to a certain extent in the ionosphere above.

Ongoing research aims to identify processes that may connect the powerful stratospheric fluctuations to the tropospheric climate under 15 km (Hood et al., 2013; Niranjankumar et al., 2011; Petrick et al., 2012; Sfîcă and Voiculescu, 2014; Veretenenko and Ogurtsov, 2014). Two possible mechanisms could be active in tandem. First, it appears that the UV warming of the ozone layer generates anomalies in the atmospheric temperature gradient, which in turn causes changes in the tropical circulation systems of the lower atmosphere and shifts in precipitation zones (Bal et al., 2011; Ineson et al., 2011; Kang et al., 2011; Kodera, 2006; Meehl et al., 2009). The increased solar radiation warms the ocean water in parallel, which, in connection with powerful trade winds, leads to fewer clouds in the tropics. This in turn allows more solar radiation to reach the surface and to warm the ocean (Meehl et al., 2009; Udelhofen and Cess, 2001). While the first process establishes the connection between the stratosphere and troposphere, the second process delivers an additional amplification contribution to the climatic total effect of solar changes. Recent studies have corroborated the existence of a link between the stratosphere and climate in the troposphere. Ozone changes due to solar activity variations have been shown to affect the winds south of Greenland and in parts the Southern Hemisphere (Reichler et al., 2012; Varma et al., 2012). UV amplification effects have not been included in current climate models so far. Moreover, changes in the individual spectral classes of solar radiation are far too insufficiently differentiated, and hence the UV effects cannot be adequately taken into account (Haigh et al., 2010).

4.2 Cloud Cover Changes Through Solar-Forced Cosmic Ray Variations

The cosmic ray amplifier concept is based on the linkage of multiple intermediate steps. The underlying concept is that average cloud coverage of the Earth changes in sync with solar activity whereby the clouds act as a parasol that cools the Earth:

- The magnetic field of the sun fluctuates with solar activity.
- The solar magnetic field encompasses the solar system and thus the Earth, therefore shielding it from incoming cosmic rays from outer space. The stronger the sun's magnetic field, the more it shields the Earth against the flux of incoming cosmic rays.
- Cosmic rays are suspected to provide part of the cloud condensation nuclei needed for forming low-level clouds for the first 3 km of atmosphere above the Earth's surface. The charged cosmic ray particles trigger the condensation of atmospheric water vapor, similar to what happens inside a condensation chamber. Particularly responsible are the highest-energy part of the secondary cosmic radiation, the so-called muons. They are the only particles that can penetrate to the lowest levels of the atmosphere. Clouds of the middle and upper atmosphere, on the other hand, are not influenced by cosmic rays, as an ample amount of cosmic radiation is always available.
- The weaker the solar activity, the weaker the sun's magnetic field, and the weaker the shield protecting the Earth from cosmic rays. This means more cosmic condensation nuclei are able to penetrate the Earth's atmosphere, which in turn lead to more condensation and thus an increase in the formation of cooling-type clouds.

The solar radiation amplifier model was developed by Danish physicist Henrik Svensmark in cooperation with Eigil Friis-Christensen beginning in the late 1990s (Svensmark, 2000; Svensmark and Friis-Christensen, 1997, 2007; Svensmark et al., 2007). The Svensmark model has attracted a lot of criticism. While a number of observations support the theory, other data seem to conflict. The IPCC discusses the model in its AR5 climate report, but assumes it to be too weak to play a significant role (IPCC, 2013a, Chapter 7, p. 614):

> Correlations between cosmic ray flux and observed aerosol or cloud properties are weak and local at best, and do not prove to be robust on the regional or global scale. Although there is some evidence that ionization from cosmic rays may enhance aerosol nucleation in the free troposphere, there is medium evidence and high agreement that the cosmic ray-ionization mechanism is too weak to influence global concentrations of CCN or droplets or their change over the last century or during a solar cycle in any climatically significant way.

The original Svensmark model may indeed have been overly simplistic. A globally homogenous relationship between solar activity and clouds via cosmic rays may not exist. Various studies indicate, however, that the mechanism may have significant climatic influence when differentiated according to latitude, season, and atmospheric altitude (Yu and Luo, 2014). Importantly, the Svensmark process appears to work best at middle latitudes, where sufficient suitable stratified clouds, liquid cloud droplets, and suitable quantities of cosmic rays exist (Kovaltsov and Usoskin, 2007; Laken et al., 2010; Marsh and Svensmark, 2000, 2003; Pallé et al., 2004; Usoskin et al., 2004). For key regions, Voiculescu and Usoskin (2012) documented a persistent response of clouds to cosmic rays over the entire studied time interval (1984—2009), indicating a real link. Cho et al. (2012) suggested that the north—south asymmetry on the sun also has to be considered when interpreting the solar—cloud effect. Further recent work providing positive evidence for the effect has been provided by, eg, Rawal et al. (2013), Kancírová and Kudela (2014), Huneeus et al. (2014), Lam et al. (2014) and Sfică and Voiculescu (2014).

Common criticism of a significant solar involvement in 20th-century climate change is that sunspots peaked in the 1960s while temperatures reached their current high plateau 40 years later. It is important to note, however, that the main 1980—2000 warming period coincides with a strong solar magnetic field that was higher than at any other time over the past 150 years (Fig. 16.5). This synchronicity provides additional evidence for the involvement of the solar magnetic field and cosmic rays in 20th/21st-century climate change.

Often overlooked in the solar—climate discussion is that the climate system is characterized by large inertia. It may take a few decades (several 11-year solar cycles) before global temperatures are fully adjusted to a new solar activity regime. A delay between solar trigger and temperature response is therefore likely. Climate time lags ranging between a few years to several 100 years have been observed (eg, Eichler et al., 2009; Perry, 2007; Ratnam et al., 2014; Tiwari et al., 2015; Wang and Zhang, 2011; Weber et al., 2004). In addition, effects of 60-year ocean cycles, such as the Pacific Decadal Oscillation (PSO) and the Atlantic Multidecadal Oscillation (AMO), are superimposed on the temperature curve. A one-to-one relation between solar activity changes and temperature curve consequently is not even expected.

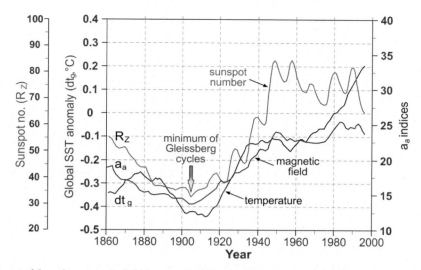

FIGURE 16.5   Development of the solar magnetic field (represented here by the a_a proxy), sunspot record, and global temperature. The 11-year solar cycle is smoothed out using a 23-year mean for the solar data. *From Mufti, S., Shah, G.N., 2011, Solar-geomagnetic activity influence on Earth's climate. Journal of Atmospheric and Solar-Terrestrial Physics 73, 1607–1615.*

5. SOLAR-FORCED MILLENNIAL CLIMATE CYCLES

For a long time, the preindustrial Holocene climate was thought to have been rather stable. This paradigm culminated in the publication of the so-called Hockey Stick Curve by Mann et al. (1999), who assumed a largely steady-state climate regime during the 850 years prior to rapid warming from 1850 onwards during the industrial period. This was a key element in the 2001 third IPCC report. However, a vast literature and subsequent studies by other authors rejected the preindustrial steady-state concept. More recent research emphasizes the cyclical climate pattern consisting of the MWP, LIA, and Modern Warm Period (Christiansen and Ljungqvist, 2012; Esper et al., 2012b; Ljungqvist, 2010; Ljungqvist et al., 2012), which shows a good degree of synchronicity with solar activity changes (see above). Nevertheless, the warm-cold-warm change of the MWP, LIA and Modern Warm Period represents only one cycle, which is thus too short to verify the validity of the solar-forced millennial climate cycle concept in a statistically robust manner.

Evidence for an older millennial cycle preceding the MWP was provided by Ljungqvist (2010) (Fig. 16.3), who described another prominent warm–cold–warm change, composed of the Roman Warm Period (RWP; 0 AD) and the Dark Ages Cold Period (DACP = Cold Phase of the Migration Period; 500 AD) leading towards the MWP. Notably, the RWP coincides with several hundred years of above-average solar activity. In contrast, the cold peak of the DACP occurred during a phase from 500 to 700 AD, when solar activity reached some of its lowest values of the last 2000 years (Fig. 16.3).

In order to test the concept of solar-driven millennial-scale climate cycles, it is important to widen the study to the last 10,000 years. This study period gives the chance to investigate up to 10 millennial-scale cycles. A further extension of the study period into the past is not attempted because the climate system during the last glacial period is thought to be significantly different than the Holocene. Yet millennial-scale climate cycles with abrupt temperature jumps are also described from this period. These are the so-called Dansgaard–Oeschger events, which some scientists interpret to be controlled by solar activity (Braun and Kurths, 2010; Muscheler and Beer, 2006; Rahmstorf, 2003).

5.1 "Bond" Events in the North Atlantic

At the end of the 1990s, a research team led by Gerard Bond examined the Holocene climate history of the North Atlantic. In the seabed deposits of the last 10,000 years, the team found unusual layers of debris deposits that repeated at regular intervals (Bond et al., 2001). The only means of transport for this kind of coarse material to this particular oceanic site was by floating icebergs, which unloaded their debris as they gradually melted, thus spreading their detritus over the ocean floor. The coarse layers were deposited during times when cool, ice-rich surface water masses from the Arctic reached far into southern regions as far south as the latitude of Great Britain.

Bond et al. (2001) compared the repetitive pattern of the debris deposits to solar activity over the same time period and found a high degree of synchronicity. Cold phases in the North Atlantic occurred predominantly during times of weak solar activity. Bond and his team numbered the Holocene cold events from 0 to 8. Cold Events 0 and 1 correspond to the LIA and the Dark Ages Cold Period, respectively (Fig. 16.6).

The synthetic mean value over all the cycles produces a period of about 1500 years (Bond et al., 2001). This, however, must be seen as a theoretical cycle length that as such may not exist. Most likely the 1500 years represent a composite value of the 2300-year Hallstatt Cycle and the 1000-year Eddy Cycle, as pointed out by Debret et al. (2009) and Obrochta et al. (2012).

Around the same time as the Bond team, a research group led by Hui Jiang studied the late Holocene sea temperature history of the north Icelandic Shelf based on diatom assemblages (Jiang et al., 2002). The study found a general late Holocene cooling trend, which is interrupted by three relatively warm periods. Reconstructed temperature changes of these millennial scale cycles ranged between 1 and 2°C. In a follow-up paper three years later, the same study group demonstrated that the north Icelandic climate development formed part of synchronous North Atlantic-wide fluctuations, implying a common forcing factor (Jiang et al., 2005). Comparison with the solar activity curve revealed solar forcing as the most likely climate driver, which according to Jiang et al., explains 30–45% of the observed sea-surface temperature (SST) variance at time scales longer than 50 years.

Oppo et al. (2003) investigated changes in the carbon-isotope composition of benthic foraminifera throughout the Holocene in a deep sea core from the subpolar northeastern Atlantic. They found that deep-water production varied on a centennial–millennial timescale, with most of the identified cold events matching well with the climate cycles reported by Bond et al. (2001).

A few years later, Berner et al. (2008) generally confirmed the findings of the Bond team, investigating a sediment core south of Iceland and west of the core locations studied by Bond et al. (2001). Based on diatom-derived palaeotemperatures, the new study found high-frequency SST variability, on the order of 1–3°C. Most of the cooling events identified by Berner et al. (2008) showed anticorrelation with ^{14}C production rate and [10]Be flux, implying solar-related changes as an important underlying mechanism for the observed ocean climate variability. Typical periodicities of the climatic oscillations were 600–1000, 1500, and 2500 years, compatible with the Eddy and Hallstatt group of solar cycles. Similar cycles were also found in a nearby North Atlantic sediment core by Thornalley et al. (2009), who agreed with the other two mentioned research teams that these oscillations may have been controlled by an external driver such as solar variability.

The North Atlantic millennial-scale climate cyclicity was once more confirmed in 2011 in a study by D'Andrea et al. (2011) from West Greenland. The authors compiled a high-resolution temperature record over the past 5600 years based on alkenone unsaturation in sediments of two lakes. They found abrupt natural temperature changes of 2°C within a few decades and 4°C within 200 years, which had a profound impact on the Saqqaq, Dorset, and Norse cultures in the region, driving settlement and subsequent abandonment. In their conclusions, D'Andrea et al. (2011) suspect that solar forcing and volcanoes are among the key drivers for the observed abrupt climate change.

A year later, Mernild et al. (2012) studied meltwater runoff of glaciers in East Greenland for the past 4000 years. The authors found characteristic centennial to submillennial variations concurrent with the RWP, the Dark Ages Cold Period (Bond Event 1), the Medieval Climate Anomaly, and the LIA (Bond Event 0). Overall, Mernild et al. (2012) found a very weak decreasing trend in runoff over the last 4000 years, which they associated with the general insolation-generated Northern Hemisphere cooling since the Holocene Thermal Maximum.

In the same year, Larsen et al. (2012) published a Holocene glacier history from central Iceland. They found a series of abrupt cooling events at 8.7–7.9, 6.4, 5.5, 4.2, 2.9, 1.4, and 0.7 kiloanna (ka) before present. The latter four events are synchronous with Bond Events 3, 2, 1, and 0. According to Larsen et al., the most plausible forcing mechanisms are solar variability and volcanic emissions superimposed on long-term insolation changes. The greatest Holocene extent of the studied Langjökull Glacier occurred in the 19th century and is coincident with peak landscape instability, followed by ice recession throughout the 20th century.

In yet another paper from that year, Moros et al. (2012) presented a late Holocene climate reconstruction from the Reykjanes Ridge southwest of Iceland. The researchers identified a series of millennial-scale cold spells centered at 5.5, 3.8, 2.8, and 1.3 ka BP and the LIA. Notably, each cooling event was increasingly colder than the previous one. A large drop in temperature, 2.5°C (Mg/Ca on *N. pachyderma* d.) and 1.5°C (alkenone SST), respectively, was observed from 2.0 to 1.3 ka. The five cold phases might correspond to Bond Events 4 to 0 when allowing for uncertainties in dating. Moros et al. (2012) also report a widespread pronounced warming at 2.0 ka, which seems to represent the RWP and reflects the warmest period of the late Holocene. The authors discuss that external/solar forcing at millennial timescale might have been the driver for the observed natural climate change.

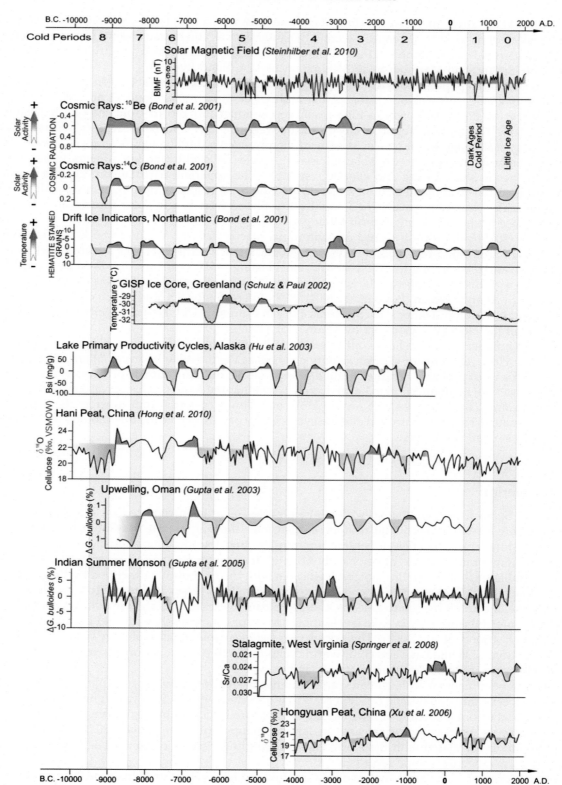

FIGURE 16.6 Solar activity (upper three curves) and climate (remaining curves) pulsated synchronously in concert according to the millennial beat during the entire 10,000-year postglacial period. Characteristic cold intervals (marked gray) occurred simultaneously over many different parts of the planet during phases of weak solar activity. In some regions the coupling between the sun and climate discontinued at times. To provide a better overview, the peaks of these phases are unmarked. Sources: numbering of North Atlantic cold phases (Bond et al., 2001); solar magnetic field (Steinhilber et al., 2010); [10]Be and ^{14}C as indicators of solar activity (Bond et al., 2001); iceberg sediment deposits North Atlantic (Bond et al., 2001); Greenland GISP ice core (Schulz, 2002); Alaska sea sediment deposit (Hu et al., 2003); China Hani peat (Hong et al., 2010); Oman stalagmite zone (Gupta et al., 2003); Indian summer monsoon (Gupta et al., 2005); West Virgina/USA stalagmite (Springer et al., 2008); and China Hongyuan peat (Xu et al., 2006).

Moffa-Sanchez et al. (2014) analyzed a sedimentary core from the Iceland basin for which they reconstructed the temperature history of the past 1000 years based on $\delta^{18}O$ and Mg/Ca ratio measurements of foraminifera shells. The reconstructed centennial-scale variations in hydrography correlate well with variability in total solar irradiance, namely the Wolf, Spörer, and Maunder Solar Minima of the LIA, Bond Event 0.

In 2015, two additional papers provided strong support for the concept of solar-driven millennial cycles. A team led by Hui Jiang notably also included one of the authors of the original Bond et al. (2001) group, Raimund Muscheler. Jiang et al. (2015) presented a high-resolution summer SST record from the North Atlantic covering the past 9300 years. The authors found a close link between the SST and solar activity records over the past 4000 years at statistically significant levels. Moreover, the presence of highly significant cycles around 90 years in the SST record suggests a connection to solar Gleissberg Cycles. A slight time lag appears when comparing the longer-term changes in solar forcing and climate responses. The apparent instantaneous climate reaction to solar forcing on multidecadal to centennial time scales, with a delayed response on longer time scales (centennial to millennial time scales), may be caused by a rapid atmospheric response to solar forcing and a delayed ocean circulation response to sustained longer-term forcing.

Balascio et al. (2015) published a record of continuous glacier activity for the last 9500 years from southeast Greenland, derived from high-resolution analysis of proglacial lake sediments. The authors documented several abrupt glacier advances beginning at 4100 years BP, each lasting ~100 years and followed by a period of retreat, superimposed on a gradual trend toward larger glacier size. Notably, the glacier advances coincided with ice rafting events in the North Atlantic Ocean (Bond et al., 2001, 1997). Interestingly, the study team consisted of two members of Bond's group at the Lamont-Doherty Earth Observatory of Columbia University, as well as Raymond Bradley, one of the coauthors of the Hockey Stick paper by Mann et al. (1999).

Summed up, a number of studies have confirmed the millennial- and centennial-scale climate cycles in the North Atlantic and their close relationship to solar activity changes as the most likely driver. Slight differences in timing, intensity, and phase most probably are due to limitations in dating, different climate proxies, and regional variations in climate response. The main question to be answered is whether the Bond Cycles are also represented outside the North Atlantic realm. Therefore, we will review Holocene climate records from other parts of the world. If an external solar climate driver has controlled climate cycles of the past 10,000 years, this should have manifested itself in climate archives on all continents.

5.2 North America

Newfoundland. Solar-driven, millennial-scale climate cycles have been reported by Marchitto and deMenocal (2003) from offshore Newfoundland. The authors reconstructed the temperature development of the last 4000 years based on magnesium/calcium ratios in benthic foraminifera and found a characteristic natural cyclicity. Results suggest that the temperature of the upper North Atlantic deep water has varied by at least 2°C during the late Holocene. This millennial-scale cooling notably coincides with the cold events described by Bond et al. (2001).

New Jersey. Li et al. (2007) studied Holocene lake sediments in New Jersey and found low lake levels at about 1.3, 3.0, 4.4, and 6.1 ka, indicated by heterogeneous, coarse, calcareous sediments showing strong magnetic intensities. The dry periods inferred from the low lake levels appear to occur concurrently with cold periods recorded in North Atlantic sediments (Bond et al., 2001). The correlation between millennial-scale dry/wet cycles inferred from lake-level fluctuations of White Lake and cold/warm cycles in North Atlantic sediments suggests sensitive moisture responses to Holocene millennial-scale climate variability.

Virginia. Holocene millennial-scale climate variability has been also described from Virgina. Willard et al. (2005) documented Holocene millennial-scale cool intervals that occurred every ~1400 years and lasted ~300—500 years, based on pollen data from Chesapeake Bay. The cool events are indicated by significant decreases in pine pollen, which we interpret as representing decreases in January temperatures of between 0.2°C and 2°C. The timing of the pine minima is correlated with the North Atlantic ice-rafting events of Bond et al. (2001) and solar minima interpreted from cosmogenic isotope records.

West Virginia. Springer et al. (2008) analyzed Sr/Ca ratios and $\delta^{13}C$ values in a Holocene stalagmite from a cave in West Virgina and found six centennial-scale droughts in the last 6000 years. The droughts correlate well to the cold events reported by Bond et al. (2001) from the North Atlantic. Spectral analysis of the West Virginia data yields coherent ~200 and ~500 year periodicities, consistent with Suess-de Vries solar irradiance data and an unnamed solar cycle. These provide convincing evidence for solar forcing of east-central North American droughts and strengthen the case for solar modulation of midcontinent climates.

North Carolina. A Holocene, high-resolution, foraminiferal-based SST reconstruction was presented by Cléroux et al. (2012) for an area offshore from Cape Hatteras, North Carolina. Wavelet transform analysis revealed a characteristic 1000-year period, pacing the $\delta^{18}O$ signal over the early Holocene. This 1000-year frequency band is significantly coherent with the 1000-year frequency band of TSI between 9.5 and 7 ka, and both signals are in phase over the rest of the studied period.

Florida. Schmidt et al. (2012) compiled an early to mid-Holocene record of SSTs and sea-surface salinities for the Florida Straits, based on Mg/Ca-paleothermometry and stable oxygen isotope measurements on planktonic foraminifera from a sediment core. The authors found that early Holocene sea surface salinity enrichments are associated with increased evaporation/precipitation ratios in the Florida Straits during periods of reduced solar forcing during North Atlantic ice-rafting events (Bond et al., 2001). Schmidt et al. interpret that variations in solar forcing over the early Holocene had a significant impact on the global tropical hydrologic cycle.

New Mexico. Based on $\delta^{18}O$ data from a cave stalagmite in New Mexico, Asmerom et al. (2007) reconstructed a high-resolution Holocene climate record for the North American monsoon region of the southwestern United States. The data show that periods of increased solar radiation correlate with decreased rainfall, indicating a strong solar influence on climate in the region.

Mexico. Marchitto et al. (2010) presented a high-resolution magnesium/calcium proxy record of Holocene SST off the west coast of Baja California Sur, Mexico. The early Holocene SSTs were characterized by millennial-scale fluctuations that correlate with cosmogenic nuclide proxies of solar variability, with inferred solar minima corresponding to El Niño-like (warm) conditions and an active sun triggering cold La Niña ocean circulation. The group observed five cold intervals between ~7000 and 11,000 years, with roughly 1000-year spacing. Coauthor of the study was Raimund Muscheler, who was a member of the Bond et al. (2001) group.

Alaska. Hu et al. (2003) analyzed lake sediment from southwestern Alaska and found cyclic variations in climate and ecosystems during the Holocene. The observed variations occurred with periodicities similar to those of solar activity and appear to be coherent with time series of the cosmogenic nuclides ^{14}C and ^{10}Be, as well as North Atlantic drift ice Bond et al. (2001). The authors conclude that small variations in solar irradiance induced pronounced cyclic changes in northern high-latitude environments.

A pioneer paper on solar-driven millennial-scale climate cycles was published by Denton and Karlén (1973). The authors described Holocene retreats and advances of Alaskan and Yukon glaciers. They noted a close correlation of glacier change with C^{14} variations measured from tree rings as proxy for galactic cosmic rays and solar activity. Already in the early 1970s, the authors suspected that the observed climate fluctuations were most probably caused by varying solar activity.

West Canada. Gavin et al. (2011) carried out multiproxy analyses on Holocene sediments of a lake in interior British Columbia. Climate development was reconstructed based on biogenic silica (BSi) abundance and changes in diatom and pollen assemblages. The authors found good correlation of the climate with solar activity changes in the early and late Holocene, but the link was less apparent during the mid-Holocene.

North America general. Viau et al. (2006) published a mean continental July temperature reconstruction based on pollen records from across North America for the past 14,000 years. Maximum values were reached between 6000 and 3000 cal year BP, after which mean July temperatures decreased. Superimposed on this trend were characteristic millennial-scale climate cycles with a period of ~1100 years, which falls reasonably close to the solar Eddy Cycle duration.

5.3 South America

Brazil. Evangelista et al. (2014) studied a sedimentary core near Rio de Janeiro and reconstructed Holocene SSTs based on alkenones. They found typical millennial-scale climate cycles, which correlate well with the cold events described by Bond et al. (2001). The authors detected typical periodic signals of 800, 1700, and 2200 years in the temperature curve. Evangelista et al. (2014) did not attempt to identify likely climate drivers for this millennial-scale climate cyclicity, but the 800-year and 2200-year periods show great affinity towards the Eddy and Hallstatt Cycles.

Millennial-scale climate cycles have been also reported by Chiessi et al. (2014), who analyzed Mg/Ca and oxygen isotopic compositions of planktonic foraminifera for the late Holocene from two marine sediment cores retrieved from the Atlantic off southern Brazil. The SSTs during this time showed fluctuations of up to 3°C and cyclicity with a period of c.730 years. While Chiessi et al. (2014) interpret the observed climate cycles primarily as related to changes in the Atlantic meridional overturning circulation, they did not rule out solar activity changes as climate driver because one of the detected periodicities of total solar irradiance had a 710-year duration.

Geologically, millennial-scale cycles suggestive of Bond Cycles and the 2400 years Hallstatt Cycles have been reported by Franco et al. (2012) from Permian–Carboniferous rhythmites in the Brazilian Paraná Basin.

Chile. Varma et al. (2011) analyzed the iron record for the past 3000 years from a sedimentary core on the Chilean continental shelf. Iron represents a proxy for the strength of Southern Hemisphere westerly winds. The data show that the winds were closely correlated to solar activity. During periods of low solar activity, the Southern Hemisphere westerly winds typically shifted equatorward, whereas they moved southward again when the sun became more active. Varma et al. (2011) conclude that the role of the sun in modifying Southern Hemisphere tropospheric circulation patterns has probably been underestimated in model simulations of past climate change.

Caniupán et al. (2014) published Holocene, alkenone-derived SST records from the southernmost Chilean fjord region. The records show consistently warmer than present-day SSTs, except for the past ~600 years. During the mid-Holocene, pronounced short-term variations of up to 2.5°C and cooling centered at 5000 years were recorded, which coincides with the first Neoglacial glacier advance in the southern Andes. The latest Holocene is characterized by two pronounced cold events centered at ~600 years ago, ie, during the Little Ice Age.

Peru. Engel et al. (2014) carried out a reconstruction of temperature and precipitation for the last 4300 years based on a peat core from the Peruvian Central Andes. They found pronounced cooling around 2800 years ago, which corresponds to the cooling of the North Atlantic Bond Event 2. Cold temperatures during the LIA coincide with Bond Event 0.

South America. Vuille et al. (2012) reviewed the South American monsoon history over the past two millennia, based on high-resolution, stable isotope proxies from speleothems, ice cores, and lake sediments. They found that the monsoon fluctuated cyclically with a period of about 1000 years, which may correspond to the solar Eddy Cycle. The South American summer monsoon weakened during the MWP and Modern Warm Period, whereas it strengthened during the Little Ice Age.

5.4 Europe

Scandinavia. Helama et al. (2010) studied mid- and late Holocene terrestrial temperature records and solar activity in stems in small lakes of northern Lapland in Finland and Norway. The tree-ring–derived temperatures exhibit persistent annual-to-millennial–scale variations, with multidecadal to multicentennial periodicities reminiscent of the sun's periodicities. Moreover, millennial and bimillennial modes of climate variability were correlative with variations in sunspot numbers on similar scales, with near-century and near-zero lags, respectively.

Ojala et al. (2015) analyzed 10,000-year-long varved sediment records from lakes in Finland and found characteristic climate oscillations at multidecadal to millennial timescales. Among the periodicities found were the solar 1000-year Eddy Cycle, the 200-year Suess-de Vries Cycle, and the 90-year Gleissberg Cycle. The authors consider solar activity changes an important solar driver that may have operated alongside other factors, such as Atlantic Ocean cycles. Ojala et al. (2015) expected that the relative influence of the various climate factors varied through time.

Esper et al. (2012a,b) reconstructed northern Scandinavian summer temperatures based on maximum latewood density data for the past 2000 years. The authors found a well-developed, millennial-scale cyclicity consisting of the RWP, MWP, LIA, and Modern Warm Period. Notably, warmth during the Roman and Medieval times was larger in extent and longer in duration than the 20th-century warmth. The two cold periods correspond to the North Atlantic Bond cold events one and 0. The authors also found a long-term cooling over the last 2000 years, which they related to orbital Milankovic changes.

Germany. Fohlmeister et al. (2012) studied cave stalagmites from western Germany, from which they derived a record of past winter climate variability for the last 10,800 years. Changes in $\delta^{18}O$, $\delta^{13}C$ values and Mg/Ca ratios were used as proxies for precipitation and temperature. The authors found a millennial-scale cyclicity, alternating between warm/wet and cold/dry, which showed a high-degree of correlation with the cycles presented by Bond et al. (2001).

Similar millennial-scale climate cyclicity was described by Müller et al. (2005) from the last interglacial, the Eemian, from southern Germany. The study team, which involved Gerard Bond, discussed analogies with the North Atlantic Holocene record and speculated that the Eemian Cycles may also have been triggered by solar activity changes. Müller et al. (2005) explained that natural cyclic changes appear to be a persistent feature of interglacial climates.

Austria. Mangini et al. (2007) carried out a climate reconstruction from the Central Alps for the last 9000 years based oxygen isotope records from stalagmites in a cave in Tyrolean Austria. The observed millennial climate cyclicity shows an excellent match with North Atlantic ice-rafting events (Bond et al., 2001).

Switzerland. Niemann et al. (2012) reconstructed the mean annual air temperature history for the last 11,000 years in southern Switzerland using specific bacterial membrane lipids in lake sediment. Climate was dominated by millennial-scale cyclicity, which according to the authors, showed a fairly good match with the record from the Austrian cave studied by Mangini et al. (2007). In particular, the cold events at 500 (LIA), 4000, 6000, and 10,000 years ago reported by Niemann et al. (2012) from Switzerland fit in timing with Bond Events 0, 3, 4, and 7. Correlation of other cold phases is more complicated, possibly because the temporal resolution of the Swiss climate record is relatively low and variable.

Magny (2004) presented a Holocene, mid-European lake-level record based on tree rings and lake sediments in the Jura Mountains, of the northern French Pre-Alps and the Swiss Plateau. The data indicate a rather unstable Holocene climate punctuated by 15 phases of higher lake level. The lake-level development shows a good correlation with the North Atlantic climate curve and the residual ^{14}C records related to cosmic rays. Magny (2004) concludes that changes in the solar activity have played a major role in Holocene climate oscillations. A decade earlier, Magny (1993) had described the solar Hallstatt Cycle from the climatic lake record.

France. Azuara et al. (2015) reconstructed the precipitation history of the past 4500 years on the Mediterranean coast in southern France based on high-resolution pollen analyses of two sediment cores recovered from lagoons. Besides a long-term aridification trend over the study period, Azuara et al. (2015) also found three superimposed arid events at around 4400, 2600 and 1200 years BP, coinciding in time with Bond Events 3, 2 and 1, respectively.

Sabatier et al. (2012) reconstructed the palaeo-storm activity for the past 7000 years on the French Mediterranean coast based on a lagoon sediment core in the Gulf of Lions. Integrating grain size, faunal analysis, clay mineralogy, and geochemistry data with radiocarbon chronology, the authors recorded seven periods of increased storm activity at 6300−6100, 5650−5400, 4400−4050, 3650−3200, 2800−2600, 1950−1400, and 400−50 years BP. These periods of increased storm activity correlate well with North Atlantic Holocene cooling events.

Jalali et al. (2015) generated a Holocene SST reconstruction for the Gulf of Lions based on sediment core material using alkenones and high-molecular-weight, odd-carbon numbered n-alkanes. Besides some Holocene longterm trends, the authors found a series of superimposed multidecadal cooling events of ∼1°C amplitude. The cold events show a fairly good match with the North Atlantic cool phases.

Spain. Pena et al. (2010) studied benthic foraminiferal stable isotopes (δ^{18}O, δ^{13}C) and molecular biomarkers of a sedimentary core taken off the northwest Spanish coast. In their climate reconstruction of the past 8200 years, the authors found rapid centennial and millennial events that were strongly linked to cold phases from the North Atlantic.

Nieto-Moreno et al. (2011) reconstructed climate variability for the last 4000 years in an offshore sediment core from the west Algerian Balearic Basin, studying changes in chemical and sediment mineralogical composition as well as grain-size distribution. The group found pronounced climatic fluctuations that in part show good correlation with the North Atlantic. The authors considered changes in solar irradiance and North Atlantic Oscillation variability as the main driving mechanisms behind the observed natural climate change.

Fletcher et al. (2013) studied a marine core from the Alboran Sea and found distinct millennial-scale climate cycles throughout the Holocene. A 900-year periodicity in the data was interpreted as an Eddy Solar Cycle, while a 1750-year periodicity was thought to relate to Atlantic Ocean cycles.

Another core from the Alboran Sea was analyzed by Rodrigo-Gámiz et al. (2014), who documented well-identified, millennial to centennial scale paleoclimate cycles for the past 20,000 years. Bond Cycle periodicities of 1000−2000 years occur. As in the previously mentioned study, these authors also suggest a mix of solar and ocean cycles as drivers for the observed natural cyclicity.

Adriatic Sea. Siani et al. (2013) presented a Holocene stable oxygen and carbon isotope SST reconstruction from a core collected in the South Adriatic Sea. Characteristic millennial-scale cyclicity of 1670 years was documented, close to the North Atlantic Cycles.

NE Mediterranean. Valsecchi et al. (2012) reported a high-resolution pollen analysis of a sediment core from the Sea of Marmara. They observed a number of climate oscillations, which they partly correlated with North Atlantic ice-rafting events.

All Mediterranean. Dermody et al. (2012) presented a reconstruction of the change in climatic humidity around the Mediterranean 3000−1000 years BP. Using a range of proxy archives and model simulations they concluded that climate during this period was typified by a millennial-scale seesaw in climatic humidity between Spain and Israel on one side and the Central Mediterranean and Turkey on the other. The Mediterranean humidity/aridity cycles show a good correlation with North Atlantic ice-rafting events.

5.5 Africa

North Africa. Jiménez-Espejo et al. (2014) reconstructed Saharan aeolian conditions and effective humidity variations during the Holocene, based on a southern Iberia alpine lacustrine record. Cyclo-stratigraphic analyses and transport mechanisms both point to solar irradiance and aridity as major triggering factors for dust supply over Western Europe during the Holocene. The observed 1500—2000 year cycles show similarities with North Atlantic ice-rafting events.

Egypt. Hennekam et al. (2014) presented high-resolution records for oxygen isotopes of the planktic foraminifer *Globigerinoides ruber* ($\delta^{18}O_{ruber}$) and bulk-sediment inorganic geochemistry for Holocene sediments from the Nile Delta. The authors found that Nile discharge was highly variable on a multicentennial time scale during the early to middle Holocene, being strongly influenced by variable solar activity.

Ethiopia. Ghinassi et al. (2012) reconstructed lake-level fluctuations of the past 3500 years for a lake in northern Ethiopia. The authors found three main highstand phases, which occurred at about 3250—3000, 2600—950 and 650—160 years BP, respectively. These wet phases may correspond to North Atlantic ice-rafting events events 2, 1, and 0, respectively. The studied lake record shows good correlation with several other East Africa lakes and therefore can be considered representative for the late Holocene climate of the region.

Mauritania. deMenocal et al. (2000) studied a sediment core extracted off Cap Blanc in Mauritania from which they reconstructed the Holocene climate based on alkenone-derived SSTs and foraminiferal faunal changes. The authors found a series of abrupt, millennial-scale cooling events, which punctuated the Holocene and correlate well with North Atlantic ice-rafting events.

5.6 Asia

Oman. Neff et al. (2001) published a high-resolution record of oxygen isotope variations, for the period from 9600 to 6100 years BP in a Th—U-dated stalagmite from Oman. The study found an excellent correlation between monsoon rainfall and changes in solar activity. The authors concluded that variations in solar radiation are one of the primary controls on centennial- to decadal- scale changes in tropical rainfall and monsoon intensity during this time.

Gupta et al. (2005) studied Holocene foraminiferal abundance patterns as a proxy for monsoons in a sediment core off Oman in the northwestern Arabian Sea. The proxy curve is characterized by strong millennium cycles in the range of 1000 to 2000 years, which correlate well with North Atlantic ice-rafting events. Gupta et al. (2005) found that weak summer monsoon winds generally correlate with reduced solar output. They conclude that solar influence plays a major role for the studied climate system and that small changes in solar irradiance can bring pronounced changes in the tropical monsoon.

India. Thamban et al. (2007) analyzed a marine core off the Indian east coast and reconstructed summer monsoon-controlled precipitation changes during the Holocene on the basis of high-resolution terrigenous proxy studies. The authors found several abrupt events in monsoon precipitation throughout the Holocene. Spectral analysis of the precipitation records reveals statistically significant periodicities that include the solar Hallstatt, Eddy, Sues-de Vries, and Gleissberg Cycles of 2200, 950, 220, and 126—92 years, respectively. Thamban et al. (2007) interpret that small changes in solar activity controlled the Indian monsoon to a large extent.

Menzel et al. (2014) carried out biogeochemical and mineralogical analyses on a sediment core of the entire Holocene from a central Indian lake. The authors documented millennial-scale climate cycles with repeated dry/wet shifts, which occurred synchronously with North Atlantic ice-rafting events. Notably, all nine Bond events are also isochronally (within dating uncertainties) reflected in the central Indian lake record. Because the studied climate record from India correlates quite well with the solar output proxy ^{14}C production rate, Menzel et al. (2014) concluded that changes in solar activity have been the cause of the observed centennial- and millennial-scale climate shifts in the Indian monsoon.

China. Wang et al. (2005) published an Asian monsoon record over the past 9000 years based on oxygen isotope data from the Dongge Cave in southern China. The authors documented eight phases when the monsoon weakened, each lasting one to five centuries. Cross-correlation of the decadal- to centennial-scale monsoon record with the atmospheric radiocarbon record as proxy for solar activity changes showed good correlation. Wang et al. (2005) concluded that much of the monsoon variability must have resulted from changes in solar output. The weak monsoon phases generally correlate well with North Atlantic ice-rafting events.

A few years later, Steinhilber et al. (2012) published a refined curve of Holocene solar activity based on a combination of ^{10}Be in ice core records from Greenland and Antarctica and the global ^{14}C tree ring record. According to

Steinhilber et al. (2012), carbon-14 that was used by Wang et al. (2005) is not fully suitable as a solar proxy alone. When Steinhilber et al. compared the isotope-based Holocene climate record of the Dongge Cave of Wang et al. with the updated solar activity curve, a very high degree of correlation was observed. The authors conclude that this "correlation is remarkable because the Earth's climate has not been driven by the Sun alone. Other forcings like volcanoes, greenhouse gas concentrations, and internal variability also have played an important role."

Donges et al. (2015) regionally expanded the Asian monsoon cave theme and studied speleothems from various caves in China, India, Indonesia, Oman, and Yemen. The data revealed pronounced regime shifts in Asian monsoon variability, including the periods around 8500—7900, 5700—5000, 4100—3700, and 3000—2400 years BP. The timing of these regime shifts is consistent with known episodes of Holocene rapid climate change (RCC) and high-latitude Bond Events. Donges et al. (2015) speculate that solar activity changes may be a key driver for the Asian monsoon and recommend future research to further substantiate this hypothesis.

Xu et al. (2015) analyzed peat deposits at a lake in southwest China and found abrupt, millennial-scale episodes when the Indian summer monsoon weakened, which correlate well with the North Atlantic Bond Events. The reconstructed phases of failed Indian summer monsoons show a good match with changes in solar activity. The authors therefore consider solar activity changes the key climate driver for the observed climate variability.

The Hai Xu group had previously reported strong evidence that solar activity changes played a major role in shaping the Holocene climate in China. Xu et al. (2002) reconstructed temperature variations of the past 6000 years based on $\delta^{18}O$ data from a peat profile in southwest China. The authors found marked temperature variations, which included periodicities that match the Eddy, Suess-de Vries, and Gleissberg solar frequency bands. Xu et al. (2002) consider solar forcing the main driving forcing of the observed climate fluctuations. A few years later, Xu et al. (2006) further corroborated these findings. In particular, it was found that the cold phases in the temperature curve from southwest China correlated well with the North Atlantic Bond cooling events 0—4.

Yu et al. (2012) reconstructed the mid-Holocene East Asian monsoon history for the period 6650 to 2150 years BP based on a sediment core extracted from an estuary in southern China. Organic carbon isotopes ($\delta^{13}C$), total carbon to total nitrogen (C/N) ratios, and total organic carbon (TOC) concentrations were used as proxies of changes in monsoonal precipitation strength. The authors documented a series of dry—wet oscillations at centennial to millennial timescales, which they interpreted as response to solar activity changes. Comparison of the climate record from south China with North Atlantic ice-rafting events yields a high degree of correspondence.

Liu et al. (2011) studied amplitudes, rates, and periodicities of temperature variations for the past 2485 years, based on tree rings on the east-central Tibetan Plateau. The results showed that extreme climatic events on the plateau, such as the MWP, LIA, and 20th-century warming appeared synchronously with those worldwide. Temperatures varied by about 1°C. Notably, the largest amplitude and rate of temperature change occurred during the Eastern Jin Event (343—425 AD), and not in the late 20th century. Cold intervals generally corresponded to sunspot minimums. The authors found various climate cycles associated with solar activity.

Liu et al. (2014) reconstructed late Holocene glacier variations on the westernmost Tibetan Plateau based on analysis of grain size and magnetic susceptibility from a glaciolacustrine lake sediment core. The authors identified four glacier expansion episodes that alternated with glacier retreat periods during the last 4000 years. Timing of the glacier expansion episodes corresponds well with late Holocene North Atlantic ice-rafting events.

Zhao et al. (2012) developed a regional vegetation and climate dynamics scenario on the basis of a high-resolution pollen record over the last 4000 years in the Taklamakan Desert in northwest China. Three periods of increased humidity, from c.4000—2620 years BP, c.1750—1260 years BP, and c.550—390 years BP, were identified, correspoding with the Bond North Atlantic ice-rafting events 2, 1, and 0 respectively.

Ruan et al. (2015) presented a record of SSTs from long-chain alkenones for the past 15,000 years in the southern Okinawa Trough in the East China Sea. The authors identified abrupt temperature drops, each lasting 200 to 1000 years and centered at ca. 8300, 7300, 4900, 4000, 3400, 2500, 1500, and 500 years BP. The cooling phases at 8600—8100, 5800—4800, 4100—3900, 1600—1300, and 600—500 years BP are synchronous with ice-rafting events in the North Atlantic. Other cooling phases at ~7300, 3400, and 2500 years BP are not identified in the sediment record of the North Atlantic, suggesting a complex sun—ocean—atmosphere coupling.

Taiwan. Liew et al. (2006) reconstructed Holocene climate variability in central Taiwan based on a subalpine pollen sequence from peat bog deposits. Abrupt and relative severe cold phases, shown by biome changes, occurred at about 11,200—11,000 years BP; 7500 years BP; 7200 years BP; 7100 years BP; 5200 years BP; 5000 years BP; and 4900 years BP. A spectral analysis of pollen of a relatively cold taxon, Salix, reveals that the time series is dominated by a 1500-year periodicity. The cold—warm cycles show close relationship to solar activity changes as reflected in the production rate of ^{10}Be.

South Korea. Lim et al. (2005) studied variations in eolian quartz over the past 6500 years as a proxy for dust in maar sediments in South Korea. The record yielded millennial- and centennial-scale fluctuations with periodicities of 1137, 739, 214, 162, 137, 127, and 111 years, implying drier conditions in the source areas in China. The detrended eolian quartz flux record correlates visually and statistically with the atmospheric Δ^{14}C record (a solar proxy). The authors suggest that the observed centennial-scale variability in eolian quartz flux may be controlled by solar activity through the Sun—East Asian monsoon linkage.

Indonesia. Oppo et al. (2009) presented a temperature reconstruction for the past 2000 years of the Makassar Strait in Indonesia using stable oxygen isotopes. The curve shows two prominent cold events centered at 700 AD and 1700 AD, corresponding to the cold phase of the Migration Period/Bond Cold Event 1 and the LIA/Bond Cold Event 0. Notably, temperatures during the MWP at 1000—1250 AD were similar to modern levels.

Tropical West Pacific. Khider et al. (2014) studied stable oxygen isotopes and magnesium/calcium records of co-occurring planktonic and benthic foraminifera from a Holocene marine sediment core collected in the western equatorial Pacific. The planktonic record exhibits millennial-scale SST oscillations of ~0.5°C over the Holocene. While the authors favor climate internal cycles as the main driver they also could not totally discount the solar forcing hypothesis. Solar forcing as an explanation for the observed millennial-scale SST variability requires (1) a large climate sensitivity and (2) a long 400-year delayed response, suggesting that if solar forcing is the cause of the variability, it would need to be considerably amplified by processes within the climate system, at least at the core location.

Subarctic North Pacific. Max et al. (2012) investigated North Pacific SSTs for the last 15,000 years based on six piston cores from the western Bering Sea, the continental slope of east Kamchatka, and the southeastern Sea of Okhotsk. The alkenone-derived temperature curve contained pronounced millennial-scale SST fluctuations that are similar to short-term climate oscillations known from Greenland isotope ice-core records. On the basis of chronology, the sea-surface variability in the northwest Pacific realm seems to be rather in-phase than out-of-phase with Greenland/North Atlantic temperature changes. Max et al. (2012), therefore, suggested a quasi-synchronicity between the north Atlantic and north Pacific SST development during the last glacial termination and argue for a strong atmospheric coupling between the north Pacific and the north Atlantic. North Atlantic ice-rafting event cyclicity and solar climate drivers have been reported from the Greenland Holocene climate record (see above). Solar influence on climate may also be assumed for the subarctic north Pacific, given the good match with Greenland reported by Max et al. (2012).

Lake Baikal. Murakami et al. (2012) analyzed temporal variation of U concentration that reflect weathering intensity associated with changes in the rainfall/moisture levels in a sediment cores in Lake Baikal. The detrended U record for the last 5200 years shows wet events at 4300—3700, 3200—2300, 1800—1200, and 800—300 years BP. These events coincide with North Atlantic ice-rafting events.

Another paper on Holocene climate change in the Lake Baikal region was published by Kravchinsky et al. (2013), who conducted a high-resolution study of wind-blown sediments and buried soils. Relative wind strength was determined by grain size analyses of different stratigraphic units. Windy and cold periods correspond to the absence of soil. Spectral analysis of the soil data demonstrates periodic changes of 1500, 1000, and 500 years of relatively warm and cold intervals during the Holocene. A reasonable correlation with North Atlantic ice-rafting events can be made. Kravchinsky et al. (2013) presumed that the 1000- and 500-year climatic cycles were driven by increased solar insolation. The 1500-year cycle associated with North Atlantic circulation appears only in the Late Holocene. Three time periods—8400—9300 years BP, 3600—5100 years BP, and the last ~250 years BP—correspond to both the highest sun spot number and the most developed soil horizons in the studied sections.

5.7 Australia and Oceania

Australia. Moros et al. (2009) studied oxygen-isotope records of two planktonic foraminifera species in a deep-sea sediment core off South Australia. A strong ~1550-year cycle was found in the *Globigerina bulloides* record with prominent cold phases centered at ca. 9200, 7300, 5800, 4300, 2700, and 1400 years BP and, possibly, the LIA, which have global counterparts. Soon et al. (2014) applied wavelet analysis to the δ18O record for the planktonic species *Globigerinoides ruber* from the same sediment core and identified prominent concentrations of power at about 2048 years, 1000 years, and 512 years, consistent with the Hallstatt, Eddy, and an unnamed solar activity cycles.

Kemp et al. (2012) reconstructed palaeosalinity changes in groundwater-influenced lakes based on ostracod assemblages, oxygen isotopes, and quartz sand counts in southeast Australia. The authors found short-lived, low-salinity events at 8800, 7200, 5900, 4800, 2400, 1300, and 400 years BP, indicative of higher effective precipitation.

These events are similar in timing and number to those recorded on Australia's southern continental shelf and also correlate well with North Atlantic ice-rafting events.

5.8 Antarctica

Masson et al. (2000) described millennial-scale climate cycles from East Antarctica based on water isotope measurements in 11 ice cores from coastal and central sites. All records exhibit nine millennial-scale oscillations. Climatic optima show a reduced pacing between warm events (typically 800 years), whereas cooler periods are associated with less-frequent warm events (pacing >1200 years). Notably, the 800−1200-year cycle duration is close to the 1000-year period of the Eddy Solar Cycle. In a subsequent publication, Masson-Delmotte et al. (2004) provided additional evidence for an 833-year periodicity that they observed in the deuterium and site temperature record.

5.9 Global

Mayewski et al. (2004) examined 50 globally-distributed Holocene paleoclimate records and identified as many as six periods of significant rapid climate change during the time periods 9000−8000, 6000−5000, 4200−3800, 3500−2500, 1200−1000, and 600−150 years BP. Most of the climate change events in these globally distributed records are characterized by polar cooling, tropical aridity, and major atmospheric circulation changes. The authors found that Holocene climate has been highly variable, following fairly regular quasi-periodic patterns. According to Mayewski et al. (2004), multiple controls must have been responsible for this variability. Of all the potential climate forcing mechanisms, long-term changes in solar insolation are considered by the authors to be the most important forcing mechanism for the rapid climate change, except perhaps those at 9000−8000 and 4200−3800 years BP.

Millennial-scale climate cyclicity and its possible solar origin has also been discussed by Soon et al. (2014), Loehle and Singer (2010), Singer and Avery (2008), and Perry and Hsu (2000).

6. COVERAGE OF MILLENNIAL CLIMATE CYCLES IN THE FIFTH IPCC CLIMATE REPORT

The objective of the IPCC reports was to cover information relevant to understanding the scientific basis of the risk of human-induced climate change. This includes a thorough literature analysis of past climate fluctuations and their possible drivers, which serve as valuable calibration data for climate models and numerical radiative forcing concepts. The IPCC's latest climate review is the Fifth Assessment Report (AR5). The final version of the Physical Science Basis (Working Group I) was published in late January 2014 and summarizes literature up to 2013.

Surprisingly, the AR5 report does not systematically discuss Holocene millennial-scale climate variability and its possible solar forcing. Unfortunately, Bond et al. (2001) is only cited once in the 1552 pages, and without any in-depth discussion (Chapter 7.4.7: "Many studies have reported observations that link solar activity to particular aspects of the climate system (eg, Bond et al., 2001)..."). No reference to the Bond et al. study appears in the Paleoclimate and Climate Forcing chapters of the AR5 report (Chapters 5 and 8).

From the 64 publications (up to 2013) on millennial-scale climate variability reviewed above, only six were cited in the AR5 report. Helama et al. (2010) and Esper et al. (2012a,b) documented regional temperatures during the MWP that reached levels comparable to those of the current Modern Warm Period. Potential solar forcing is omitted in the IPCC report even though it is in the title of Helama et al. (2010). Also, Vuille et al. (2012) is only discussed with respect to the MWP, ignoring important cyclicity patterns that were reported by the authors. Moros et al. (2009) reported a strong 1550-year climate cycle that is ignored in the IPCC report, and refocused on an anomaly between 5000 and 4000 years ago (Chapter 5.5.1.3: "Increased amplitude of millennial-to-centennial scale SST variability between 5 ka and 4 ka is recorded in several locations, possibly due to variations in the position and strength of the westerlies (Moros et al., 2009; Euler and Ninnemann, 2010; Shevenell et al., 2011))."

Lastly, Steinhilber et al. (2012) is only cited by the AR5 as a solar activity reference curve. The good match with climate proxies in a Chinese cave reported by Steinhilber et al. is not mentioned in AR5. Of the vast majority of papers (up to 2013) on millennial-scale climate variability that we reviewed above, 58 out of 64, or 91%, were not considered in the IPCC AR5 report. Another 20 papers, which we discussed above, were published in 2014/2015, after AR5 had been completed.

The IPCC appears to have underestimated the importance of Holocene millennial-scale climate variability and possible solar forcing in its AR5 report.

7. CONCLUSIONS

Millennial-scale climate variability is a well-established phenomenon in the Holocene development across the globe. Climate oscillations have been described from all oceans and continents. Millennial-scale climate cycles are known from upper, middle, and low latitudes, and encompass all climate zones from the Arctic to the tropics. The amplitude of the observed temperature fluctuations are often more than 1°C and thus have a similar or even greater range than the warming that the world has experienced since the LIA. Furthermore, many of these Holocene natural climate fluctuations show the same level of abruptness as the 20th-century warming.

A common characteristic of many of the documented millennial climate fluctuations is the good match with solar activity changes (Beer and van Geel, 2008). Besides solar activity changes, internal millennial ocean cycles may have contributed to the observed climate oscillations. Both solar and internal climate system autocyclic drivers are not yet implemented in current climate models. Notably, climate models still do not manage to reproduce the variable Holocene climate (Bothe et al., 2012; Crook and Forster, 2011; Lohmann et al., 2012). Yet successful hindcast capability is generally considered a prerequisite that qualifies models to be used for modeling of future climate.

The last millennial-scale climate cycle comprising the MWP, LIA, and Modern Warm Period needs to be compared to previous Holocene climate oscillations. That the general climate pattern of the past 1000 years may represent mostly a continuation of the natural millennial cyclicality that characterized the postglacial climate development cannot be excluded. This would imply that only a limited amount of warming since the end of the LIA can be attributed to CO_2. Consequently, CO_2 climate sensitivity might be at the lower end of the range indicated by the IPCC in its reports. Recent research points into this direction (eg, Lewis and Curry, 2015; Loehle, 2014; Masters, 2014; Mauritsen and Stevens, 2015; Spencer and Braswell, 2014; van der Werf and Dolman, 2014; von der Heydt et al., 2014).

Future research needs to attempt detailed correlations of Holocene climate curves, complemented by additional data filling gaps in currently poorly covered regions. The correlations need to use the latest solar activity reconstructions (eg, Steinhilber et al., 2012), which have updated earlier versions. Due to complexity and inertia of the climate system, regionally opposing trends and time lags are likely to occur, even if triggered by the same external process, eg, solar activity changes. Correlations also need to take into consideration coarse sampling resolution as well as limitations in the various age-dating techniques employed. Superimposition of climate internal autocycles and events such as volcanic eruptions further complicate the resulting overall climate record. A good understanding of global Holocene millennial- and centennial-scale climate variability and its possible solar forcing is the calibration basis for new-generation climate models that reliably can reproduce past climate change and provide more robust future climate prognoses.

References

Abreu, J.A., Beer, J., Ferriz-Mas, A., 2010. Past and future solar activity from cosmogenic radionuclides. In: Cranmer, S.R., Hoeksema, J.T., Kohl, J.L. (Eds.), SOHO-23: Understanding a Peculiar Solar Minimum, ASP Conference Series, vol. 428, pp. 287–295.

Asmerom, Y., Polyak, V., Burns, S., Rassmussen, J., 2007. Solar forcing of Holocene climate: new insights from a speleothem record, southwestern United States. Geology 35 (1), 1–4.

Azuara, J., Combourieu-Nebout, N., Lebreton, V., Mazier, F., Müller, S.D., Dezileau, L., 2015. Late Holocene vegetation changes in relation with climate fluctuations and human activities in Languedoc (Southern France). Climate of the Past Discussions 11 (5), 4123–4157.

Bal, S., Schimanke, S., Spangehl, T., Cubasch, U., 2011. On the robustness of the solar cycle signal in the Pacific region. Geophysical Research Letters 38, 1–5.

Balascio, N.L., D'Andrea, W.J., Bradley, R.S., 2015. Glacier response to North Atlantic climate variability during the holocene. Climate of the Past Discussions 11 (3), 2009–2036.

Bard, E., Frank, M., 2006. Climate change and solar variability: What's new under the sun? Earth and Planetary Science Letters 248, 1–14.

Beer, J., van Geel, B., 2008. Holocene climate change and the evidence for solar and other forcings. In: Battarbee, R.W., Binney, H.A. (Eds.), Natural Climate Variability and Global Warming: A Holocene Perspective. Wiley-Blackwell, Chichester, pp. 138–162.

Berner, K.S., Koç, N., Divine, D., Godtliebsen, F., Moros, M., 2008. A decadal-scale Holocene sea surface temperature record from the subpolar North Atlantic constructed using diatoms and statistics and its relation to other climate parameters. Paleoceanography 23 (2), PA2210.

Bond, G., Kromer, B., Beer, J., Muscheler, R., Evans, M.N., Showers, W., Hoffmann, S., Lotti-Bond, R., Hajdas, I., Bonani, G., 2001. Persistent solar influence on North Atlantic climate during the holocene. Science 294, 2130–2136.

Bond, G., Showers, W., Cheseby, M., Lotti, R., Almasi, P., deMenocal, P., Priore, P., Cullen, H., Hajdas, I., Bonani, G., 1997. A pervasive millennial-scale cycle in North Atlantic holocene and glacial climates. Science 278, 1257–1266.

Bothe, O., Jungclaus, J.H., Zanchettin, D., Zorita, E., 2012. Climate of the last millennium: ensemble consistency of simulations and reconstructions. Climate of the Past Discussions 8, 2409–2444.

Braun, H., Kurths, J., 2010. Were Dansgaard-Oeschger events forced by the sun? The European Physical Journal Special Topics 191, 117–129.

Caniupán, M., Lamy, F., Lange, C.B., Kaiser, J., Kilian, R., Arz, H.W., León, T., Mollenhauer, G., Sandoval, S., De Pol-Holz, R., Pantoja, S., Wellner, J., Tiedemann, R., 2014. Holocene sea-surface temperature variability in the Chilean fjord region. Quaternary Research 82 (2), 342–353.

Chiessi, C.M., Mulitza, S., Groeneveld, J., Silva, J.B., Campos, M.C., Gurgel, M.H.C., 2014. Variability of the Brazil current during the late holocene. Palaeogeography, Palaeoclimatology, Palaeoecology 415, 28–36.

Cho, H., Ho, C.-H., Choi, Y.-S., 2012. The observed variation in cloud-induced longwave radiation in response to sea surface temperature over the Pacific warm pool from MTSAT-1R imagery. Geophysical Research Letters 39 (18), L18802.

Christiansen, B., Ljungqvist, F.C., 2012. The extra-tropical Northern Hemisphere temperature in the last two millennia: reconstructions of low-frequency variability. Climate of the Past 8, 765–786.

Cléroux, C., Debret, M., Cortijo, E., Duplessy, J.-C., Dewilde, F., Reijmer, J., Massei, N., 2012. High-resolution sea surface reconstructions off Cape Hatteras over the last 10 ka. Paleoceanography 27 (1), PA1205.

Crook, J.A., Forster, P.M., 2011. A balance between radiative forcing and climate feedback in the modeled 20th century temperature response. Journal of Geophysical Research 116 (D17), D17108.

D'Andrea, W.J., Huang, Y., Fritz, S.C., Anderson, N.J., 2011. Abrupt Holocene climate change as an important factor for human migration in West Greenland. PNAS Early Edition 1–5. May 30.

Damon, P.E., Sonett, C.P., 1991. Solar and terrestrial components of the atmospheric ^{14}C variation spectrum. In: Sonett, C.P., Giampapa, M.S., Mathews, M.S. (Eds.), The Sun in Time. University of Arizona Press, Tuscon, pp. 360–388.

Debret, M., Sebag, D., Crosta, X., Massei, N., Petit, J.-R., Chapron, E., Bout-Roumazeilles, V., 2009. Evidence from wavelet analysis for a mid-Holocene transition in global climate forcing. Quaternary Science Reviews 28, 2675–2688.

deMenocal, P., Ortiz, J., Guilderson, T., Sarnthein, M., 2000. Coherent high- and low-latitude climate variability during the holocene warm period. Science 288, 2198–2202.

Denton, G.H., Karlén, W., 1973. Holocene climatic variations-their pattern and possible cause. Quaternary Research 3, 155–205.

Dermody, B.J., de Boer, H.J., Bierkens, M.F.P., Weber, S.L., Wassen, M.J., Dekker, S.C., 2012. A seesaw in Mediterranean precipitation during the Roman Period linked to millennial-scale changes in the North Atlantic. Climate of the Past 8, 637–651.

Donges, J.F., Donner, R.V., Marwan, N., Breitenbach, S.F.M., Rehfeld, K., Kurths, J., 2015. Non-linear regime shifts in Holocene Asian monsoon variability: potential impacts on cultural change and migratory patterns. Climate of the Past 11 (5), 709–741.

Dreschhoff, G.A.M., 2008. Paleo-Astrophysical data in relation to temporal characteristics of the solar magnetic field. In: Caballero, R., D'Olivo, J.C., Medina-Tanco, G., Nellen, L., Sánchez, F.A., Valdés-Galicia, J.F. (Eds.), Proceedings of the 30th International Cosmic Ray Conference. Universidad Nacional Autónoma de México, Mexico City, pp. 541–544, 2008.

Eichler, A., Olivier, S., Henderson, K., Laube, A., Beer, J., Papina, T., Gäggeler, H.W., Schwikowski, M., 2009. Temperature response in the Altai region lags solar forcing. Geophysical Research Letters 36, 1–5.

Engel, Z., Skrzypek, G., Chuman, T., Šefrna, L., Mihaljevič, M., 2014. Climate in the Western Cordillera of the Central Andes over the last 4300 years. Quaternary Science Reviews 99, 60–77.

Ermolli, I., Matthes, K., Dudok de Wit, T., Krivova, N.A., Tourpali, K., Weber, M., Unruh, Y.C., Gray, L., Langematz, U., Pilewskie, P., Rozanov, E., Schmutz, W., Shapiro, A., Solanki, S.K., Woods, T.N., 2013. Recent variability of the solar spectral irradiance and its impact on climate modelling. Atmospheric Chemistry and Physics 13 (8), 3945–3977.

Esper, J., Büntgen, U., Timonen, M., Frank, D.C., 2012a. Variability and extremes of northern Scandinavian summer temperatures over the past two millennia. Global and Planetary Change 88–89, 1–9.

Esper, J., Frank, D.C., Timonen, M., Zorita, E., Wilson, R.J.S., Luterbacher, J., Holzkämpe, S., Fischer, N., Wagner, S., Nievergelt, D., Verstege, A., Büntgen, U., 2012b. Orbital forcing of tree-ring data. Nature Climate Change 2, 862–866.

Euler, C., Ninnemann, U.S., 2010. Climate and Antarctic intermediate water coupling during the late Holocene. Geology 38, 647–650.

Evangelista, H., Gurgel, M., Sifeddine, A., Rigozo, N.R., Boussafir, M., 2014. South tropical Atlantic anti-phase response to holocene Bond events. Palaeogeography, Palaeoclimatology, Palaeoecology 415, 21–27.

Fletcher, W.J., Debret, M., Goñi, M.F.S., 2013. Mid-Holocene emergence of a low-frequency millennial oscillation in western Mediterranean climate: implications for past dynamics of the North Atlantic atmospheric westerlies. The Holocene 23 (2), 153–166.

Fohlmeister, J., Schröder-Ritzrau, A., Scholz, D., Spötl, C., Riechelmann, D.F.C., Mudelsee, M., Wackerbarth, A., Gerdes, A., Riechelmann, S., Immenhauser, A., Richter, D.K., Mangini, A., 2012. Bunker Cave stalagmites: an archive for central European Holocene climate variability. Climate of the Past 8, 1751–1764.

Franco, D.R., Hinnov, L.A., Ernesto, M., 2012. Millennial-scale climate cycles in Permian–Carboniferous rhythmites: permanent feature throughout geologic time? Geology 40 (1), 19–22.

Gavin, D.G., Henderson, A.C.G., Westover, K.S., Fritz, S.C., Walker, I.R., Leng, M.J., Hu, F.S., 2011. Abrupt Holocene climate change and potential response to solar forcing in western Canada. Quaternary Science Reviews 30 (9–10), 1243–1255.

Ghinassi, M., D'Oriano, F., Benvenuti, M., Awramik, S., Bartolini, C., Fedi, M., Ferrari, G., Papini, M., Sagri, M., Talbot, M., 2012. Shoreline fluctuations of Lake Hayk (northern Ethiopia) during the last 3500 years: Geomorphological, sedimentary, and isotope records. Palaeogeography, Palaeoclimatology, Palaeoecology 365–366, 209–226.

Gray, L.J., Beer, J., Geller, M., Haigh, J.D., Lockwood, M., Matthes, K., Cubasch, U., Fleitmann, D., Harrison, G., Hood, L., Luterbacher, J., Meehl, G.A., Shindell, D., van Geel, B., White, W., 2010. Solar influences on climate. Reviews of Geophysics 48, 1–53.

Gupta, A.K., Anderson, D.M., Overpeck, J., 2003. Abrupt changes in the Asian southwest monsoon during the holocene and their links to the North Atlantic ocean. Nature 421, 354–357.

Gupta, A.K., Das, M., Anderson, D.M., 2005. Solar influence on the Indian summer monsoon during the Holocene. Geophysical Research Letters 32, L17703.

Haigh, J.D., Winning, A.R., Toumi, R., Harder, J.W., 2010. An influence of solar spectral variations on radiative forcing of climate. Nature 467, 696–699.

Helama, S., Fauria, M.M., Mielikäinen, K., Timonen, M., Eronen, M., 2010. Sub-Milankovitch solar forcing of past climates: mid and late Holocene perspectives. Geological Society of America Bulletin.

Hennekam, R., Jilbert, T., Schnetger, B., de Lange, G.J., 2014. Solar forcing of Nile discharge and sapropel S1 formation in the early to middle Holocene eastern Mediterranean. Paleoceanography 29 (5), 343–356.

Hong, B., Uchida, M., Leng, X.T., Hong, Y.T., 2010. Peat cellulose isotopes as indicators of Asian monsoon variability. PAGES News 18, 18–20.

Hood, L., Schimanke, S., Spangehl, T., Bal, S., Cubasch, U., 2013. The surface climate response to 11-Yr solar forcing during northern winter: observational analyses and comparisons with GCM simulations. Journal of Climate 26 (19), 7489–7506.

Hu, F.S., Kaufman, D., Yoneji, S., Nelson, D., Shemesh, A., Huang, Y., Tian, J., Bond, G., Clegg, B., Brown, T., 2003. Cyclic variation and solar forcing of holocene climate in the Alaskan subarctic. Science 301, 1890–1893.

Huneeus, N., Boucher, O., Alterskjær, K., Cole, J.N.S., Curry, C.L., Ji, D., Jones, A., Kravitz, B., Kristjánsson, J.E., Moore, J.C., Muri, H., Niemeier, U., Rasch, P., Robock, A., Singh, B., Schmidt, H., Schulz, M., Tilmes, S., Watanabe, S., Yoon, J.-H., 2014. Forcings and feedbacks in the GeoMIP ensemble for a reduction in solar irradiance and increase in CO_2. Journal of Geophysical Research: Atmospheres 119 (9), 5226–5239.

Ineson, S., Scaife, A.A., Knight, J.R., Manners, J.C., Dunstone, N.J., Gray, L.J., Haigh, J.D., 2011. Solar forcing of winter climate variability in the Northern Hemisphere. Nature Geoscience 4, 753–757.

IPCC, 2013a. Climate Change 2013: The Physical Science Basis. Contribution of Working Group I to the Fifth Assessment Report of the Intergovernmental Panel on Climate Change. Cambridge University Press, Cambridge, United Kingdom and New York, NY, USA, 1535 p.

IPCC, 2013b. Summary for Policymakers. In: Stocker, T.F., Qin, D., Plattner, G.-K., Tignor, M., Allen, S.K., Boschung, J., Nauels, A., Xia, Y., Bex, V., Midgley, P.M. (Eds.), Climate Change 2013: The Physical Science Basis. Contribution of Working Group I to the Fifth Assessment Report of the Intergovernmental Panel on Climate Change. Cambridge University Press, Cambridge, United Kingdom and New York, NY, US.

Jalali, B., Sicre, M.A., Bassetti, M.A., Kallel, N., 2015. Holocene climate variability in the north-Western Mediterranean sea (Gulf of Lions). Climate of the Past Discussions 11 (4), 3187–3209.

Jiang, H., Eiríksson, J., Schulz, M., Knudsen, K.-L., Seidenkrantz, M.-S., 2005. Evidence for solar forcing of sea-surface temperature on the North Icelandic Shelf during the late Holocene. Geology 33 (1), 73–76.

Jiang, H., Muscheler, R., Björck, S., Seidenkrantz, M.-S., Olsen, J., Sha, L., Sjolte, J., Eiríksson, J., Ran, L., Knudsen, K.-L., Knudsen, M.F., 2015. Solar forcing of Holocene summer sea-surface temperatures in the northern North Atlantic. Geology.

Jiang, H., Seidenkrantz, M.-S., Knudsen, K.L., Eríksson, J., 2002. Late-Holocene summer sea-surface temperatures based on a diatom record from the north Icelandic shelf. The Holocene 12 (2), 137–147.

Jiménez-Espejo, F.J., García-Alix, A., Jiménez-Moreno, G., Rodrigo-Gámiz, M., Anderson, R.S., Rodríguez-Tovar, F.J., Martínez-Ruiz, F., Giralt, S., Delgado Huertas, A., Pardo-Igúzquiza, E., 2014. Saharan aeolian input and effective humidity variations over western Europe during the Holocene from a high altitude record. Chemical Geology 374–375, 1–12.

Kancírová, M., Kudela, K., 2014. Cloud cover and cosmic ray variations at Lomnický štít high altitude observing site. Atmospheric Research 149, 166–173.

Kang, S.M., Polvani, L.M., Fyfe, J.C., Sigmond, M., 2011. Impact of polar ozone depletion on subtropical precipitation. Science 332, 951–954.

Kemp, J., Radke, L.C., Olley, J., Juggins, S., De Deckker, P., 2012. Holocene lake salinity changes in the Wimmera, southeastern Australia, provide evidence for millennial-scale climate variability. Quaternary Research 77 (1), 65–76.

Khider, D., Jackson, C.S., Stott, L.D., 2014. Assessing millennial-scale variability during the Holocene: a perspective from the western tropical Pacific. Paleoceanography 29 (3), 143–159.

Kodera, K., 2006. The role of dynamics in solar forcing. Space Science Reviews 125, 319–330.

Kovaltsov, G.A., Usoskin, I.G., 2007. Regional cosmic ray induced ionization and geomagnetic field changes. Advances in Geosciences 13, 31–35.

Kravchinsky, V.A., Langereis, C.G., Walker, S.D., Dlusskiy, K.G., White, D., 2013. Discovery of Holocene millennial climate cycles in the Asian continental interior: has the sun been governing the continental climate? Global and Planetary Change 110 (Part C), 386–396.

Laken, B.A., Kniveton, D.R., Frogley, M.R., 2010. Cosmic rays linked to rapid mid-latitude cloud changes. Atmospheric Chemistry and Physics 10, 10941–10948.

Lam, M.M., Chisham, G., Freeman, M.P., 2014. Solar wind-driven geopotential height anomalies originate in the Antarctic lower troposphere. Geophysical Research Letters 41 (18), 6509–6514.

Larsen, D.J., Miller, G.H., Geirsdóttir, Á., Ólafsdóttir, S., 2012. Non-linear Holocene climate evolution in the North Atlantic: a high-resolution, multi-proxy record of glacier activity and environmental change from Hvítárvatn, central Iceland. Quaternary Science Reviews 39, 14–25.

Lewis, N., Curry, J., 2015. The implications for climate sensitivity of AR5 forcing and heat uptake estimates. Climate Dynamics 45 (3–4), 1009–1023.

Li, Y.-X., Yu, Z., Kodama, K.P., 2007. Sensitive moisture response to Holocene millennial-scale climate variations in the Mid-Atlantic region, USA. The Holocene 17 (1), 3–8.

Liew, P.M., Lee, C.Y., Kuo, C.M., 2006. Holocene thermal optimal and climate variability of East Asian monsoon inferred from forest reconstruction of a subalpine pollen sequence, Taiwan. Earth and Planetary Science Letters 250, 596–605.

Lim, J., Matsumoto, E., Kitagawa, H., 2005. Eolian quartz flux variations in Cheju Island, Korea, during the last 6500 yr and a possible Sun–monsoon linkage. Quaternary Research 64, 12–20.

Liu, X., Herzschuh, U., Wang, Y., Kuhn, G., Yu, Z., 2014. Glacier fluctuations of Muztagh Ata and temperature changes during the late Holocene in westernmost Tibetan Plateau, based on glaciolacustrine sediment records. Geophysical Research Letters 41 (17), 6265–6273.

Liu, Y., Cai, Q., Song, H., An, Z., Linderholm, H.W., 2011. Amplitudes, rates, periodicities and causes of temperature variations in the past 2485 years and future trends over the central-eastern Tibetan Plateau. Chinese Science Bulletin 56 (28–29), 2986–2994.

Ljungqvist, F.C., 2010. A new reconstruction of temperature variability in the extra-tropical northern hemisphere during the last two millennia. Geografiska Annaler: Series A 92 (3), 339–351.

Ljungqvist, F.C., Krusic, P.J., Brattström, G., Sundqvist, H.S., 2012. Northern Hemisphere temperature patterns in the last 12 centuries. Climate of the Past 8, 227–249.

Loehle, C., 2014. A minimal model for estimating climate sensitivity. Ecological Modelling 276, 80–84.

Loehle, C., Singer, S.F., 2010. Holocene temperature records show millennial-scale periodicity. Canadian Journal of Earth Sciences 47 (10), 1327–1336.

Lohmann, G., Pfeiffer, M., Laepple, T., Leduc, G., Kim, J.-H., 2012. A model-data comparison of the Holocene global sea surface temperature evolution. Climate of the Past Discussion 8, 1005–1056.

Ma, L.H., 2007. Thousand-year cycle signals in solar activity. Solar Physics 245, 411–414.

Magny, M., 1993. Solar influences on Holocene climatic changes illustrated by correlations between past lake-level fluctuations and the atmospheric ^{14}C record. Quaternary Research 40, 1–9.

Magny, M., 2004. Holocene climate variability as reflected by mid-European lake-level fluctuations and its probable impact on prehistoric human settlements. Quaternary International 113, 65–79.

Mangini, A., Verdes, P., Spötl, C., Scholz, D., Vollweiler, N., Kromer, B., 2007. Persistent influence of the North Atlantic hydrography on central European winter temperature during the last 9000 years. Geophysical Research Letters 34 (2) n/a–n/a.

Mann, M.E., Bradley, R.S., Hughes, M.K., 1999. Northern hemisphere temperatures during the past millennium: inferences, uncertainties, and limitations. Geophysical Research Letters 26 (6), 759–762.

Marchitto, T.M., deMenocal, P.B., 2003. Late holocene variability of upper North Atlantic deep water temperature and salinity. Geochemistry, Geophysics, Geosystems 4 (12).

Marchitto, T.M., Muscheler, R., Ortiz, J.D., Carriquiry, J.D., Geen, A.v, 2010. Dynamical response of the tropical Pacific ocean to solar forcing during the early holocene. Science 330, 1378–1381.

Marsh, N., Svensmark, H., 2000. Cosmic rays, clouds, and climate. Space Science Reviews 94 (1/2), 215–230.

Marsh, N., Svensmark, H., 2003. Galactic cosmic ray and El Nino–southern oscillation trends in International Satellite cloud Climatology Project D2 low-cloud properties. Journal of Geophysical Research 108, 6-1–6-11.

Masson-Delmotte, V., Stenni, B., Jouzel, J., 2004. Common millennial-scale variability of Antarctic and Southern Ocean temperatures during the past 5000 years reconstructed from the EPICA Dome C ice core. The Holocene 14 (2), 145–151.

Masson, V., Vimeux, F., Jouzel, J., Morgan, V., Delmotte, M., Ciais, P., Hammer, C., Johnsen, S., Lipenkov, V.Y., Mosley-Thompson, E., Petit, J.-R., Steig, E.J., Stievenard, M., Vaikmae, R., 2000. Holocene climate variability in Antarctica based on 11 ice-core isotopic records. Quaternary Research 54 (3), 348–358.

Masters, T., 2014. Observational estimate of climate sensitivity from changes in the rate of ocean heat uptake and comparison to CMIP5 models. Climate Dynamics 42 (7–8), 2173–2181.

Mauritsen, T., Stevens, B., 2015. Missing iris effect as a possible cause of muted hydrological change and high climate sensitivity in models. Nature Geoscience 8 (5), 346–351.

Max, L., Riethdorf, J.-R., Tiedemann, R., Smirnova, M., Lembke-Jene, L., Fahl, K., Nürnberg, D., Matul, A., Mollenhauer, G., 2012. Sea surface temperature variability and sea-ice extent in the subarctic northwest Pacific during the past 15,000 years. Paleoceanography 27 (3) n/a–n/a.

Mayewski, P.A., Rohling, E.E., Stager, J.C., Karlén, W., Maasch, K.A., Meeker, L.D., Meyerson, E.A., Gasse, F., van Kreveld, S., Holmgren, K., Lee-Thorp, J., Rosqvist, G., Rack, F., Staubwasser, M., Schneider, R.R., Steig, E.J., 2004. Holocene climate variability. Quaternary Research 62, 243–255.

Meehl, G.A., Arblaster, J.M., Matthes, K., Sassi, F., van Loon, H., 2009. Amplifying the Pacific climate system response to a small 11-Year solar cycle forcing. Science 325, 1114–1118.

Menzel, P., Gaye, B., Mishra, P.K., Anoop, A., Basavaiah, N., Marwan, N., Plessen, B., Prasad, S., Riedel, N., Stebich, M., Wiesner, M.G., 2014. Linking holocene drying trends from Lonar Lake in monsoonal central India to North Atlantic cooling events. Palaeogeography, Palaeoclimatology, Palaeoecology 410, 164–178.

Mernild, S.H., Seidenkrantz, M.-S., Chylek, P., Liston, G.E., Hasholt, B., 2012. Climate-driven fluctuations in freshwater flux to Sermilik Fjord, East Greenland, during the last 4000 years. The Holocene 22 (2), 155–164.

Moffa-Sanchez, P., Born, A., Hall, I.R., Thornalley, D.J.R., Barker, S., 2014. Solar forcing of North Atlantic surface temperature and salinity over the past millennium. Nature Geoscience 7 (4), 275–278.

Moros, M., De Deckker, P., Jansen, E., Perner, K., Telford, R.J., 2009. Holocene climate variability in the Southern Ocean recorded in a deep-sea sediment core off South Australia. Quaternary Science Reviews 28 (19–20), 1932–1940.

Moros, M., Jansen, E., Oppo, D.W., Giraudeau, J., Kuijpers, A., 2012. Reconstruction of the late-holocene changes in the Sub-Arctic Front position at the Reykjanes Ridge, North Atlantic. The Holocene 22 (8), 877–886.

Mufti, S., Shah, G.N., 2011. Solar-geomagnetic activity influence on Earth's climate. Journal of Atmospheric and Solar-Terrestrial Physics 73, 1607–1615.

Müller, U.C., Klotz, S., Geyh, M.A., Pross, J., Bond, G.C., 2005. Cyclic climate fluctuations during the last interglacial in central Europe. Geology 33 (6), 449–452.

Murakami, T., Takamatsu, T., Katsuta, N., Takano, M., Yamamoto, K., Takahashi, Y., Nakamura, T., Kawai, T., 2012. Centennial- to millennial-scale climate shifts in continental interior Asia repeated between warm–dry and cool–wet conditions during the last three interglacial states: evidence from uranium and biogenic silica in the sediment of Lake Baikal, southeast Siberia. Quaternary Science Reviews 52, 49–59.

Muscheler, R., Beer, J., 2006. Solar forced Dansgaard/Oeschger events? Geophysical Research Letters 33, L20706.

Neff, U., Burns, S.J., Mangini, A., Mudelsee, M., Fleitmann, D., Matter, A., 2001. Strong coherence between solar variability and the monsoon in Oman between 9 and 6 kyr ago. Nature 411, 290–293.

Niemann, H., Stadnitskaia, A., Wirth, S.B., Gilli, A., Anselmetti, F.S., Sinninghe Damsté, J.S., Schouten, S., Hopmans, E.C., Lehmann, M.F., 2012. Bacterial GDGTs in Holocene sediments and catchment soils of a high-alpine lake: application of the MBT/CBT-paleothermometer. Climate of the Past 8, 889–906.

Nieto-Moreno, V., Martinez-Ruiz, F., Giralt, S., Jimenez-Espejo, F., Gallego-Torres, D., Rodrigo-Gamiz, M., Garcia-Orellana, J., Ortega-Huertas, M., de Lange, G.J., 2011. Tracking climate variability in the western Mediterranean during the Late Holocene: a multiproxy approach. Climate of the Past 7, 1395–1414.

Niranjankumar, K., Ramkumar, T.K., Krishnaiah, M., 2011. Vertical and lateral propagation characteristics of intraseasonal oscillation from the tropical lower troposphere to upper mesosphere. Journal of Geophysical Research 116, 1–10.

Obrochta, S.P., Miyahara, H., Yokoyama, Y., Crowley, T.J., 2012. A re-examination of evidence for the North Atlantic "1500-year cycle" at Site 609. Quaternary Science Reviews 55, 23–33.

Ojala, A.E.K., Launonen, I., Holmström, L., Tiljander, M., 2015. Effects of solar forcing and North Atlantic oscillation on the climate of continental Scandinavia during the Holocene. Quaternary Science Reviews 112, 153−171.

Oppo, D.W., McManus, J.F., Cullen, J.L., 2003. Palaeo-oceanography: deepwater variability in the holocene epoch. Nature 422 (6929), 277.

Oppo, D.W., Rosenthal, Y., Linsley, B.K., 2009. 2,000-year-long temperature and hydrology reconstructions from the Indo-Pacific warm pool. Nature 460, 1113−1116.

Pallé, E., Butler, C.J., O'Brien, K., 2004. The possible connection between ionization in the atmosphere by cosmic rays and low level clouds. Journal of Atmospheric and Solar-Terrestrial Physics 66, 1779−1790.

Pena, L.D., Francés, G., Diz, P., Esparza, M., Grimalt, O., Nombela, M.A., Alejo, I., 2010. Climate fluctuations during the Holocene in NW Iberia: high and low latitude linkages. Continental Shelf Research 30, 1487−1496.

Perry, C.A., 2007. Evidence for a physical linkage between galactic cosmic rays and regional climate time series. Advances in Space Research 40, 353−364.

Perry, C.A., Hsu, K.J., 2000. Geophysical, archaeological, and historical evidence support a solar-output model for climate change. Proceedings of the National Academy of Sciences 97 (23), 12433−12438.

Petrick, C., Matthes, K., Dobslaw, H., Thomas, M., 2012. Impact of the solar cycle and the QBO on the atmosphere and the ocean. Journal of Geophysical Research: Atmospheres 117, D17 n/a−n/a.

Rahmstorf, S., 2003. Timing of abrupt climate change: a precise clock. Geophysical Research Letters 30 (10).

Ratnam, M.V., Santhi, Y.D., Kishore, P., Rao, S.V.B., 2014. Solar cycle effects on Indian summer monsoon dynamics. Journal of Atmospheric and Solar-Terrestrial Physics 121 (Part B), 145−156.

Rawal, A., Tripathi, S.N., Michael, M., Srivastava, A.K., Harrison, R.G., 2013. Quantifying the importance of galactic cosmic rays in cloud microphysical processes. Journal of Atmospheric and Solar-Terrestrial Physics 102, 243−251.

Reichler, T., Kim, J., Manzini, E., Kröger, J., 2012. A stratospheric connection to Atlantic climate variability. Nature Geoscience 5, 783−787.

Rodrigo-Gámiz, M., Martínez-Ruiz, F., Rodríguez-Tovar, F.J., Jiménez-Espejo, F.J., Pardo-Igúzquiza, E., 2014. Millennial- to centennial-scale climate periodicities and forcing mechanisms in the westernmost Mediterranean for the past 20,000 yr. Quaternary Research 81 (1), 78−93.

Ruan, J., Xu, Y., Ding, S., Wang, Y., Zhang, X., 2015. A high resolution record of sea surface temperature in southern Okinawa Trough for the past 15,000 years. Palaeogeography, Palaeoclimatology, Palaeoecology 426, 209−215.

Sabatier, P., Dezileau, L., Colin, C., Briqueu, L., Bouchette, F., Martinez, P., Siani, G., Raynal, O., Von Grafenstein, U., 2012. 7000 years of paleostorm activity in the NW Mediterranean Sea in response to Holocene climate events. Quaternary Research 77 (1), 1−11.

Schmidt, M.W., Weinlein, W.A., Marcantonio, F., Lynch-Stieglitz, J., 2012. Solar forcing of Florida Straits surface salinity during the early Holocene. Paleoceanography 27 (3), PA3204.

Schulz, M., 2002. On the 1470-year pacing of Dansgaard-Oeschger warm events. Paleoceanography 17 (2).

Sfică, L., Voiculescu, M., 2014. Possible effects of atmospheric teleconnections and solar variability on tropospheric and stratospheric temperatures in the Northern Hemisphere. Journal of Atmospheric and Solar-Terrestrial Physics 109, 7−14.

Shevenell, A.E., Ingalls, A.E., Domack, E.W., Kelly, C., 2011. Holocene Southern Ocean surface temperature variability west of the Antarctic Peninsula. Nature 470, 250−254.

Siani, G., Magny, M., Paterne, M., Debret, M., Fontugne, M., 2013. Paleohydrology reconstruction and holocene climate variability in the south Adriatic sea. Climate of the Past 9 (1), 499−515.

Singer, S.F., Avery, D.T., 2008. Unstoppable Global Warming—Every 1500 Years. Rowan & Littlefield Publishers, Lanham, MD, 278 p.

Solanki, S.K., Usoskin, I.G., Kromer, B., Schüssler, M., Beer, J., 2004. Unusual activity of the Sun during recent decades compared to the previous 11,000 years. Nature 431, 1084−1087.

Sonett, C.P., Finney, S.A., 1990. The spectrum of radiocarbon. Philosophical Transactions of Royal Society of London A330, 413−426.

Soon, W., Velasco Herrera, V.M., Selvaraj, K., Traversi, R., Usoskin, I., Chen, C.-T.A., Lou, J.-Y., Kao, S.-J., Carter, R.M., Pipin, V., Severi, M., Becagli, S., 2014. A review of Holocene solar-linked climatic variation on centennial to millennial timescales: physical processes, interpretative frameworks and a new multiple cross-wavelet transform algorithm. Earth-Science Reviews 134, 1−15.

Spencer, R., Braswell, W., 2014. The role of ENSO in global ocean temperature changes during 1955−2011 simulated with a 1D climate model. Asia-Pacific Journal of Atmospheric Sciences 50 (2), 229−237.

Springer, G.S., Rowe, H.D., Hardt, B., Edwards, R.L., Cheng, H., 2008. Solar forcing of Holocene droughts in a stalagmite record from West Virginia in east-central North America. Geophysical Research Letters 35, 1−5.

Steinhilber, F., Abreu, J.A., Beer, J., Brunner, I., Christl, M., Fischer, H., Heikkilä, U., Kubik, P.W., Mann, M., McCracken, K.G., Miller, H., Miyahara, H., Oerter, H., Wilhelms, F., 2012. 9400 years of cosmic radiation and solar activity from ice cores and tree rings. Proceedings of the National Academy of Sciences 109 (16), 5967−5971.

Steinhilber, F., Abreu, J.A., Beer, J., McCracken, K.G., 2010. Interplanetary magnetic field during the past 9300 years inferred from cosmogenic radionuclides. Journal of Geophysical Research 115, A01104.

Steinhilber, F., Beer, J., Fröhlich, C., 2009. Total solar irradiance during the Holocene. Geophysical Research Letters 36, L19704.

Suess, H.E., 1980. The radiocarbon method in tree rings of the last 8000 years. Radiocarbon 20, 200−209.

Svensmark, H., 2000. Cosmic rays and earth's climate. Space Science Reviews 93, 155−166.

Svensmark, H., Friis-Christensen, E., 1997. Variation of cosmic ray flux and global cloud coverage - a missing link in solar-climate relationships. Journal of Atmospheric and Solar-Terrestrial Physics 59 (11), 1225−1232.

Svensmark, H., Friis-Christensen, E., 2007. Reply to Lockwood and Fröhlich − the persistent role of the Sun in climate forcing. Danish National Space Center, Scientific Report 3 (2007).

Svensmark, H., Pedersen, J.O.P., Marsh, N.D., Enghoff, M.B., 2007. Experimental evidence for the role of ions in particle nucleation under atmospheric conditions. Proceedings of the Royal Society of London A 463, 385−396.

Thamban, M., Kawahata, H., Rao, V., 2007. Indian summer monsoon variability during the holocene as recorded in sediments of the Arabian Sea: timing and implications. Journal of Oceanography 63 (6), 1009−1020.

Thornalley, D.J.R., Elderfield, H., McCave, I.N., 2009. Holocene oscillations in temperature and salinity of the surface subpolar North Atlantic. Nature 457, 711−714.

Tiwari, M., Nagoji, S.S., Ganeshram, R.S., 2015. Multi-centennial scale SST and Indian summer monsoon precipitation variability since the mid-Holocene and its nonlinear response to solar activity. The Holocene 25 (9), 1415–1424.

Udelhofen, P.M., Cess, R.D., 2001. Cloud cover variations over the United States: an influence of cosmic rays or solar variability? Geophysical Research Letters 28 (13), 2617–2620.

Usoskin, I.G., Marsh, N., Kovaltsov, G.A., Mursula, K., Gladysheva, O.G., 2004. Latitudinal dependence of low cloud amount on cosmic ray induced ionization. Geophysical Research Letters 31, 1–4.

Valsecchi, V., Sanchez Goñi, M.F., Londeix, L., 2012. Vegetation dynamics in the Northeastern Mediterranean region during the past 23,000 yr: insights from a new pollen record from the Sea of Marmara. Climate of the Past 8 (6), 1941–1956.

van der Werf, G.R., Dolman, A.J., 2014. Impact of the Atlantic Multidecadal Oscillation (AMO) on deriving anthropogenic warming rates from the instrumental temperature record. Earth System Dynamics 5 (2), 375–382.

Varma, V., Prange, M., Lamy, F., Merkel, U., Schulz, M., 2011. Solar-forced shifts of the southern hemisphere westerlies during the holocene. Climate of the Past 7, 339–347.

Varma, V., Prange, M., Spangehl, T., Lamy, F., Cubasch, U., Schulz, M., 2012. Impact of solar-induced stratospheric ozone decline on southern hemisphere westerlies during the late Maunder minimum. Geophysical Research Letters 39 (20), L20704.

Vasiliev, S.S., Dergachev, V.A., 2002. The ~2400-year cycle in atmospheric radiocarbon concentration: bispectrum of ^{14}C data over the last 8000 years. Annales Geophysicae 20, 115–120.

Veretenenko, S., Ogurtsov, M., 2014. Stratospheric polar vortex as a possible reason for temporal variations of solar activity and galactic cosmic ray effects on the lower atmosphere circulation. Advances in Space Research 54 (12), 2467–2477.

Viau, A.E., Gajewski, K., Sawada, M.C., Fines, P., 2006. Millennial-scale temperature variations in north America during the holocene. Journal of Geophysical Research: Atmospheres 111 (D9) n/a–n/a.

Voiculescu, M., Usoskin, I., 2012. Persistent solar signatures in cloud cover: spatial and temporal analysis. Environmental Research Letters 7.

von der Heydt, A.S., Köhler, P., van de Wal, R.S.W., Dijkstra, H.A., 2014. On the state dependency of fast feedback processes in (paleo) climate sensitivity. Geophysical Research Letters 41 (18), 6484–6492.

Vuille, M., Burns, S.J., Taylor, B.L., Cruz, F.W., Bird, B.W., Abbott, M.B., Kanner, L.C., Cheng, H., Novello, V.F., 2012. A review of the South American monsoon history as recorded in stable isotopic proxies over the past two millennia. Climate of the Past 8, 1309–1321.

Wang, X., Zhang, Q.-B., 2011. Evidence of solar signals in tree rings of Smith fir from Sygera Mountain in southeast Tibet. Journal of Atmospheric and Solar-Terrestrial Physics 73, 1959–1966.

Wang, Y., Cheng, H., Edwards, R.L., He, Y., Kong, X., An, Z., Wu, J., Kelly, M.J., Dykoski, C.A., Li, X., 2005. The holocene Asian monsoon: links to solar changes and North Atlantic climate. Science 308, 854–857.

Weber, S.L., Crowley, T.J., van der Schrier, G., 2004. Solar irradiance forcing of centennial climate variability during the Holocene. Climate Dynamics 22, 539–553.

Willard, D.A., Bernhardt, C.E., Korejwo, D.A., Meyers, S.R., 2005. Impact of millennial-scale Holocene climate variability on eastern North American terrestrial ecosystems: pollen-based climatic reconstruction. Global and Planetary Change 47, 17–35.

Xu, H., Hong, Y., Lin, Q., Hong, B., Jiang, H., Zhu, Y., 2002. Temperature variations in the past 6000 years inferred from δ ^{18}O of peat cellulose from Hongyuan, China. Chinese Science Bulletin 47 (18), 1578–1584.

Xu, H., Hong, Y., Lin, Q., Zhu, Y., Hong, B., Jiang, H., 2006. Temperature responses to quasi-100-yr solar variability during the past 6000 years based on d^{18}O of peat cellulose in Hongyuan, eastern Qinghai–Tibet plateau, China. Palaeogeography, Palaeoclimatology, Palaeoecology 230, 155–164.

Xu, H., Yeager, K.M., Lan, J., Liu, B., Sheng, E., Zhou, X., 2015. Abrupt Holocene Indian Summer Monsoon failures: a primary response to solar activity? The Holocene 25 (4), 677–685.

Yu, F., Luo, G., 2014. Effect of solar variations on particle formation and cloud condensation nuclei. Environmental Research Letters 9.

Yu, F., Zong, Y., Lloyd, J.M., Leng, M.J., Switzer, A.D., Yim, W.W.-S., Huang, G., 2012. Mid-Holocene variability of the East Asian monsoon based on bulk organic δ^{13}C and C/N records from the Pearl River estuary, southern China. The Holocene 22 (6), 705–715.

Zhao, K., Li, X., Dodson, J., Atahan, P., Zhou, X., Bertuch, F., 2012. Climatic variations over the last 4000 yr BP in the western margin of the Tarim Basin, Xinjiang, reconstructed from pollen data. Palaeogeography, Palaeoclimatology, Palaeoecology 321–322, 16–23.

CHAPTER

17

The New Little Ice Age Has Started

H.I. Abdussamatov
Pulkovo Observatory of the RAS, St. Petersburg, Russia

1. INTRODUCTION

The long-term equilibrium state of the average annual energy balance of the Earth between the total solar irradiance (TSI) coming into outer layers of the atmosphere of the Earth and going out from the Earth into space of the total energy radiation from the top of the atmosphere to the surface determines the stability of the climate so that the Earth's temperature does not change. If in the long term the Earth keeps in more solar energy than it expels, then our planet will warm, and in contrast, if the Earth radiates more energy to space than it receives from the Sun, the Earth will cool. The main reasons for long-term deviations in the average annual energy balance of the Earth from the equilibrium state are a quasi-bicentennial cyclical variation in incoming TSI [equal to up to ~0.5% (Shapiro et al., 2011)] and the portion absorbed by the Earth, and the solar energy absorbed by the Earth remaining uncompensated by the energy of the long-wave radiation emitted by the planet into space over a time interval of 20 ± 8 years, which is controlled by the thermal inertia of the world oceans (Abdussamatov, 2009b, 2012a, 2013a,b, 2015a). As a result, the average annual energy balance of the Earth due to the quasi-bicentennial variations in TSI long term deviates from the equilibrium, which is the basic state of the surface—atmosphere climate system. Long-term deviation in the average annual energy balance of the Earth from the equilibrium state (excess of incoming TSI accumulated by the ocean or its deficiency) dictates a corresponding change in the Earth's energy state and climate (Abdussamatov, 2009b, 2012a, 2013a,b, 2015a). The variations in the TSI quasi-bicentennial cycle, together with important successive influences of a chain of subsequent causal feedback effects can explain all climate changes. Long-term positive (negative) deviation in the average annual energy balance of the Earth from the equilibrium state [excess (deficiency) of incoming TSI accumulated by the ocean] first will gradually warm (cool) the upper layers of water of the oceans in the tropics. The basic features of the Earth's climate variations are connected, in particular, with fluctuations in the power of both the atmospheric circulation and the ocean currents, including the thermal current of the Gulf Stream, which is driven by the heat accumulated by ocean water in the tropics. They are determined by the direct action of the quasi-bicentennial cyclic variation of the TSI and important successive influences of a chain of subsequent causal feedback effects caused by climate changes. Secondary mechanisms have additional influences in the form of subsequent climatic causal feedback effects in the process of cooling, and cause additional significant (depending on the duration of the cooling) temperature decline (with some time lag) due to the gradual, nonlinear rise in the Earth's Bond albedo and the natural decrease in atmospheric concentration primarily of water vapor (according to the Clausius—Clapeyron relation) and other greenhouse gases (according to Henry's law), and the opposite effect in the process of warming. Significant climate variations during the past 7.5 millennia indicate that bicentennial quasi-periodic TSI variations define the corresponding cyclic mechanism of climatic change from warming to a Little Ice Age and set the timescales of practically all physical processes that occur in the Sun—Earth system (Herschel, 1801; Eddy, 1976; Climate Oscillations, 1988; Abdussamatov, 2009b, 2012a, 2013a,b, 2015a; Section 6). The global nature of climate change during the past thousand years confirms an ikaite record: a medieval climatic optimum and a Little Ice Age Maunder minimum also spread to the Antarctic Peninsula (Lu et al., 2012). My definition of the Little Ice Age differs from that in the literature in that a long period of global cooling during the 14th— 19th centuries has been interrupted by several warming periods (Abdussamatov, 2013a, 2014, 2015a). Deep cooling was associated with Wolf (approximately from 1280 to 1340), Spörer (approximately from 1450 to 1550), Maunder (approximately from 1645 to 1715), and Dalton (approximately from 1790 to 1830) grand minima and should not be seen as single little ice ages (Abdussamatov, 2013a, 2014, 2015a). Quasi-bicentennial cyclic climate changes are always the response of the climate system to corresponding cyclic external influences of the Sun.

The observed decline in TSI since 1990 and the upcoming grand minimum in 2043 ± 11 and expected beginning of the deep cooling phase of the Earth in 2060 ± 11 will certainly have a significant impact on the global economy, society, and national security of countries. The upcoming deep global cooling will dictate the direction of variations in different natural processes on the Earth's surface and in the atmosphere as well as the change for worse conditions for creating material and financial resources in society. For practical purposes, the most important task is to determine the tendencies of expected climate change for the next 50—100 years. Nowadays, the problem of future deep global cooling is not only a major and important scientific problem of planetary scale facing humankind, but also a serious economic, social, demographic, and political problem whose solution determines further prospects of human civilization development. A deep cooling will directly influence the development of scientific, technical, and economic potentials of modern civilization. The effects of the recent deep cooling in the period of the Maunder minimum are a warning of the serious threat to the future of energy security for all humanity. Early understanding of the reality of the forthcoming deep global cooling and the physical mechanisms responsible for it will directly determine the choice of adequate and reliable measures that will allow humankind, in particular, the populations of countries situated far from the equator, to adapt to the future conditions, which will be especially strong in the zone of influence of a weakened Gulf Stream on western Europe and the eastern parts of the United States and Canada. After the middle of the current century humankind will meet with the same very difficult conditions of the period of the Maunder minimum.

2. MILANKOVITCH CYCLES AND INTERRELATED VARIATIONS IN CLIMATE AND ABUNDANCE OF CARBON DIOXIDE IN THE ATMOSPHERE

Within the past million years, global cyclical glaciation has occurred with temperature drops of ~10°C, during which the ice cover spread to much lower latitudes than now. What causes these deep changes in the Earth's climate? Any profound climate change is a response to long-term external effects on the Earth's climate system, the world's oceans and land, the cryosphere (snow and ice), and the biosphere. The main factor in profound change of climate has always been a long-term fluctuation of the TSI, taking into account its direct and important subsequent secondary feedback influences. The relative positions of the planets and the Sun influence the distribution and strength of gravitational fields. Indeed, the position and shape of the Earth's orbit and the axis orientation of its rotation in space have experienced over the centuries slow changes that influence the irradiance of the Earth by the Sun. A detailed theoretical description of the mechanisms of the astronomical cycles and their influence on the Earth's climate was proposed by the Serbian astrophysicist Milutin Milankovitch, who developed the astronomical theory of deep climate change with duration of tens of thousands of years. Insignificant, more long-term (more than 10,000 years) variations in the annual average TSI entering the Earth's upper atmosphere because of cyclical changes in the shape of the Earth's orbit, inclination of the Earth's axis relative to its orbital plane, and its precession, known as the astronomical Milankovitch cycles (Fig. 17.1), together with very important (due to long-term variations in TSI and temperature) subsequent nonlinear feedback effects, lead to the big glacial periods (with a period of about 100,000 years) with glacial/interglacial cycles (Fig. 17.2) (Milankovitch, 1941; Astronomical theory). The astronomical Milankovitch cycles lead to long-term changes in the average annual

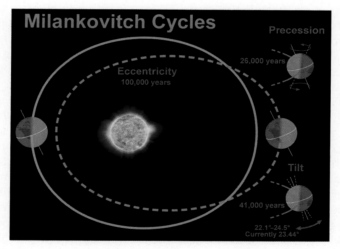

FIGURE 17.1   Astronomical Milankovitch cycles (Milankovitch, 1941; Astronomical theory).

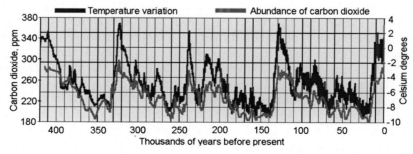

FIGURE 17.2   Variations in the Earth's climate and volumetric concentrations of carbon dioxide for a period of 420,000 years (according to the ice core data near Vostok, Antarctica) (Petit et al., 1999; Climate change, 2000).

energy radiated by the Sun that is absorbed by the whole planet and by different regions of the Earth [taking into account the uneven distribution of land vs ocean in the Northern (approximately 39%) and Southern (approximately 19%) Hemispheres] due to variations in orbital forcing. As a result, considerable changes mostly in the average annual distance between the Sun and the Earth with the period of about 100,000 years lead to more long-term variations in the annual average TSI entering the Earth's upper atmosphere. These variations in the TSI together with many chains of subsequent secondary feedback effects cause significant temperature fluctuations from warming to the Big Glacial Period, as well as in the atmospheric concentrations of water vapor, carbon dioxide, and other greenhouse gases (Fig. 17.2). Significant achievements of modern methods of measurement are determining the climatic characteristics (not only the temperature but also the composition of the atmosphere) in the periods of the last glacial/interglacial cycles. It turns out that the Antarctic ice cores provide clear evidence of a close coupling between variations in temperature and the atmospheric concentration of carbon dioxide during the glacial/interglacial cycles of at least the past 800,000 years.

Precise information on relative temporal changes in the temperature and the atmospheric concentration of carbon dioxide can assist in refining our understanding of the physical processes involved in this coupling. According to the ice core data drilled from a depth of over 3600 m near the Vostok site, Antarctica, during the glacial/interglacial cycles a rise in concentrations of greenhouse gases has begun every time after warming begins and ended after the warming was replaced by cooling (Fig. 17.2). It is worth emphasizing that the temperature starts to decrease, after reaching its highest values in the glacial/interglacial cycles, despite the fact that the concentration of greenhouse gases continues to grow (Petit et al., 1999; Climate change, 2000; Fischer et al., 1999; Pedro et al., 2012). The peaks of the carbon dioxide concentration have never preceded the warming, but on the contrary always took place 800 ± 400 years after it, being its consequence, ie, they have always been a natural consequence of the temperature increase caused by long-term growth of the incoming average annual solar energy (Fig. 17.2) (Petit et al., 1999; Climate change, 2000; Fischer et al., 1999; Pedro et al., 2012). According to Henry's law, warm water absorbs less gas because the solubility of a gas in a liquid is directly proportional to the partial pressure of the gas above the liquid, and hence more carbon dioxide remains in the atmosphere. So, the analysis of ice cores shows that higher or lower levels of carbon dioxide concentration are observed after warming or cooling, respectively, ie, considerable changes in the atmospheric concentration of carbon dioxide are always determined by the corresponding temperature fluctuations of the world oceans.

Thus, long-term cyclic variations in the annual average total energy of the solar radiation entering the upper layers of the Earth's atmosphere caused by the astronomical Milankovitch cycles, taking into account their direct and long-term important subsequent secondary feedback influences, are the main fundamental cause of corresponding significant (glacial) climate variations on the Earth. Long-term Milankovitch cycles cause changes in TSI, which are the causal factor in climate change (Milankovitch, 1941; Astronomical theory). Long-term changes in TSI cause changes in the temperature, which cause long-term subsequent secondary feedback influences and together they are enough to cause climate change in glacial/interglacial cycles. It should be emphasized that the amount of natural flows of carbon dioxide, water vapor, and dust from the world ocean and land to the atmosphere (M_{in}) and from the atmosphere (M_{out}) to the world's oceans and land exceed many times the anthropogenic discharges of these substances into the atmosphere (M_{ant}) (Nigmatulin, 2010). The overall content of carbon dioxide in the world ocean is ~50 times higher than in the atmosphere and as it warms or cools according to the Sun's intensity, it releases or absorbs these gases, respectively. Even a weak "breath" of the ocean can change dramatically the carbon dioxide level in the atmosphere. Changes in atmospheric carbon dioxide are not tracking changes in human emissions (Nigmatulin, 2010). The impact of these gases in driving climate change is secondary and minor compared to the power of the Sun. Natural causes play the most important role in climate variations and in the level of carbon dioxide in the atmosphere rather than human activity because natural factors are substantially more powerful. Therefore, there is no evidence that carbon dioxide is a major factor in the warming of the Earth. Although carbon dioxide has some warming influence, but the Sun plays a far greater role in the whole scheme of things. In the modern era, using data series on the concentrations of atmospheric carbon dioxide and global temperatures, these were also investigated as to the phase relation (leads/lags) between them for the period January 1980 to December 2011 (Humlum et al., 2013). Humlum et al. concluded that the change in atmospheric carbon dioxide from January 1980 is natural, rather than human induced, ie, changes in the amount of atmospheric carbon dioxide in the modern era also are always lagging behind the corresponding changes in temperature. At the same time carbon dioxide is a key component of the life cycle of the biosphere, and the increase in its concentration—a major factor in plant growth and development, increases agricultural productivity.

3. INTERRELATED VARIATIONS IN THE CLIMATE, TOTAL SOLAR IRRADIANCE, AND SOLAR ACTIVITY

The correlation between variations in sunspot activity and climate was first announced by English astronomer William Herschel in 1801 after he had discovered the inverse interrelation between wheat prices and cyclic variations in the solar activity before and during a cool period known as the Dalton minimum (Herschel, 1801). During high levels of the solar activity, wheat production increased, resulting in a drop in prices. When the number of sunspots significantly dropped, the wheat production decreased—and the prices went up. Herschel assumed that the change in wheat price was due to corresponding climate changes but he was not able to explain the physical nature of the phenomenon. Later American astronomer John Eddy discovered the interconnection between periods of significant variations in solar activity during the past millennium and corresponding deep climatic changes of both phase and amplitude (Eddy, 1976). Russian geophysicist Evgeniy Borisenkov discovered that each of the 18, Maunder-type minima of solar activity over the past 7500 years was associated with deep cooling, whereas periods of high solar activity (maxima) corresponded to warming (Climate Oscillations, 1988).

The lower envelope curve in Fig. 17.3, connecting the values of TSI at three successive minima levels between 11-year cycles 21 and 22, 22 and 23, and 23 and 24, is the total level of its quasi-bicentennial variation (Abdussamatov, 2003, 2004, 2005, 2007a,b, 2009a,b, 2012a, 2013a) relative to which its 11-year cyclic variations occur. We used PMOD composite data (Fröhlich, 2013) because ACRIM composite data show an incorrect increase in the TSI from 1986 to 1996 (Krivova et al., 2009). The quasi-bicentennial cycle of the Sun is one of the most intense solar cycles. The 11-year

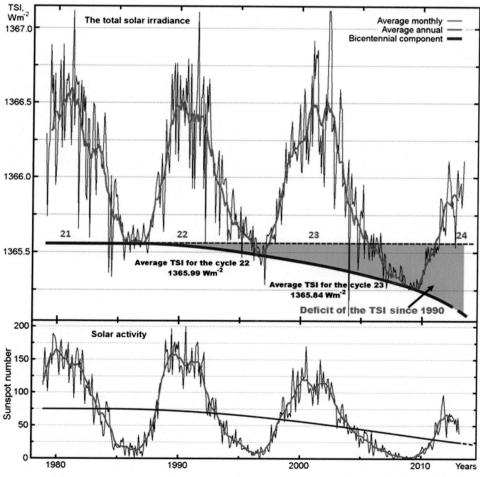

FIGURE 17.3   Variations in the solar activity based on monthly data (SunSpot Data, 2014) and the total solar irradiance (*TSI*) based on daily data (Fröhlich, 2013) between 1978 and 2013 and the deficit in TSI since 1990.

and quasi-bicentennial cyclic variations in solar activity and TSI, being a manifestation of the same processes in the solar interior, are interconnected and synchronized in both phase and amplitude (Fig. 17.3) (Abdussamatov, 2003, 2004, 2005, 2007a,b, 2009a,b, 2012a, 2013a). This allows extrapolating a relatively short series (since 1978) of highly accurate extra-atmospheric measurements of the TSI onto long periods of past time using a long series of solar activity (Avdyushin and Danilov, 2000; Lean, 2000; Solanki and Krivova, 2004; Shapiro et al., 2011; SunSpot Data, 2014). It enables one to study the course of TSI during the past centuries and even millennia to match it to the corresponding climate changes in the past and to study future variations, taking into account their direct and important subsequent secondary feedback influences.

The studies (Bal et al., 2011; McPhaden et al., 2011) confirm our earlier results (Abdussamatov, 2009a,b) indicating the direct joint effect (with some lag) of the 11-year (11 $\pm$ 3 years) and quasi-bicentennial (200 $\pm$ 70 years) cyclic variations in TSI on the changes in the state of the tropical Pacific Ocean surface layer (tens to hundreds of meters deep), resulting in El Niño and La Niña events associated with the appearance respectively of warm or cool water. The changes in the observed parameters of El Niño over the past 31 years did not correspond to the changes predicted by the climate models that are suggesting the dominant role of greenhouse gases (Bal et al., 2011; McPhaden et al., 2011). Thus, the oscillations in El Niño parameters were mostly due to natural causes, namely, cyclical variations in the TSI, taking into account their direct and important subsequent secondary feedback influences. It should be emphasized that cyclical variations in solar activity are the accompanying phenomena of physical processes occurring in the interior of the Sun and do not substantially affect either the TSI and the Earth's climate (Abdussamatov, 2005, 2007a, 2009b, 2013a).

4. QUASI-BICENTENNIAL VARIATIONS IN TOTAL SOLAR IRRADIANCE AND MECHANISMS OF ITS SECONDARY ADDITIONAL INFLUENCES

Quasi-bicentennial variations in TSI are about 0.5% from the latest reconstructed data (Shapiro et al., 2011) and their direct impact alone is insufficient to account for all corresponding quasi-bicentennial cyclic changes in the Earth's temperature, from the periods of warming to the Little Ice Age. An additional "amplifier" is needed to enhance the direct impact of the variations in TSI on the observed global climate changes (Abdussamatov, 2009b, 2012a, 2013a,b). Such amplifiers of the direct impact of variations in TSI on climate variation are the indirect subsequent additional secondary influences of TSI in the form of a successive chain of causal feedback effects: natural changes in the global albedo of the Earth as a planet, the albedo (additional changes in the absorbed part of TSI), and changes in the concentrations of greenhouse gases in the atmosphere (water vapor, carbon dioxide, methane, etc.), additional variations in the influence of the greenhouse effect. The albedo of the Earth is defined as the ratio of the flux of a portion of the solar radiation reflected and scattered by the spherical surface—atmosphere system back into space in all directions to the flow of the total solar radiation coming into the upper atmosphere. The albedo is defined by the global optical properties of the Earth as a whole, with its air and water envelopes averaged over the total vertical, starting from the surface up through the atmosphere. The albedo of the Earth is increased to its maximum level during deep cooling (in particular, because of increasing coverage by snow and ice) and reduced to its minimum level during warming, whereas the concentrations of greenhouse gases in the atmosphere (mostly water vapor at the surface layers of the atmosphere, as well as carbon dioxide and other gases) vary inversely because their variations are mostly defined by the temperature of the ocean and land. The albedo is a particularly important physical parameter of the energy budget of the Earth as a planet. Successive variations in the parameters of the Earth's surface and atmosphere due to variations in the TSI quasi-bicentennial cycle will lead to the generation of numerous nonlinear changes in temperature due to multiple repetitions of such causal cycles of the subsequent secondary feedback effects. The impact of the causal feedback effects in the period of cooling, caused by changes in the parameters of the Earth's surface and atmosphere due to decreasing of the TSI quasi-bicentennial cycle, leads to a gradual nonlinear increase in the value of the albedo and a decrease in the content of greenhouse gases in the atmosphere. It will lead to an additional reduction in the absorbed portion of solar energy and reduce the influence of the greenhouse effect. The multiple repetitions of these changes will lead to a chain of successive drops in the Earth's temperature, which can surpass the influence of the direct effect of the TSI decrease in the bicentennial cycle, even if the TSI will subsequently remain unchanged over a certain period of time. A similar pattern was observed in the late 20th century. Unfortunately, the dynamics describing the rate of increase in the total area of snow and ice covering the Earth's surface, as well as the rate of decrease in the concentration of greenhouse gases in the atmosphere, is a nonlinear function of the temperature drop and difficult to predict. The natural concentration of carbon dioxide in the atmosphere is known to have been, during the glacial periods in the Earth's history, more than two times lower than today (Fig. 17.2) (Petit et al., 1999; Climate change, 2000).

5. THE QUASI-BICENTENNIAL SOLAR CYCLE DETERMINES VARIATIONS IN BOTH THE DURATION AND THE POWER OF THE 11-YEAR SOLAR CYCLE

Long-term successive changes from the impacts of 11-year solar cycles are accompanied by significant changes in climate. The duration and power of the 11-year cycles of TSI and solar activity are the most important characteristics of variations in the physical processes occurring in the deep interior of the Sun (Fig. 17.4). Quasi-bicentennial solar cycles are the primary cycles that govern variations in the 11-year subsidiary cycles in TSI and solar activity (Abdussamatov, 2006, 2009a,b, 2013a). Identification of some common features among the last 14 11-year solar activity cycles is very important. Fig. 17.5 demonstrates variations in the duration of the 11-year solar cycles of activity, *P*, as a function of phases of the quasi-bicentennial cycle. The length of the 11-year cycle depends on the phase of the quasi-bicentennial cycle and increases gradually from the growing phase to the maximum and descending phase of the quasi-bicentennial solar cycle (Fig. 17.5) (Abdussamatov, 2006, 2009b, 2013a). It is obvious that such relationship exists also for the 11-year cyclic variations in the TSI because the cyclic variations in solar activity and the TSI are interconnected and synchronized (Abdussamatov, 2003, 2004, 2005, 2007a,b, 2009a,b, 2012a, 2013a). Eleven-year

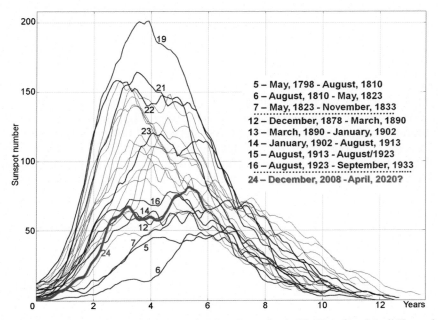

FIGURE 17.4 Variations in the relative sunspot number in solar cycles 1–24. *Data from http://sidc.oma.be/sunspot_data/.*

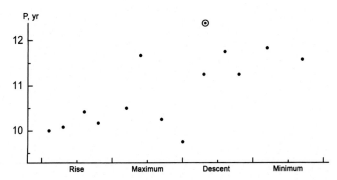

FIGURE 17.5 The duration of an 11-year cycle depends on the phase of the quasi-bicentennial cycle and consequently increases from the rising phase to the maximum and descending phase of the quasi-bicentennial cycle (⊙, cycle 23) (Abdussamatov, 2006, 2009b, 2013a).

cycles of solar activity developing in the descending phase of the quasi-bicentennial cycle have a duration of about $P = 11.7 \pm 0.8$ years, and they are ordinarily longer than in the cycle developing during the rising phase and maximum of the quasi-bicentennial cycle.

These data can be explained by a decrease in the average duration of the last eight cycles of solar activity, from 15 to 22, which developed during the rising phase and maximum of the quasi-bicentennial cycle, down to $P = 10.4$ years, compared to the average duration of $P = 10.9$ years for the last 14 cycles of solar activity. This relationship allows us to predict not only the duration of the current 11-year cycle, but also the durations of the subsequent cycles 25 and 26, which will be formed during the descent of the current quasi-bicentennial cycle. The succeeding cycles 25, 26, and 27 in the phase of the decline of the quasi-bicentennial cycle are expected to begin in approximately 2020.7 ± 0.6, 2032.2 ± 1.2, and 2043.7 ± 1.8, respectively.

It is very important to establish the extent to which such changes in the laws of 11-year cycle duration lead to changes in the power of their average cyclic energy. Only the average weighted level of the TSI and solar activity during the cycle may allow objective and quantitative determination of the average level of the power of the 11-year cycle, as well as prediction of its impact on processes occurring in the Sun–Earth system. The relative power of the 11-year cycle of solar activity is the average weighted level of the index of solar activity throughout the cycle:

$$\overline{W}(\text{cycle}) = \frac{\sum W \Delta t}{\sum \Delta t}$$

Here W is the sunspot number and Δt is the time interval between successive observations throughout the cycle. The absolute average weighted power of the energy of an 11-year cycle is defined as

$$\overline{S}_\odot(\text{cycle}) = \frac{\sum S_\odot \Delta t}{\sum \Delta t}$$

where $S_\odot$ is the TSI (Abdussamatov, 2015a,b).

To establish the relationship between the length of 11-year solar cycles and their relative average cyclical radiative forcing a graph was plotted for all of the 24 cycles of solar activity (Fig. 17.6). The results show the presence of a generally inverse relationship between the length of 11-year cycles and their relative radiative forcing, ie, the lower the relative radiative forcing of a solar cycle, the greater the cycle length, and vice versa. It is obvious that such an interconnection exists also between the absolute average cyclic radiative forcing, the average weighted absolute value of the TSI 11-year cycle, and the solar cycle length. High absolute radiative forcing of 11-year cycles and of their

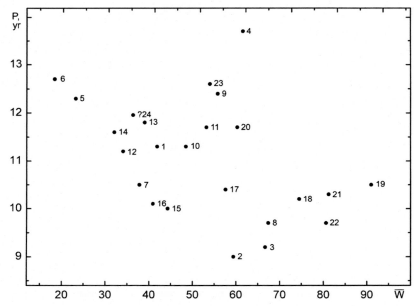

FIGURE 17.6 Dependence of the relative solar activity cycle forcing of all 11-year solar cycles on their length, *P*. Numerals indicate the cycle numbers (Abdussamatov, 2015a,b).

small length (below average with respect to all 24 cycles) during the growth phase of the quasi-bicentennial cycle and the opposite pattern during its phase decline might serve as an indicator of climate change only during these phases of the bicentennial solar cycle (Abdussamatov, 2015a,b). The inverse relationship between the length of 11-year solar cycles and their radiative forcing is attributed to the influence of the quasi-bicentennial solar cycle, which determines the total pattern of development of 11-year cycles. These facts prove once again that the 11-year cycles are genetically related to the quasi-bicentennial cycle, which determines the regular development of their duration and power as the filial 11-year cycles (Abdussamatov, 2006, 2009a,b, 2013a).

In the growth phase of the Sun's quasi-bicentennial cycle, the length of the solar 11-year cycles typically is shorter relative to the average for all 24 cycles, and the height of their maximum and the absolute power of the average cyclic energy sequentially are increased, which with some delay leads to a temperature rise of the Earth. Therefore, previous studies showed that the length of the 11-year solar cycle might possibly be an indicator of climate change, because a correlation between the solar cycle length and the temperature of the surface layer in the Northern Hemisphere over 130 years was detected (Friis-Christensen and Lassen, 1991). The dependence observed by Friis-Christensen and Lassen (1991) can be easily explained by an established inverse interconnection between the duration of the 11-year solar cycle and its power. However, the observed (Friis-Christensen and Lassen, 1991) dependence can exist only sporadically and only in the phase of rise or decline of the solar quasi-bicentennial cycle. Thus the gradual decline or increase in the duration of the 11-year cycle is generally an indicator of the onset phase of growth or decline of the solar quasi-bicentennial cycle, respectively, and an indicator of the beginning of future climate change, taking into account the thermal inertia of the oceans.

6. THE AVERAGE ANNUAL ENERGY BALANCE OF THE EARTH

When the absorbed average annual energy (TSI) during a long-term period is equal to the average annual energy long-wave radiation emitted into space, the Earth's temperature does not change—the energy budget of the Earth is in balance. To understand climate change, we must determine what drives long-term deviation from the equilibrium state of the average annual energy balance of the Earth. Therefore, we must first identify the components of the average annual energy balance of the Earth and the reasons for their long-term changes. The annual average difference between the energy of the total solar radiation coming into the outer layers of the Earth's atmosphere,

$$E_{in} = (S_\odot + \Delta S_\odot)/4 \tag{17.1}$$

and the energy portion of the reflected and scattered solar radiation and the energy long-wave radiation of the Earth going out into space,

$$E_{out} = (A + \Delta A)(S_\odot + \Delta S_\odot)/4 + \varepsilon\sigma(T_p + \Delta T_p)^4 \tag{17.2}$$

determine the energy balance of the budget of the surface—atmosphere system, E. The difference between the incoming E_{in} and outgoing E_{out} radiation is described by Eq. (17.3) (Fig. 17.7) (Abdussamatov, 2012a,b, 2013a,b, 2015a),

$$E = (S_\odot + \Delta S_\odot)/4 - (A + \Delta A)(S_\odot + \Delta S_\odot)/4 - \varepsilon\sigma(T_p + \Delta T_p)^4 \tag{17.3}$$

or by the difference between the portion of the TSI energy absorbed by the Earth and the long-wave radiation energy emitted into space,

$$E = (S_\odot + \Delta S_\odot)(1 - A - \Delta A)/4 - \varepsilon\sigma(T_p + \Delta T_p)^4. \tag{17.4}$$

Here, $S_\odot$ is TSI, $\Delta S_\odot$ is the increment of TSI, A is the global albedo of the Earth (albedo), ΔA is the increment of the albedo, ε is the emissivity of the surface—atmosphere system, σ is the Stefan—Boltzmann constant, T_p is the planetary thermodynamic temperature, ΔT_p is the increment of the planetary thermodynamic temperature, and E is the specific power of the enthalpy change of the active oceanic and atmospheric layer (Wm^{-2}), which can be considered the energy balance of the annual average budget in the debit and credit of the thermal power of our planet. Planetary thermodynamic temperature is the average temperature over the entire surface of the planet (the Earth's surface and atmosphere). The factor 1/4 on the right side of Eqs. (17.1)—(17.4) reflects the fact that the solar radiation flux is projected onto the cross-sectional area of the terrestrial sphere (circle), whereas the Earth emits from the entire sphere surface, which is four times as large. The specific power of the Earth's enthalpy change E is a particular indicator of

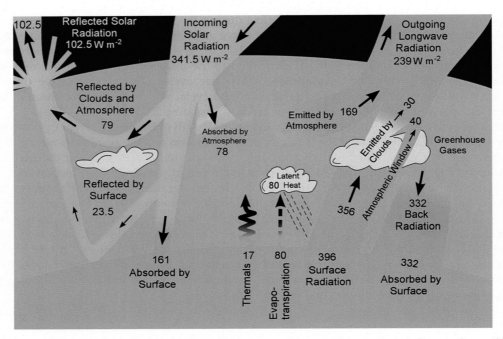

FIGURE 17.7   The average annual values of components of the total energy balance of the Earth as a planet in the equilibrium state (Trenberth et al., 2009; Abdussamatov, 2013a).

the deficit or excess of the thermal energy, which can be considered the energy balance of the annual mean budget in the debit and credit of the thermal power of the planet.

At the same time, the increment of the Earth's effective temperature is involved in the radiation balance immediately after the change in absorbed power, in contrast to the planetary thermodynamic temperature that is involved in the heat balance. The relative impact of variations in TSI and albedo on the change in the effective temperature of the Earth can be determined from the radiation balance of the Earth as a planet:

$$S_\odot/4 = \sigma T^4_e + A S_\odot/4, \tag{17.5}$$

where T_e is the Earth's effective temperature. Let us introduce the long-term increment of the effective temperature $\Delta T_e = T_e - T_{eo}$, where T_e is the current value of the Earth's effective temperature and T_{eo} is its initial value. We believe that the increment in the effective temperature is the result of increments in TSI $\Delta S_\odot$ and albedo ΔA. In this case, the radiative balance equation (Eq. (17.5)) takes the form

$$(S_\odot + \Delta S_\odot)/4 = \sigma(T_e + \Delta T_e)^4 + (A + \Delta A)(S_\odot + \Delta S_\odot)/4. \tag{17.6}$$

or

$$(S_\odot + \Delta S_\odot)(1 - A - \Delta A)/4 = \sigma(T_e + \Delta T_e)^4, \tag{17.7}$$

Because the increments in the effective temperature are small, $\Delta T_e \ll T_e$, the following equality is fulfilled with a high degree of accuracy:

$$(S_\odot + \Delta S_\odot)(1 - A - \Delta A)/4 = \sigma T^4_e + 4\sigma T^3_e \Delta T_e. \tag{17.8}$$

Subtracting Eq. (17.5) from Eq. (17.8) yields

$$\Delta S_\odot(1 - A - \Delta A) - S_\odot \Delta A = 16\sigma T^3_e \Delta T_e. \tag{17.9}$$

The formula for the increment of the Earth's effective temperature due to the increment of the TSI and albedo can be obtained from Eq. (17.9):

$$\Delta T_e = [\Delta S_\odot(1 - A - \Delta A) - S_\odot \Delta A]/(16\sigma T^3_e). \tag{17.10}$$

If the TSI does not change ($\Delta S_\odot = 0$), we obtain from Eq. (17.10) that

$$\Delta T_e = -S_\odot \Delta A / \left(16\sigma T^3_e\right). \tag{17.11}$$

Using well-known values of the effective temperature of the Earth and the TSI for the 21st and 22nd solar cycles, which are $T_e = 254.8$K and $S_\odot = 1366\,\text{Wm}^{-2}$, respectively, we obtain from Eq. (17.11) for $\Delta S_\odot = 0$:

$$\Delta T_e = -91\,\Delta A. \tag{17.12}$$

For the albedo being constant ($\Delta A = 0$) we obtain from Eq. (17.10)

$$\Delta T_e = \Delta S_\odot \left(1 - A\right) / \left(16\sigma T^3_e\right). \tag{17.13}$$

By using the known now value of the Earth's global albedo, which is $A = 0.3$ according to the latest data (Trenberth et al., 2009), we obtain from Eq. (17.13) for $\Delta A = 0$:

$$\Delta T_e = 0.047\,\Delta S_\odot. \tag{17.14}$$

Estimation of the ratio of the relative contributions of the increments $\Delta S_\odot$ and ΔA to the increment ΔT_e can be done by adopting the conditions of their mutual compensation while maintaining the energy balance (Abdussamatov, 2012a, 2013a)

$$\Delta S_\odot \left(1 - A - \Delta A\right) - S_\odot \Delta A = 0 \tag{17.15}$$

in Eq. (17.10). From Eq. (17.15) one can get the ratio of the relative contribution of the increments $\Delta S_\odot$ and ΔA to the increment ΔT_e,

$$\Delta S_\odot / S = \Delta A / \left(1 - A - \Delta A\right) \tag{17.16}$$

or if $S_\odot = 1366\,\text{Wm}^{-2}$ and $A = 0.3$, then

$$\Delta S_\odot = 1366 \cdot \Delta A / \left(0.7 - \Delta A\right). \tag{17.17}$$

Eq. (17.14) indicates that if the albedo of the Earth remains unchanged ($\Delta A = 0$) and only the TSI bicentennial decreases at $\Delta S_\odot = -6.4\,\text{Wm}^{-2}$ (Shapiro et al., 2011), the global effective temperature of the Earth decreases at $\Delta T_e = -0.3$K [estimates of the difference between the increment in the temperature of the global surface air (in view of the time of its delay) and the effective temperature are insignificant when their values are small]. Such gradually decline in the Earth's temperature about -0.3K may cause an increase in the Earth's albedo of about 1%, and, according to Henry's law and the Clausius–Clapeyron relation, a substantial reduction in the greenhouse gas concentrations in the atmosphere. Such increase in the albedo will result in an additional drop in the temperature of the Earth as a planet, which in turn will cause a further additional increase in the albedo and substantial reduction in the greenhouse gas concentrations in the atmosphere. A long succession chain of these cycles leads to successive additional chains of cooling and to a Little Ice Age, taking into account direct and important subsequent secondary feedback influences of TSI. The effective temperature of the surface–atmosphere system describes the inertia-free process of radiative heat exchange in the equilibrium thermal regime (Eqs. (17.6) and (17.7)). The increment in the effective temperature of the Earth would be carried out immediately with the change in the power to absorb solar radiation and the albedo of the Earth (Eq. (17.10)), unlike the planetary thermodynamic temperature, which determines the energy balance based on the thermal inertia of the Earth.

The change in albedo appreciably affects the temperature of the Earth and is, along with TSI, the most important factor in determining future climate change. The thermal inertia of the oceans is slowly changing in the climate system. However, the close connection of the atmosphere and the oceans leads to a significant delay of response of the atmospheric climate to external stimuli. Therefore, the Earth's thermodynamic temperature does not change immediately because of variations in the TSI and albedo. There is an appreciable lag in time determined using the constant thermal inertia of the planet (Abdussamatov et al., 2010):

$$t = 0.095(1 + 0.42 \cdot l)\,\text{year}, \tag{17.18}$$

where l is the depth of the active layer of the oceans. If the depth of the active layer of the ocean is 300–700 m, the constant thermal inertia is

$$t = 20 \pm 8\,\text{year}. \tag{17.19}$$

7. QUASI-BICENTENNIAL VARIATION IN THE TOTAL SOLAR IRRADIANCE LEADS TO AN ENERGY IMBALANCE OF THE SURFACE–ATMOSPHERE SYSTEM

The temporal changes in long-wave radiation emitted to space from the surface–atmosphere system always lag behind changes in short-wave solar radiation absorbed by the Earth because the enthalpy of the world's oceans changes slowly. Variation of temperature lags relative to absorbed solar radiation by 20 ± 8 years owing to the large heat capacity and thermal inertia of the oceans (Abdussamatov, 2009b, 2012a, 2013a, 2015a; Abdussamatov et al., 2010). Any long-term change in average annual short-wave solar radiation absorbed by the Earth due to the quasi-bicentennial variation of the TSI is not compensated by a corresponding change the emission of long-wave radiation from the Earth into space. That is why the debit and credit parts of the average annual energy balance of the Earth (Eqs. (17.3) and (17.4)) always deviate from the equilibrium state of the surface–atmosphere climate system ($E \neq 0$). Therefore, in the growth phase of the quasi-bicentennial solar cycle, the Earth receives more solar energy than is emitted by radiation into space, and its average annual energy balance is positive ($E > 0$), and vice versa in the recession phase of the quasi-bicentennial cycle—it is then negative ($E < 0$). As a result, the average annual energy balance of the Earth oscillates around the quasi-bicentennial equilibrium state. Therefore, long-term monitoring of the deviation of the average annual energy balance of the planet from the equilibrium state will define the trend and magnitude of the energy excess or deficit accumulated by the oceans. In the decay phase of the quasi-bicentennial cycle, the average annual energy balance of the Earth is negative ($E < 0$) and a long-term deficit in incoming solar energy can lead to cooling of the planet. The total quantity of stored solar energy (ΣE) (or its deficit) over a long period of monitoring can determine the corresponding depth of the impending climate variations. Quasi-bicentennial variations in the TSI determine the mechanism of cyclic alternations in climate change and set the timescales for physical processes taking place in the Sun–Earth system. That is why the Earth's climate changes every 200 ± 70 years.

8. CURRENT TOTAL SOLAR IRRADIANCE DECREASE IN THE QUASI-BICENTENNIAL CYCLE HAS LED TO A LONG-TERM DEFICIT IN THE EARTH'S ENERGY BALANCE AND THE BEGINNING OF A NEW LITTLE ICE AGE

Since 1990, TSI has been gradually decreasing. The annual rate of decreasing quasi-bicentennial TSI increased from cycle 22 to cycles 23 and 24 (Figs. 17.3 and 17.8). The smoothed value of TSI in the minimum between cycles 23 and 24 ($1365.27 \pm 0.02\ \mathrm{Wm^{-2}}$) was lower by approximately 0.23 and $0.30\ \mathrm{Wm^{-2}}$ than at the minima between cycles 22 and 23 and cycles 21 and 22, respectively. The average cyclical values of TSI were also lower approximately by $0.15\ \mathrm{Wm^{-2}}$ in cycle 23 relative to cycle 22. An average annual decrease rate in the TSI during cycle 22 was approximately $0.007\ \mathrm{Wm^{-2}\ year^{-1}}$, whereas in cycle 23 it became approximately $0.02\ \mathrm{Wm^{-2}\ year^{-1}}$. Note that the level of the maximum 11-year component of TSI has decreased within 6 years of the current cycle 24 by approximately $0.5\ \mathrm{Wm^{-2}}$ to the level of cycle 23 (Figs. 17.3 and 17.8). The current increasing decline in TSI (with an abrupt drop in cycle 24) is about $0.1\ \mathrm{Wm^{-2}\ year^{-1}}$ (Fig. 17.3) and its decrease will continue in cycle 25. The proportion of solar energy absorbed by the Earth since 1990 decreases at practically the same rates (Abdussamatov, 2009b, 2012a,b, 2013a,b, 2015a). However, the observed decrease in the average annual TSI portion absorbed by the Earth since 1990 has not been compensated for by a decrease in the average annual energy emitted into space because of the thermal inertia of the oceans (Abdussamatov et al., 2010). Since 1990 the Earth has radiated more energy back into space than the solar energy it has absorbed. Such gradual expenditure of all solar energy accumulated by the oceans during the 20th century led to the beginning of a new Little Ice Age after the maximum phase of cycle 24 (Fig. 17.8).

The Earth will continue to have a negative average annual energy balance ($E < 0$) in cycles 25–28 because the Sun is moving into a grand minimum. The observed trend of the increasing decline in TSI (Figs. 17.3 and 17.8) suggests that this decline will correspond to the analogous TSI decline in the period of the Maunder minimum (Shapiro et al., 2011). This decrease may reach approximately $1363.4 \pm 0.8, 1361.0 \pm 1.6$, and a deep minimum of $1359.6 \pm 2.4\ \mathrm{Wm^{-2}}$ (with gradual decreasing accuracy) at the minima between cycles 24 and 25, 25 and 26, and 26 and 27, respectively (Figs. 17.8 and 17.9). The maximum number of sunspots smoothed over 13 months could reach 50 ± 15 and 30 ± 20 in cycles 25 and 26, respectively (Abdussamatov, 2003, 2004, 2005, 2007a,b, 2009a,b, 2012a,b, 2013a, 2015a). Based on studies of variations in TSI for three consecutive minima between cycles 21 and 22, 22 and 23, and 23 and 24

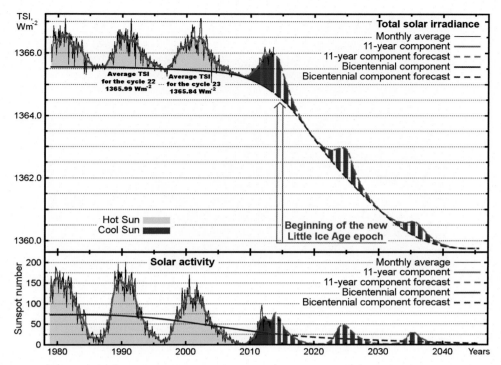

FIGURE 17.8   Cyclic variations in the total solar irradiance (*TSI*) *(data are taken from Fröhlich, C., 2013, Solar Constant. www.pmodwrc.ch/pmod. php?topic=tsi/composite/SolarConstant)* and sunspot number *(data are taken from SunSpot Data, 2014. SIDC—Solar Influences Data Analysis Center. http://sidc.oma.be/sunspot-data/)* and a forecast of their variations to cycles 24—26 until 2045 (the hot Sun is marked by *yellow* and the cool Sun is marked by *red*). The *red arrow* indicates the beginning of the new Little Ice Age epoch.

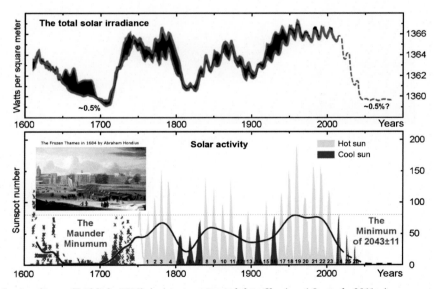

FIGURE 17.9   Total solar irradiance (Fröhlich, 2013) *(using reconstructed data Shapiro, A.I., et al., 2011. A new approach to the long-term reconstruction of the solar irradiance leads to large historical solar forcing. Astronomy & Astrophysics 529, A67.)* and solar activity from 1611 (SunSpot Data, 2014) and the prognosis of their variations until the end of the 21st century *(dashed lines)*: the hot Sun is marked by *yellow* and the cool Sun is marked by *red*.

(Figs. 17.3 and 17.8), the projected start of the grand minimum phase of the Maunder-type quasi-bicentennial TSI is in cycle 27 (±1) in 2043 ± 11, with a possible duration of 45—65 years. The start of the deep cooling phase of the new Little Ice Age is expected in 2060 ± 11 (Figs. 17.10 and 17.11). The expected course of lowering temperature is shown in Fig. 17.12. What we are seeing now in solar cycle 24 and bicentennial cycle was predicted by me in 2003—2007,

FIGURE 17.10   The frost fair on the frozen River Thames (WorldGallery.CO.UK).

FIGURE 17.11   Frost fair on the frozen River Thames (Tom de Castella).

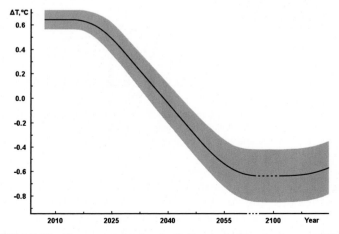

FIGURE 17.12   The prognosis of natural climate changes for the next 100 years.

long before cycle 24 began (Abdussamatov, 2003, 2004, 2005, 2007a,b, In the middle of the XXI century, 2006; Scientist predicts, 2006; The price of sensations, 2006; Russian scientist, 2006; Global cooling, 2006; People, 2007; Russian academic, 2007).

However, it should be borne in mind that the Earth is a large and complex closed system with a large number of simultaneously acting internal structural elements. Therefore, global climate characteristics depend on a combination of many factors, acting both individually and comprehensively. Natural causes of climate change include cloud cover and the surface area of snow and ice cover, ongoing interactions between the atmosphere and the ocean, cryosphere and land surface changes, and shifts in some regional oceanic and atmospheric currents, taking into account, in particular, the uneven distribution of land on the surface of the Northern and Southern Hemispheres, ~39% and ~19%, respectively. Natural internal causes of climate change can lead to short-term fluctuations in the Earth's temperature. However, natural fluctuations in temperature with only internal causes may usually be negligible, only within ±0.1°C. Intrasystem causes of temperature fluctuations are a part of the climate system. However, their long-term cyclical trends are caused by external factors related to bicentennial variations in the TSI in view of the thermal inertia of the oceans.

Long-term changes in the Sun's energy output can account for almost all the climate changes on planets of the Solar system. Even insignificant, long-term TSI variations may have serious consequences for the climate of the Earth and other planets of the Solar system. Warming on Mars (Odyssey, 2005; Ravilious, 2007) and other planets was observed in the last quarter of the 20th century. That was attributed to an action of season "solar summer" and the quasi-bicentennial alternation in climate conditions throughout the Solar system. At the end of 2015, after the maximum phase of solar cycle 24, the season "solar summer" ended and the season "solar autumn" began, and then in 2060 ± 11 will be the coming season "solar winter." "Spring" in the Solar system will come only at the beginning of the 22nd century. Geologists call past warm epochs "climatic optimums," and cold times, dark ages; yet governments across the world are preparing only for warming.

9. SENSITIVITY OF CLIMATE TO WATER VAPOR AND CARBON DIOXIDE

The concentration of water vapor in the atmosphere is the dominant factor in the greenhouse effect (Abdussamatov, 2009b, 2012a,b, 2013a, 2015a). Water vapor absorbs about 68% of long-wave emissions from the Earth's surface, whereas carbon dioxide absorbs only approximately 12% (Fig. 17.13). This proportion is due to the partial overlap of spectral absorption bands of carbon dioxide and water vapor and the constancy of moisture content in the atmosphere at low pressure and temperature variations. Therefore, the absorption by the atmosphere of thermal radiation of the Earth's surface is mainly determined by the concentration of water vapor. Altitude is an important factor in temperature change caused by water vapor in the atmosphere. The concentration of water vapor in

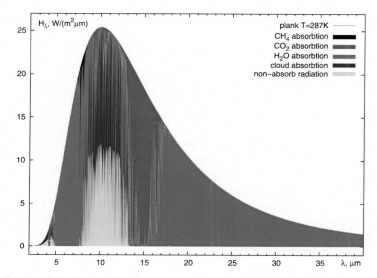

FIGURE 17.13 The spectral density of the thermal flow long-wave radiation of the Earth's surface (as a blackbody).

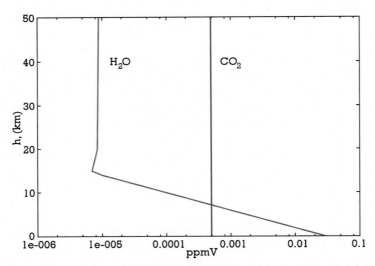

FIGURE 17.14   The changes in the concentrations of water vapor and carbon dioxide with height (Abdussamatov, 2009b, 2010, 2013a, 2014).

the atmosphere depends strongly on altitude (Fig. 17.14). Carbon dioxide is homogeneously distributed to a height of about 80—100 km (it), but water vapor has its maximum concentration at the surface, drops abruptly with height in the troposphere, and remains on some constant level in the stratosphere (Fig. 17.14). Even small growth of the average concentration of water vapor in the atmosphere, together with a simultaneous rise in the average concentration of carbon dioxide caused by warming can lead to a significant increase in water vapor concentration at the very bottom of the surface layers of the air relative to the carbon dioxide concentration. This leads to substantial changes in the transfer of thermal flow of the long-wave radiation of the Earth's surface by the water vapor, because of the significant overlapping bands of the absorption of water vapor and carbon dioxide in the wide regions of the emission spectrum within the windows of atmospheric transparency (Fig. 17.13). A significant increase in the concentration of water vapor in the lowest surface layers of the atmosphere significantly increases the proportion of the absorption by water vapor of the long-wave radiation of the Earth's surface in the overlapping spectral absorption bands of carbon dioxide and water vapor within the windows of the atmospheric transparency. As a result, the climate sensitivity to increasing concentrations of carbon dioxide decreases with significant growth of water vapor concentration in the surface layer caused by warming (Abdussamatov, 2009b, 2010, 2013a, 2014). Therefore, with an increase in the concentration of carbon dioxide in the atmosphere at its current high level, it is impossible to expect a significant increase in the absorption by the carbon dioxide of the radiation of the Earth's surface.

10. CONVECTION, EVAPORATION, AND CONDENSATION IN TRANSFER OF THERMAL FLOW AT THE EARTH'S SURFACE

As early as 1908, the American physicist Robert Wood made two identical boxes (minigreenhouses) of black cardboard: one of them was covered with a glass plate and the other, with a plate made of rock salt crystals, which are almost transparent in the infrared part of the spectrum (Wood, 1909). The temperature in both greenhouses simultaneously reached approximately 130°F (approximately 54.4°C). However, the plate made of rock salt is transparent at long wavelengths and, according to the commonly adopted theory of the greenhouse effect, this cover should not produce it at all. Robert Wood established that in the greenhouse, where the heat is blocked from all sides and there is no air exchange with the atmosphere, the radiative component is negligibly small compared to the convective component. Hence, heat that accumulates in the greenhouse only slightly depends on its cover transparency to infrared radiation, ie, absorption of infrared radiation by the glass is not the main reason for the heat to accumulate in the greenhouse. Thus, convection, evaporation, and condensation, together with the greenhouse effect, participate in the transmission of thermal flow long-wave radiation of the Earth's surface to the atmosphere (for example, see Fig. 17.7).

11. POWERFUL VOLCANIC ERUPTIONS LEAD ONLY TO SHORT-TERM COOLING PERIODS

Relatively powerful volcanic eruptions increase the amounts of solid particles and gases in the lower stratosphere. Their scattering, screening, and partial absorption of solar radiation decrease the portion of the TSI reaching the surface, which can result in short-term cooling, which was observed after the Pinatubo volcano. Volcanic microparticles in the atmosphere contribute to cloud formation, which also prevents solar radiation from reaching the surface. They simultaneously absorb infrared radiation, but their antigreenhouse effect is more pronounced than the greenhouse effect. However, these changes are not long term because of the limited lifetime of volcanic particles in the atmosphere. The impact of volcanic eruptions on climate depends not only on the mass of exhaust gases and microparticles, but also the altitude to which they are ejected, since at an altitude of 15—20 km, they can linger for several years. The atmosphere is self-cleaning and gradually increases its transparency to its previous level over a time span of 6 months to a few years (Abdussamatov, 2009b, 2010, 2013a, 2014). That is why the role of volcanic eruptions in climate variations cannot be long term and cannot be a determining factor. When Mt. Pinatubo erupted in the Philippines in 1991, about 20 million tons of sulfur dioxide was ejected into the atmosphere, and the global temperature dropped by about 0.5°C from 1991 to 1993. However, later the atmosphere cleaned itself of these additives and finally returned to its initial state.

12. FUTURE DEEP COOLING CAN BECOME A MAJOR PROBLEM FOR THE DEVELOPMENT OF PETROLEUM IN THE ARCTIC

The Arctic is a unique territory, primarily because of its incredible oil and gas resources and almost half the world's fish production. The Arctic Ocean, according to geological surveys, is the richest in petroleum of all the oceans. Intense interest in the development of the Arctic has been stimulated by UN experts who have predicted further melting of Arctic ice because of global warming. This could open up new areas of the shelf, making them available for deep-water drilling. However, deep cooling in the new Little Ice Age in the middle of this century would make it almost impossible to exploit offshore fields and pump oil and gas tens to hundreds of kilometers from the coast at depths of hundreds of meters (Abdussamatov, 2009b, 2013a). Because the Arctic Ocean will be covered with thicker floating ice and the ice will move, drilling will be extremely difficult and dangerous. In the future, a fuel and energy complex will not have easy access to oil and gas in the Arctic. Deep cooling can become one of the major risks in the development of hydrocarbon deposits in the Arctic. So long-term forecasts portend for the fuel and energy complex even more complex and difficult working conditions, and not only in the Arctic. In this way, the upcoming new Little Ice Age will have a very serious impact also on energy security. In addition, the possibility of exploitation of the northern sea route in the future seems to the author also impossible.

13. INCREASING GLOBAL TEMPERATURE ON THE EARTH HAS STOPPED SINCE 1997

TSI has gradually decreased since 1990. The Sun has not heated our planet as much as previously since 1990 (Figs. 17.3 and 17.8). However, because of the thermal inertia of the world's oceans, we are basking in warm additional solar energy stored in the oceans during the 20th century. As a result of the TSI fall since 1990, the warming, contrary to predictions by many climatologists, instead of accelerating, has stopped since 1997, and the average temperature on Earth is stabilized, not growing, and there are already signs of deep cooling (Fig. 17.15). At the same time, since 1997, the concentration of carbon dioxide in the atmosphere has continued to grow.

A rise in the world ocean's level is the most reliable indicator of the rate of temperature growth and one of the problems of our time. More than 40 scientists in 20 groups participating in Arctic research combined their efforts to estimate the contribution of ice melting in Greenland and Antarctica to the global sea level. Since 1992 the global rise in the level of the world ocean has been 0.59 ± 0.2 mm/year on average (Shepherd et al., 2012). This means that the current ocean level is not rising, which would reflect a current state of warming—it has stopped, and the global temperature has not grown during this entire period.

The decline in TSI since 1990 and, correspondingly, lower amount of solar energy coming into the tropical part of the world ocean will gradually cause a weakening of the power of the atmospheric and oceanic circulation and, first

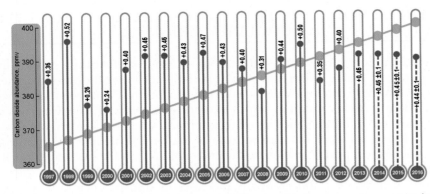

FIGURE 17.15   The trends in both average annual global temperature (with respect to the average temperature for the time interval 1961—90) *(data are taken from the HadCRUT3 (2015))* and carbon dioxide concentration (ppmv).

of all, a decrease in power of the warm current of the Gulf Stream because the amount of heat provided by ocean currents from the tropical areas to the Gulf of Mexico will decrease. This will lead to stronger cooling of the climate in the zone of its action in western Europe and the eastern parts of North America. The temperature in Greenland has shown no increase for decades (Booker, 2015). A study of snow variability in the Swiss Alps in 1864—2009 also shows a reversal of the temperature trend in the Swiss Alps since 2000 (Scherrer et al., 2013). The average temperature around the globe will fall by about at 1.5°C when we enter the deep cooling phase of the Little Ice Age, expected in the year 2060 ± 11. The cooling phase will last for about 45—65 years, for four to six 11-year cycles of the Sun, after which on the Earth, at the beginning of the 22nd century, will begin the new, next quasi-bicentennial cycle of warming—Little Ice Age. Natural causes play the most important role in climate variations, rather than human activity.

14. CONCLUSION

The Sun is the main factor controlling the climatic system and even slight long-term TSI variations may have serious consequences for the climate of the Earth and other planets of the Solar system. The total sign and value of the deviation of the average annual energy balance of the Earth from the equilibrium state over a long time (excess of incoming TSI accumulated by the ocean, or its deficiency) determine a corresponding change in the energy (thermal) state of the surface—atmosphere system and, hence, a forthcoming climate variation and its amplitude. That is, variations in the Earth's climate are a function of long-term deviations in the average annual energy balance of the Earth from the equilibrium state between the total incoming solar radiation energy in the upper layers the Earth's atmosphere and the total outgoing energy from the Earth back into space. Significant climate variations during at least the past 800,000 years indicate that quasi-bicentennial and 100,000-year cyclic variations in the TSI entering the upper layers of the Earth's atmosphere (taking into account the direct and subsequent nonlinear influences of secondary feedback effects) are the main fundamental cause of corresponding alternations in climate from warming to the Little Ice Age and Big Glacial Period. The quasi-bicentennial cyclic variations in the TSI control and practically totally determine the mechanism of quasi-bicentennial cyclic alternations in climate change and set corresponding timescales of practically all physical processes taking place in the Sun—Earth system and also are key to understanding cyclic changes in both nature and society. In 2003—2007, I warned that the world should prepare for imminent global cooling. Long before the beginning of the 24th solar cycle, when the anthropogenic nature of warming was the most commonly adopted, I predicted the onset of the grand minimum of both TSI and solar activity in approximately 2040 ± 11 and the beginning of the corresponding decrease in global temperature—the beginning of the epoch of the 19th (over the past 7500 years) Little Ice Age—the period from 2012 to 2015 (after the maximum phase of the 24th solar cycle) (Abdussamatov, 2003, 2004, 2005, 2007a,b; In the middle of the XXI century, 2006; Scientist predicts, 2006; The price of sensations, 2006; Russian scientist, 2006; global cooling, 2006; People, 2007; Russian academic, 2007; Trimble et al., 2007). And the deep minimum phase of the Little Ice Age will begin in the period from 2055 to 2060 (±11). The most obvious reason for deep global cooling is the upcoming grand minimum of the Sun. These predictions in the course past years are finding more and more practical confirmations, and by the Sun itself (the significant drop in levels of TSI and solar activity at the levels of my predictions) and also by the current lack of warming and of a practical rise the level of the world ocean since 1997, which are under the direct control of the decreasing-phase quasi-bicentennial TSI cycle.

Long-term negative deviation of the Earth's average annual energy balance from the equilibrium state since 1990 has led to corresponding variations in its energy state and thermal mode. On our planet, after the maximum phase of the 24th solar cycle at the end of 2015, we began the descent into the epoch of the quasi-centennial 19th Little Ice Age in the past 7500 years (Abdussamatov, 2016a). Variations in the parameters of the Earth's surface and atmosphere, caused by cooling, will be generating a long chain of causal cycle of the subsequent secondary feedback effects. The increase in the albedo and the decrease in the greenhouse gas concentrations in the atmosphere by the upcoming cooling according Henry's law and the Clausius—Clapeyron relation will lead to an additional reduction in the absorbed solar energy and reduce the greenhouse effect influence. They will lead to an additional drop in temperature, which can surpass the influence of the direct effect of the quasi-bicentennial TSI decrease (Eq. (17.14)). The Earth has experienced such major cooling occurrences five times over the past 1000 years and not less than 18 times in the past 7500 years, and a global freeze will come regardless of whether or not industrialized countries put a cap on their greenhouse gas emissions, because the amounts of natural flows of carbon dioxide from the oceans and land to the atmosphere (M_{in}) and from the atmosphere (M_{out}) to the oceans and land exceed many times the anthropogenic discharges of these substances into the atmosphere (M_{ant}) (Nigmatulin, 2010). The common view that human industrial activity is a deciding factor in warming has emerged from a misinterpretation of cause and effect. Pictures of the frozen Thames (Figs. 17.10 and 17.11) and a historical study of the effects of recent deep cooling in the period of the Maunder minimum are warnings about the serious threat also to the future of energy security for humanity. Humankind by the middle of the current century will meet with the same very difficult times as well as a change for the worse in conditions for creating material and financial resources of the society.

Earlier in the last quarter of the 20th century the simultaneous warming of the Earth, Mars, and other planets and satellites of planets in the Solar system was established. In 2005, data from NASA's Mars Global Surveyor and Odyssey missions revealed that the carbon dioxide "ice caps" near Mars's south pole had been diminishing for three summers in a row (Odyssey, 2005; Ravilious, 2007). Is there anything in common for all the planets of the Solar system whose action could result in their simultaneous warming during the same time period? Because there are no human-made emissions on Mars, this warming must be due to other things, such as a warming Sun, and these same causes are responsible for the warming observed on Earth throughout practically all the 20th century. This common factor affecting simultaneously all the bodies of the Solar system is a long-term high TSI level during practically the whole 20th century. That is why simultaneous warming on Earth, Mars, and the whole Solar system has a natural solar origin and confirms the action of the solar summer throughout the Solar system and alternation of climatic conditions in it. In general, by analogy with the seasons on Earth there is also a similar alternation of climatic conditions in the Solar system, dictated by the quasi-bicentennial cyclic variation in the TSI (Abdussamatov, 2013a,b, 2015a). From this point of view, now the whole of our Solar system after the solar summer came the season of the solar autumn and then in 2060 ± 11 we will reach season of the solar winter of the quasi-bicentennial cycle of climate variation. Only in the early 22nd century will come the season "sunny spring" in the entire region of the Solar system. The observed long-term decline in TSI and forthcoming deep cooling will, first of all, essentially affect climate-dependent natural resources and influence, in the first place, economics closely connected with the state of the climate. Temperature to the mid-21st century may be reduced to the level of the Maunder minimum, which took place in the years 1645—1715. Thus climate change is a natural process, beyond human control, and is not practically connected with our activities. The new Little Ice Age could cause long-term and extremely cold winters characterized by the freezing of normally ice-free rivers as well as by snow fields in areas that have for several centuries not witnessed such climate conditions.

Simultaneous gradual weakening of the Gulf Stream leads to a stronger cooling, which will be most felt in western Europe and the eastern parts of United States and Canada. The world must start preparing for the new Little Ice Age right now. Politicians and business leaders must make full economic calculations of the impact of the new Little Ice Age on everything—industry, agriculture, living conditions, development. The most reasonable way to a fight against the new Little Ice Age is a complex of special steps aimed at support of economic growth and energy-saving production to adapt mankind to the forthcoming period of deep cooling, which will last approximately until the beginning of the 22nd century. Early understanding of the reality of the forthcoming global cooling and physical mechanisms responsible for it directly determines the choice of adequate and reliable measures that will allow humankind, in particular populations of countries situated far from the equator, to adapt in advance to the deep global

FIGURE 17.16   The Lunar Observatory is a system of two identical optical telescopes.

cooling. The upcoming climate change will be the most important challenge and a priority issue for the world and define the main events in politics, the economy, and the most important areas of the whole of humanity in the coming decades. Monitoring and research of long-term variations in global climate parameters are extremely important tasks in a thorough understanding of the physical mechanisms of global climate change and determining the patterns of its formation, which will allow the development of reliable methods predicting future deep climate changes. The most reliable way to accurately predict the depth and exact time of the beginning phase of a deep minimum of the upcoming Little Ice Age is to study the long-term variations in the most effective global parameter: the deviation of the Earth's average annual energy balance from the equilibrium state. Therefore, we have developed a very important new space project, the Lunar Observatory, for direct monitoring of both the albedo and the long-wave radiation of the Earth going out into space, as well as the state of the surface, clouds, vegetation, cryosphere, concentration aerosols, and ozone around the globe for investigation of the deviation of the average annual energy balance of the Earth from the equilibrium state and the physical reasons for climatic changes, with the Moon's surface having the ideal conditions of the lunar night (Fig. 17.16) (Abdussamatov, 2015c,d, 2016b,c). The Lunar Observatory is a system of two identical optical telescopes working in series to measure the complex of global climatic parameters of the Earth. Comprehensive measurements will be taken of variations in radiation emanating from the Earth, reflected and scattered TSI to space in all directions in the wavelength ranges $\Delta\lambda = 0.2-4$ µm, the thermal radiation of the Earth $\Delta\lambda = 4-100$ µm, and the main atmospheric transparency window $\Delta\lambda = 8-13$ µm, as well as remote sensing of the Earth from the surface of the Moon in 10 narrow intervals of the spectrum range $\Delta\lambda = 0.2-3$ µm. These measurements will be carried out consecutively during ~94% of the lunar day continuously only in the lunar night in places installation of the telescopes. It is a scientific issue of fundamental importance and a unique tool for the best and deeper learning of the physics of solar—terrestrial relations and the establishment of the physical causes of climate change and, therefore, their most reliable forecasting. Establishing physical causes of global climate change and the timely creation of the most reliable methods for predicting the future of deep cooling will prevent and weaken maximally the impact of global deterioration of the natural environment and of climate, social, and economic situations around the world, as well as climate-driven changes in the geopolitical situation.

Acknowledgments

I sincerely express deep gratitude to Professor Easterbrook for his great work in editing my complex and difficult English text.

References

Abdussamatov, H.I., Bogoyavlenskii, A.I., Khankov, S.I., Lapovok, E.V., 2010. Modeling of the Earth's planetary heat balance with electrical circuit analogy. Journal of Electromagnetic Analysis and Applications 2, 133–138.

Abdussamatov, H.I., 2003. Long-term correlated variations of the solar activity, radius, irradiance and climate. In: Proc. Int. Conf. on Climatic and Ecological Aspects of the Solar Activity, St. Petersburg, pp. 3–10 (in Russian).

Abdussamatov, H.I., 2004. About the long-term coordinated variations of the activity, radius, total irradiance of the Sun and the Earth's climate. In: Proceedings of IAU Symposium No 223 Cambridge. Cambridge University Press, pp. 541–542.

Abdussamatov, H.I., 2005. Long-term variations of the integral radiation flux and possible temperature changes in the solar core. Kinematics and Physics of Celestial Bodies 21, 328–332.

Abdussamatov, H.I., 2006. The time of the end of the current solar cycle and the relationship between duration of 11-year cycles and secular cycle phase. Kinematics and Physics of Celestial Bodies 22, 141–143.

Abdussamatov, H.I., 2007a. Decrease of the solar radiation flux and drastic fall of the global temperature on the Earth in the middle of the XXI century. Izvestiya Krymskoi Astrofizicheskoi Observatorii 103 (4), 292–298 (in Russian). http://www.gao.spb.ru/russian/cosm/astr/bull_crim_103%284%29_292.pdf.

Abdussamatov, H.I., 2007b. Optimal prediction of the peak of the next eleven-year activity cycle and of the peaks of several succeeding cycles on the basis of long-term variations in the solar radius or solar constant. Kinematics and Physics of Celestial Bodies 23, 97–100.

Abdussamatov, H.I., 2009a. The Sun defines the climate. Russian Journal 'Nauka i Zhizn' ('Science and Life') 34–42. http://www.gao.spb.ru/english/astrometr/abduss_nkj_2009.pdf.

Abdussamatov, H.I., 2009b. The Sun Dictates the Climate of the Earth. Published by Logos, St Petersburg, 197 pp. (in Russian).

Abdussamatov, H.I., 2010. The Sun dictates the climate. In: Fourth International Conference on Climate Change (ICCC-4), ppt-presentation, Chicago.

Abdussamatov, H.I., 2012a. Bicentennial decrease of the solar constant leads to the Earth's unbalanced heat budget and deep climate cooling. Kinematics and Physics of Celestial Bodies 28, 62–68.

Abdussamatov, H.I., 2012b. Bicentennial decrease of the total solar irradiance leads to unbalanced thermal budget of the Earth and the Little Ice Age. Applied Physics Research 4, 178–184.

Abdussamatov, H.I., 2013a. Grand Minimum of the Total Solar Irradiance Leads to the Little Ice Age. Printed by Nestor-Istoriya, St. Petersburg, 246 pp. (in Russian).

Abdussamatov, H.I., November 25, 2013b. Grand Minimum of the Total Solar Irradiance Leads to the Little Ice Age (Summary for Policy Makers). Science & Public Policy Institute (SPPI), pp. 1–7 (in English).

Abdussamatov, H.I., 2014. 2014 – the beginning of the new Little Ice Age. In: Ninth International Conference on Climate Change (ICCC-9), ppt-presentation, Las Vegas.

Abdussamatov, H., 2015a. Current long-term negative average annual energy balance of the Earth leads to the new Little Ice Age. Thermal Science 19 (Suppl. 2), S279–S288. http://dx.doi.org/10.2298/TSCI140902018A.

Abdussamatov, H.I., 2015b. Power of the energy of 11-year solar cycle and its dependence on solar cycle length. Kinematics and Physics of Celestial Bodies 31, 193–196.

Abdussamatov, H.I., 2015c. Lunar Observatory (in Russian). http://www.gao.spb.ru/russian/project/lunar_observatory.pdf.

Abdussamatov, H.I., 2015d. Lunar Observatory: Program 9 of the Presidium of Russian Academy of Sciences. Report for 2015. Direction 9. Research Methods of the Solar System, pp. 312–321. http://pr9.cosmos.ru/sites/pr9.cosmos.ru/files/report2015/9-Gerasimov.pdf (in Russian).

Abdussamatov, H.I., 2016a. The began quasi-centennial epoch of the new Little Ice Age. In: Abstract Book, Eleventh Annual Conference "Plasma Physics in the Solar System", February 15–19, 2016. Space Research Institute of the RAS, Moscow, p. 187 (in Russian). http://plasma2016.cosmos.ru/docs/Plasma2016-AbstractBook.pdf.

Abdussamatov, H.I., 2016b. Lunar observatory for investigations of deviation energy balance of the Earth from the equilibrium state and reasons changes of the climate. In: Abstract Book, Eleventh Annual Conference "Plasma Physics in the Solar System", February 15–19, 2016. Space Research Institute of the RAS, Moscow, p. 188 (in Russian). http://plasma2016.cosmos.ru/docs/Plasma2016-AbstractBook.pdf.

Abdussamatov, H.I., 2016c. Lunar Observatory for Investigations of Deviation Energy Balance of the Earth From the Equilibrium State and Reasons Changes of the Climate: Study of the Earth From Space, No. 5 (in Russian).

Astronomical Theory of Climate Change, NOAA Paleoclimatology, NOAA, http://www.ncdc.noaa.gov/paleo/milankovitch.html.

Avdyushin, S.I., Danilov, A.D., 2000. The Sun, weather, and climate: a present-day view of the problem (Review). Geomagnetism and Aeronomy 40, 545–555.

Bal, S., Schimanke, S., Spangehl, T., Cubasch, U., 2011. On the robustness of the solar cycle signal in the Pacific region. Geophysical Research Letters 38, L14809–L14814.

Booker, C., July 26, 2015. How Arctic ice has made fools of all those poor warmists. The Sunday Telegraph. UK newspaper. http://www.telegraph.co.uk/comment/11763272/How-Arctic-ice-has-made-fools-of-all-those-poor-warmists.html.

Borisenkov, E.P. (Ed.), 1988. Climate Oscillations of the Last Millenium. Gidrometeoizdat, Leningrad, 408 p. (in Russian).

Climate Change: New Antarctic Ice Core Data, 2000. http://www.daviesand.com/Choices/Precautionary_Planning/New_Data/.

Eddy, J.A., 1976. The Maunder minimum. Science 192, 1189–1202.

Fischer, H., Wahlen, M., Smith, J., Mastroianni, D., Deck, B., 1999. Ice core records of atmospheric CO_2 around the last three glacial terminations. Science 283, 1712–1714.

Friis-Christensen, E., Lassen, K., 1991. Length of the solar cycle: an indicator of solar activity closely associated with climate. Science 254, 698–700.

Fröhlich, C., 2013. Solar Constant. www.pmodwrc.ch/pmod.php?topic=tsi/composite/SolarConstant.

Herschel, W., 1801. Observations tending to investigate the nature of the Sun, in order to find the causes or symptoms of its variable of light and heat; with remarks on the use that may possibly be drawn from solar observations. Philosophical Transactions of the Royal Society of London 91, 265–318.

Humlum, O., Stordahl, K., Solheim, J.E., 2013. The phase relation between atmospheric carbon dioxide and global temperature. Global and Planetary Change 100, 51–69.

In the middle of the XXI century, 2006. Scientists Predict Global Cooling. www.rian.ru/science/20060206/43366505.html.

Krivova, N.A., Solanki, S.K., Wenzler, T., 2009. ACRIM-Gap and total solar irradiance revisited: is there a secular trend between 1986 and 1996? Geophysical Research Letters 36, L20101.

Lean, J.L., 2000. Short term, direct indices of solar variability. Space Science Reviews 94, 39—51.

Lu, Z., Rickaby, R.E.M., Kennedy, H., Kennedy, P., Pancost, R.D., et al., 2012. An ikaite record of late Holocene climate at the Antarctic Peninsula. Earth and Planetary Science Letters 325, 108—115.

McPhaden, M.J., Lee, T., McClurg, D., 2011. El Nino and its relationship to changing back-ground conditions in the tropical Pacific ocean. Geophysical Research Letters 38, L15709—L15712.

Milankovitch, M., 1941. Kanon der Erdbestrahlungen und seine Anwendung auf das Eiszeitenproblem. In: 1998, Canon of Insolation and the Ice Age Problem. With Introduction and Biographical Essay by Nikola Pantic, Hardbound, Alven Global, Belgrade, 636 pp. (in English).

Nigmatulin, R.I., 2010. The ocean: climate, resources, and natural disasters. Herald of the Russian Academy of Sciences 80, 338—349.

Odyssey studies changing weather and climate on Mars, July 13, 2005. The Changing South Polar Cap of Mars: 1999-2005. MGS MOC Release No. MOC2-1151. http://www.msss.com/mars_images/moc/2005/07/13/.

Pedro, J.B., Rasmussen, S.O., van Ommen, T.D., 2012. Tightened constraints on the time-lag between Antarctic temperature and CO2 during the last deglaciation. Climate of the Past 8, 1213—1221.

People Do Not Affect Global Warming, Say Scientists, 2007. www.rian.ru/science/20070108/58645571.html.

Petit, J.R., et al., 1999. Climate and atmospheric history of the past 420,000 years from the Vostok ice core, Antarctica. Nature 399, 429—436.

Ravilious, K., 2007. Mars Melt Hints at Solar, Not Human, Cause for Warming, Scientist Says. National Geographic News. http://news.nationalgeographic.com/news/2007/02/070228-mars-warming.html.

Russian Academic Says CO2 Not to Blame for Global Warming, 2007. http://en.rian.ru/russia/20070115/59078992.html.

Russian Scientist Issues Global Cooling Warning, 2006. http://en.rian.ru/russia/20060825/53143686.html.

Russian Scientist Predicts Global Cooling, 2006. http://upi.com/NewsTrack/view.php?StoryID=20060825-091321-7556r.

Scherrer, S.C., Wüthrich, C., Croci-Maspoli, M., Weingartner, R., Appenzeller, C., 2013. Snow variability in the Swiss Alps 1864—2009. International Journal of Climatology. http://dx.doi.org/10.1002/joc.3653. Published online: 4 February 2013.

Scientist predicts, 2006. Mini Ice Age. www.upi.com/NewsTrack/Science/2006/02/07/scientist_predicts_mini_ice_age/2345/.

Shapiro, A.I., et al., 2011. A new approach to the long-term reconstruction of the solar irradiance leads to large historical solar forcing. Astronomy & Astrophysics 529, A67.

Shepherd, A., Ivins, E.R., et al., 2012. A reconciled estimate of ice-sheet mass balance: Science 338 (6111), 1183—1189.

Solanki, S.K., Krivova, N.A., 2004. Solar irradiance variations: from current measurements to long-term estimates. Solar Physics 224, 197—208.

SunSpot Data, 2014. SIDC—Solar Influences Data Analysis Center. http://sidc.oma.be/sunspot-data/.

The Price of Sensations About the Global Warming and the Global Ice Age, 2006. http://english.pravda.ru/science/earth/08-02-2006/75628-climate-0/. http://english.pravda.ru/science/earth/08-02-2006/75628-climate-1/.

The HadCRUT3, 2015. http://www.metoffice.gov.uk/research/monitoring/climate/surface-temperature/.

Trenberth, K.E., Fasullo, J.T., Kiehl, J., 2009. Earth's global energy budget. Bulletin of the American Meteorological Society 90, 311—323.

Trimble, V., Aschwanden, M.J., Hansen, C.J., 2007. Astrophysics in 2006. Space Science Reviews 132, 1—182.

Tom de Castella. Frost fair: When an elephant walked on the frozen River Thames. http://www.bbc.com/news/magazine-25862141; http://thames.me.uk/s00051.htm.

Wood, R.W., 1909. Note on the theory of the greenhouse. Philosophical Magazine 17, 319—320.

WorldGallery.CO.UK. http://www.worldgallery.co.uk/art-print/jan-griffier-the-elder-thames-frost-fair-435965#435965-31-40-1.

18

Aspects of Solar Variability and Climate Response

D. Archibald[1], E.L. Fix[2]

[1]Rhaetian Management, City Beach, WA, Australia; [2]Avionics Fix, Beavercreek, OH, United States

1. SOLAR ACTIVITY IN THE LITTLE ICE AGE AND MODERN WARM PERIOD

Fig. 18.1 shows the flux of ^{10}Be atoms at the NGRIP location on the Greenland ice sheet. ^{10}Be atoms are created by the impact of galactic cosmic rays on oxygen and nitrogen atoms in the upper atmosphere. During periods of high solar activity, the Sun's magnetosphere pushes galactic cosmic rays away from the inner planets of the solar system. The converse occurs during low solar activity, so high ^{10}Be values indicate low solar activity. Solar activity varied through the Little Ice Age and into the modern warm period. For example, the Maunder Minimum has a defined period of consistently higher ^{10}Be values resulting from lower solar activity. This period includes the coldest decade, from 1690 to 1700, of the last 1000 years. That decade has the most consistently high ^{10}Be values of this 600-year record. Similarly, the Dalton Minimum has high ^{10}Be values and a colder climate. There is a change of character into the Modern Warm Period, with low values indicating high solar activity.

The only visual indication of varying solar activity is the changing incidence of sunspots in the normally 11-year solar cycle. Good records have been kept of sunspot activity from the mid-17th century. Solar cycles have been numbered from the mid-18th century. Fig. 18.2 shows the solar record from 1693 with a projection of the amplitude of the next solar cycle, Solar Cycle 25, based on the current strength of the solar poloidal fields. The solar amplitude data is sourced from the Solar Influences Data Analysis Center. What is evident from this record is the rise in activity through the 18th century and the change in character into the Dalton Minimum with Solar Cycles 5 and 6. The Little Ice Age continued up to 1930, when there was a jump in activity to the larger solar cycles of the Modern Warm Period. The modern warm period was caused by the highest level of solar activity for 8000 years (Usoskin et al., 2005). The peak of Solar Cycle 24 occurred in April 2014.

The Solar Cycle 24—25 Minimum is expected in 2021. This will also mark the end of the Modern Warm Period. Schatten and Tobiska (2003) predicted a return to a Maunder Minimum-like level of solar activity. From their abstract:

The surprising result of these long-range predictions is a rapid decline in solar activity, starting with cycle #24. If this trend continues, we may see the Sun heading towards a "Maunder" type of solar activity minimum—an extensive period of reduced levels of solar activity.

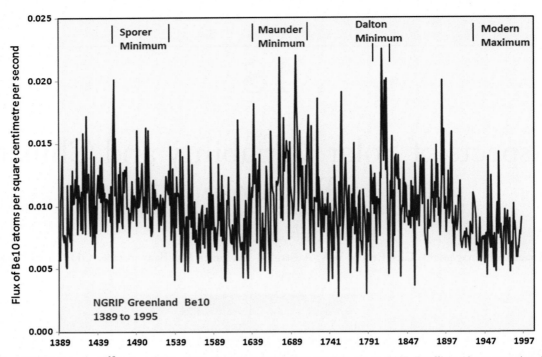

FIGURE 18.1   NGRIP Greenland ¹⁰Be flux, 1389–1995. *Data from Finkel, R.C., Nishiizumi, K., 1997, Beryllium-10 concentrations in the Greenland Ice Sheet Project 2 ice core from 3-40 ka. Journal of Geophysical Research Oceans 102(C12), 26699–26706. http://dx.doi.org/10.1029/97JC01282.*

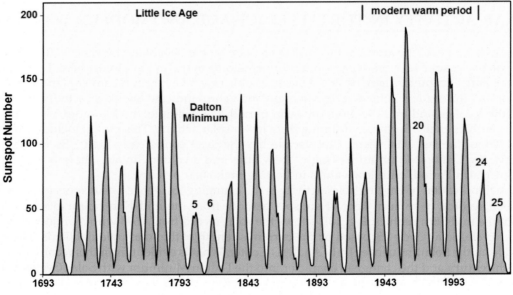

FIGURE 18.2   Sunspot numbers 1693 to 2034.

One of the strongest correlations between solar activity and climate is the level of Lake Victoria in East Africa, as shown by Mason (2006) Fig. 18.3. The correlation was recognized early in the 20th century but the relationship broke down in 1930. There was a 2 m increase in lake level from 1962 to 1964, after which the relationship between solar activity and lake level resumed.

The correlation between solar activity and the level of Lake Victoria in East Africa is striking (Fig. 18.4). What is also significant is that the relationship broke down in 1930 and then resumed in the 1960s. The 30-year hiatus coincides with the rise of solar activity in the modern warm period, culminating in the high amplitude of Solar Cycle 19.

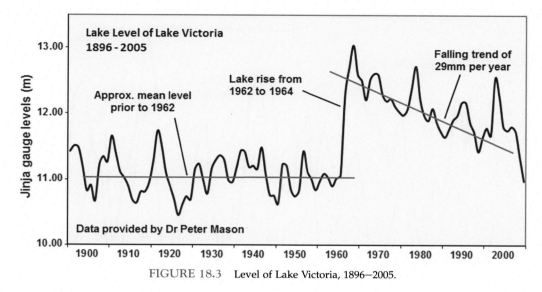

FIGURE 18.3   Level of Lake Victoria, 1896–2005.

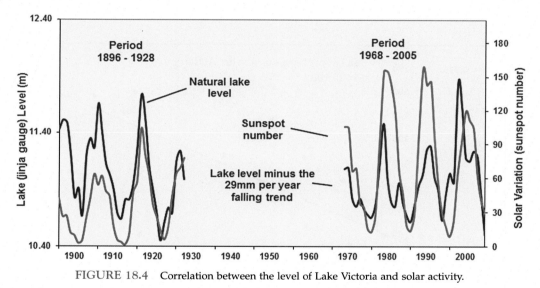

FIGURE 18.4   Correlation between the level of Lake Victoria and solar activity.

Sea-level rise has not been monotonic during the Modern Warm Period, with sea level falling during four solar cycle minima (Fig. 18.5). As with the level of Lake Victoria, the relationship between sea level and solar activity broke down about 1930 but then resumed. The relationship is strongest over the 40 years to 1987.

The relationship between solar activity and sea level has enabled the break-over level between warming and cooling in the modern warm period to be quantified. What Fig. 18.6 shows is that sea level falls below a sunspot number of 40 and rises above that number. The effect is linear. A sunspot number of 40 corresponds to an F10.7 flux of 100.

The relationship between solar activity and the rate of change of sea level shown in Fig. 18.6 enables a forecast of sea-level change to be made. Fig. 18.7 shows sea-level change derived from coastal tide gauge records from 1870 combined with satellite data from 2001 and a projection based on lower solar activity over Solar Cycle 25. What is significant from this graph is the inflection point in 1930 in rate of sea-level rise. Prior to 1930, sea level was rising at 1.0 mm per annum on average. After 1930, the rate nearly doubled to 1.9 mm per annum.

One of the authors (Fix, 2012), modeled solar activity as an oscillatory process excited by the movement of the major gas planets around the barycenter of the solar system. Fig. 18.8 shows the match between the output of a model of solar activity and monthly sunspot number from 1914. The sunspot cycle is depicted using the Hale Cycle, rather than the more usual Schwabe Cycle, and successive cycles are shown with alternating polarity, denoting the alternating magnetic polarity of the sunspots that define the Hale Cycle. The match is quite good, including

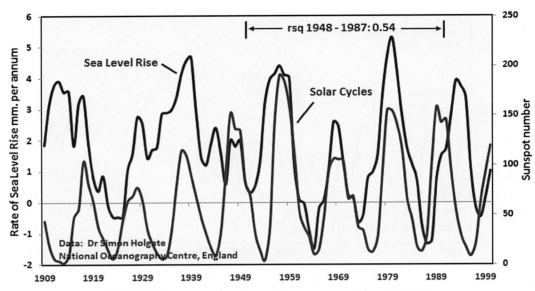

FIGURE 18.5   Sea-level rise and fall and solar activity, 1909—2000.

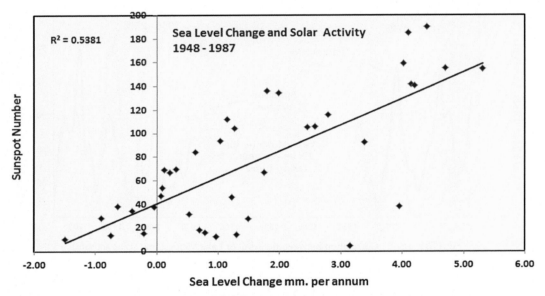

FIGURE 18.6   Sea-level change and solar activity, 1948—1987.

predicting the shape of the decline of Solar Cycle 23. The model indicates that Solar Cycles 24 and 25 will each be eight years long. This would make Solar Cycle 24 similar to Solar Cycle 2, which was nine years long and had a rapid drop in activity in the last two years of the cycle. A succession of such short cycles has not been seen in the solar record yet. Solar Cycle 25 is projected to be a small, short cycle, which may be the start of a period of very low solar activity as per the predictions of Schatten and Tobiska.

The simple model simulation depicted above, on its own, places the peak of Cycle 18 too late, affecting several subsequent solar cycles. To correct this, the model's increase toward the peak of Cycle 18 was stopped in March 1947, making the simulated Cycle 18 peak match the data much more closely. The model was allowed to continue from that point. For subsequent comparisons between full and partial models (Figs. 18.9—18.11), that correction was not made, in order to allow accurate model-to-model comparisons; thus the "full-model" depictions following do not match the model depicted in Fig. 18.9. All versions of the model are started with initial condition 0 in June 1912, when the actual sunspot numbers were near 0 for an extended time.

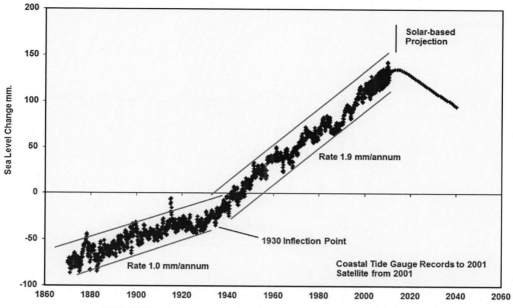

FIGURE 18.7   Sea-level rise from 1870 and future projection.

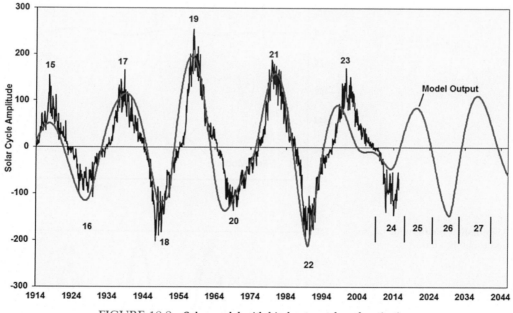

FIGURE 18.8   Solar model with hindcast match and projection.

The planets that cause bulk of the sun's chaotic orbit about the barycenter, and thus excite the solar cycle, are the four gas planets: Jupiter, Saturn, Uranus, and Neptune. To determine the relative contribution of the individual planets, the model was run with each planet in turn and compared to the full model. Fig. 18.9 shows that Jupiter by itself has little effect on solar variability.

Similarly, running the model with only Saturn shows a similarly low amplitude response, though with Saturn's 29 year orbital period instead of Jupiter's 11.86 years.

The combination of Jupiter and Saturn is more than additive and allows the model to simulate most of the sunspot cycle. In contrast, the full model shows the influence of including Uranus and Neptune. In Solar Cycles 18 and 22, Neptune and Saturn increased the amplitude of the cycle while they decreased the amplitude of Solar Cycles 20

FIGURE 18.9 Simulation model with Jupiter only compared to the full model.

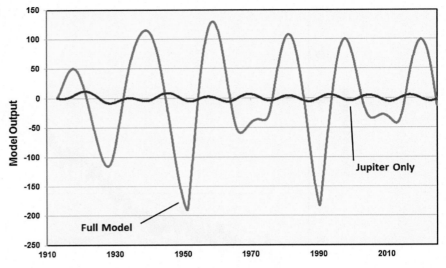

FIGURE 18.10 Simulation model with Saturn only compared to the full model.

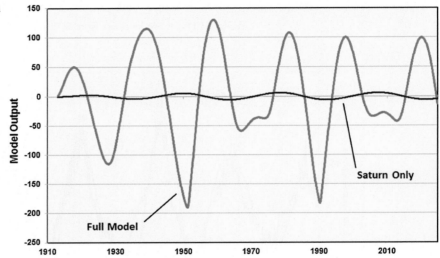

FIGURE 18.11 Simulation model with Jupiter and Saturn compared to the full model.

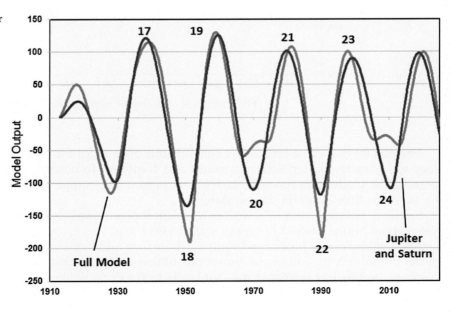

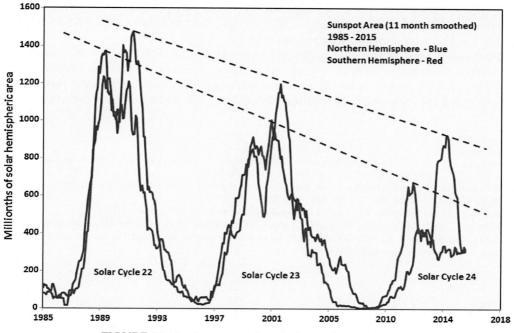

FIGURE 18.12   Sunspot area by solar hemisphere, 1985–2015.

and 24. Thus the cold period of the 1970s cooling period associated with Solar Cycle 20 may have been due to the influence of Uranus and Neptune.

The significance of the plot in Fig. 18.12 is the peaks of activity for both solar hemispheres have held a trend over at least three solar cycles, as shown by the dashed lines. The separate peaks of activity of each hemisphere result in some solar cycles being double-topped. In Solar Cycle 24, the northern hemisphere sunspot area peaked in December 2011 and the southern hemisphere sunspot area peaked two and a half years later in June 2014.

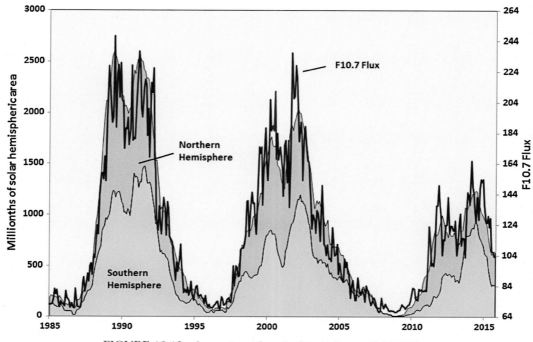

FIGURE 18.13   Sunspot area by solar hemisphere with F10.7 flux.

The fact that the Sun's hemispheres have had separate trends in sunspot activity for over 30 years suggests that the longer orbital periods of Uranus (84 years, 6.48° inclination to the Sun's equator) and Neptune (165 years, 6.43° inclination to the Sun's equator) may be responsible.

The sum of the sunspot areas of the solar hemispheres equates to the F10.7 flux of the Sun. The F10.7 flux is the most commonly used measure of solar activity (Fig. 18.13). The significance of this is that solar activity, and thus the effect on the Earth's climate, varies by solar hemisphere. A further refinement of the solar model may be possible by modeling each solar hemisphere separately in three dimensions. The solar model is currently two-dimensional.

References

Finkel, R.C., Nishiizumi, K., 1997. Beryllium-10 concentrations in the Greenland Ice Sheet Project 2 ice core from 3-40 ka. Journal of Geophysical Research Oceans 102 (C12), 26699–26706. http://dx.doi.org/10.1029/97JC01282.

Fix, E., 2012. The relationship of sunspot cycles to gravitational stresses on the Sun: results of a proof-of-concept simulation. In: Easterbrook, D. (Ed.), Evidence-Based Climate Science. Elsevier, pp. 335–352.

Mason, P.J., 2006. Lake Victoria: A Predictably Fluctuating Resource, Hydropower and Dams, Issue Three.

Schatten, K.H., Tobiska, W.K., 2003. Solar activity heading for a Maunder minimum? Bulletin of the American Astronomical Society 35, 3, 6.03.

Usoskin, I.G., Schuessler, M., Solanki, S.K., Mursula, K., 2005. Solar activity, cosmic rays, and the Earth's temperature: a millennium-scale comparison. Journal of Geophysical Research 110, A10102.

CHAPTER

19

The Notch-Delay Solar Hypothesis

D.M.W. Evans

Science Speak, Perth, Australia

OUTLINE

1. INTRODUCTION

By correcting problems in the conventional basic climate model used to compute sensitivity, the equilibrium climate sensitivity (the surface warming in response to a doubling of the atmospheric CO_2 concentration) was recently shown to be most likely less than 0.5°C (Evans, 2016). However this implies that the rising CO_2 of recent decades is insufficient to account for the observed surface warming: it most likely caused less than 20% of the surface warming (Evans, 2016).

If CO_2 is not the 80% cause of global warming, then what is? This chapter presents a search for that cause or causes.

We look for a *nonanthropogenic cause*. There was considerable temperature variation during the current interglacial period before industrialization: globally, the depth of the Little Ice Age around 1640 AD was ∼1.5°C cooler than today, while the Medieval Warm Period around 1000 AD was about as warm as the modern period (Christiansen and Ljungqvist, 2012). The last century, which has seen ∼94% of human emissions of CO_2 in all of history, has warmed only ∼0.8°C.

We seek a nonanthropogenic source of *albedo modulation*. Albedo from any source other than feedbacks in response to surface warming is omitted from the general circulation climate models, but appears to be crucial. We need a name for it: externally driven albedo (EDA) (see Appendix A) is defined here as albedo due to something external, such as the Sun—the total change in albedo is the change in EDA plus the change in albedo in response to surface warming. By comparing the fractional variation in albedo to the fractional variation in total solar irradiance (TSI), and accounting for albedo feedbacks to surface warming, the effect on surface temperature of changes in EDA was shown to be at least twice (and possibly many times greater than) the direct heating effect of changes in TSI (Evans, 2016). Thus EDA is strong enough to be the leading suspect as the cause of global warming.

We search for a natural source of albedo modulation related *to the Sun*. The Sun is the most obvious source of EDA, or is at least likely be involved if the source is something in our solar system, such as the Jovian planets.

We explore *the link between TSI and surface temperature on Earth*. The only solar parameter measured for long enough to be significant for climate is the number of sunspots, which have been recorded since 1610. All the other solar parameters are either inferred from proxies or were only measured after 1947—which leaves only a relatively short period of three full solar cycles (or Hale cycles, each ∼22 years), which are characterized by an unusually high level of sunspot activity. TSI, which used to be known as the "solar constant" because it appears constant to surface-based observers, has been measured since satellite observations began in late 1978. The relationship between sunspots and TSI since 1978 has been studied and used to estimate TSI back to 1610; although we use TSI in this search, bear in mind that much of the TSI record is deduced from the sunspot record. The relationship between sunspots and TSI is contentious and there are competing TSI reconstructions whose backgrounds or low-frequency components differ significantly. However, note as the evidence and hypothesis are presented later in the chapter that the precise reconstruction makes little or no difference, except to the prediction in Section 7—where predictions are made for each of the two main reconstructions.

Here is a roadmap of this chapter. The formal system from TSI to the surface temperature T_S is a linear invariant system, so it is amenable to Fourier analysis. Careful analysis using newly developed low-noise Fourier techniques reveals an apparently new feature: a notch in the transfer function of the system, centered on the frequency corresponding to 11 years, the average length of the sunspot cycle; the peaks in TSI every ∼11 years as part of the sunspot cycle are not reflected in the surface temperature record. The notch suggests the possibility of, but does not mandate, a delay of several years between a change in TSI and the corresponding change in T_S. Evidence for a delay of ∼11 years is found in several disparate sources. This leads to the force-X hypothesis, that a warming influence called force X affects the Earth's albedo, and that force X is a smoothed version of the TSI delayed by one sunspot cycle. Thus tremors in TSI foretell what force X will be ∼11 years later. Also force X is at a minimum when the Sun's magnetic field flips polarity, which causes notching—the flips coincide with the peaks in TSI. Fitting the TSI and temperature records to the resulting notch-delay solar model, we find that force X has between 10 and 20 times as much influence on surface temperatures as the direct heating effect of changes in TSI. There was a significant fall in TSI around 2004, suggesting a fall in force X and significant surface cooling from ∼2017 (the current sunspot cycle is 13 years, 2004 + 13 = 2017). This prediction leads to a falsifiability condition on the notch-delay hypothesis.

We employ some ideas perhaps unfamiliar to many climate scientists, chiefly linear invariant systems and low-noise Fourier analysis. They are mainly relegated, along with other subsidiary details, to the four supplementary files:

- *The Notch-Delay Solar Theory—Supplementary Information*. General purpose supplementary information.
- *Systems, Sinusoids, the Fourier Transform, and Filters*. Frequency-domain knowledge required by this chapter, explained from scratch, including linear invariant systems; sinusoids; the Fourier transform; simple low-pass, delay, and notch filters; transfer functions; and step responses.
- *The Optimal Fourier Transform (OFT)*. The OFT is a low-noise version of the Fourier transform for time series, pioneered for this chapter, which is superior to the standard Discrete Fourier Transform (DFT) and Fast Fourier Transform (FFT) for finding signals in noisy data sets.
- *climate.xlsm*. Spreadsheet containing all the data and code.

2. THE NOTCH IN THE EMPIRICAL TRANSFER FUNCTION

This section estimates the amplitude of the transfer function from TSI to surface temperature. We use the main TSI and surface temperature data sets, and discover the notch.

2.1 The Formal System

For simplicity while searching for a relationship, we assume that the TSI is the *only* influence on T_S. As discussed in the Introduction, it is plausible that TSI is a dominant cause—or more precisely, contains information about a dominant cause. If TSI mostly predicts T_S, and there is a strong and obvious relationship between them, then this assumption is adequate for the exploratory analysis here.

Formally, consider the system whose input is the TSI anomaly at 1 AU, denoted by ΔS, and whose output is the mean global surface temperature anomaly, ΔT_S, both functions of time, as shown in Fig. 19.1.

2.2 Some Background System Theory

Our system is assumed to be linear, because we are only dealing with small perturbations to a stable climate near steady state, and any classical physical system is approximately linear for sufficiently small perturbations. The climate is widely assumed to be linear for warming influences during an interglacial. The system is also assumed to be (time-) invariant—its properties have not changed significantly during the current interglacial.

Linear invariant systems have a special property. If the input is a sinusoid, then the output is also a sinusoid at the same frequency, and for a given system at a given frequency, the ratio of the amplitude of the output sinusoid to the amplitude of the input sinusoid is constant and the difference between the phases of the output and input sinusoids is also a constant. Linear invariant systems are amenable to Fourier analysis: the input can be expressed as a sum of sinusoids using the Fourier transform, and then, by the linearity of the system, each of the input sinusoid maps to the same output sinusoid that it would if the input sinusoid were the only input present—that is, what happens at each frequency is independent of what happens at other frequencies. This is the only significance of sinusoids in analyzing systems.

A linear invariant system is completely described by its transfer function, which is a function of frequency whose value at each frequency consists of the amplitude multiplier and the phase shift caused by the system. Complex numbers are used to represent sinusoids: both have amplitudes and phases, and a sinusoid is simply represented by the complex number with the same amplitude and phase. At a given frequency, the value of the transfer function is the complex number whose:

- Amplitude is the amplitude of the output sinusoid at the frequency divided by the amplitude of the input sinusoid at the frequency.
- Phase is the phase of the output sinusoid at the frequency less the phase of the input sinusoid at that frequency.

The value of the transfer function at a given frequency is the ratio of the complex number representing the output sinusoid to the complex number representing the input sinusoid, using complex division. The Fourier transform of the system output is the complex product of the transfer function and the Fourier transform of the system input.

2.3 The Data

The most prominent public data sets, on all time spans available, were used in this study. The data sets are noisy and sometimes contradictory, so we did not pick a "best" data set or combine them into a composite data set, but instead combined their spectra into a single spectrum that best fitted them all.

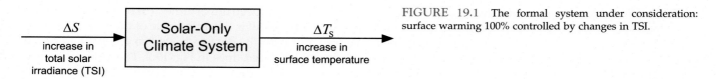

ΔS
increase in total solar irradiance (TSI)

Solar-Only Climate System

ΔT_S
increase in surface temperature

FIGURE 19.1 The formal system under consideration: surface warming 100% controlled by changes in TSI.

The TSI data sets used are Lean's reconstruction from sunspots from 1610 to 2008 with the Wang, Lean, and Sheeley background correction, the PMOD satellite observations from late 1978, the Steinhilber reconstructions from [10]Be in ice cores going back 9300 years, Delaygue and Bard's reconstruction from ^{14}C and [10]Be from 695 AD, the f10.7 solar radio flux from 1947, and the SIDC/SILSO (V1) sunspot counts from 1749. ACRIM satellite data from 1978 were omitted because they disagree with the other data before 1992, which then leaves only two sunspot cycles—too short to establish its spectrum.

The temperature data sets used are the satellite records from late 1978 (UAH and RSS), the surface thermometer records from 1850 or 1880 (HadCrut4, GISTEMP, and NCDC), the two comprehensive proxy time series of Christiansen and Ljungqvist (2012) going back to 1500 with 91 proxies and 1 AD with 32 proxies, the Dome C ice cores going back 9300 years (to match the period for which there is TSI data), and Moberg's 18-proxy series from one AD.

2.4 Low-Noise Fourier Analysis

Two functions of time—the TSI anomaly ΔS and surface warming—are considered here. If those functions were known perfectly for all time, then we could calculate their Fourier transforms and learn the amplitude and phase of all of their constituent sinusoids. Instead our data samples these functions, imperfectly, to form a number of overlapping time series of limited extents, from which we estimate the main constituent sinusoids in the functions.

The signals we are potentially looking for are small, at about the noise level in the data. Consider the direct heating effect of the TSI peaks that occur at the maximum of each sunspot cycle. From Evans (2016), the direct surface warming due to a change in TSI of ΔS is:

$$\Delta T_{S,A} = \frac{(1-\alpha)}{4} \frac{\lambda_{SB}M}{(1-f_\alpha\lambda_{SB}M)}\Delta S \simeq 0.12[0.07, 0.36]\frac{°C}{W\ m^{-2}}\Delta S, \tag{19.1}$$

where α is the albedo (~ 0.30), λ_{SB} is the Stefan−Boltzmann sensitivity ($0.267°C\ W^{-1}\ m^2$), M is the ARTS multiplier ($2.0\ [1.5, 2.7]$), and f_α is the albedo feedback to surface warming ($0.4 \pm 0.5\ W\ m^{-2}°C^{-1}$). TSI typically varies from the trough to the peak of a sunspot cycle by $\sim 0.8\ W\ m^{-2}$ out of $1361\ W\ m^{-2}$, thus causing $\sim 0.1°C$ of surface warming, about the same as the $0.1°C$ typical error margin in modern temperature records. However processing techniques like Fourier analysis that correlate an expected signal against the data can easily see beneath the noise floor—for example, the global positioning system signal is typically one four hundredth (-26 dB, power) of the noise floor of the Earth.

There are many ways of applying Fourier analysis to estimate the constituent sinusoids. We took care to use methods that minimized the introduction of noise. In particular, we did not arbitrarily change the data in the time series by adding data points whose values are zero in order to "pad" the data series to a convenient length, or use "windowing"—both often done automatically by software packages that apply a FFT. We originally discovered the notch (described later in the chapter) using the standard Discrete Fourier transform (DFT), by matching a TSI time series to a temperature time series covering the same period with the same number of regularly spaced data points. But while it is possible to detect the notch using the DFT, we wanted to be surer; so we developed a superior low-noise version of the DFT, called the Optimal Fourier transform (OFT). The OFT has far greater sensitivity and frequency resolution than the DFT, mainly because it is not confined to the preset frequencies of the DFT—but it takes much longer to compute. All the results here use the OFT.

To be properly confident in a Fourier analysis, it must not vary significantly if minor changes are made to the processing methods or to the data—for instance, removing the initial or final 5% or 10% of a data series. The results here are all robust with respect to minor changes in processing technique or data.

2.5 The Empirical Transfer Function

The various data sets each contributed points to a combined amplitude spectrum of either TSI or temperature. The combined spectra were both smoothed, and then their ratio found. The result is the amplitude of the empirical transfer function, the black line in Fig. 19.2. Repeating the procedure, but with the data restricted to pre-1910, to pre- or post-1945, to pre-1970, or to instrumental data (no proxies), gives the colored lines in Fig. 19.2. The TSI data are at 1 AU and so they are deseasonalized, whereas the eccentric orbit of the Earth means the actual TSI incident on the Earth is seasonalized, so we ignore frequencies greater than one cycle per year.

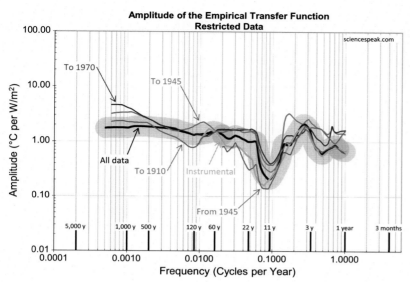

FIGURE 19.2 The amplitude of the empirical transfer function, when the data are restricted as marked. The *black* line is for all data (unrestricted); the *gray* area is a zone around the *black* line. The second horizontal scale is the period of the sinusoids. All scales are logarithmic.

The data were only sufficient to estimate the amplitudes of the constituent sinusoids in the TSI or temperature (or equivalently, their power spectra—the power of a sinusoid is the square of its amplitude). Estimations of phases were not robust. Although the amplitude spectra of physical phenomena like radiation tend to be relatively smooth functions of frequency, the phase spectra are often highly discontinuous, so smoothing and averaging is not appropriate.

2.6 The Notch

The first robust feature of the empirical transfer function is the notch, a relatively narrow region of lower response centered on a period of ∼ 11 years. Sinusoids in ΔS with periods around 11 years are severely attenuated when transferred by the system to ΔT_S, relative to sinusoids at other frequencies. Peaks in TSI and sunspots, which occur ∼ 11 years apart on average, do *not* result in corresponding peaks in the surface temperature. In electronic audio equipment, a filter that removes the hum due to mains power is called a notch filter, because it removes the sinusoids in a narrow range of frequencies around the mains frequency. It appears that something is removing the 11-year "solar hum" from ΔT_S.

The notch is a curious fact. Solar radiation warms the Earth, providing all the heat as incoming radiation—visible light, UV, infrared, and so on. So we would expect the peaks in TSI from the Sun every ∼ 11 years to produce small but detectable corresponding peaks in surface temperature; yet they do not. (This observation gave rise to this chapter.)

The notch occurs because the temperature amplitude spectrum is flat—also found by Eschenbach (2014) independently via different means—while the TSI amplitude spectrum has, of course, a pronounced peak around 11 years.

If low-altitude cloud cover troughed at every sunspot maxima, as suggested by Figs. 2—12 of Lockwood (2004), perhaps in response to troughs in galactic cosmic rays at sunspot maxima, then presumably albedo would also trough, and surface temperature peak, during sunspot maxima. This would make the empirical fact of the notch even more remarkable.

The notch is *not* intrinsic to the main part of the notch-delay hypothesis. The force ND hypothesis (described in Section 5 later in the chapter) allows for force D, the most consequential part of the hypothesis, which does not depend on the notch's existence (data sets are subject to revision) or meaningfulness (under the notch-delay hypotheses below, the formal system is not quite invariant because the delay is one sunspot cycle and the duration of a sunspot cycle varies with time).

2.7 Implications of the Notch for Climate Influences

The TSI peak every sunspot cycle fails to cause a corresponding peak in the Earth's surface temperature record—therefore a countervailing cooling influence is present at precisely the times when TSI peaks.

The duration of the sunspot cycle varies considerably, from 9 to 14 years, yet notching implies that the countervailing influence is always synchronized to the TSI peaks—therefore the timing of the countervailing influence is controlled by the Sun.

The countervailing influence completely counters the warming influence of the TSI peaks—therefore the countervailing influence is as at least as strong as the direct heating effect of the changes in TSI.

2.8 Indirect Solar Sensitivity

The second robust feature of the empirical transfer function is that it is remarkably flat for periods over 20 years. For slower TSI fluctuations the sensitivity of surface warming to increases in TSI, herein called the indirect solar sensitivity (ISS), appears to be relatively constant. For periods over 200 years, the ISS is $\sim 1.7 \pm 0.2$°C W^{-1} m^2 (the value of the black line in Fig. 19.2 as the period increases). This is well above the direct solar sensitivity of 0.12 [0.07, 0.36]°C W^{-1} m^2 in Eq. (19.1), implying that the long term influence of the TSI on ΔT_S is ~ 14 [4.2, 27] times larger than the direct heating effect of TSI.

Hence changes in TSI signal something that has an effect on ΔT_S that is much larger than the direct heating effect of changes in TSI—which is compatible with the finding reported in the Introduction at the start of the chapter about the relatively large influence of EDA.

3. THE DELAY

This section makes a suggestive case for a delay between a change in smoothed TSI and the corresponding change in surface temperature, of ~ 11 years or one sunspot cycle.

3.1 A Notch Might Mean a Delay

We are interested in all possible systems as per Fig. 19.1 that are compatible with the empirical transfer function in Fig. 19.2: a notch in the amplitude transfer function, centered on a period of ~ 11 years, with no constraints on phases.

Assuming the system is describable by a linear differential equation like typical physical systems with continuous variables, the simplest filter that could produce a notch is second order, corresponding to a second-order linear differential equation—which only has terms in the input and output functions, their derivatives, and their second derivatives. Higher-order notch filters are cascades of second-order notch filters, corresponding to higher derivatives—for example, a cascade of two second-order filters is described by a fourth-order linear differential equation. Invoking Occam's razor, we assume the system contains a single second-order filter.

Second-order notch filters come in four "classes"—filters within a class may have different values of the real-valued parameters but are qualitatively identical, while the classes differ only by the values of two binary parameters.

The step response of a system is the output of the system when the input is a unit step function—namely 0 until time zero and 1 thereafter (it "steps up" from 0 to 1 at time zero). A causal step response is 0 before time zero—it obeys the "law of cause and effect," the response comes *after* the cause or stimulus. But a noncausal step response is nonzero before time zero, which is impossible in our universe because the response starts *before* the input steps up. Two of the four classes of second-order notch filters have causal step responses, while two have noncausal step responses, as shown in Fig. 19.3.

Clearly the causal step responses are possibilities for the Sun—Earth relationship, but what about the noncausal ones? Their noncausality dies out exponentially with decreasing time, so simply delaying the step response by a few years by combining the notch filter with a delay filter makes the step response of the combined filter causal, to a good approximation. Note that a delay filter only affects the phases of the transfer function, not the amplitudes.

If the Sun—Earth relationship involves only the causal step responses then a delay is compatible with the observed empirical transfer function, while if it involves the noncausal step responses then a delay of several years is mandatory. This suggests that there *might* be a delay of several years, which motivated us to look for evidence of a delay.

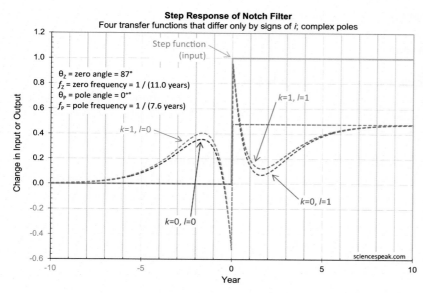

Step Response of Notch Filter
Four transfer functions that differ only by signs of i; complex poles

θ_Z = zero angle = 87°
f_Z = zero frequency = 1 / (11.0 years)
θ_P = pole angle = 0·°
f_P = pole frequency = 1 / (7.6 years)

Step function (input)

$k=1, l=1$

$k=1, l=0$

$k=0, l=1$

$k=0, l=0$

sciencespeak.com

FIGURE 19.3 The step responses of the four classes of second-order notch filter, for realistic parameter values for the Sun—Earth relationship of Fig. 19.1 (values determined in Section 8.2 later in the chapter). We characterize each class by the values of two binary parameters k and l: when l is 1, the step response is casual, but when l is 0, the step response is noncausal.

3.2 Observational Evidence for a Delay

A delay of ~11 years from changes in smoothed TSI to corresponding changes in surface temperature has been found independently several times, though apparently mostly interpreted as delays in the propagation of heat around the Earth. Few, if any, appear to have considered the delay might be in the Sun itself.

3.2.1 A 10-Year Delay to Tropical Atlantic Sea-Surface Temperatures

Soon (2009) found a good correlation between changes in 10-year-delayed TSI to changes in the tropical Atlantic sea surface temperature from 1870 (see his Fig. 19.4), and ascribed it to delays in heat propagation in the oceans:

> The chosen delay time of 10 years is only a rough estimate for the thermal-cryospheric-salinity and mechanical wind stress effects occurring within the Arctic and northern North Atlantic basins to propagate southward. But it is clear from both empirical evidence...and careful ocean modeling...that a physical delay of some 5—20 years is reasonable.

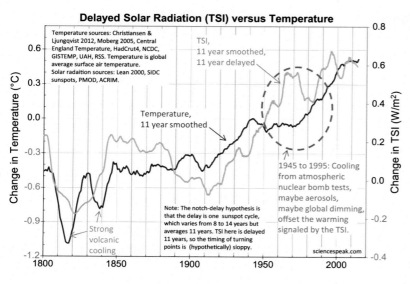

Delayed Solar Radiation (TSI) versus Temperature

Temperature sources: Christiansen & Ljungqvist 2012, Moberg 2005, Central England Temperature, HadCrut4, NCDC, GISTEMP, UAH, RSS. Temperature is global average surface air temperature.
Solar radaition sources: Lean 2000, SIDC sunspots, PMOD, ACRIM.

TSI, 11 year smoothed, 11 year delayed

Temperature, 11 year smoothed

Strong volcanic cooling

Note: The notch-delay hypothesis is that the delay is one sunspot cycle, which varies from 8 to 14 years but averages 11 years. TSI here is delayed 11 years, so the timing of turning points is (hypothetically) sloppy.

1945 to 1995: Cooling from atmospheric nuclear bomb tests, maybe aerosols, maybe global dimming, offset the warming signaled by the TSI.

sciencespeak.com

FIGURE 19.4 Global temperature and 11-year delayed TSI, both 11-year smoothed, have mainly trended together.

3.2.2 A 12.42-Year Delay to Sea-Surface Temperatures Near Iceland

Moffa-Sanches et al. (2014) found a lag of ~12.42 years from changes in TSI to correlated changes in North Atlantic surface temperatures derived from a marine sediment core in the Iceland Basin, from 900 AD.

3.2.3 A 12-Year Delay to Northern Hemispheric Ground Temperatures

Usoskin et al. (2004) found that the correlation coefficient between the Northern Hemisphere ground temperature from Mann and Jones (2003) and sunspot numbers reconstructed from ^{10}Be, from 850 AD, was greatest when the temperature lagged the sunspot numbers by ~12 years (see their Fig. 19.3).

3.2.4 Delay of One Sunspot Cycle to Northern Hemispheric Ground Temperatures

The correlation between temperature and the length of the previous sunspot cycle (solar cycle) is one of the strongest correlations in climate science, unexplained to date and largely disregarded, but the notch-delay hypothesis offers support and explanation.

Friis-Christensen and Lassen (1991) found that the length of a sunspot cycle correlates well with the Northern Hemispheric surface temperature on land during the *following* sunspot cycle—the longer a sunspot cycle, the cooler the Earth during the following sunspot cycle—from 1861. The correlation is strong to 1970 in their data, and then there is a dispute. Damon and Laut (2004) claim they mishandled their data and that the correlation from 1970 instead predicted level temperatures while in fact they went up strongly, thereby breaking the correlation and supporting the CO_2 theory. However, this is strongly disputed by Friis-Christensen and Svensmark (2004).

Butler and Johnson (1994) found the correlation applied to temperatures at the Armagh observatory in Northern Ireland from 1795.

Archibald (2010) showed the correlation applied to the 350-year Central England temperature record, the De Bilt data from Holland, and temperature records at a number of places in the northeastern United States: "in the latter, the relationship is that each 1-year increase in solar cycle length corresponds to a 0.7°C decline of atmospheric temperature during the following cycle." Archibald also proposed using the correlation as a predictive tool.

The duration of the ascending part of a sunspot cycle (roughly its first half) is anti-correlated with the peak sunspot number of the cycle, which is known as the Waldmeier effect. However the strength of this negative correlation depends strongly on the measure of the rise time and which index of sunspot numbers is used (Dikpati et al., 2008). Higher sunspot numbers correlate with a higher peak of TSI, so from the Waldmeier effect we deduce that a longer sunspot cycle correlates with lower levels of TSI during the cycle, which correlates with lower surface temperatures during the following sunspot cycle.

Thus lower TSI during one sunspot cycle correlates with lower surface temperatures during the next sunspot cycle.

The delay implied by the correlation is roughly one sunspot cycle, or ~11 years. Note also that the existence of the correlation supports the notion that the Sun has a major influence on temperatures.

3.2.5 Delay of 10–12 Years to Surface Temperatures in Norway and the North Atlantic

Solhiem et al. (2012) found that a lag of 10–12 years gives the maximum correlation between sunspot cycle length (SCL) and surface temperatures in Norway and the North Atlantic, from 1880:

> This points to the Atlantic currents as reinforcing a solar signal…It is reasonable to expect a time lag for the locations investigated, since heat from the Sun, amplified by various mechanisms, is stored in the ocean mainly near the Equator, and transported into the North Atlantic by the Gulf Stream to the coasts of Northern Europe…They also found that temperatures shifted 11 years back in time, correlated better with SCL measured between minima than between maxima.

3.3 Recent History Suggests a Delay

Lockwood and Froehlich (2007) found that four measures of solar activity—sunspots, TSI, coronal source flux, and neutron count due to high-energy cosmic rays—all peaked around 1986 and 1987 after rising since at least 1970, once the usual fluctuations of the sunspot cycle were removed by a smoothing process. Global surface temperature rose until peaking in 1998 (or maybe 1997 if the effect of the 1998 El Niño is smoothed out), before leveling off.

This suggests a delay of ~11 years from changes in TSI to corresponding changes in surface temperatures. Indeed, without a delay it is difficult to see how TSI could be signaling the major influence on T_S.

3.4 Observations Suggestive of a Delay

A composite TSI record and a composite temperature record was constructed by splicing together the data mentioned in Section 2.2 earlier in the chapter. Fig. 19.4 shows global temperature versus 11-year-delayed TSI, back to 1800, where the TSI is 11-year smoothed to remove most of the effect of the sunspot cycle (the smoother simply averages the values in a centered 11-year window; if the sunspot cycle was exactly 11 years such a smoother would remove all cyclic behavior). With the obvious exception of the 1950s through the early 1980s, which we discuss in Section 8.2, the temperature and 11-year-delayed TSI trend up and down mainly in unison—which is suggestive of an $\sim$11-year delay. Be aware that the data are from proxies before 1850 for temperatures and before 1979 for TSI.

3.5 Implications of the Delay for Climate Influences

In the reasoning and observations described, the magnitude of the warming is great enough to be easily observed—so something either amplifies the direct heating effect of a change in TSI, or is a force in its own right. In either case, there is a warming influence that lags TSI by $\sim$11 years, and its magnitude is much greater than the direct heating effect of changes in TSI.

Note also the observed delay of $\sim$11 years cannot be simply due to propagation of heat around the Earth because:

- The delayed warming influence just mentioned is too large.
- The time constant of the low-pass filter that mimics the thermal inertia of the Earth is $\sim$5 years (Section 8.1)—so the global temperature reflects the new level of direct heating by the TSI much sooner than the $\sim$11 years of the delay (Fig. 19.10).

4. THE FORCE-X HYPOTHESIS

This section puts together numerous clues about strong influences on surface warming into a working hypothesis about a single warming influence, "force X."

4.1 Discussion

We have now identified four strong manifestations of one or more climate influences not included in conventional climate models, where "strong" means to have at least as much effect on surface temperature as the direct heating effect of changes in TSI:

1. EDA is causing albedo modulation (see the Introduction).
2. The notch implies a countervailing influence during TSI peaks (see Section 2.6).
3. The empirical transfer function implies an ISS (see Section 2.7).
4. The delay implies an influence that lags TSI by $\sim$11 years (see Section 3.4).

The EDA and notch are new pieces of the puzzle; the ISS and delay have long been known but not necessarily connected.

It is well known that variation in direct heating by TSI is too small to explain global warming: the 11-year-smoothed TSI rose $\sim$0.7 W m^{-2} from 1900 to 2000, but that only caused $\sim$0.08 [0.05, 0.25]°C (Eq. 19.1) of the observed 0.8°C of observed surface warming, or $\sim$10%.

How many independent strong influences can there be (IPCC, 2013)? Table TS.6 lists only greenhouse gases, aerosols, and albedo changes due to land-use change as long-term strong influences, and volcanoes can be strong but are transitory; yet the four manifestations listed earlier in this section are none of these. The simplest explanation is that they are not manifestations of four separate and previously unknown influences, but of one. Here we will be guided by Occam's razor.

Let us assume there is only one influence. We call it "force X"—because although the outline and some properties of the influence can be deduced at this stage, the exact mechanism is unknown. (There is a historical precedent for the "X": Wilhelm Röntgen named X-rays thusly in 1895 because although he could demonstrate their presence and effects, he did not know what they were exactly. Hopefully when and if force X becomes completely known, it will be renamed.)

The notch shows that force X is synchronized to the Sun—if the delay was of constant duration then force X would get out of synch with the TSI peaks. Thus the delay is not a propagation delay on Earth. It is difficult to see how the puny relative changes in TSI during a TSI maximum could alter anything of significance on Earth, so it would appear that force X originates in the Sun, though it may act via agents on Earth. The EDA finding indicates force X acts via albedo modulation. The delay indicates that force X acts about ∼11 years after a corresponding change in smoothed TSI, implying that changes in TSI occur ∼11 years before the corresponding changes in force X. The duration of the delay is suggestive of one sunspot cycle (a Schwabe cycle, ∼11 years), or half of one full solar cycle (a Hale cycle, ∼22 years)—a half-cycle delay in the Sun's dynamo is perhaps the simplest and most natural lag in a rotating system. The ISS indicates that force X has an order of magnitude more influence on surface temperatures than the direct heating effect of TSI over the longer term.

We will assume without loss of generality that force X is a warming influence, rather than cooling. At the sunspot maximum each cycle, the TSI peaks, and so force X must trough to counteract the effect of the peaking TSI in the surface temperature record. These TSI peaks occur just when the Sun's magnetic field flips polarity and the solar polar field goes through zero (though many other aspects of the Sun's magnetic field do not go through zero).

There exists an influence on the Earth's mean surface temperature, called force X, such that:

1. When force X increases, the surface starts warming immediately and becomes warm after a delay determined only by the thermal inertia of the Earth.
2. Force X modulates the Earth's albedo.
3. Changes in force X occur half of a full solar cycle, or ∼11 years on average, after corresponding changes in smoothed TSI (smoothed or averaged over at least an entire sunspot cycle).
4. Force X is weaker when the Sun's magnetic field is reversing its polarity.
5. On decadal and centennial time scales, force X is such that the surface warming associated with a change in TSI is an order of magnitude more than the direct heating effect of that change in TSI.

The observational evidence for a delay (Sections 3.2–3.4) is of a statistical nature, consistent with underlying changes in TSI found by smoothing TSI over a sunspot cycle or longer, sufficient to be independent of the TSI peak in the middle of the sunspot cycle. The delay is not sharply defined; TSI does not precisely foretell every small change in force X and surface temperatures that occur ∼11 years later. Instead, force X lags the underlying or smoothed TSI. Hence there is no contradiction in force X being at its weakest during a sunspot cycle just when TSI is peaking, which is of course about ∼11 years after TSI last peaked. Force X appears to change on a decadal scale only in response to decadal changes in TSI. Also bear in mind that TSI is not force X, only an imperfect predictor of force X, and we have only the sunspot record and estimated sunspot-TSI relationships to work with.

The coincidence between peaks in TSI and the flipping of the solar magnetic field could explain the notching. As TSI peaks, force X is in a trough (see Fig. 19.5), and these countervailing influences cancel out sufficiently closely as to

FIGURE 19.5   When TSI peaks, the solar magnetic field is at its weakest because it is reversing polarity. This figure merely illustrates the timing; the solar polar field is but one aspect of the Sun's magnetic field.

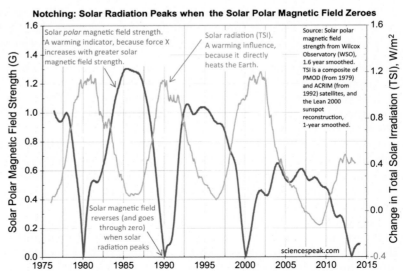

leave no trace in the surface temperature record. This begs the question of whether there might be some feedbacks or some other principle behind such a precise cancellation.

Many solar phenomena are related to the power delivered by the Sun's electromagnetic field, and thus to the product of its electric field strength and magnetic flux—which correlate, so the power is roughly proportional to the square of the magnetic flux. The number of sunspots is such a phenomenon; so because the *square* of the magnetic flux is indifferent to its polarity, the sunspots follow an apparent ~11-year cycle even though the full solar cycle is ~22 years. Some phenomena, such as hydrological cycles on Earth, are also sensitive to the full 22-year cycles.

The duration of the hypothesized delay is one sunspot cycle, or ~11 years on average. But, starting from the moment of a significant change in underlying or smoothed TSI, is the delay to the corresponding change in force X the duration of the current, previous, or next sunspot cycle, or maybe a weighted average of all of them?

An analogy may help to understand the delay. A four-stroke combustion engine has four phases: "suck, squeeze, bang, and blow." If you know how much fuel and air is inhaled during the "suck" phase then you know how much power will be produced in the "bang" phase, which comes half a full cycle (two phases) later. Apparently something similar is happening with the Sun: the sunspots, or the tiny changes in TSI, tell us how much force X there will be half of a full solar cycle later.

5. THE FORCE-ND HYPOTHESIS

This section considers an alternative hypothesis, in which there are *two* indirect warming influences: "force N" causes notching, while "force D" explains the delay, the ISS, and the EDA finding. Although we prefer the force-X hypothesis, this is presented for completeness or in case the notch is found not to exist or not to be meaningful.

5.1 Discussion

The force X hypothesis is based on the assumption that the four strong influences listed near the beginning of Section 4.1 are all manifestations of the *same* influence, namely force X. At least two possible problems exist with this. The first is that the cloudiness fraction, available from 1983, shows no peaking during the TSI peaks of 1990 and 2001, and if anything shows a *decrease* in cloud fraction around 2001 (Cloud Analysis Part 1: Cloud Amount). Low-altitude cloud cover underwent a distinct *trough* around 1990, but there was no particular feature in 2001 (Lockwood, 2004, Figs. 2—12; and Climate and Clouds: Cloud data). But force X acts by albedo modulation and produces a cooling peak to counteract the TSI peak at the sunspot maxima, suggesting it creates an *increase* in cloud cover around sunspot maxima.

However, the increase in cloud cover fraction required to counteract the extra TSI at a sunspot maximum is ~0.05%, too small to detect. (0.8 W m^{-2} of extra TSI at 1 AU is $0.8 \times (1 - 0.3)/4$ or 0.14 W m^{-2} of extra absorbed solar radiation (ASR), which is countered by an increase in cloud fraction of $0.14/239$ or 0.05% since the average ASR is 239 W m^{-2}).

So either force X is affecting albedo by something other than clouds, or the small countervailing increase in cloud fraction goes undetected among noise and larger moves.

The second potential problem is that the increase in TSI during a sunspot maximum implies *increased* force X one sunspot cycle later, which may well be during the next sunspot maximum, just when force X *decreases* in order to counteract the direct heating by the extra TSI. This could be explained by the changes in TSI that foretell changes in force X ~11 years later needing to be changes in underlying or trend TSI, while the temporary changes in force X at sunspot maxima are due instead to the reversal of the Sun's magnetic field. (Each step response in Fig. 19.3 is slightly complicated; obviously there exists a step response corresponding to any empirical transfer function for the solar-only system, so a single force X explanation is possible.)

Another explanation is that there are two separate influences, one that manifests itself around sunspot maxima and causes notching, and another that changes in delayed response to changes in underlying TSI and is responsible for the delay, the ISS, and the EDA finding.

5.2 Hypothesis ND

Let us go to the next simplest alternative after the one-influence assumption of force X, and assume there are two influences. We call them "force N," which causes notching, and "force D," the delayed force, which acts about one sunspot cycle after being signaled by a change in smoothed TSI and is the same as force X except not responsible for the notching. We assume both are warming influences. Schematically:

$$\text{force X} = \text{force N} + \text{force D}. \tag{19.2}$$

We often make statements that apply under either notch-delay hypothesis: "force X/D" means "force X or force D."

5.3 Force N and Force D

If there are indeed two separate significant influences on the climate, beyond those currently considered by the IPCC, then it makes the climate puzzle much harder to solve than if there was only one.

Force N does not necessarily work through albedo modulation, though it could. It could even work by cloud modulation that is too small to be detected (see Section 5.1). It causes notching so it is synchronized to the Sun.

Force D is also synchronized to the Sun, because (1) the correlation between temperature and the length of the previous sunspot cycle in Section 3.2.4 is synchronized to the Sun, and (2) it is not simply propagation of heat, as discussed in Section 3.4. It works by externally driven albedo modulation.

Interestingly, the force D transfer function (of the system from TSI, which signals force D, to surface warming), which is the transfer function of force X in Fig. 19.2 without the notch, is compatible with the transfer function of a simple accumulator or first-order low-pass filter as shown in Fig. 19.6. (An example of a first-order low-pass filter is a capacitor fed through a resistor, which charges or discharges depending on the voltage applied across the combination of resistor and capacitor.) The fall-off in amplitude for frequencies above three or four cycles per year suggested by the empirical transfer function in Fig. 19.2 implies a low-pass filter with a break frequency of ~ 5 years, which indeed is what we get by curve fitting such a model to the data in Section 8.1.

Note that, as shown by the ISS, force D operates with a large amplification factor over the direct heating effect of TSI, so while force D is *proportional* to the accumulation of TSI it is *not due* to the cumulative effect of the direct heating of TSI (see Eq. 19.1).

A simple integral of TSI over time is an accumulator of TSI, so the time-integral of TSI is roughly proportional to force D. The transfer function of an integrator depends on its details (no integrator goes back forever in time), but all are characterized by the downward-sloping amplitude line on the right of Fig. 19.6. For example, a simple integrator circuit implemented with an op-amp is a low-pass filter exactly as per Fig. 19.6. It has been widely observed that time-integrals of TSI roughly fit the shape of the surface warming over the last few centuries. See also Fig. 19.10.

Researchers who have found a high sensitivity of temperature to TSI may have found a high sensitivity to force X/D. For example:

- Shaviv (2008) looked at three independent ocean records (net heat flux, sea level changes from tide gauges, and sea-surface temperatures) and found forcings associated with solar cycle variations that are 5−7 times those associated with TSI variations in the current climate models.

FIGURE 19.6   Transfer function of a low-pass filter. A low-pass filter "passes" sinusoids with frequencies well below f_B but "blocks" those well above f_B (and the higher the frequency, the more it is attenuated).

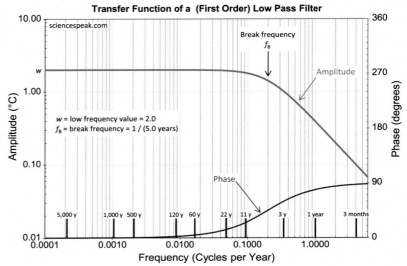

- Douglass and Clader (2002) found that the sensitivity to TSI is twice that of the no-feedback Stefan–Boltzmann radiation model balance, from satellite observations of TSI and temperature.
- Scafetta and West (2009) argue for high sensitivity to TSI and cite paleolithic temperature reconstructions and glacial epochs induced by Milankovitch astronomical cycles, in response to Duffy et al. (2009), who argue that solar variability does not explain late-20th-century warming.

6. MECHANISMS

We do not know the mechanisms behind forces X, N, or D.

6.1 Possibilities

Among others:

- Solar stimulation of ozone via UV or energetic electron or particle precipitation—which changes the relative proportions of ozone and the relative heights of the tropopause at the poles and equator, which in turn affects the degree of north–south extent in the jet streams, which affects the amount of air mass mixing at boundaries of climate zones, which determines cloudiness and albedo (Wilde, 2010, 2015; Woollings et al., 2010).
- Cosmic rays are suspected of encouraging cloud formation and thus affecting albedo, and are influenced by the Sun's magnetic field, so they may be involved in force D. Cosmic rays decrease during TSI peaks, presumably decreasing clouds and albedo and warming the Earth's surface, so they are *not* responsible for force N.
- Solar stimulation of plankton—which produce aerosols that affect clouds (McCoy et al., 2015).
- Meteoritic dust influences albedo, depositing particles large enough to reflect and scatter light but small enough to persist in the stratosphere for months. Meteor rates vary inversely with sunspot numbers (Ellyet, 1977), so, like cosmic rays, they might explain force D but not force N. The dust contains minerals that catalyze plankton growth (see previous point).
- The interplanetary electric field affects cloud cover Voiculescu et al. (2013).
- The Jovian planets may influence solar activity (Sharp, 2013; Wilson I. R., 2013; McCracken et al., 2014), and might also be responsible for changes in force X/D half of a full solar cycle afterwards.
- Asymmetries in the motion of the Sun about the center of mass of solar system are correlated with deviations in the Earth's length of day (LOD). The time rate of change of the LOD correlates with the phase of the North Atlantic Oscillation, while deviations of the LOD from its long-term trend correlate with the phase of the Pacific Decadal Oscillation (Wilson, 2011). These ocean oscillations are correlated with decadal changes in surface temperature, so may be responsible for or related to force D.

Or there may be solar influences that are not yet explained, eg, Stober (2010). Force X/D may involve combinations of the factors just listed.

6.2 Possible Clue to Force X/D?

There is a faint chance that the Nimbus-7/ERB measurements of TSI from 1979 to 1993 may have inadvertently measured (some aspect of) force X/D. These TSI measurements are notable both for being the earliest and for disagreeing with later TSI measurements by being notably higher.

Yoshimura (1996) found that the ERB-TSI lagged the sunspots by 10.3 years (pp. 606–607). Force X/D lags sunspots by that duration, so perhaps the difference between whatever Nimbus-7/ERB measured and what later TSI instruments measured is related to force X/D. Yoshimura concluded (p. 601):

> We argue that the time lags between the TSI and magnetic field variations demand us to consider the influences of the Sun on the Earth and on the space environment through two channels which are physically linked together but their variations may not necessarily be in phase in time. One channel is through the irradiance variations and the other is through the magnetic field variations. Time evolution of a phenomenon on the Earth that is influenced by the Sun can be in phase as well as out of phase with the solar magnetic cycle if this phenomenon is mainly caused by the irradiance variations of the Sun.

7. A PREDICTION

This section predicts an upcoming global cooling, based on the large fall in underlying TSI in 2004 and either of the notch-delay hypotheses. Some caveats: there is no satisfactory instrument for measuring TSI even today; much TSI "data" and all TSI before 1979 are based on reconstruction via questionable models; it is not known which solar parameter is best for predicting force X/D in ~ 11 years and it might not be TSI; and the observations of a delay in Section 3 are based on a variety of TSI measures.

7.1 The Recent Fall in TSI

We constructed a composite TSI by combining all of the sources of TSI mentioned in Section 2.2, so it relies mainly on PMOD and Lean's reconstruction with the background correction of Wang, Lean, and Sheeley. It is shown in Fig. 19.7 to give historical perspective: the recent fall in TSI in ~ 2004 is one of the three largest falls in TSI ever recorded, with records starting in 1610. It is almost the same magnitude as the fall from 1610 to 1645 that led to the Maunder Minimum and the depth of the Little Ice Age, or the fall from 1795 to 1810 that led to the Dalton Minimum.

Fig. 19.8 shows a closer view of the recent fall, with sunspots and several measures of TSI, including PMOD/ Lean's reconstruction, our composite TSI, and recent reconstructions.

If the TSI measures based on PMOD are the better predictor of force X/D, then force X/D will decline significantly in response to the fall in underlying TSI from around 2004. The fall in underlying TSI is ~ 0.3 W m^{-2}, suggesting the associated long-term (200 + years) surface cooling is $\sim 0.5°$C (from Section 2.7, the ISS is $\sim 1.7 \pm 0.2°$C W^{-1} m^2) and the associated cooling after 20 years is $\sim 0.3°$C (from Fig. 19.2, the amplitude of the empirical transfer function is $\sim 1.0°$C W^{-1} m^2 after 20 years).

If the more recent TSI reconstructions such as SORCE/TIM are the better predictors of force X/D, then force X/D has been falling in response to a slide in TSI beginning ~ 1995, which accelerated from ~ 2004. The total fall in underlying TSI is ~ 0.4 W m^{-2}, suggesting an associated long-term surface cooling of $\sim 0.7°$C and cooling after 20 years of $\sim 0.4°$C.

The fall in underlying TSI around 2004 has been noted elsewhere by different methods. For example, Herrera et al. (2014) deduce that a grand minimum in solar activity began in 2004 (using PMOD data) or 2002 (ACRIM).

7.2 The Delay

The current solar cycle looks to be ~ 13 years long, suggesting the delay from change in TSI to corresponding change in force X is most likely ~ 13 years.

FIGURE 19.7   Our composite TSI and composite temperature since 1610. The recent fall in TSI is the one of the three largest and steepest on record.

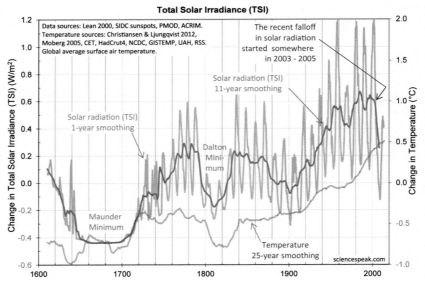

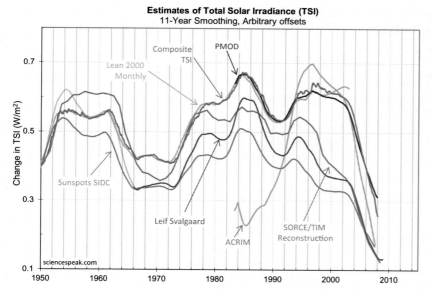

Estimates of Total Solar Irradiance (TSI)
11-Year Smoothing, Arbitrary offsets

FIGURE 19.8 Various measures of sunspots or TSI, all 11-year smoothed (averaged over a centered 11-year window, to eliminate the effect of an 11-year cycle). The TIM is the latest instrument for measuring TSI, but only started in January 2003, too late to cast smoothed light on 2004. The SORCE/TIM reconstruction by Kirvova is used by the IPCC in AR5, and the reconstruction from Leif Svalgaard in mid-2014 is similar.

If the notch is best modeled with the noncausal step responses, the cooling will start about one sunspot cycle after the TSI fall. But if the causal notch step responses are more appropriate, most of the cooling impact might not begin until up to another ∼5 years after that (Fig. 19.10). At the stage we do not which notch model most resembles reality.

The fall in force X and surface temperatures corresponding to the fall in TSI that started around 2004 will most likely start sometime around 2017 (=2004 + 13), or up to 5 years after that.

7.3 A Prediction

If the PMOD TSI is the better predictor of force X/D (more likely): Global temperatures should come off the current plateau into a sustained and significant cooling starting sometime from 2017 to 2021. The cooling will be ∼0.3°C in the medium term (the 2020s), taking the planet back to the global temperature that prevailed in the 1980s.

If the SORCE/TIM reconstructed TSI is the better predictor of force X/D (less likely): There has been mild global cooling due to force X/D starting sometime from 2008 (=1995 + 13) to 2012, and it will accelerate from sometime in 2017−2021. The total medium-term cooling (the 2020s) will be about 0.4°C. However, there has been no mild cooling since 2008 to 2012, so either the SORCE/TIM reconstruction is inappropriate or the notch-delay hypotheses is wrong (in which case there is no prediction), or that mild cooling was counteracted by warming from extra CO_2 (in which case there will be mild cooling starting sometime from 2017 to 2021, as the fall in force X accelerates).

In summary: The notch-delay hypothesis predicts sustained and significant global cooling starting sometime from 2017 to 2022, of ∼0.3°C but perhaps milder. If the predicted cooling does not occur then the notch-delay hypothesis is false.

8. THE NOTCH-DELAY SOLAR MODEL

The notch-delay hypothesis is a physical model that describes a solar influence on the Earth's surface temperatures by its vital properties. It is not fully fleshed-out because the mechanics of force X are unknown, but, like X-rays when they were first discovered, enough critical properties are known or hypothesized to be useful.

This section creates a numerical model from the physical model of force X, finds its parameters by fitting it to the TSI and surface temperature records, and demonstrates that it can explain the global warming of the last few centuries entirely in terms of force X.

The only significance of this is that it demonstrates the existence of a physically based alternative to the CO_2 theory of global warming. The model is primitive and needs more work.

A schematic diagram of the notch-delay solar model is shown in Fig. 19.9.

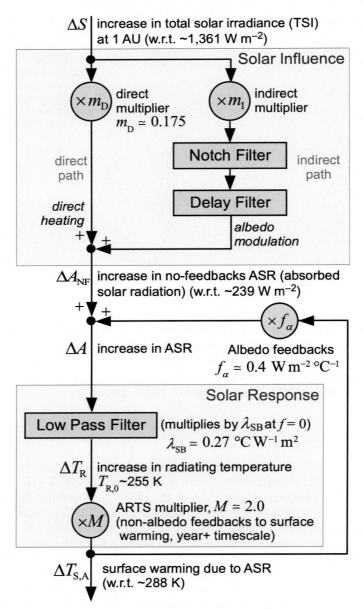

FIGURE 19.9   The notch-delay solar model.

By the notch-delay hypothesis, TSI "controls" EDA in a modeling sense—changes in EDA are due to force X or to forces N and D, force X/D is signaled ~11 years in advance by the tiny changes in the TSI, and force N is signaled by the TSI peaks. These properties give rise to the "indirect path" in Fig. 19.9, for the influence of TSI on surface temperatures via albedo modulation. There is of course also a "direct path," for the direct heating effect of TSI.

The input to each path is the record of TSI changes, and the output of each path contributes a corresponding record of changes in no-feedbacks ASR. Each path has a multiplier that expresses how much change in ASR is produced for a given change in TSI.

The indirect path has a notch filter, to simulate the notch in the empirical transfer function. As discussed in Section 3.1, the model uses the simplest possible notch filter, a second-order filter with two zeroes and two poles. Although there are four classes of such filter and the poles may be a complex conjugate pair or real, in all the simulations here we only used a noncausal notch filter ($k = 0$, $l = 0$) with complex poles—we have not yet explored the other options. The indirect path also has a simple fixed-length delay (a variable delay would be a prime aim of future research).

After the solar influence, the notch-delay solar model is as per the sum-of-warmings model from Evans, 2016, which has the same solar response as the conventional basic climate model. Fig. 19.9 shows parameter values derived from AR5, and m_D is $(1 - \alpha)/4$. The simulations reported here omit albedo feedbacks due to surface warming other than by extra ASR, but these are small.

8.1 Finding the Model Parameters

The notch-delay solar model has eight independent parameters, which were found by curve fitting. A multivariate optimization was performed to find the parameter values that best fitted the composite temperature and TSI since 1630, but particularly since 1900, and also fitted the empirical transfer function. The optimization took into account surface warming due to volcanoes, black carbon, snow albedo, and land use as per the GISS Climate Model E from 2011, and also the megatons of nuclear bombs exploded in the atmosphere. Global warming contributions from CO_2, aerosols, CFCs, and other greenhouse gases were ignored—recall that the aim of this exercise is only to demonstrate the existence of an alternative to the CO_2 theory.

The optimization surface was complicated, with many local minima. The data are uncertain, more so further back in time. No set of model parameters was clearly optimum, but certain parameter values were clearly better and we were able to find a representative "best" parameter set adequate for this exercise:

- The delay was found to be between 10 and 20 years, but the fits were better around 11 years. (Delays less than 10 years were rejected because that did not match the nonfall of temperature to date after the fall in TSI in 2004.) Note that the notch-delay model here has a fixed delay for simplicity, whereas in the notch-delay hypothesis the delay is one sunspot cycle length and thus variable.
- The low-pass filter was found to have a break period (time constant) of about five years, which agrees with what others have found. From (Schwartz, 2012): "The time constant characterizing the response of the upper ocean compartment of the climate system to perturbations is estimated as about 5 years, in broad agreement with other recent estimates, and much shorter than the time constant for thermal equilibration of the deep ocean, about 500 years."
- The direct-path multiplier m_{DLM} was found to be ~ 0.1, agreeing with Eq. (19.1).

That these three parameters have reasonable values gives us some confidence that the model and curve fitting are consistent with reality.

Fig. 19.10 shows the step responses of the model with the parameter values thus found. It includes the step response actually optimized, SR00, and a step response based on a causal notch filter with the same parameter values SR01—though because the parameter values are not optimized for SR01, this is *not* an optimized step response. Notice in both step responses the rise to $\sim 0.1°C$ after two years—this is the effect of the direct path (without the low-pass filter, which simulates the thermal inertia of the Earth, all of this eventual rise would occur immediately at zero years). The remainder of each step response, much larger and shapelier, is due to the indirect path. The

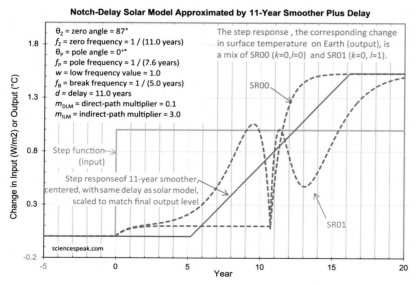

FIGURE 19.10 Two of the possible step responses of the notch-delay solar model: The parameter values were found by optimizing SR00, and SR01 is merely shown using those same values. A simple 11-year smoother-with-delay (*green line*) is a crude approximation of the model.

indirect-path multiplier m_{ILM} was found to be 30 times greater than the direct-path multiplier m_{DLM}, and while the notch attenuates about half of this, the effect of the indirect path is clearly much larger than the direct heating effect of changes in TSI.

Fig. 19.10 also shows the step response of a simple 11-year smoother—whose value at a given time is the average of the values over the 11 years centered on that time, which eliminates any cyclic variation if the period of the cycle is exactly 11 years. Crudely, it approximates any of the possible model step responses. To a first approximation, the notch-delay solar model is an 11-year smoother delayed by one sunspot cycle—as implicitly used in the prediction of Section 7.

8.2 Running the Model

Fig. 19.11 shows a climate simulation whose inputs are the record of TSI changes, the record of forcings due to volcanoes, black carbon, snow albedo, and land use from 1880 as per GISS Climate Model E, and the record of megatons of atmospheric nuclear testing. The simulation adds the surface warming predicted by the notch-delay solar model with the parameter values found earlier, the warmings due to the GISS factors when scaled so as to best fit the temperature data, and the warmings due to the nukes when scaled and lagged (by a decay rate) so as to best fit the observed temperatures. Note again that the aim of this exercise is only to demonstrate the existence of a nongreenhouse-gas explanation for recent global warming, so the warming due to greenhouse gases was entirely omitted from the simulation.

The fit of the simulated temperature to the observed temperature is adequate, matching the general trends but not the short-term variations—given that the model is using a fixed-length delay but the hypothesized delay is the somewhat variable length of the sunspot cycle, which is what we would expect if the model was working well. This demonstrates that a solar influence, based on a physical model, can explain the recent global warming.

The solar model predicts a warming in the 1950s through early 1980s (especially the 1960s) to reflect the rise in TSI from the 1940s (and especially in the 1950s), but that warming is counteracted in this simulation by a substantial cooling due to the atmospheric bomb tests. That cooling peaked at over 0.3°C in 1963, the year of the Limited Test Ban Treaty between the USSR, USA, and UK. Over 500 nuclear bombs were detonated in the atmosphere from 1945 to 1980; they put up fine dust that stayed in the atmosphere for years, reflecting sunlight back into space and lowering the incoming radiation (Fujii, 2011). Perhaps the nuclear winter hypothesis is partly correct. Alternative causes of cooling in this period include pollutant aerosols and global dimming (Pinker et al., 2005) and the Pacific Decadal Oscillation (while it may be an internal mechanism rather than properly regarded as an exogenous cause, it might reflect an exogenous influence). In any case, with only the limited data to hand, the climate model here found the nuclear bomb data fitted reasonably well.

FIGURE 19.11 Surface temperatures hindcast from TSI data by the notch-delay solar model, with contributions from volcanoes, black carbon, snow albedo, and land use after 1880, and from atmospheric nuclear bomb tests.

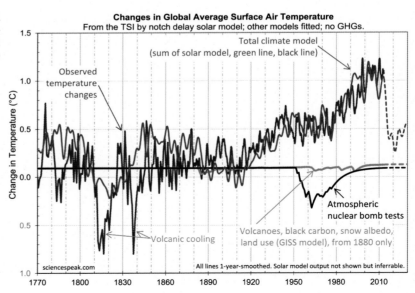

Fig. 19.11 includes a dotted red line predicting the upcoming cooling. It is premature because the model here uses a fixed-length delay, ignoring the extra length of the current sunspot cycle. It is too large, because the model simplistically puts all of the long-term effect of force X/D into play immediately, whereas the empirical transfer function of Fig. 19.2 indicates the medium-term ISS (20—50 years) is ~60% of the long-term ISS, suggesting the cooling is more likely going to be around 0.3°C.

9. CONCLUSIONS

The notch-delay hypothesis provides an alternative explanation for the recent global warming to the CO_2 hypothesis. It is a physical model that can explain the warming of the last two centuries. While force X/D is only partially known, enough critical properties can be deduced to form a working description.

All the bedrock arguments in favor of the CO_2 hypothesis have now been dismantled: recent warming can be explained by something other than CO_2, namely force X/D and albedo modulation; the basic physics, correctly applied, shows that the sensitivity of surface temperature to a doubling of CO_2 is less than 0.5°C, much too low to account for the warming of the last century (Evans, 2016); the Earth has seen greater temperature variation over the last 2000 years than it has since the industrial revolution started (eg, Christiansen and Ljungqvist, 2012); several solar indicators including TSI peaked in ~1986 but surface temperatures kept rising until ~1998, but this is explained by the delay between TSI and force X/D (Table 19.1).

Science is about testable hypotheses. Both the CO_2 and the notch-delay solar hypotheses hindcast warming or stasis for the past few decades, but in the near future they part company dramatically: strong warming (CO_2) versus strong cooling (solar).

- The IPCC in 1990 predicted warming of 0.2—0.5°C per decade for the ensuing decades (IPCC, First Assessment Report, 1990, p. xi), whereas it warmed at most 0.17°C per decade since then (nearly all in the 1990s). Their warming predictions have been tempered somewhat since then, but the CO_2 hypothesis can only predict strong warming in line with the relentlessly rising CO_2 concentration.
- The notch-delay solar hypothesis predicts sustained and significant global cooling starting sometime in the period 2017 to 2022, of ~0.3°C but perhaps milder.

If no cooling of at least 0.1°C (on a 1-year smoothed basis) in global average surface air temperature occurs by 2022, then the notch-delay solar model is falsified. On the other hand, if the cooling should occur, then it appears that Nature, not Man, has been the main cause of the recent global warming.

TABLE 19.1 CO_2 Theory Versus the Notch-Delay Solar Theory

Aspect	CO_2 hypothesis	Notch-delay solar hypothesis
Compatible with "missing hotspot" observation	No	Yes
Explains warming trend 1800—1998	Yes	Yes
Explains temperature changes 1 AD—1800	No	Maybe
Explains temperature stasis 1998—2015	Not really	Yes
Temperature prediction for next decade	+0.2°C	−0.3°C
Explains "natural variation" in temperatures	No	Maybe
Influence of direct heating effect of TSI changes	Minor	Minor
Influence of Sun other than by direct heating	None	Major
Influence of albedo	Minor	Major
Equilibrium climate sensitivity (ECS) to CO_2	1.5—4.5°C	Less than 0.5°C
Indirect solar sensitivity (ISS)	Zero	$1.7 \pm 0.2°C\,W^{-1}\,m^2$
Requires amplification by feedbacks	Yes	No
Detailed mechanism hypothesized	Yes	No

APPENDIX A: ACRONYMS

ASR	A	Absorbed solar radiation
CO_2		Carbon dioxide
EDA		Externally-driven albedo
ISS		Indirect solar sensitivity
TSI	S	Total solar irradiance

Acknowledgments

A big thank-you to numerous readers at the *joannenova.com.au* blog for commenting on the work in progress in mid-2014, making many useful suggestions. Peter Sinclair proposed the possibility of meteoritic dust. Bernard Hutchins Jr. rightly insisted that a notch filter can be causal. "Sun-Sword" found the Yoshimura paper. Joanne Nova noticed the coincidence of notching with the reversal of the Sun's magnetic field, and realized its significance. I am grateful to Joanne Nova, David Stockwell, Bob Carter, Geoff Sharp, Garth Paltridge, David Archibald, Craig Loehle, Christopher Monckton, Stephen Farish, Michael Cejnar, Nir Shaviv, Michael Hammer, Tim Channon, and Luboš Motl for discussions or advice. Professor Ronald Bracewell, late of Stanford University, put me on the path of simplifying and decluttering the Fourier transform, without which this project might have failed.

The author has no conflict of interest. He receives no monies or employment from any government or other organization, only donations to joannenova.com.au.

References

Archibald, D., 2010. Retrieved 10 1, 2013, from David Archibald .info: http://www.davidarchibald.info/papers/Past-and-Future-of-Climate.pdf.

Butler, C.J., Johnston, D.J., 1994. The link between the solar dynamo and climate — the evidence from a long mean air temperature series from northern Ireland. Infrared Astronomical Journal J.21, 251–254.

Christiansen, B., Ljungqvist, F.C., 2012. The extra-tropical Northern Hemisphere temperature in the last two millennia: reconstructions of low-frequency variability. Climate of the Past. http://dx.doi.org/10.5194/cp-8-765-2012.

Climate and Clouds: Cloud data., n.d. Retrieved from Climate4You: http://www.climate4you.com/ClimateAndClouds.htm.

Cloud Analysis Part 1: Cloud Amount., n.d. Retrieved 11 8, 2014, from International Satellite Cloud Climatology Project: http://isccp.giss.nasa.gov/climanal1.html.

Damon, P.E., Laut, P., 2004. Pattern of strange errors plagues solar activity and terrestrial climate data. Eos, Transactions American Geophysical Union 85 (39), 370–374.

Dikpati, M., Gilman, P., de Toma, G., 2008. The Waldmeier Effect: an artifact of the definition of wolf sunspot number? Astrophysical Journal 673, L99–L101.

Douglass, D.H., Clader, D.B., 2002. Climate sensitivity of the Earth to solar irradiance. Geophysical Research Letters 29 (16). http://dx.doi.org/10.1029/2002GL015345.

Duffy, P.B., Santer, B.D., Wigley, T.W., 2009. Solar variability does not explain late-20th-century warming. Physics Today.

Ellyet, C., 1977. Solar influence on meteor rates and atmospheric density variations at meteor heights. Geophysical Research. http://dx.doi.org/10.1029/JA082i010p01455.

Eschenbach, W., April 10, 2014. Solar Periodicity. Retrieved April 10, 2014, from Watts Up With That: http://wattsupwiththat.com/2014/04/10/solar-periodicity/.

Evans, D.M., 2016. Correcting Problems with the Conventional Basic Calculation of Climate Sensitivity.

Friis-Christensen, E., Lassen, K., 1991. Length of the solar cycle; an indicator of solar activity closely associated with climate. Science 254 (2032), 698–700.

Friis-Christensen, E., Svensmark, H., 2004. Comments on Damon and Laut. Retrieved January 15, 2014, from dtu.dk: http://wwwx.dtu.dk/upload/institutter/space/forskning/05_afdelinger/sun-climate/full_text_publications/comment%20to%20eos_28_sept_04.pdf.

Fujii, Y., April 2011. The role of atmospheric nuclear explosions on the stagnation of global warming in the mid 20th century. Journal of Atmospheric and Solar-Terrestrial Physics 73 (5–6), 643–652.

Herrera, V.V., Mendoza, B., Herrera, G.V., 2014. Reconstruction and prediction of the total solar irradiance: from the Medieval Warm Period to the 21st century. New Astronomy 34 (2015), 221–233.

IPCC, 1990. First Assessment Report. Cambridge University Press.

IPCC, 2013. Fifth Assessment Report. Cambridge University Press.

Lockwood, M., October 27, 2004. Earth-Sun-Heliosphere Interactions Experiment: The Earthshine Mission, A. Science Case. Retrieved October 7, 2013, from. ftp://ftp.ukssdc.ac.uk/pub/lockwood/ESHINE_Team/EARTHSHINEscience_case_Oct04.pdf.

Lockwood, M., Froehlich, C., 2007. Recent oppositely directed trends in solar climate forcings and the global mean surface air temperature. Proceedings of the Royal Society. http://dx.doi.org/10.1098/rspa2007.1880.

Mann, M.E., Jones, P.D., 2003. Global surface temperatures over the past two millennia. Geophysical Research Letters 30, 1820–1824. http://dx.doi.org/10.1029/2003GL017814.

McCoy, D.T., Burrows, S.M., Wood, R., Grosvenor, D.P., Elliot, S.M., Ma, P.-L., et al., 2015. Natural aerosols explain seasonal and spatial patterns of Southern Ocean cloud albedo. Science Advances. http://dx.doi.org/10.1126/sciadv.1500157.

McCracken, K.G., Beer, J., Steinhilber, F., 2014. Evidence for planetary forcing of the cosmic ray intensity and solar activity throughout the past 9400 years. Solar Physics. http://dx.doi.org/10.1007/s11207-014-0510-1.

Moffa-Sanchez, P., Born, A., Hall, I.R., Thornalley, D.J., Barker, S., 2014. Solar forcing of North Atlantic surface temperature and salinity over the past millennium. Nature Geoscience, Supplementary Information.

Pinker, R.T., Zhang, B., Dutton, E.G., May, 06, 2005. Do satellites detect trends in surface solar radiation? Science 38, 850–854.

Scafetta, N., West, B.J., 2009. Interpretations of climate-change data. Physics Today 62 (11), 8.

Schwartz, S.E., 2012. Determination of Earth's transient and equilibrium climate sensitivities from observations over the twentieth century: strong dependence on assumed forcing. Surveys in Geophysics (Special Issue).

Sharp, G.J., 2013. Are Uranus & Neptune responsible for solar grand minima and solar cycle modulation? International Journal of Astronomy and Astrophysics 260–273. http://dx.doi.org/10.4236/ijaa.2013.33031.

Shaviv, N.J., 2008. Using the oceans as a calorimeter to quantify the solar radioactive forcing. Journal of Geophysical Research 113, A11.

Solheim, J.-E., Stordahl, K., Humlum, O., 2012. The long sunspot cycle 23 predicts a significant temperature decrease in cycle 24. Journal of Atmospheric and Solar-Terrestrial Physics.

Soon, W.W.-H., 2009. Solar Arctic-mediated climate variation on multidecadel to centennial timescales: empirical evidence, mechanistic explanation, and testable consequences. Physical Geography 30 (2), 144–184.

Stober, D., 2010. The Strange Case of Solar Flares and Radioactive Elements. Retrieved April 11, 2014, from Stanford News. http://news.stanford.edu/news/2010/august/sun-082310.html.

Usoskin, I.G., Schuessler, M., Solanki, S.K., Mursula, K., 2004. Solar activity over the last 1150 years: does it correlate with climate? In: Favata, F., Hussain, G., Battrick, B. (Eds.), Proc. The 13th Cambridge Workshop on Cool Stars, Stellar Systems and the Sun, pp. 19–22. Hamburg: ESA SP-560, Jan. 2005.

Voiculescu, M., Usoskin, I., Condurache-Bota, S., 2013. Clouds blown by the solar wind. Environmental Research Letters.

Wilde, S., 2010, 2015. Is the Sun Driving Ozone and Changing the Climate? Retrieved from JoNova: http://joannenova.com.au/2015/01/is-the-sun-driving-ozone-and-changing-the-climate/.

Wilson, I.R., 2011. Are changes in the Earth's rotation rate externally driven and do they affect climate? General Science Journal.

Wilson, I.R., 2013. The Venus–Earth–Jupiter spin–orbit coupling model. Pattern Recogn. Phys. 1, 147–158.

Woollings, T., Lockwood, M., Masato, G., Bell, C., Gray, L., 2010. Enhanced signature of solar variability in Eurasian winter climate. Geophysical Research Letters 37, L20805. http://dx.doi.org/10.1029/2010GL044601.

Yoshimura, H., 1996. Coupling of total solar irradiance and solar magnetic field variations with time lags: magneto-thermal pulsation of the Sun. In: Astronomical Society of the Pacific, ASP Conference Series, vol. 95, pp. 601–608.

CLIMATE MODELS

20

Correcting Problems With the Conventional Basic Calculation of Climate Sensitivity

D.M.W. Evans
Science Speak, Perth, Australia

OUTLINE

Evidence-Based Climate Science, Second Edition
http://dx.doi.org/10.1016/B978-0-12-804588-6.00020-3

1. INTRODUCTION

The conventional basic climate model is used to calculate the sensitivity of surface temperature to carbon dioxide (CO_2) (see the appendix for acronyms), expressed as the equilibrium climate sensitivity (ECS)—the surface warming due to a doubling of the concentration of atmospheric CO_2.

Predating computer simulations, the conventional basic climate model applies "basic physics" to the climate. The idea that "it's the physics" makes the carbon dioxide theory impregnable in the minds of many. The basic model ignited concern about carbon dioxide; without it we probably would not be too worried. It is distinct from the global circulation models (GCMs). Dating back to 1896 with Arrhenius (Weart, 2015), it is described in some detail in the Charney Report (Charney, et al.,1979, pp. 7—9), the seminal document that ushered in the current era of concern about CO_2. We start this chapter with a complete presentation of the basic model in its modern form.

The fundamental predictions of modern climate science are failing; there is the stubborn fact of the "pause," the water vapor emissions layer did not ascend when the surface was warming in the 1980s and 1990s (the "missing hotspot"), and temperature does not follow CO_2 in the ice cores. Yet despite the empirical evidence, many climate scientists are convinced that the current theory is correct. This certainty does not rest on opaque computer simulations but rather on the conventional basic climate model. The basic model is the cornerstone of concerns about CO_2 and the understanding that CO_2 is the main cause of modern warming.

The basic climate model, like any model, simplifies reality by making approximations that seem reasonable. It appears plausible that it captures the essence of the situation—the model relies on the same techniques used in countless physical science models. But note the survivorship bias—those models have been tested against reality and found to be useful.

In modeling, it is difficult to know before testing whether a model will happen to work well enough. It is often impossible to know in advance the impact of the errors introduced by the inevitable approximations, or whether anything vital has been omitted. In fields where experiments can be performed quickly, such as chemistry or electronics, a model is tested within hours or days and quietly discarded if it turns out not to work. But climate changes slowly, so testing the basic climate model has taken decades.

The usual approach of those questioning establishment climate science has been to dispute the values of the parameters in the model but accept the model architecture. Here we do the converse—we agree with the physics of modern climate science and accept all the Intergovernmental Panel on Climate Change's (IPCC) values such as feedback values, and instead we critique the *architecture* of the conventional basic climate model, how the parameters are combined to estimate the ECS.

This chapter states the basic model in full, examines it for flaws, finds two major problems, develops an alternative model without those problems, and estimates the ECS.

2. THE CONVENTIONAL CALCULATION OF CLIMATE SENSITIVITY

This section describes the conventional basic climate model and the calculation of the ECS. This account is based on less complete descriptions in Held and Soden (2000; hereafter HS00), Pierrehumbert (2010, pp. 163—165; Ph10), and IPCC, Fifth Assessment Report (2013; AR5).

The models in this chapter are for the passage of Earth from one steady state to another. A variable (eg, X) in the initial steady state has a "0" subscript (eg, X_0), while the change from the initial to the final steady state is prefixed with a "Δ" (eg, ΔX). In steady state the outgoing long-wave radiation (OLR) R matches the absorbed solar radiation (ASR) A,

$$A = R, \tag{20.1}$$

in what is known as "energy balance" or "radiation balance" (both are $\sim 239 \, W/m^2$). Thus

$$\Delta A = \Delta R. \tag{20.2}$$

2.1 Balance Between Temperature and CO_2

Consider the simplest of hypothetical situations, where only the CO_2 concentration and the temperature can change, other climate drivers are held constant, and there are no feedbacks.

The CO_2 concentration is represented by C. The amount of OLR blocked by CO_2 rises logarithmically with C, so we represent the CO_2 concentration by its base-2 logarithm L:

$$\Delta L = L - L_0 = \log_2 C - \log_2 C_0 = \log_2 \left(\frac{C}{C_0} \right) = \log_2 \left(\frac{C_0 + \Delta C}{C_0} \right). \tag{20.3}$$

The mean tropospheric temperatures are all considered to change uniformly (HS00), so if T is any mean tropospheric temperature then ΔT is equal to ΔT_S, where T_S is the mean surface temperature.

R depends on T_S and L, and everything else is held constant, so schematically R is $R(T_S,L)$. Thus

$$dR = \frac{\partial R}{\partial T_S} dT_S + \frac{\partial R}{\partial L} dL. \tag{20.4}$$

The ratio of ΔR to ΔT_S for small changes when everything is held constant, except OLR and the mean tropospheric temperatures, is $\partial R/\partial T_S$. AR5 (p. 818) calls this value the Planck feedback (beware: it is not a "feedback" in the way that the term is used below or generally used, as something that affects what caused it) and gives its value as $3.2 \pm 0.1 \, (W/m^2)/°C$. We represent it here by its reciprocal, the Planck sensitivity:

$$\lambda_0 = \frac{1}{\partial R/\partial T_S} \simeq \frac{1}{3.2 \pm 0.1} = 0.31 \pm 0.01 \, °C/(W/m^2). \tag{20.5}$$

The ratio of $-\Delta R$ to ΔL for small changes when everything else is held constant, except OLR and the CO_2 concentration, is $-\partial R/\partial L$. Because ΔL is unity when CO_2 doubles, it is the decrease in OLR emitted by CO_2 molecules per doubling of the CO_2 concentration when everything else is held constant, which AR5 (p. 8SM-7) gives as

$$D_{R,2X} = -\frac{\partial R}{\partial L} = 3.7[3.5, 4.1] \, W/m^2. \tag{20.6}$$

With these values for the partial derivatives, Eq. (20.4) now describes the change in OLR due to changes in T_S and L. Consider a move between two steady states: assuming Earth stays close to steady state as CO_2 changes, the small changes dX become small changes ΔX. A is independent of T_S and L (recall that there are no feedbacks), so ΔA is zero, so ΔR is also zero, by Eq. (20.2), so Eq. (20.4) becomes

$$0 = \frac{\partial R}{\partial T_S} \Delta T_S + \frac{\partial R}{\partial L} \Delta L. \tag{20.7}$$

Plugging in the values of the partial derivatives, the surface warming is thus

$$\Delta T_S = \lambda_0 D_{R,2X} \Delta L. \tag{20.8}$$

The increase in OLR due to tropospheric warming is balanced by the decrease in OLR due to extra CO_2. In this situation the doubling of CO_2 can be viewed as a two-stage process: (1) CO_2 doubles, holding everything else constant, which reduces OLR by $D_{R,2X}$; then (2) the troposphere warms, holding everything else constant, exactly enough to restore the OLR back to its original level. Or the two stages can be considered as two processes that occur simultaneously without interfering with each other, and we add the changes in OLR they cause (the climate is linear for small changes, so processes and changes superpose).

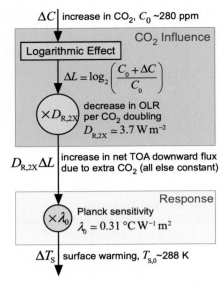

FIGURE 20.1   Conventional basic climate model for the hypothetical situation where only temperature and CO_2 can vary.

The "no feedbacks ECS," the warming per doubling of CO_2 when there are no other drivers and no feedbacks, is thus

$$\text{ECS}_{\text{NF}} = \Delta T_{\text{S}}|_{\Delta L=1} = \lambda_0 D_{R,2X} \simeq (3.2 \pm 0.1)^{-1} \times 3.7[3.5, 4.1] = 1.16[1.08, 1.29] \ ^\circ\text{C}. \tag{20.9}$$

For extra clarity about connections and operations, this chapter makes liberal use of diagrams of computations, which provide more insight than bunches of simultaneous equations, as long known by those dealing with electrical circuits. In these diagrams arrows indicate the direction of computation or information flow. Fig. 20.1 illustrates the analysis of this situation, wherein Eqs. (20.3) and (20.8) provide the computational path from ΔC to ΔT_{S}.

2.2 Balance Between Temperature and Multiple Drivers, With Feedbacks

This generalizes the previous subsection by considering any number of independent climate drivers that do not depend directly on temperature. It also adds feedbacks, which, while independent of all the variables under consideration, are dependent on temperature—so as the Earth warms they further affect ASR or OLR, which affects warming, which further affects these feedback variables, and so on. This is the full version of the conventional basic climate model.

Consider the hypothetical situation where only the temperature and the climate variables $V_1,\ldots,V_n$ and $U_1,\ldots,U_m$ can change, $n, m \in \{1, 2, \ldots\}$, with other drivers held constant and no other feedbacks. Each variable V_i or U_i is not directly dependent on any of the other variables. Each driver variable V_i is not directly dependent on the temperature, but each feedback variable U_i depends directly on the temperature. Some variables might change the relativities between various mean atmospheric temperatures, but without loss of generality we will take the temperature to be T_{S}. Variables usually considered to be drivers include the solar constant, externally driven albedo (EDA), and the CO_2 concentration. Variables usually thought of as feedbacks include surface albedo, lapse rate, humidity, the average height of the water vapor emissions layer, and the average height of cloud tops. The previous subsection is the special case where n equals 1, m is zero, and V_1 is L.

The net top-of-atmosphere (TOA) downward flux is

$$G = A - R. \tag{20.10}$$

G depends on T_{S} as well as on each of $V_1,\ldots,V_n, U_1,\ldots,U_m$ while everything else is held constant, so schematically G is $G(T_{\text{S}},V_1,\ldots,V_n, U_1,\ldots,U_m)$. Thus

$$dG = \frac{\partial G}{\partial T_{\text{S}}}dT_{\text{S}} + \sum_{i=1}^{n} \frac{\partial G}{\partial V_i}dV_i + dT_{\text{S}} \sum_{i=1}^{m} \frac{\partial G}{\partial U_i} \frac{dU_i}{dT_{\text{S}}}. \tag{20.11}$$

Consider a move between two steady states: assuming Earth stays close to steady state as CO_2 changes, the small changes dX become small changes ΔX. By radiation balance Eq. (20.2) ΔG is zero and $\partial G/\partial T_S$ is $-\lambda_0^{-1}$ because $\partial A/\partial T_S$ is zero and by Eqs (20.5) and (20.10), so

$$\frac{\Delta T_S}{\lambda_0} = \sum_{i=1}^{n} \frac{\partial G}{\partial V_i} \Delta V_i + \Delta T_S \sum_{i=1}^{m} \frac{\partial G}{\partial U_i} \frac{dU_i}{dT_S}. \tag{20.12}$$

Ignoring the feedbacks, this says that the increase in OLR due to warming is balanced by an increase in net downward flux (ie, increases in ASR plus decreases in OLR) due to changes in the drivers. Changes in $V_1,...,V_n$ can be viewed as an $n+1$-stage process: in stage i variable V_i changes by ΔV_i holding everything else constant, which increases the net TOA downward flux G by $(\partial A/\partial V_i - \partial R/\partial V_i)\Delta V_i$, then, in stage $n+1$, T_S increases by ΔT_S holding everything else constant, which increases OLR by exactly enough to restore the net change in G to zero. These $n+1$ stages can be viewed as simultaneous processes that do not interfere with each other, adding their changes in G because the climate is linear for small changes (which thus superpose).

Feedbacks complicate the balance by adding more downward flux due to changing values of $U_1,...,U_m$ in response to changes in T_S, which have to be balanced by more OLR due to a further change in T_S, and so on in a feedback loop. Solving Eq. (20.12) for ΔT_S,

$$\Delta T_S = \frac{\sum_{i=1}^{n} \frac{\partial G}{\partial V_i} \Delta V_i}{\frac{1}{\lambda_0} - \sum_{i=1}^{m} \frac{\partial G}{\partial U_i} \frac{dU_i}{dT_S}} = \frac{\lambda_0}{1 - f\lambda_0} \sum_{i=1}^{n} \frac{\partial G}{\partial V_i} \Delta V_i \tag{20.13}$$

where the total feedback is

$$f = \sum_{i=1}^{m} \frac{\partial G}{\partial U_i} \frac{dU_i}{dT_S}. \tag{20.14}$$

This expression for ΔT_S is the same as it is without feedbacks (ie, when f is zero) except that it is multiplied by $1/(1 - f\lambda_0)$, which is the distinctive form of a feedback f around a multiplier λ_0; apply Fig. 20.2 with a equal to λ_0 and b equal to f.

In Eq. (20.13) the contribution to ΔT_S of every driver shares a part in common: the feedback-modified Planck sensitivity $\lambda_0/(1 - f\lambda_0)$. This inspires the concept of the "forcing" of a driver, as the part peculiar to the driver variable. The (instantaneous radiative) forcing of a change ΔV_i in the driver V_i is the resulting change in net TOA downward flux, namely

$$\Delta F_i = \frac{\partial G}{\partial V_i} \Delta V_i = \left(\frac{\partial A}{\partial V_i} - \frac{\partial R}{\partial V_i} \right) \Delta V_i \tag{20.15}$$

(AR5, Section 8.1.1.1 and Fig. 8.1; for changes between steady states, this is the appropriate form of forcing). The contribution ΔV_i makes to surface warming is the product of its forcing and the feedback-modified Planck sensitivity:

$$\Delta T_{S,i} = \frac{\lambda_0}{1 - f\lambda_0} \Delta F_i. \tag{20.16}$$

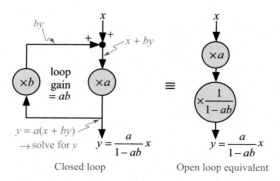

FIGURE 20.2 The closed-loop and open-loop forms of a feedback circuit are equivalent.

The radiation imbalance is the total change in downward flux before feedbacks, or the sum of forcings due to the drivers before feedbacks (ie, when f is zero):

$$\Delta I = \sum_{i=1}^{n} \Delta F_i = \sum_{i=1}^{n} \frac{\partial G}{\partial V_i} \Delta V_i = \sum_{i=1}^{n} \left(\frac{\partial A}{\partial V_i} - \frac{\partial R}{\partial V_i} \right) \Delta V_i. \tag{20.17}$$

(Incidentally, most drivers directly affect either ASR or OLR but not both, so usually either $\partial A/\partial V_i$ or $\partial R/\partial V_i$ is zero.)

The situation is illustrated in Fig. 20.3. Here the first driver V_1 is the no-feedbacks ASR A_{NF}, the portion of ASR unaffected by feedbacks to surface warming. (Moving between steady states, changes in ASR are due either to ΔA_{NF} or to albedo feedbacks to surface warming.) The second driver V_2 is the logarithm of the CO_2 concentration, namely L. The computational paths from the drivers to ΔT_S are provided by Eq. (20.13), noting that $\partial A/\partial A_{\text{NF}}$ is unity, $\partial R/\partial A_{\text{NF}}$ is zero, $\partial A/\partial L$ is zero, and $-\partial R/\partial L$ is $D_{R,2X}$ [by Eq. (20.6)].

Adding the forcings and applying the feedbacks in Fig. 20.3, the surface warming is

$$\Delta T_S = \frac{\lambda_0}{1 - f\lambda_0} \Delta I \tag{20.18}$$

where

$$\Delta I = \Delta A_{\text{NF}} + D_{R,2X}\Delta L + \sum_{i=3}^{n} \frac{\partial G}{\partial V_i} \Delta V_i. \tag{20.19}$$

Each driver adds its forcing to the input of the Planck sensitivity.

AR5 (Table 9.5 and Fig. 9.43, and p. 591) reports the individual and total feedbacks from the CMIP5 as. in $(\text{W/m}^2)/°C$: water vapor $+1.6 \pm 0.3$, lapse rate -0.6 ± 0.4, water vapor and lapse rate combined $+1.1 \pm 0.2$, surface albedo $+0.3 \pm 0.1$, cloud $+0.3 \pm 0.7$, and total

$$f = 1.7[0.97, 2.43] \ (\text{W/m}^2)/°\text{C}. \tag{20.20}$$

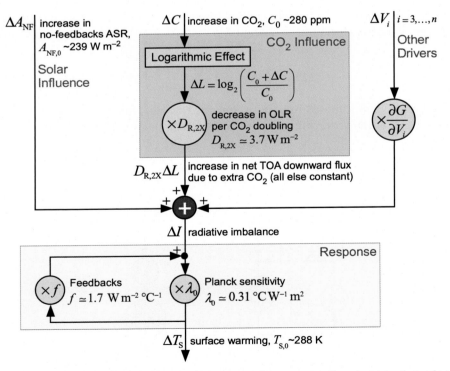

FIGURE 20.3 Conventional basic climate model, for changes from one steady state to another. Assumes that only temperature, the drivers (no-feedbacks ASR, CO_2, and others) and the feedbacks can vary, and that they do not depend directly on one another except that the feedbacks depend only on surface temperature. This is a radiation-balance architecture: the radiation imbalances (or forcings) due to the drivers are added at the purple circle, then balanced by the increase in feedbacks-adjusted OLR due to surface warming.

(Incidentally, the loop gain, the total amplification going once around the feedback loop, is

$$f\lambda_0 \simeq 1.7[0.97, 2.43] \times (3.2 \pm 0.1)^{-1} = 0.53[0.30, 0.76]. \tag{20.21}$$

It is less than unity. If it were greater than or equal to unity, then the infinite sum implicit in the denominator in Eq. (20.18) would not converge and the computed ΔT_S would be infinite.)

2.3 The Conventional Calculation of the Equilibrium Climate Sensitivity

The ECS is the surface warming ΔT_S when the CO_2 concentration doubles (ie, ΔL is unity) and other drivers are unchanged:

$$\text{ECS} = \frac{\lambda_0}{1 - f\lambda_0}D_{R,2X} = \frac{D_{R,2X}}{\lambda_0^{-1} - f} \simeq \frac{3.7[3.5, 4.1]}{3.2 \pm 0.1 - 1.7[0.97, 2.43]} \simeq 2.5[1.24, 3.7] \, °\text{C}. \tag{20.22}$$

This accords with AR5 (p. 1033), which finds the ECS as likely to be 1.5–4.5°C for a doubling of equivalent CO_2 concentration. Setting f to zero in Eq. (20.22) gives the no-feedbacks ECS of Eq. (20.9).

3. PROBLEMS WITH THE CONVENTIONAL BASIC CLIMATE MODEL

This section presents three problems with the conventional basic model used to estimate climate sensitivity. Noting that there are problems is not new, but apparently itemizing them is unusual: from Sherwood et al. (2015), "While the forcing—feedback paradigm has always been recognized as imperfect, such discrepancies have previously been attributed to variations in 'efficacy' (Hansen et al., 1984), which did not clarify their nature."

3.1 Analysis by Partial Derivatives Is Problematic

The ensemble of climate variables forms a rich web of feedbacks and indirect interconnections: "in climate, everything depends on everything." Consequently it is not possible to vary only one variable and allow the OLR or ASR to vary, while holding all other feedbacks constant, as required for the partial derivatives in the model to exist; see Eqs. (9.4) and (9.11). The partial derivatives in question include λ_0 and $D_{R,2X}$ [Eqs. (20.5) and (20.6)].

The conventional calculation of ΔT_S must therefore be treated with suspicion. One might argue that the partial derivatives are good approximations, but this is an unknowable assertion because the partial derivatives are hypothetical quantities—so they cannot be empirically verified and using them incurs an unknown amount of error.

3.1.1 The Required Partial Derivatives Do Not Exist

When partial differentiation is taught, the variables are nearly always independent. The archetypical example is a function of orthogonal spatial coordinates x and y. One can move independently in the x or y directions; although x and y might be tied to together by a constraint in some situations, in general one can hold x constant while changing y and vice versa.

When a quantity depends on a set of mutually dependent variables, a partial derivative of the quantity "has no definite meaning" (Auroux, 2010, who gives a worked example), because of ambiguity over which variables are truly held constant and that change because they depend on the variable allowed to change. The problem in the basic climate model is not ambiguity—Eqs. (20.4) and (20.11) are unambiguous about which variables are to be held constant—and no amount of mathematical notation or redefinition of functions can hide the physics [note that Eq. (20.4) applies in a hypothetical situation where all feedbacks are held constant]. One simply cannot hold all the feedbacks constant as required by the basic climate model while allowing only, say, surface temperature and OLR to vary.

Consider the situation of Section 2.2, wherein the net TOA downward flux G is a function of T_S, n drivers V_i, and m feedbacks U_i. For brevity, let $\mathbf{W}$ denote $V_1,...,V_n,U_1,...,U_m$, so schematically G is $G(T_S,\mathbf{W})$. By definition, the partial derivative of G with respect to T_S at the initial steady state is

$$\frac{\partial G}{\partial T_S}(T_{S,0}, \mathbf{W}_0) = \lim_{h \to 0} \frac{G(T_{S,0} + h, \mathbf{W}_0) - G(T_{S,0}, \mathbf{W}_0)}{h}. \tag{20.23}$$

But there is no such climate state as $(T_{S,0} + h, \mathbf{W}_0)$, where the surface has warmed by h but all the other variables are as per the initial steady state, because each feedback variable U_i depends directly on T_S and so has also changed. Thus, technically $\partial G / \partial T_S$ does not exist.

3.1.2 *Cannot Hold Everything Constant Except for One Variable*

The notion of "holding everything else constant" can be ambiguous or arbitrary in climate. For example, when holding all else but T_S and the OLR constant, the mean tropospheric temperatures all change uniformly (HS00), but are stratospheric temperatures constant?

- If stratospheric temperatures remain constant, as per Soden and Held (2006, p. 3356), there is ambiguity at the tropopause—can temperature change on one side of the tropopause but not on the other? Is the height of the tropopause constant?
- If stratospheric temperatures are not constant but change uniformly with the tropospheric temperatures (because "everything else" includes lapse rates), at what height do the changes stop? Outer space does not warm.

Another example: when holding all else but T_S and OLR constant, in the troposphere the temperatures change with T_S so the water-holding ability of the air changes—are the specific or the relative humidities held constant?

3.2 Omitting Feedbacks That Are Not Temperature Dependent

The architecture of the conventional model is a radiation balance. For each climate driver V_i and the surface temperature, the increase in net TOA downward flux is computed, and of course these sum to zero. This arrangement is symmetric in the climate drivers and the surface temperature. There is asymmetry in the status of these variables, however, because the surface warming is unknown while the forcings for the other variables are known, as illustrated by the arrows in Fig. 20.3, which indicate what is computed from what.

When feedbacks are introduced, the conventional model only applies them to surface temperature: all the conventional feedbacks are in response to surface warming. This breaks the symmetry of the radiation balance. The unconventional but symmetric introduction of feedbacks is shown in Fig. 20.4—each driver, not just the surface

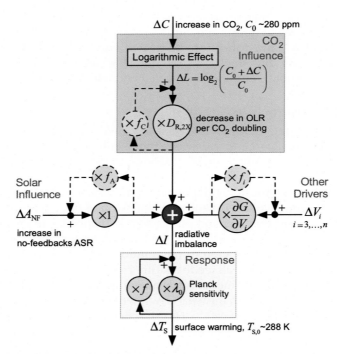

FIGURE 20.4   The symmetric application of feedbacks to the radiation-balance architecture: each driver and the surface temperature has its own specific feedbacks. The conventional model (Fig. 20.3) only has feedbacks in response to surface warming—it omits the dashed feedbacks. (The no-feedbacks ASR has no feedbacks, so f_A is zero.)

temperature, has its own feedbacks (which go in the opposite direction to existing information flow, because they feed *back*).

In the conventional model all the feedbacks are in response to changes in T_S: they are directly dependent on T_S but not on the climate drivers or other feedbacks. If there exists a feedback that responds to a climate driver, but is not triggered by some other driver or the surface warming it causes, then there is literally no place for it in the conventional architecture and it is omitted from the conventional basic climate model.

One might argue that any climate driver affects T_S, so modeling all feedbacks as responses to surface warming is adequate. However, if there is a driver that does *not* trigger a given feedback, then that feedback cannot be a response only to surface warming. It also implies the feedbacks do not "know" which driver caused them; for example, a surface warming of 0.2°C could be due to extra CO_2 (which leaves OLR constant, ignoring albedo feedbacks) or an increase in total solar irradiance (TSI; which increases OLR), but in the conventional model the feedback is identical in either case, a serious oversimplification physically.

These omissions could be remedied by adding driver-specific feedbacks to each driver, as in Fig. 20.4. A driver's feedbacks might then significantly change the "ultimate" forcing due to the driver. A feedback that responds only to changes in CO_2 is proposed in Section 4.

3.3 Unrealistic Physical Features

While no model is perfectly realistic, the clashes between the conventional model and physical reality are severe. These clashes are a direct result of the radiation-balance architecture, which suggests that something more than a radiation balance is going to be required to more realistically model the effect of increased CO_2.

3.3.1 *Interchangeability*

The conventional model computes the radiation imbalance ΔI, from which it calculates the surface warming $(1 - f\lambda_0)^{-1}\lambda_0\Delta I$. All of the information about the influences of the drivers is encapsulated in ΔI; the model is blind to anything about the drivers that is not in ΔI.

The various climate drivers are thus all interchangeable (or "fungible")—the influences of any two drivers that cause the same radiation imbalance (have the same forcing) are treated identically in the conventional model. Interchangeability is an inevitable consequence of the radiation balance architecture (Figs. 20.3 and 20.4).

Therefore the conventional model is structurally unable to distinguish warming due to extra ASR from warming due to extra CO_2: same forcing, same contribution to ΔI, so same ΔT_S, and same feedbacks. But these drivers are fundamentally quite different:

- Extra ASR increases the input of energy into the climate system. The extra energy is mainly directed to the surface, and it increases OLR.
- Extra CO_2 impedes the loss of energy from the climate system. The decreasing loss occurs in the higher atmosphere and does not change OLR (by energy balance, because CO_2 does not affect ASR, ignoring the minor effect of albedo feedbacks in response to surface warming).

The feedbacks in the conventional model only respond to surface warming, so the feedbacks to extra ASR and extra CO_2 are identical in the conventional model—they drive identical changes in the average height of the water vapor emission layer and cloud tops, changes in average lapse rate, etc. Given the physical differences involved, this seems implausible.

The concept of forcing relies on interchangeability to be useful, otherwise why bother adding forcings? The semantics of the word *forcing* obscure important differences: while increased CO_2 is obviously a "forcing," in that it forces the climate to change, it is not the same type of "forcing" as extra ASR; the latter changes OLR when steady state resumes, while the former does not (except through albedo feedbacks).

3.3.2 *Solar Response Applied to the CO_2 Influence*

In the conventional model all forcings are interchangeable, so they are all equivalent to extra ASR. Defining the "solar response" as the response to extra ASR, measured in °C per W/m² of forcing, the conventional model applies the solar response to every forcing. Increased ASR is the one forcing whose effects on OLR, before feedbacks, Arrhenius could be relatively certain of in 1896 when he estimated sensitivity to CO_2, via the Stefan–Boltzmann equation.

The conventional model expresses the Stefan–Boltzmann equation through the Planck feedback, the increase in OLR per unit of surface warming under the Planck conditions—namely that all else besides tropospheric temperature and OLR are held constant, there are no feedbacks, all tropospheric temperatures move uniformly, and stratospheric temperatures are unchanged (Soden and Held, 2006, pp. 3355−56; Supplementary Section 3). The Planck conditions are motivated by the use of partial derivatives in the conventional model (Section 3.1).

While 80% of the Planck feedback's value is due to the Stefan−Boltzmann equation, most of the rest is because the CO_2 and ozone emission layers are mainly in the stratosphere, whose temperatures are constant under the Planck conditions. Thus the amount of OLR they emit is less sensitive to changes in T_S than if they were in the troposphere at all relevant wavelengths. There is also a small effect due to the nonuniform distribution of surface temperature by time and latitude.

3.3.2.1 Natural Solar Response

To explore the solar response and the Stefan−Boltzmann equation *without* partial derivatives and the Planck conditions, let us consider the more natural application of the Stefan−Boltzmann equation to Earth. The Earth radiates all its heat to space as the OLR R. We *define* a quantity T_R such that

$$R = \sigma \varepsilon T_R^4, \tag{20.24}$$

where σ is the Stefan−Boltzmann constant (5.67×10^{-8} W/m^2 K^4) and ε is the Earth's emissivity (~ 0.995). This being the Stefan−Boltzmann equation, T_R is a temperature (~ 255K), hereafter called the "radiating temperature." While the Stefan−Boltzmann equation cannot be literally applied to Earth because there is no solid, uniform, isothermal surface that emits all the OLR, this definition of T_R effectively applies it. See Supplementary Section 3 for more detail.

T_R is numerically similar to the Earth's effective temperature T_e, the temperature of a *black* body that emits the same OLR as the Earth: R equals σT_e^4, so T_e is $\varepsilon^{1/4} T_R$ or $\sim 0.999 T_R$. The numerical difference between T_R and T_e is insignificant, but here we are concerned with OLR from the real Earth so it is more natural to use radiating temperature.

The Stefan−Boltzmann sensitivity (SBS) is defined as the slope of the T_R curve as a function of R in the Earth's current neighborhood, found by differentiating in Eq. (20.24):

$$\lambda_{SB} = \frac{dT_R}{dR} = \frac{1}{4\sigma\varepsilon T_R^3} = \frac{T_R}{4R} \simeq \frac{255}{4 \times 239} = 0.267\,°\text{C}/(\text{W/m}^2). \tag{20.25}$$

The SBS is the ratio of ΔT_R to the corresponding ΔR, in all circumstances. The slope of the Stefan−Boltzmann curve at the Earth's current state is regarded here as a constant because the Earth does not stray far from this point— the effect of a change in λ_{SB} is second order in the modeling here.

To discover ΔR and ΔT_R in the conventional model, we perform two rearrangements of Fig. 20.3: the first shows ΔR, and the second reveals ΔT_R. These rearrangements also suggest how to develop a better model.

3.3.2.2 First Rearrangement

ΔR is equal to ΔA by Eq. (20.2), and ΔA is the sum of ΔA_{NF} and the increase in ASR due to feedbacks in response to surface warming. So let us partition the feedbacks in response to surface warming into those that affect albedo, denoted by f_α (all the "surface albedo" and some of the "cloud" feedbacks), and those that do not, denoted by $f_{\bar\alpha}$:

$$f = f_\alpha + f_{\bar\alpha}, \tag{20.26}$$

where, from AR5 and sources referenced by AR5 (see Supplementary Section 1),

$$f_\alpha = 0.4 \pm 0.5\,(\text{W/m}^2)/°\text{C}, \tag{20.27}$$

$$f_{\bar\alpha} = 1.3 \pm 0.5\,(\text{W/m}^2)/°\text{C}. \tag{20.28}$$

Then Fig. 20.3 becomes Fig. 20.5.

(To compute ΔT_S in Fig. 20.5, first apply Fig. 20.2 with a equal to λ_0 and b to $f_{\bar\alpha}$ to form a multiplier $\lambda_0/(1 - f_{\bar\alpha}\lambda_0)$, then move the output of the albedo feedbacks to between the output of the purple adder and the Planck sensitivity, then reapply Fig. 20.2 with a equal to this multiplier and b to f_α:

$$\Delta T_S = \frac{\frac{\lambda_0}{1 - f_{\bar\alpha}\lambda_0}}{1 - f_\alpha \frac{\lambda_0}{1 - f_{\bar\alpha}\lambda_0}}\Delta I = \frac{\lambda_0}{1 - f\lambda_0}\Delta I. \tag{20.29}$$

This agrees with Eq. (20.18), so Figs. 20.3 and 20.5 are identical for computing ΔT_S.)

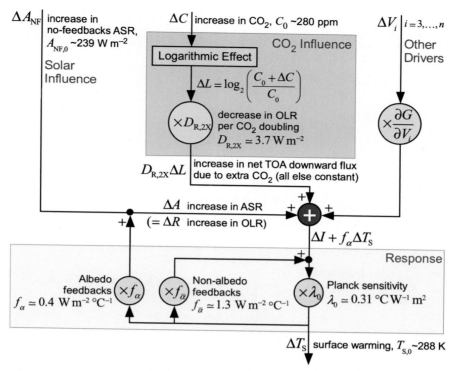

FIGURE 20.5 Conventional basic climate model, rearranged to explicitly show the increase in ASR and thus OLR. As per Fig. 20.3 except feedbacks in response to surface warming are partitioned by whether or not they affect albedo.

3.3.2.3 Second Rearrangement

To reveal ΔT_R, two changes are required. First, replace the Planck sensitivity λ_0 with λ_{SB}, because (1) we are interested in ΔT_R but λ_0 involves ΔT_S while λ_{SB} involves ΔT_R, and (2) λ_0 only applies under the Planck conditions while λ_{SB} always applies. So we define the sensitivity ratio as

$$\eta = \frac{\lambda_0}{\lambda_{SB}} = \frac{(3.2 \pm 0.1)^{-1}}{0.267} \simeq 1.17 \pm 0.04 \tag{20.30}$$

and replace λ_0 by $\eta\lambda_{SB}$.

Second, by its definition in Eq. (20.25), multiplication by the SBS only produces ΔT_R when it multiplies ΔR. So we convert the nonalbedo feedbacks into an equivalent open-loop multiplier e, so that they will not add to the input like in Fig. 20.5. Applying Fig. 20.2 to the loop consisting of the nonalbedo feedbacks and the Planck sensitivity in Fig. 20.5, we set

$$e = \frac{1}{\lambda_0} \frac{\lambda_0}{1 - f_{\bar{\alpha}}\lambda_0} = \frac{1}{1 - f_{\bar{\alpha}}\eta\lambda_{SB}} \simeq \frac{1}{1 - (1.3 \pm 0.5) \times (1.17 \pm 0.04) \times 0.267} = 1.7[1.3, 2.3]. \tag{20.31}$$

Then Fig. 20.5 becomes Fig. 20.6.

(To compute ΔT_S in Fig. 20.6, move the output of the albedo feedbacks to between the output of the purple adder and the SBS, then apply Fig. 20.2 with a equal to $\lambda_{SB}\eta e$ and b equal to f_α:

$$\Delta T_S = \frac{\lambda_{SB}\eta e}{1 - f_\alpha\lambda_{SB}\eta e}\Delta I = \frac{\lambda_0}{1 - f\lambda_0}\Delta I. \tag{20.32}$$

This agrees with Eqs. (20.18) and (20.29), so Figs. 20.3, 20.5, and 20.6 are identical for computing ΔT_S.)

3.3.2.4 Remarks

The Stefan–Boltzmann equation relates R to T_R, and its slope, λ_{SB}, relates ΔR to ΔT_R. Applied to Earth, it only says something about R, ΔR, T_R, and ΔT_R; it says nothing about other amounts of radiation or other temperatures. It is our primary means of converting between temperature and radiation.

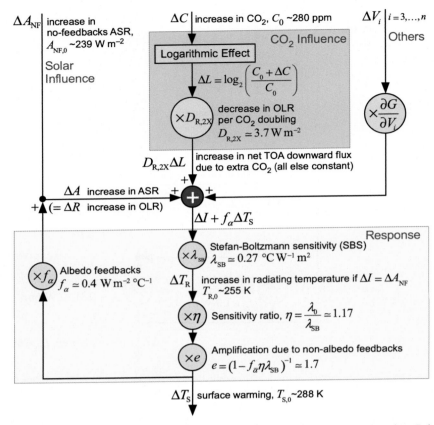

FIGURE 20.6   Conventional basic climate model, rearranged to explicitly show the OLR input to the Stefan–Boltzmann sensitivity.

For the move between two steady states, ΔA is equal to ΔR, which multiplication by λ_{SB} converts to ΔT_R, which in turn depends on the temperatures of the physical emission layers that emit OLR to space. Thus the SBS λ_{SB} relates the response of temperatures on Earth to the energy from the Sun; it describes the solar response of the Earth, before feedbacks.

In the conventional model, the influence of extra CO_2 is fed into the SBS (purple adder, Fig. 20.6). If there is no solar influence and no influence from the drivers marked "other," the radiation imbalance ΔI is $D_{R,2X}\Delta L$, which is input into the SBS along with the albedo feedback $f_\alpha \Delta T_S$. Fig. 20.6 shows that the conventional basic climate model applies the solar response to all climate influences, including CO_2.

While it is the inevitable result of the radiation-balance architecture, how realistic can it be to apply the Earth's solar response, its response to an increase in absorbed sunlight, to an increase in the amount of OLR blocked by a greenhouse gas? Shouldn't a response specific to the greenhouse gas be applied instead? Applying the solar response to nonsolar influences seems to be inviting problems, yet the conventional model allows only the one response to any influence, "one size fits all."

Notice that if the input to the SBS was only the solar influence, ΔA, then the output of the SBS would be the radiating temperature. Only under this condition does the SBS relate the increase in OLR to the increase in radiating temperature, which is the only thing it is qualified to do by the Stefan–Boltzmann equation. This tells us how to improve the model—do not apply the SBS or Planck sensitivity to the nonsolar forcings.

That the conventional model necessarily applies a specifically solar response to the influence of extra CO_2 may have tended to be overlooked because the usual view of the conventional model (Fig. 20.3) obscures ΔR and ΔT_R while entangling the albedo and nonalbedo feedbacks.

4. PROPOSED FEEDBACK: "REROUTING"

This section proposes a feedback in response to extra CO_2, which is omitted from the basic sensitivity calculation because it is not a response to surface warming. It is also omitted from current GCMs. Part of f_C in Fig. 20.4, it potentially reduces the ultimate radiation imbalance and surface warming due to increased CO_2.

4.1 Background

OLR has four main routes to space: emitted on the wavelengths of CO_2 emission and absorption by the CO_2 emission layer (CO_2EL), in the atmospheric window from the (near) surface or from the cloud tops, or on the wavelengths of water vapor emission and absorption from the water vapor emissions layer (WVEL). For brevity we use *pipe* here to mean a combination of wavelengths and emission layer, and call these four routes the "CO_2 pipe," "surface pipe," "cloud top pipe," and "water vapor pipe." For this section, all heat escapes the Earth through one of these four pipes (see Fig. 20.8 and Supplementary Section 3.4).

Increasing CO_2 impedes the flow of heat through the CO_2 pipe. When steady state is resumed the total OLR is the same as it was originally because the ASR is the same (ignoring the minor albedo feedbacks to surface warming). Increasing CO_2 redistributes the flow of OLR through the pipes but does not increase the total OLR.

For minor excursions from the climate of the current interglacial, a pipe's OLR is solely determined by the temperature of its emitting layer. The OLR in the surface pipe is determined by T_S, the OLR in the water vapor pipe is determined by the average temperature of the WVEL, which in turn is determined by its average height and the lapse rate, and so on. Knowing the rearrangement of OLR between the pipes gives the change in OLR in the surface pipe, and thus surface warming and the ECS.

In the conventional model, increasing CO_2 causes a sympathetic *decrease* in OLR in the water vapor pipe, due to amplification by water vapor feedbacks. Extra CO_2 is represented as a forcing, equivalent to extra ASR, which warms the surface, causing more evaporation and more water vapor, causing the WVEL to ascend because there is more water vapor coming from a warmer surface, whereupon the WVEL is cooler, which reduces the OLR in the water vapor pipe. Hence the surface and cloud top pipes must compensate for the OLR decreases through the CO_2 *and* water vapor pipes, by increasing their combined OLR by a matching amount. This requires much more surface warming than if the water vapor pipe also carried more OLR in response to the decreased OLR in the CO_2 pipe.

4.2 The Feedback

From the point of view of heat in the upper troposphere, increasing CO_2 makes it harder for photons to escape to space from CO_2 molecules because there are more CO_2 molecules above, and therefore *relatively* easier to escape in photons fired from water vapor molecules. Increased CO_2 thus increases the relative propensity of OLR to come from water vapor molecules. The energy has to escape to space somehow, the relative attractiveness of the CO_2 pipe has decreased compared to the water vapor pipe, and the heat is essentially available to all molecules because they swap energy back and forth by thermal collisions. If more OLR is being emitted from the water vapor molecules, the WVEL must have warmed.

Increasing the CO_2 concentration warms the upper troposphere because the emissions spectrum changes and there is more warming by downward emissions from the extra CO_2. This heats neighboring molecules, including water vapor molecules in the WVEL and some cloud tops, so more OLR is emitted by water vapor molecules and cloud tops. The WVEL emits more so it must be at a higher average temperature, due to a combination of warming by increased CO_2 and a decline in average height moving it to a warmer altitude.

4.3 Causes the WVEL to Descend

How does increased CO_2 affect the average height of the WVEL? Atmospheric water vapor is dynamic, so a possible mechanism involves meteorology. (In contrast, CO_2 is relatively static and well mixed, so radiative concerns are usually sufficient to explain its behavior.)

Upper tropospheric warming by increased CO_2 distorts the local lapse rate, which becomes less steep (less cooling per km of rise). The atmosphere around the WVEL altitude becomes warmer and more stable. The moist air rising by convection thus rises less vigorously and not as high, and so the average height of the WVEL declines. Because increasing CO_2 lowers the vigor of convection in the upper troposphere, humidity builds up and clouds condense at lower levels, suggesting the average height of the cloud tops declines.

This explanation of the lowering of the WVEL by rerouting relies only on the altered movements of water vapor due to increased CO_2 rather than on radiation transfer.

Note that it is possible for the WVEL to descend despite increased evaporation from the surface, if the extra water vapor is mainly confined to the lower troposphere and the consequent greater stability at low altitudes leads to less overturning and less transport of water vapor to the upper troposphere; indeed, this seems to be happening as reported by Paltridge et al. (2009) from a study of the better radiosonde data from 1973.

4.4 Comments

It is called the "rerouting feedback" because some fraction of the OLR that is blocked by rising CO_2 levels from escaping to space from CO_2 molecules is *rerouted* to space via emission from water vapor and cloud tops instead.

This rerouting takes place high in the atmosphere, far from the surface, so there is no place for it in the conventional basic climate model—it is in the blind spot of that model, which contains only feedbacks in response to surface warming. Perhaps a suitable variable to describe the strength of the feedback is the height of the CO_2EL plus the height of the WVEL.

The heat rerouted to space via water vapor molecules is not available to travel down and warm the surface, as in the conventional models. Thus the rerouting feedback reduces the impact of increasing CO_2 on surface warming. If this feedback is real and significant, it could help explain why CO_2 is not as potent as the IPCC supposes.

4.5 A Negative Feedback?

The rerouting feedback reduces the ultimate radiation imbalance due to extra CO_2, so it is a negative feedback in terms of its effect on the CO_2 forcing, so f_C is negative. Applying Fig. 20.2 with a equal to $D_{R,2X}$ and b to f_C, the rerouting feedback changes the radiation imbalance due to increasing CO_2 from $D_{R,2X}\Delta L$ to

$$\frac{1}{1 - f_C D_{R,2X}} D_{R,2X}\Delta L.$$

For example, if f_C was -0.6 then $\left(1 - f_C D_{R,2X}\right)^{-1}$ would be $\sim 30\%$, and the influence of extra CO_2 would be reduced by 70%.

Although the rerouting feedback reduces the sensitivity of T_S to changes in CO_2, and although f_C is negative, it is not a feedback in response to ΔT_S so it is not a "negative feedback" as that term is understood in the conventional paradigm.

Only CO_2 enrichment triggers the rerouting feedback, illustrating why the influence of extra CO_2 is not interchangeable with extra ASR (Section 3.3.1).

4.6 Energy Considerations

Consider in broad terms how the climate might adjust to a decrease in OLR in the CO_2 pipe. The blocked OLR has to find its way to space somehow. The resistance of the surface pipe to carrying more OLR is exceptionally high in the tropics because heat loss from the surface via evaporation rises exponentially with T_S (Kininmonth, 2010, elaborates on this). The resistance of the water vapor pipe to carrying more OLR might be relatively low because it requires only that the average height of the WVEL to decline by a few tens of meters. Like the WVEL, the cloud tops might descend slightly with little apparent energy requirement.

The energy required to warm the surface on a sustained basis, with the ocean warming that would entail, might be much greater than the energy required to change the average height of the WVEL or cloud tops sufficiently to change OLR by the same amount. This would suggest that the bulk of the response to the decrease in OLR escaping via the CO_2 pipe would come as more OLR from the WVEL or cloud tops, rather than from the surface, which is consistent with the proposed rerouting feedback and a lower ECS.

5. EXTERNALLY DRIVEN ALBEDO COULD BE SIGNIFICANT

Externally driven albedo (EDA) is albedo that is not attributable to feedbacks in response to surface warming. EDA includes any changes in albedo due to modulation by the Sun. The total change in albedo is the change in EDA plus the change in albedo due to feedbacks in response to surface warming.

The proportional-variation argument that follows suggests that changes in EDA have been a larger influence on T_S than the direct heating effect of changing TSI—the proportional variation in observed albedo, even after taking out the variation caused by albedo feedbacks to surface warming, is much larger than the proportional variation in TSI. Hence we will include EDA in our alternative model; EDA is omitted from conventional climate models.

5.1 Theory

Let ΔA_{EDA} be the increase in ASR due to increasing EDA; let ΔA_{TSI} be the increase in ASR due to the direct heating effect of increasing TSI. Then the increase in the no-feedbacks ASR, the increase in ASR that is independent of surface warming, is

$$\Delta A_{\text{NF}} = \Delta A_{\text{EDA}} + \Delta A_{\text{TSI}}. \tag{20.33}$$

Let the albedo α and the TSI S be α_0 ($\sim 30\%$) and S_0 ($\sim 1361 \text{ W/m}^2$) in the initial steady state, and let the change in EDA to the final steady state be $\Delta \alpha_E$. Then

$$\Delta A_{\text{NF}} = [1 - (\alpha_0 + \Delta \alpha_E)] \frac{(S_0 + \Delta S)}{4} - (1 - \alpha_0) \frac{S_0}{4}$$

$$\simeq \frac{\Delta S}{4} - \frac{1}{4}(\alpha_0 \Delta S + S_0 \Delta \alpha_E) \tag{20.34}$$

to first order. By comparison with Eq. (20.33),

$$\Delta A_{\text{EDA}} = -\frac{1}{4} S_0 \Delta \alpha_E$$

$$\Delta A_{\text{TSI}} = \frac{1}{4}(1 - \alpha_0) \Delta S. \tag{20.35}$$

Thus

$$\frac{\Delta A_{\text{EDA}}}{\Delta A_{\text{TSI}}} = -\frac{\Delta \alpha_E}{1 - \alpha_0} \bigg/ \frac{\Delta S}{S_0}$$

$$= -\frac{\alpha_0}{1 - \alpha_0} \left(\frac{\Delta \alpha_E}{\alpha_0} \bigg/ \frac{\Delta S}{S_0} \right) \tag{20.36}$$

$$\simeq -\frac{3}{7} \times \frac{\text{proportional variation in EDA}}{\text{proportional variation in TSI}}.$$

5.2 Data

TSI, known as "the solar constant" until 1979 when satellite measurements began, has varied less than 1.8 W/m^2, or 0.13%, over the 400 years since sunspot records began, according to the reconstruction of Lean (2000) with the background corrections of Wang et al. (2005). Some reconstructions favored by the IPCC (eg, by Kirvova or by Svalgaard) say the variation is even smaller.

Albedo data is sketchy but sufficient for a lower bound on proportional variation.

Palle et al. (2008) found some agreement between earthshine, the ISCCP FD product, and satellite (CERES) observations: from 1984 to 1998 the first two (CERES started in 2000) showed a fall in smoothed reflected solar radiation (upwelling SW) of $\sim 1\%$ or $\sim 1.0 \text{ W/m}^2$, then a rise by almost as much to mid-2000, then a roughly constant level to 2005 when the data stops. Some confirmation comes from Pinker et al. (2005), who found that ASR increased from 1984 to 1998 by $\sim 0.16 \text{ W/m}^2$ per year, suggesting a fall in reflected solar radiation of $\sim 2.2 \text{ W/m}^2$ (the 11-year smoothed TSI rose $\sim 0.3 \text{ W/m}^2$ in that period, accounting for only $0.3 \div 4 \times 0.7$ or $\sim 0.05 \text{ W/m}^2$), before leveling off then decreasing after 2000. We conclude there was a fall in reflected radiation of at least $\sim 1.0 \text{ W/m}^2$, which, since total reflected radiation is $\sim 100 \text{ W/m}^2$, corresponds to a change in albedo of $\sim 1\%$ (from say

30.0–29.7%). This is being conservative as even the CERES data from 2000 to 2004 (Stephens, O'Brien, Webster, Pilewski, Kato and Li, 2015, p. 4) shows variation of $\sim 2\,\mathrm{W/m^2}$ in reflected radiation, or albedo variation of $\sim 2\%$.

Albedo feedback from 1984 to 1998 due to surface warming is responsible for $f_\alpha \Delta T_S$, or $\sim (0.4 \pm 0.5) \times 0.35$ or $0.14 \pm 0.18\,\mathrm{W/m^2}$, leaving a fall of at least $0.86 \pm 0.18\,\mathrm{W/m^2}$ in reflected solar radiation.

Thus the proportional variation in EDA from 1984 to 1998 was at least $\sim 0.86 \pm 0.18\%$, from say 30.0% to $29.74 \pm 0.05\%$. These figures can only understate the relative variation of EDA, due to the relative shortness of the period from 1984 to 1998 compared to the last 400 years for TSI.

5.3 Conclusions

Plugging these proportional variations into Eq. (20.36),

$$\left| \frac{\Delta A_{\mathrm{EDA}}}{\Delta A_{\mathrm{TSI}}} \right| \geq \left| \frac{3}{7} \times \frac{0.86 \pm 0.18\%}{0.13\%} \right| \simeq 2.8[2.3, 3.3]. \tag{20.37}$$

ASR directly drives surface temperature, so the effect of changes in EDA on surface warming is at least twice as great as the direct effect of changes in TSI, and possibly much more. If the reconstructions that say the TSI variations are smaller are correct, then the relative influence of EDA is even larger.

6. ALTERNATIVE MODEL

This section develops an alternative model that fixes the conventional problems with omitting feedbacks other than to surface warming and with applying the solar response to all climate influences, and it ameliorates the partial derivative problem. It comes at the cost of requiring climate data to calculate the ECS.

6.1 Strategy

6.1.1 A New Organizing Principle

The conventional model is organized around a radiation balance; it adds the radiation imbalance ("forcing") caused by each climate driver to produce a total radiation imbalance. Due to the conventional interchangeability of climate drivers, this total radiation imbalance is equivalent, before feedbacks, to a net increase in ASR. The surface warming is then calculated, as that required under the Planck conditions to eliminate this imbalance with a net increase in OLR, after feedbacks, are applied. This leads to the architecture in Fig. 20.3 or Fig. 20.6; its nub is the adder (or summation mode) shown as a big purple circle, which sums the radiation imbalances due to the drivers.

That is, the essence of the conventional model is that it sums the radiation imbalances (forcings) due to the various climate influences.

The alternative model employs a different organizing principle. It adds the perturbation in surface temperature ("warming") caused by each climate driver, to produce a total surface warming. Its nub is a sum of warmings. The climate is widely assumed to be linear for the small temperature perturbations involved in global warming, so the temperature perturbations due to the various climate drivers are independent and superpose; that is, we can calculate the warming due to each driver independently of what else is happening, and the total warming is just the sum of the individual warmings. By the assumed linearity, the effects of one climate driver on other drivers are only second order. Warmings are additive if the various climate influences each warm the land and sea in approximately similar proportions.

That is, the essence of the alternative model is that it sums the surface temperature perturbations (warmings) due to the various climate influences.

6.1.2 A Sum-of-Warmings Model

The major advantage of the sum-of-warmings approach is that it allows tailored responses to different climate influences:

- It applies a specific CO_2 response to the influence of CO_2.
- It can contain CO_2-specific feedbacks, which respond to increasing CO_2 rather than to surface warming, such as the rerouting feedback.

6.1.3 Radiation Balance

Energy must balance between steady states, so the alternative model also applies a radiation balance. The radiation in a sum-of-warmings model is balanced simply by setting the increase in ASR in the model equal to the increase in OLR. (Thus the alternative basic climate model, like the conventional model, can only be applied between steady states.)

6.1.4 Solvability

A sum-of-warmings model results in one equation, in which the total surface warming is equated to the sum of the warmings due to the individual drivers.

The surface warming due to increasing CO_2 is characterized by a single CO_2 sensitivity parameter, λ_C, because that warming is assumed proportional to the forcing produced by the extra CO_2.

The increase in ASR for the last few decades is unknown because EDA is unknown, but it equals the increase in OLR, so instead we need to know the change in OLR.

Hence the single equation from the sum-of-warmings model, over an observed period in the last few decades, contains two unknown quantities: (1) the CO_2 sensitivity λ_C, and (2) the increase in OLR. With two unknowns, the equation is not yet solvable.

6.1.5 An OLR Model

The alternative model includes an OLR (sub) model that estimates the increase in OLR from changes in the physical emissions layers, due to surface warming, the increase in CO_2, and changes in the heights of the emissions layers, the lapse rate, and the cloudiness fraction. This drags a lot more data into the calculation, but it is perhaps the simplest way of determining the OLR, or at least bounding it.

The OLR data sets are not used directly because the data either isn't of sufficient resolution or sufficient length. The NOAA OLR data set, from 1974 to 2013, reads low (~ 232 W/m^2, while the true value is ~ 239 W/m^2), and that data is gridded and interpolated so to construct a sufficiently sensitive global mean that would require the assistance of the people who originally managed the data set. The CERES global OLR data set starts in 2000 and so is too short (though later, for short and recent observation periods, the CERES OLR figures are agreeable). A major advantage to using an OLR model, however, is that it gives more insight.

6.1.6 The Alternative Model

The alternative model combines a sum-of-warmings model with an OLR model. The OLR model provides the increase in OLR for the sum-of-warmings model, yielding a single equation in which the only unknown over a period of observations is λ_C. So we can estimate λ_C, and the model is complete, ready to estimate the ECS.

Unfortunately, the climate data is not good enough to form an estimate of the change in OLR over the last few decades; however, it is good enough to put meaningful bounds on the ECS. Essentially the emission-layer data puts a lower bound on the increase in OLR, which by energy balance puts a lower bound on the increase in ASR, which puts a lower bound on the component of surface warming due to the solar response, which (given the observed surface warming) puts an upper bound on the surface warming due to extra CO_2, which puts an upper bound on the ECS (see Section 8.2.1).

6.1.7 Comparison

The conventional model is based solely on a radiation balance. It calculates the ECS via a simplistic analysis primarily using laboratory physics: the Stefan–Boltzmann equation (modified for the Planck conditions), the OLR blocked by a doubling of CO_2 (spectroscopy), and the feedbacks to surface warming (mainly the properties of moist air).

The alternative model is based on a sum-of-warmings model *and* an OLR model. Using radiation balance to equate the changes in OLR and ASR, these two models are combined into a joint model, from which the CO_2 sensitivity λ_C can be estimated over observed periods. The ECS can then be estimated as the CO_2 response to a doubling of CO_2. Thus it calculates the ECS using a mix of laboratory physics and observed climate data.

6.2 The Sum-of-Warmings Model

In the sum-of-warmings model each climate driver has its own "response"— a combination of a specific sensitivity and specific feedbacks.

6.2.1 Solar Response

The solar response describes how the surface temperature responds to changes in ASR. The Earth's rate of warming is proportional to the difference between ASR and OLR, the Stefan–Boltzmann equation describes the OLR in terms of the Earth's radiating temperature T_R, and together they give a first-order linear differential equation that identifies the relationship between the ASR A and T_R as functions of time (Supplementary Section 4). This equation has the same structure as that of a simple low-pass filter, so the thermal inertia of the Earth is mimicked by a low-pass filter: T_R is a smoothed version of A, slow changes in A are faithfully reproduced in T_R, but faster changes are attenuated. For moves between steady states, the differential equation simplifies to just

$$\Delta T_R = \lambda_{SB}\Delta A. \tag{20.38}$$

How do we then get from ΔT_R to $\Delta T_{S,A}$, the surface warming due to an increase in ASR? The dependence of $T_{S,A}$ on T_R as a function of time is complicated: T_R depends on the temperatures of the various emission layers, one of which is the surface, and the rest of which are suspended in the atmosphere. The relationship between T_R and $T_{S,A}$ is thus mainly mediated by the atmosphere. The atmosphere is complicated and has many feedbacks. However, it acts and reacts quickly—usually within days, always within weeks—which allows a great simplification: on timescales of a year or more (such as moves between steady states, as the CO_2 concentration rises) and for small perturbations, $\Delta T_{S,A}$ is (presumably) proportional to ΔT_R. Accordingly, we model $\Delta T_{S,A}$ as

$$\Delta T_{S,A} = M\Delta T_R, \tag{20.39}$$

where M is the amplification of radiating temperature to surface multiplier. M describes the effects of all the feedbacks in response to surface warming except those that influence albedo, and is the open-loop form of the nonalbedo feedbacks in response to surface warming. Thus the solar response is

$$\Delta T_{S,A} = M\lambda_{SB}\Delta A. \tag{20.40}$$

This is the same form as the solar response in the conventional model, so by comparison with Fig. 20.6 when all inputs are zeroed except ΔA_{NF},

$$M = \eta e = (1.17 \pm 0.04) \times 1.7[1.3, 2.3] = 2.0[1.5, 2.7]. \tag{20.41}$$

In other words, the solar response is as per the conventional model with values from AR5, though we have replaced the Planck sensitivity with the Stefan–Boltzmann sensitivity.

6.2.2 CO₂ Response

The CO_2 response describes how the surface temperature responds to changes in CO_2 forcing. The conventional model applies the solar response to the CO_2 forcing (Figs. 20.3 and 20.6), which leads to various unrealistic physical features (Section 3.3). So, as noted in the remarks of Section 3.3.2, the CO_2 forcing should not add to the input of the Planck or Stefan–Boltzmann sensitivities. Also, the CO_2 forcing should not affect ΔT_R (and thus also OLR, because T_R is a proxy for OLR) except via surface warming and albedo feedbacks, because by linearity the various climate drivers do not significantly interfere with each other and ΔT_R is part of the solar response, and because increasing CO_2 merely redistributes OLR between the various pipes without changing the total amount of OLR or T_R (ignoring albedo feedbacks).

There being no existing appropriate modeling structure, we create one. The surface warming due to increased CO_2 is presumably proportional to the radiation imbalance $D_{R,2X}\Delta L$, so for small temperature perturbations let us define the CO_2 sensitivity λ_C by

$$\Delta T_{S,C} = \lambda_C D_{R,2X}\Delta L, \tag{20.42}$$

where $\Delta T_{S,C}$ is the surface warming due to the increase in CO_2. λ_C includes the rerouting feedback and any other CO_2-specific feedbacks. λ_C is positive.

6.2.3 Diagram

The complete sum-of-warmings model is shown in Fig. 20.7. Its essence is the adder shown as a big red circle, which sums the warmings caused by the various climate drivers. It also shows EDA as a separate input that, along with TSI, contributes to the no-feedbacks ASR input. This is *not* an energy balance model—energy balance is imposed later when we join it with the OLR model (Section 6.4) and set ΔA to. ΔR.

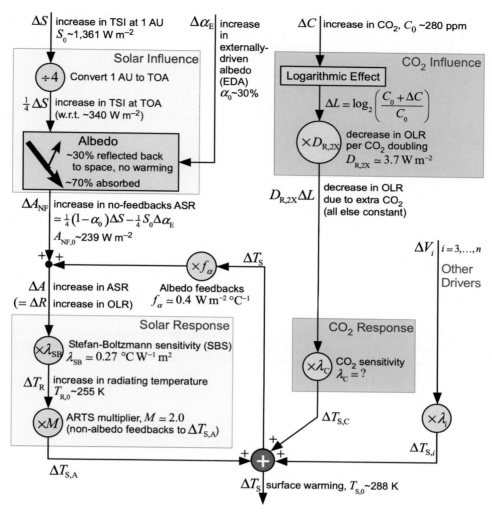

FIGURE 20.7 Sum-of-warmings model of the climate system, for changes from one steady state to another. It adds the warmings due to each climate driver, calculated independently of each other.

Consider the situation where there are only solar and CO_2 inputs ($n = 2$). The sum-of-warmings model (Fig. 20.7) is identical to the conventional model (Fig. 20.6) *except* that the CO_2 forcing is applied to its own specific CO_2 response instead of added to the input of the solar response along with the increase in no-feedbacks ASR and albedo feedbacks. Hence the sum-of-warmings is just one reconnection away from the conventional model, and it is not obvious that they are different architectures rather than just a mistaken connection! The essential difference is that in the sum-of-warmings model the solar and CO_2 influences are separate all the way to the surface warming, rather than entangled at the radiative stage as in the conventional model. The alternative model becomes the conventional model if the CO_2 sensitivity λ_C is set equal to the solar sensitivity $\lambda_{SB}M$.

6.3 The OLR Model

The OLR model, illustrated in Fig. 20.8 and developed in Supplementary Section 2, estimates the change in OLR as

$$\Delta R = \tau \Delta T_S + \theta_W \Delta h_W + \theta_U \Delta h_U + g \Delta \Gamma + D_\beta \Delta \beta + \theta_M \Delta h_M - D_{R,2X} \Delta L \qquad (20.43)$$

where

- h_W, h_U, and h_M are the average heights of the WVEL, the cloud tops, and the methane emission layer, respectively; $h_{W,0} \simeq 8$ km, $h_{U,0} \simeq 3.3$ km, $h_{M,0} \simeq 3$ km
- Γ is the average lapse rate; $\Gamma_0 \simeq 6.5°C/km$
- β is the cloud fraction; $\beta_0 \simeq 62\%$

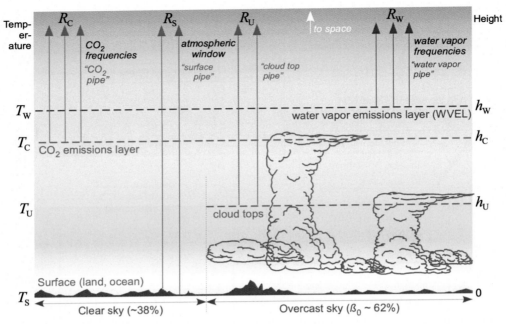

FIGURE 20.8   The OLR model estimates the change in OLR from properties of the main emission layers. (Although the CO_2 emissions layer is in the stratosphere around the center of its blockage at 15 μm, averaging by wavelength across the whole CO_2 blockage gives an average height around 7 km, out in the wings of the blockage, which also happens to be where the main changes due to increasing CO_2 are occurring. Hence this depiction.)

and

$$\tau \simeq 3.2 \ (\text{W/m}^2)/^\circ\text{C}$$

$$\theta_W \simeq -8.7 \ \text{W/m}^2 \ \text{km}$$

$$\theta_U \simeq -4.6 \ \text{W/m}^2 \ \text{km}$$

$$g \simeq \begin{cases} -13.5 \ \text{W/m}^2\text{per}^\circ\text{C/km} & \text{uniform} \\[2ex] -9.4 \ \text{W/m}^2\text{per}^\circ\text{C/km} & \text{partial} \end{cases} \tag{20.44}$$

$$D_\beta \simeq -0.42 \ \text{W/m}^2\text{per} \ 1\%$$

$$\theta_M \simeq -0.5 \ \text{W/m}^2\text{km}.$$

The value of g depends on whether the lapse rate is assumed to change uniformly across the troposphere (conventional assumption) or only in the lower 5 km (in line with radiosonde data, as noted in Section 4.3, or Fig. 20.10). Note that τ is $\partial R/\partial T_S$; its value agrees with the Planck feedback from AR5 [Eq. (20.5)].

6.4 Joining the Sum-of-Warmings and OLR Models to Form the Alternative Model

Start with the sum-of-warmings model with only the solar and CO_2 drivers because more drivers introduce too many unknowns (omit the drivers marked "Other" in Fig. 20.7). The analysis is between steady states, so ignore transitory effects of influences such as volcanoes.

Adding the temperature perturbations due to ASR and CO_2 in the sum-of-warmings model,

$$\Delta T_S = \Delta T_{S,A} + \Delta T_{S,C} = M\lambda_{SB}\Delta A + \lambda_C D_{R,2X}\Delta L. \tag{20.45}$$

Applying energy balance [Eq. (20.2)],

$$\Delta T_S = M\lambda_{SB}\Delta R + \lambda_C D_{R,2X}\Delta L. \tag{20.46}$$

Now join the models by using the OLR model [Eq. (20.43)] to replace ΔR,

$$\omega\Delta T_S = \theta_W\Delta h_W + \theta_U\Delta h_U + g\Delta\Gamma + D_\beta\Delta\beta + \theta_M\Delta h_M + \left(\frac{\lambda_C}{M\lambda_{SB}} - 1\right)D_{R,2X}\Delta L \tag{20.47}$$

where

$$\omega = \frac{1}{M\lambda_{SB}} - \tau \simeq -1.34 \text{ W/m}^{2\circ}\text{C}. \tag{20.48}$$

This equation, called the joint-model equation (JME), gives the surface warming in terms of the emission layer parameters, λ_C, and the increase in CO_2.

Hence, for a period between two steady states, the JME estimates CO_2 sensitivity as

$$\lambda_C = M\lambda_{SB}\left[\frac{\omega\Delta T_S - \left(\theta_W\Delta h_W + \theta_U\Delta h_U + g\Delta\Gamma + D_\beta\Delta\beta + \theta_M\Delta h_M\right)}{D_{R,2X}\Delta L} + 1\right]. \tag{20.49}$$

The fraction of global warming due to extra CO_2 can then be estimated as

$$\mu = \frac{\Delta T_{S,C}}{\Delta T_S} = \frac{\lambda_C D_{R,2X}\Delta L}{\Delta T_S}, \tag{20.50}$$

and the ECS as

$$\text{ECS} = \Delta T_{S,C}\big|_{\Delta L=1} = \lambda_C D_{R,2X}. \tag{20.51}$$

(Incidentally, the sum-of-warmings and OLR models cannot be portrayed together in a single diagram like Fig. 20.7. In the sum-of-warmings model the CO_2 influence $D_{R,2X}\Delta L$ *cannot* add to the ASR going into the solar response $M\lambda_{SB}$. But in the OLR model, $D_{R,2X}\Delta L$ *must* add (negatively) to the OLR, which by energy balance is also the ASR.)

6.5 Comparisons With the Conventional Model

The alternative model in Eq. (20.47) solves or ameliorates each of the problems of the conventional model described in Section 3.

The alternative model has less reliance on partial derivatives and "holding everything else constant":

- The sum-of-warmings architecture relies only on linearity to independently compute the warming due to each driver and add them—no partial derivatives. In contrast, the radiation-balance architecture of the conventional model inherently depends on partial derivatives—in particular, the balancing radiation computed by the Planck sensitivity relies on the Planck conditions, which raise tropospheric temperatures uniformly while holding everything else except OLR constant.
- The OLR model uses partial derivatives to account for the main factors that affect OLR [see Eq. (20.43)]. For example, it computes the change in OLR in the CO_2 pipe by separately considering the effect of changing the CO_2 concentration, lapse rate, or surface warming while holding the other two factors constant. However, the alternative model [Eqs. (20.47) and (20.49)] is less dependent on the value of the partial derivatives, and less vulnerable to errors in any one partial derivative, than the conventional model [Eqs. (20.9) and (20.22)], because the conventional ECS is proportional to the Planck sensitivity λ_0, a partial derivative.
- The sum-of-warmings model employs the Stefan–Boltzmann sensitivity, which applies under all conditions. The heart of the conventional model is the Planck sensitivity, which only applies under the Planck conditions.

An advantage of the sum-of-warmings approach is that, with data gathered over enough time, it should be possible to statistically disentangle the warmings due to various drivers. In principle, the feedbacks and sensitivity to each driver could be verified by real-world data. In contrast, partial derivative estimates are intrinsically hypothetical and not empirically verifiable.

The sum-of-warmings approach allows for separate feedbacks for each individual driver. For example, it allows for a CO_2 sensitivity λ_C that includes feedbacks that only apply to the influence of CO_2, such as the rerouting feedback. The conventional model applies the same solar response to each driver, so it is structurally unable to accommodate the rerouting feedback.

In the sum-of-warmings model the drivers are not interchangeable. For example, it applies a different response to surface warming due to more ASR than it does to decreased OLR from the upper atmosphere due to CO_2 enrichment. This undermines the notion of forcing—in the sum-of-warmings model the same forcing (radiation imbalance) from different drivers could cause very different surface warmings.

6.6 Massive Negative Feedback Omitted From the Conventional Model?

Many who suspect that the ECS is substantially less than 3°C currently accept the conventional model for calculating the ECS (Fig. 20.3). In the conventional model the only way the ECS can be dramatically lower than 3°C is if the feedbacks f are much less positive than in AR5, so that those feedbacks do not amplify the effect of the CO_2 forcing so much. Those critics have therefore searched for a massive missing negative feedback in response to surface warming. However, the theoretical case for net-positive feedback to surface warming seems strong, and after 30 years of searching it seems unlikely that a massive negative feedback in response to surface warming has been overlooked.

The alternative model of Fig. 20.7 provides a possible answer for those critics—do not apply the solar sensitivity and feedbacks to the influence of CO_2, but instead apply a separate CO_2 sensitivity and feedbacks. Solar input mainly heats the bottom of the atmosphere, while CO_2 enrichment affects outgoing energy from the higher atmosphere. So might not they have different responses?

Although the rerouting feedback is not a "negative feedback" as that term is understood in the conventional model, if the CO_2 sensitivity λ_C is much lower than the solar sensitivity $M\lambda_{SB}$ then it would have the same effect on ECS estimates as if a massively negative feedback had been omitted from the conventional model.

7. THE "HOTSPOT"

The "hotspot" is the informal name for the warming of the upper troposphere caused by an ascent of the water vapor emission layer (WVEL).

In the conventional model, surface warming for any reason causes a hotspot—an ascending WVEL *is* the water vapor amplification mechanism—and we assume here that the solar response in the conventional models is correct.

In the alternative model, although the solar response causes the WVEL to ascend, and although the surface warming caused by the CO_2 response causes the WVEL to ascend, the CO_2-specific feedbacks within the CO_2 response could cause the WVEL to either ascend or descend; before turning to the data there is no way of knowing, though if the rerouting feedback is the dominant CO_2-specific feedback then it might be expected to descend.

Given the apparent lack of a hotspot over the last few decades, coincident with a rapid rise in CO_2 concentration, it would appear that the CO_2 response causes the WVEL to descend.

7.1 The Hotspot Is Caused by an Ascending WVEL

The hotspot is the warming of a volume of air in the upper troposphere, at around the average height of the WVEL, which is ~8 km or ~360 hPa, and higher in the tropics (Supplementary Section 2.2).

The WVEL is the upper optical boundary of the water vapor in the upper atmosphere, on average. The air above the WVEL is dry, but the air below the WVEL is moist and therefore warmer because water vapor is condensing and releasing its latent heat. If the WVEL ascends, it creates the hotspot by warming the volume that was dry and cool when just above the WVEL, which becomes moist and warmer as the WVEL ascends above it.

Conversely, a falling WVEL would produce a volume of cooling (a "coolspot"?).

Water vapor is quite dynamic in the upper troposphere, its upper boundary often moving up and down several kilometers over a week at a given longitude and latitude, so the "instantaneous WVEL" moves up and down. The WVEL is the average of the instantaneous WVEL. Because of this dynamism, hotspot warming can extend for a couple of kilometers in height; as the WVEL moves up the instantaneous WVEL tends to move up, warming volumes over a couple of kilometers of vertical extent to some degree.

The traditional illustration of the hotspot comes in a diagram of atmospheric warming (color) by latitude (x-axis) and height (y-axis). If the instantaneous WVEL stayed very close to the WVEL (ie, no dynamism), the hotspot would be a strip of strong warming in the upper troposphere at 300—400 hPa, and the height of the strip would be the amount by which the hotspot ascended. But the dynamism of the water vapor causes a cloud of instantaneous WVEL heights clustering around the WVEL, ensuring that the hotspot is smeared out over a couple of kilometers in height.

The hotspot is distinct from warming in the upper troposphere due to a change in lapse rate. As the surface warms due to increased ASR, more evaporation causes a moister atmosphere and thus a lower lapse rate, which causes the atmosphere at a given height to warm. While this surface warming also causes the WVEL to ascend (next section), warming due to increased lapse rate is broadly and diffusely spread through the atmosphere with a shallow gradient, in contrast to the hotspot, which is a smear of warming centered on the WVEL.

7.2 The Solar Response Causes the WVEL to Ascend

The conventional explanation for the hotspot is that surface warming causes more evaporation, and the greater volume of water vapor in the atmosphere is assumed to push up the WVEL, by virtue of greater volume and more vigorous upward convection. However, as noted in Section 4.3, more water vapor in the atmosphere does not necessarily lead to more water vapor in the upper troposphere if the extra water vapor is mainly confined to a more stable lower troposphere with less overturning, as appears to be the case according to the radiosonde data.

Within the solar response (Fig. 20.7), $\Delta T_{S,A}$ is about twice ΔT_R because of amplification by the nonalbedo solar feedbacks [Eqs. (20.39) and (20.41)]. Because ΔT_R is roughly the average of the warmings of the various emission layers, and all the emission layers warm by $\sim \Delta T_S$ before the effects of changes to lapse rates and emission layer heights, one or more of the main nonsurface emission layers *must* warm significantly less than the surface. The height of the CO_2EL is determined by the concentration of CO_2, so its height cannot change. The lapse rate changes, but that works *against* the amplification of warming (the lapse rate feedback is negative). That leaves just the heights of the WVEL and the cloud tops, one or both of which must ascend to cooler places in the troposphere. Clouds are not well understood, but it is generally thought that WVEL provides the bulk of the accommodation. Thus the WVEL ascends significantly in response to surface warming due to more ASR.

In the conventional model all climate drivers cause the solar response (Fig. 20.3), and all surface warming is due to the solar response, so ΔT_S is $\Delta T_{S,A}$. Thus, the conventional explanation is that all warming influences cause the WVEL to ascend, thereby causing a hotspot.

7.3 The WVEL Has Not Ascended in the Last Few Decades

The only instruments with sufficient vertical resolution to measure the change in height of the WVEL over the last few decades (Δh_W) are the radiosondes. Satellites are much less suitable because they aggregate information from several vertical kilometers into each data point.

Radiosonde-derived temperature and humidity data is used here. It is accepted that the latter especially must be treated with great caution, particularly at altitudes above the 500 hPa pressure level. Following the discussion in Paltridge et al. (2009), the humidity data is restricted to tropical and midlatitude data at least ~ 0.5 g/kg, from 1973. While the data is not good enough to estimate small changes in Δh_W, it is presumed sufficient to at least distinguish the *direction* of movement.

Surface temperatures here are the midpoints of UAH and HadCrut4, 5-year smoothed and centered.

7.3.1 Temperature Data

The temperatures measured by the radiosondes are shown in Fig. 20.9, for 1979 to 1999 (the only image as a function of height and latitude ever publicly released, apparently). Over those two decades ΔT_S was $\sim 0.12°C$ per decade, which would have caused a similar warming at all levels of the troposphere had there been no change in the lapse rate.

Assuming lapse rate change was uniform at all heights, the warming it caused can be estimated from the lapse rate feedback f_{LR} (AR5, see Supplementary Section 1) and g from the OLR model as $\Delta T_S f_{LR} g^{-1}$, or $0.12 \times (-0.6 \pm 0.4)/(-13.5)$, or $\sim 0.0053 \pm 0.0036$ (°C/km) per decade. At the WVEL height of ~ 8 km this causes warming of $\sim 0.04 \pm 0.03°C$ per decade, so the WVEL warmed by $\sim 0.16 \pm 0.03°C$ per decade just from ΔT_S and the attendant lapse rate change.

Hence, *when looking for changed atmospheric structure* in Fig. 20.9, the meaning of the colors in the key all move one to two steps cooler; for example, the lightest yellow means $0.06-0.16°C$ of cooling. If all the atmospheric structures stayed in the same place, Fig. 20.9 would show the second or third lightest yellows throughout most of the

FIGURE 20.9   Atmospheric warming 1979 to 1999, as measured by radiosondes. The horizontal axis shows latitude, the vertical axis height (km on the right, hPa on the left). From the US CCSP report of 2006, Fig. 5.7E in Section 5.5 on page 116 (Santer, 2006), see also (Singer, 2011).

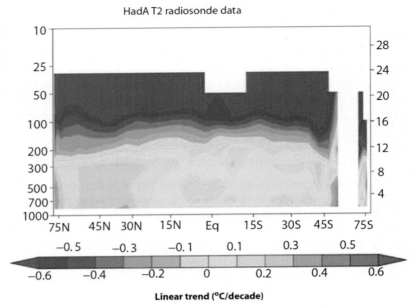

troposphere, including around heights of 8 km. Thus the presence of the lightest yellow around 8 km suggests cooling occurred *due to changes in atmospheric structure* (offsetting the surface warming and associated lapse rate changes).

The cooling strips above 12 km are due to ozone depletion and are too high to be of interest here. The most important part of the WVEL is in the tropics, where it is warmest and most of the radiation to space is occurring, and the WVEL, like the tropopause, is somewhat higher. Fig. 20.9 indicates no warming or perhaps cooling in the tropics around the WVEL height, after discounting the effect of ΔT_S. Thus the WVEL did not ascend, and may have descended.

Dr. Roy Spencer used a different mix of microwave channels to specifically look for the hotspot using the satellite data in May 2015 and concluded: "But I am increasingly convinced that the hotspot really has gone missing."

7.3.2 Humidity Data

Consider the radiosonde specific humidity data, shown in Fig. 20.10. The more reliable data only goes to 400 hPa, but above the 500 hPa pressure level the trend is one of drying. This agrees with the model in Figs. 1B, 2B, 3, and 5 of Paltridge et al. (2009). The same trends are shown by the earlier radiosonde data from 1948 to 1973. Again this is not compatible with an ascending WVEL and suggests a descent.

7.3.3 Conclusion

The WVEL has not ascended in the last few decades:

$$\Delta h_W \leq 0. \tag{20.52}$$

7.4 The CO_2 Response Causes the WVEL to Descend

In the last few decades there was surface warming and the WVEL did not ascend, so the conventional model is incorrect.

In the alternative model, the warming influences of ASR and CO_2 are considered. The albedo data discussed in Section 5 indicates a small fall in reflected solar radiation from 1984 that is larger than the smoothed changes in TSI occurring in that period, so ASR presumably increased from 1984, which caused some surface warming and thereby caused the WVEL to ascend. Yet the WVEL was observed to descend. Therefore the WVEL descended due to the CO_2 response (to the increasing CO_2), and this outweighed the ascent due to the solar response (to the increased ASR).

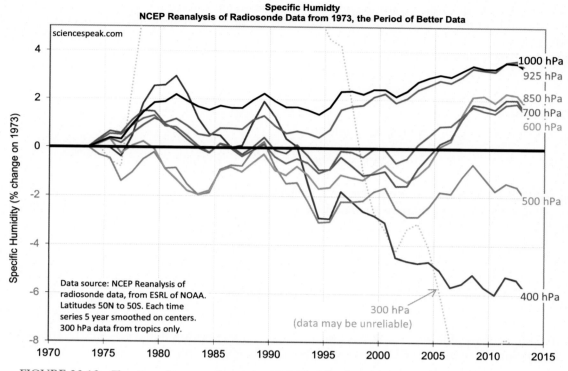

FIGURE 20.10 The atmosphere near the average WVEL height of 360 hPa shows a drying trend since 1973.

Hence the CO_2 response to increasing CO_2 causes the WVEL to descend. This is supporting evidence for the rerouting feedback.

In other words, the strong rise in CO_2 concentration since the 1970s and the lack of a hotspot during periods of surface warming together suggest that the effect of the CO_2 response is to cause the WVEL to descend, and that this descent was only partly offset by the ascent caused by extra ASR and the surface warming.

8. CALCULATING THE ECS USING THE ALTERNATIVE MODEL

This section employs the alternative model to quantitatively analyze the climate of the last few decades, estimating the CO_2 sensitivity, the ECS, and the fraction of global warming due to increasing CO_2. The numerical calculations here are all in the supplementary spreadsheet.

8.1 Simple Case

The "simple case" is defined by

$$f_\alpha = \Delta h_U = \Delta \Gamma = \Delta \beta = \Delta h_M = 0. \tag{20.53}$$

It ignores the complications of the minor albedo feedback to surface warming, so $\Delta T_{S,A}$ and the OLR are independent of the CO_2 concentration (Fig. 20.7). It ignores changes in the less critical emission layer parameters, leaving a simple tradeoff between what are presumably the main influences. The JME [Eq. (20.47)] becomes just

$$\omega \Delta T_S = \theta_W \Delta h_W + \left(\frac{\lambda_C}{M \lambda_{SB}} - 1 \right) D_{R,2X} \Delta L. \tag{20.54}$$

8.1.1 CO_2 Constant

During a period of constant CO_2, ΔL is zero so

$$\omega \Delta T_S = \theta_W \Delta h_W. \tag{20.55}$$

The WVEL ascends with surface warming and descends when the surface cools, because only the solar response is active. For example, if the surface warms by 1.0°C due to TSI or EDA changes, then T_R increases by ~ 0.5°C [Eqs. (20.39) and (20.41)] and the WVEL ascends by ~ 150 m. Or if the WVEL descends by 100 m, then the associated surface cooling is ~ 0.65°C and the drop in radiating temperature is ~ 0.33°C.

8.1.2 CO$_2$ Doubles

During a period when CO_2 doubles, ΔL is one so

$$\omega \Delta T_S = \theta_W \Delta h_W + \left(\frac{\lambda_C}{M\lambda_{SB}} - 1 \right) D_{R,2X}. \tag{20.56}$$

$\Delta T_{S,C}$ is the ECS, by definition. It is equal to $\mu \Delta T_S$, where μ is the fraction of warming caused by the CO_2 increase. It is also equal to $\lambda_C D_{R,2X}$. Using these two relationships to replace ΔT_S and λ_C by ECS$/\mu$ and ECS$/D_{R,2X}$ respectively in Eq. (20.56),

$$\text{ECS} = \phi \left(\theta_W \Delta h_W - D_{R,2X} \right) \tag{20.57}$$

where

$$\phi = \left(\frac{\omega}{\mu} - \frac{1}{M\lambda_{SB}} \right)^{-1}. \tag{20.58}$$

Notice that for a given value of μ, the ECS is linear in Δh_W. If one supposes that μ is $\sim 80\%$ (according to Table TS.6 of the Technical Summary in AR5, CO_2 provided $\sim 80\%$ of the change in radiative forcing since 1750) then

$$\text{ECS} \simeq \left(2.5 \frac{\Delta h_W}{\text{km}} + 1.0 \right) \text{°C}. \tag{20.59}$$

The greater the ECS, the more the WVEL ascends; conversely, the more the WVEL ascends, the higher must be the ECS. However, given that the WVEL appears not to have ascended, the ECS cannot exceed ~ 1.0°C by Eq. (20.59), but such an ECS is insufficient to explain 80% of the recent warming, so the fraction of warming due to the ECS must be lower. Lowering μ to say 50%,

$$\text{ECS} \simeq \left(1.9 \frac{\Delta h_W}{\text{km}} + 0.8 \right) \text{°C}, \tag{20.60}$$

so if the WVEL has not ascended then the ECS is constrained to be less than 0.8°C, which is not enough to explain 50% of the recent warming. And so on. See Fig. 20.11.

8.1.3 Recent Decades

In recent decades there are periods whose endpoints are without undue volcanic interference, when we can estimate ΔT_S and ΔL, and bound Δh_W. We assume the climate has stayed sufficiently close to steady state through recent decades to apply Eq. (20.54), from which

$$\lambda_C = M\lambda_{SB} \left[\frac{\omega \Delta T_S - \theta_W \Delta h_W}{D_{R,2X} \Delta L} + 1 \right]. \tag{20.61}$$

Eqs. (20.50) and (20.51), for the fraction of global warming due to increasing CO_2 and for the ECS, can then be applied. Estimates for several scenarios are shown in Table 20.1.

The estimate of ECS for the period of more reliable radiosonde data, from 1973, is 0.5°C if the WVEL stayed at the same height. But if the WVEL ascended 25 m, the ECS would have to be 0.96°C, while if the WVEL fell 25 m then the ECS must be less than 0.1°C, which demonstrates how important the WVEL is to climate sensitivity.

In the last few decades, CO_2 has been steadily increasing, while temperature moved around. Our emission-layer data is not good enough to track OLR, so a range of ECS estimates is unavoidable. However, the radiosondes point to a nonascending WVEL, which unambiguously implies that the ECS is much lower than conventionally believed and that the rising CO_2 concentration was not the main factor warming the surface.

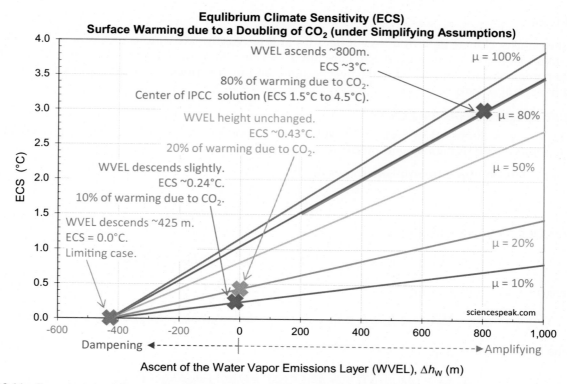

FIGURE 20.11 For a given warming-causation-fraction μ, the ECS is linear in the WVEL ascent. The green cross marks the center of the IPCC position (AR5); the red and orange crosses mark the position of this paper (below). If the WVEL has not ascended, as per the radiosondes, then the ECS must be less than 1.2°C even if all the recent warming is due to increasing CO_2 ($\mu = 1$), but an ECS of 1.2°C is insufficient to explain all of the recent warming so $\mu < 1$!

TABLE 20.1 Simple-Case Scenarios. The A and B Scenarios Match the Period of Radiosonde Data Back to 1973 (More Reliable) and 1948 (Less Reliable), During Which the Radiosondes Indicate the WVEL Did Not Ascend: The Effect of Several Values of WVEL Ascent Are Shown. Surface Warming Averages UAH and HadCrut4, Both 5-Year Smoothed. CO_2 From Mauna Loa (Law Dome Prior to 1959). Details and More Scenarios in the Supplementary Spreadsheet

Scenario	Start	End	ΔT_0 (°C)	ΔC (ppm)	Δh_W (m)	λ_C [°C/(W/m^2)]	μ (%)	ECS (°C)
A1	1973	2011	0.514	62	0	0.13	24	0.50
A2	1973	2011	0.514	62	25	0.26	47	0.96
A3	1973	2011	0.514	62	−25	0.01	1	0.03
B1	1948	2011	0.488	81	0	0.25	64	0.93
B2	1948	2011	0.488	81	50	0.44	111	1.63
B3	1948	2011	0.488	81	−50	0.06	16	0.23
D1	1973	2001	0.400	41	0	0.08	12	0.28
D2	1973	2006	0.475	52	0	0.10	16	0.37
D3	1963	2011	0.456	73	0	0.24	57	0.88
D4	1968	2011	0.458	69	0	0.22	49	0.80
D5	1978	2011	0.434	57	0	0.16	31	0.59
D6	1983	2011	0.428	49	0	0.10	17	0.39

8.2 Full Case

The full case uses climate data from recent decades in Eqs. (20.49)–(20.51) to estimate λ_C, μ, and the ECS. The climate data is insufficient to form good estimates but is sufficient to draw interesting conclusions. The data about the climate parameters are considered in the next sections, and then various combinations of parameter values are evaluated in several scenarios.

8.2.1 WVEL Height Bounds the ECS

WVEL height was discussed in Section 7.3. While we do not know how its height varied over time, merely knowing that the WVEL did not ascend produces a useful upper bound:

1. Radiosonde data on upper troposphere shows the WVEL descended [Eq. (20.52)]:

$$\Delta h_W \leq 0.$$

2. So by the OLR model [Eq. (20.43), $\theta_W < 0$],

$$\Delta R \geq \tau \Delta T_S + \theta_U \Delta h_U + g \Delta \Gamma + D_\beta \Delta \beta + \theta_M \Delta h_M - D_{R,2X} \Delta L.$$

3. So by energy balance [Eq. (20.2)],

$$\Delta A \geq \tau \Delta T_S + \theta_U \Delta h_U + g \Delta \Gamma + D_\beta \Delta \beta + \theta_M \Delta h_M - D_{R,2X} \Delta L.$$

4. So by the solar response [Fig. 20.7 or Eq. (20.40), $M\lambda_{SB} > 0$],

$$\Delta T_{S,A} \geq M\lambda_{SB} \left[\tau \Delta T_S + \theta_U \Delta h_U + g \Delta \Gamma + D_\beta \Delta \beta + \theta_M \Delta h_M - D_{R,2X} \Delta L \right].$$

5. So by the sum-of-warmings model [Fig. 20.7 or Eq. (20.45)],

$$\Delta T_{S,C} \leq \Delta T_S - M\lambda_{SB} \left[\tau \Delta T_S + \theta_U \Delta h_U + g \Delta \Gamma + D_\beta \Delta \beta + \theta_M \Delta h_M - D_{R,2X} \Delta L \right].$$

6. So by the definition of ECS as $\Delta T_{S,C}$ when ΔL is one,

$$\text{ECS} \leq (1 - M\lambda_{SB}\tau)\Delta T_S - M\lambda_{SB} \left[\theta_U \Delta h_U + g \Delta \Gamma + D_\beta \Delta \beta + \theta_M \Delta h_M \right] + M\lambda_{SB} D_{R,2X}.$$

8.2.2 Cloud Height

Davies and Molloy (2012) report a decrease in the global effective height of cloud tops from March 2000 to February 2010, using the Multi-angle Imaging SpectroRadiometer (MISR) on the Terra satellite. The linear trend was of -44 ± 22 m/decade; the difference between the first and last years was -31 ± 11 m. The annual mean height is measured with a sampling error of 8 m. Detected regional height anomalies correlate well with changes in the Southern Oscillation Index.

However, Evan and Norris,(2012) claim that the decrease reported by the MISR is an artifact due to a systematic reduction in the number of retrievals of clouds at lower elevations during the early years of the MISR mission, apparently due to "satellite orbit inclination maneuvers" causing "erroneous co-registration of the nine MISR cameras." But they also note that "there is no obvious reason why the camera co-registration issues should affect cloud height retrievals at one height in the atmosphere more or less strongly than retrievals at another height in the atmosphere." Using a post hoc method for removing the bias, they report an ascending trend of $+54$ m/decade, which agrees with the MODIS-Terra data showing increasing cloud height of $+61$ m/decade. The MODIS-Terra cloud height data is of distinctly lesser quality than the MISR for measuring cloud-top height; both begin in 2000.

There does not appear to be any other cloud height data of note. Unfortunately the cloud height data is conflicted, and is after the period of warming from the 1970s to the 1990s. We explore both ascending and descending cloud-top scenarios in the following discussion.

8.2.3 Lapse Rate

The published radiosonde data on lapse rate trends only seems to extend to 700 hPa. Behavior in the upper troposphere might be quite different (Fig. 20.10). Gaffen et al. (2000) report that observed surface-to-700-hPa lapse rates fluctuated less than 1.5% either way about an average value from 1960 to 1998, and there might have been no overall trend (the trend might have decreased from 1960 to 1979 then increased from 1979 to 1998).

In lieu of empirical data on changes in lapse rate $\Delta\Gamma$, we estimate it from the lapse rate feedback f_{LR} in AR5 (Supplementary Section 1). Though this feedback is only for the solar response (Fig. 20.7), we assume it applies for any surface warming because it is intended as such, the effect is theoretically straightforward, and we have no better information. Assuming a uniformly changing lapse rate as per the conventional model, the extra OLR due to $\Delta\Gamma$ is $-\Delta T_S f_{LR}$ from the lapse rate feedback while it is $g_{uniform}\Delta\Gamma$ by the OLR model, so

$$\Delta\Gamma \simeq -\frac{\Delta T_S f_{LR}}{g_{uniform}}. \tag{20.62}$$

To apply this in the alternative model, we assume the lapse rate only changes in the lower troposphere in line with the radiosonde data; the increase in OLR due to lapse rate changes is estimated to be $g_{partial}\Delta\Gamma$.

8.2.4 Cloud Fraction

The International Satellite Cloud Climatology Project indicates that the cloud fraction rose by $\sim$2% from 1984 to 1987, then fell $\sim$4% to 2000, and then rose $\sim$0.5% to 2010 (Climate and Clouds: Cloud data). Marchand (2012) reports cloud fraction from 2001 to 2011 as measured by MISR rising $\sim$0.1% and by MODIS-Terra and MODIS-Aqua rising $\sim$0.3%. The total change from 1984 to 2010 was $\sim -1.5\%$ according to the ISCCP, but that is exaggerated by a factor of 2–4 by comparison to MISR and MODIS over 2001 to 2011, so perhaps the cloud fraction fell by $\sim$0.5% from 1984 to 2011. There does not seem to be prior data.

8.2.5 Scenarios

Table 20.2 shows several scenarios. In any scenario, λ_C, μ, and the ECS must all be positive, which constrains the input values.

The C scenarios are for 2000 to 2010, where we have cloud-top height data. Suppose the WVEL remained at the same height. If the cloud tops descended between 42 and 20 m as per the MISR observations, the ECS is likely between 0.7°C and 1.4°C, and μ is from 125% to 250% (C1, C2). But if the cloud tops ascended between 54 and 61 m in line with the MODIS observations, then the ECS is $\sim$3.8°C and μ is $\sim$650% (so high because the CO_2 warming is much larger than the warming that actually occurred, which requires the existence of an unknown cooling influence that does not affect ASR) (C3, C4). The unrealistically high values of μ suggest that the cloud tops more likely descended than ascended and that the MISR observations are more likely to be correct. If the WVEL descended, then estimates of μ and ECS decrease: a MISR-average cloud-top descent of 31 m and a WVEL descent of 18 m requires an ECS of zero (C5). A WVEL descent of $\sim$50 m is required to bring μ down to $\sim$100% if the cloud tops rise $\sim$50 m (C6).

TABLE 20.2 Full-Case Scenarios. As per Table 20.1. The C Scenarios Are for the Period of Cloud-Top Height Data

Scenario	Start	End	ΔT_S (°C)	ΔC (ppm)	Δh_W (m)	Δh_U (m)	$\Delta\Gamma$ (°C/km)	$\Delta\beta$ (%)	Δh_M (m)	λ_C [°C/(W/m²)]	μ (%)	ECS (°C)	ΔR (W/m²)
A4	1973	2011	0.514	62.0	0	0	−0.023	−0.50	0	−0.11	−20	−0.42	1.16
A5	1973	2011	0.444	62.0	0	0	−0.020	−0.25	0	0.02	4	0.07	0.80
A6	1973	2011	0.514	62.0	−25	100	−0.023	−0.50	0	0.03	5	0.11	0.91
A7	1973	2011	0.514	62.0	0	200	−0.023	−0.50	0	0.42	76	1.57	0.23
B4	1948	2011	0.488	81.0	0	0	−0.022	−0.50	0	0.07	18	0.27	0.75
B5	1948	2011	0.488	81.0	0	200	−0.022	−0.50	0	0.47	120	1.75	−0.18
C1	2000	2010	0.045	21.0	0	−42	−0.002	0.20	0	0.19	125	0.71	−0.02
C2	2000	2010	0.045	21.0	0	−20	−0.002	0.20	0	0.38	246	1.39	−0.12
C3	2000	2010	0.045	21.0	0	54	−0.002	0.20	0	0.99	653	3.68	−0.47
C4	2000	2010	0.045	21.0	0	61	−0.002	0.20	0	1.05	691	3.90	−0.50
C5	2000	2010	0.045	21.0	−18	−31	−0.002	0.20	0	0.00	0	0.00	0.08
C6	2000	2010	0.045	21.0	−50	50	−0.002	0.20	0	0.18	115	0.65	−0.01

In the A scenarios with the better radiosonde data from 1973 to 2011, there is cloud fraction data from 1984, but no cloud-top height data before 2000. If the cloud tops do not ascend (in line with their probable behavior after 2000), the WVEL does not ascend (as per the radiosondes), and the cloud fraction change was $\sim -0.5\%$ (in line with observations from 1984), then the ECS estimate is negative (A4). The ECS must be positive, so this indicates that on the basis of the most likely changes the ECS is very small, putting no lower bound on the estimate. Perhaps the pre-satellite warming and the cloud fraction change were exaggerated twofold: this would increase μ to $\sim 4\%$ and the ECS estimate to $\sim 0.07°C$ (A5). Even if the cloud tops ascended 100 m (twice the MODIS figures for 2000−2010), and the WVEL descended 25 m, λ_C is ~ 0.03, μ is $\sim 5\%$ and the ECS is $\sim 0.1°C$ (A6). If the cloud tops rose by 200 m (difficult to reconcile with the MISR observations, particularly as the clouds tops average only ~ 3.3 km) and the WVEL did not change, estimates approach the conventional: $\mu \sim 76\%$ and ECS $\sim 1.6°C$ (A7).

The longest scenarios are the B scenarios, back to 1948 but with less-reliable or missing data. If the WVEL and cloud tops remained at the same heights, and cloud fraction changed by $\sim -0.5\%$ (the net change observed from 1984 to 2011), then λ_C is ~ 0.07, μ is $\sim 18\%$, and the ECS is $\sim 0.27°C$ (B4).

8.2.6 Conclusions

There is no strong basis in the data for favoring any scenario in particular, but the A4, A5, A6, and B4 scenarios are the ones that best reflect the input data over longer periods. Hence we conclude that the ECS might be almost zero, is likely less than 0.25°C, and most likely less than 0.5°C, that μ is likely less than 20%, and that λ_C is likely less than 0.15°C/(W/m^2). Given a descending WVEL, it is difficult to construct a scenario consistent with the observed data in which the influence of CO_2 is greater than this.

9. CONCLUSIONS

The conventional basic model is based on just a radiation balance, adding the forcings due to the various influences on climate. We developed an alternative model based instead on adding the temperature perturbations due to the various influences, and using a radiation balance as well. The alternative model solves the two main problems with the conventional architecture, albeit at the expense of requiring more climate data.

If we just confine attention to climate influences that change ASR or CO_2, the two models are the same, except that the conventional model applies the solar response to the CO_2 influence while the alternative model applies its own CO_2 response to the CO_2 influence. The conventional adds the CO_2 forcing to the ASR; the alternative allows separate feedbacks in response to extra ASR and extra CO_2, and adds the temperature perturbations due to extra ASR and extra CO_2. Because the architectures only differ by a single connection, they cannot both be correct.

Putting data from the last few decades into the alternative model shows that the ECS estimates depend critically on whether the WVEL and cloud tops ascended or descended. Both data items are disputed, but in both cases the more appropriate, higher-resolution data says they did not ascend. If they did not ascend, the data is likely only compatible with an ECS less than $\sim 0.5°C$, and the fraction of recent warming due to rising CO_2 levels is less than 20%. As recommended by Elliot and Gaffen (1991) and by Paltridge et al. (2009), there needs to be a detailed effort to extract useful data from the radiosonde records similar to the effort given to extracting surface temperature trends from the individual stations of the international meteorological network.

The solar response was found to be at least three times stronger than the CO_2 response (both are measured in °C of surface warming per unit of forcing). The conventional basic climate model incorrectly applies the solar response to the CO_2 influence, which made CO_2 appear much more potent than it is.

The rerouting feedback is omitted by the conventional architecture because it occurs in response to extra CO_2 but not to surface warming. The proposed rerouting mechanism suggests that an increasing CO_2 concentration exerts a downward force on the WVEL and average cloud top height; this could have outweighed the upward force due to surface warming, which is compatible with the evidence of a missing hotspot. The rerouting feedback is compatible with the low climate sensitivity found by the alternative model—extra CO_2 might have little effect on surface temperatures simply because most of the heat blocked from escaping to space by extra CO_2 reroutes, emitted to space from water vapor or cloud tops instead.

APPENDIX: ACRONYMS

ARTS	M	Amplification of Radiating Temperature to Surface (Multiplier)
ASR	A	Absorbed solar radiation
CO_2		Carbon dioxide
CO_2EL		Carbon dioxide emission layer
ECS		Equilibrium climate sensitivity
EDA		Externally driven albedo
GCM		General circulation model
JME		Joint-model equation
OLR	R	Outgoing long-wave radiation
SBS		Stefan–Boltzmann sensitivity
TOA		Top of (the) atmosphere
TSI	S	Total solar irradiance
WVEL		Water vapor emission layer

Acknowledgments

I am grateful to Joanne Nova, Garth Paltridge, Christopher Monckton, and William Kininmonth for helpful feedback. Thanks also to Michael Hammer (for spectroscopic advice that led to the OLR model) and to Stephen Wilde (for suggesting the rerouting mechanism). Thank you to numerous readers at the joannenova.com.au blog for the donations that made this work possible.

Disclosure of Potential Conflict of Interests.

The author declares that he has no conflict of interest. He receives no monies or employment from any government or other organization, only donations to his wife's blog at joannenova.com.au.

References

Auroux, D., 2010. Partial Differentiation with Non-independent Variables. http://ocw.mit.edu/courses/mathematics/18-02sc-multivariable-calculus-fall-2010/2.-partial-derivatives/part-c-lagrange-multipliers-and-constrained-differentials/session-43-clearer-notation/MIT18_02SC_MNotes_n1_2.pdf.

Charney, J., Arakawa, A., Baker, J., Bolin, B., Dickinson, R., Goody, R., et al., 1979. Carbon Dioxide and Climate: A Scientific Assessment. National Academy of Sciences, Washington, DC.

Climate and Clouds: Cloud data. http://www.climate4you.com/ClimateAndClouds.htm.

Davies, R., Molloy, M., 2012. Global cloud height fluctuations measured by MISR on Terra from 2000 to 2010. Geophysical Research Letters L03701.

Elliot, W.P., Gaffen, D.J., 1991. On the utility of radiosonde humidity archives for climate studies. Bulletin of the American Meteorological Society 72, 1507–1519.

Evan, A.T., Norris, J.R., 2012. On global changes in effective cloud height. Geophysical Research Letters 39, L19710.

Gaffen, D.J., Santer, B.D., Boyle, J.S., Christy, J.R., Graham, N.E., Ross, R.J., 2000. Multidecadal changes in the vertical temperature structure of the tropical troposphere. Science 287, 1242–1245.

Hansen, J., Lacis, A., Rind, D., Russell, G., Stone, P., Fung, I., et al., 1984. Climate sensitivity: analysis of feedback mechanisms. In: Climate Processes and Climate Sensitivity. Geophysical Monograph 29 vol. 5.

Held, I.M., Soden, B.J., 2000. Water vapor feedback and global warming. Annual Review of Energy and the Environment 25, 441–475.

IPCC, 2013. Fifth Assessment Report. Cambridge University Press.

Kininmonth, W., 2010. A natural constraint to anthropogenic global warming. Energy and Environment 21 (4), 225–236.

Lean, J., 2000. Evolution of the Sun's spectral irradiance since the Maunder Minimum. Geophysical Research Letters 2425–2428.

Marchand, R., 2012. Trends in ISCCP, MISR, and MODIS cloud-top-height and optical-depth histograms. Journal of Geophysical Research: Atmospheres 118. http://dx.doi.org/10.1002/jgrd.50207.

Palle, E., Goode, P.R., Montanes-Rodriguez, P., 2008. Inter-annual variations in Earth's reflectance 1999–2007. Journal of Geophysical Research 114.

Paltridge, G., Arking, A., Pook, M., 2009. Trends in middle- and upper-level tropospheric humidity from NCEP reanalysis data. Theoretical and Applied Climatology 98, 351–359.

Pierrehumbert, R.T., 2010. Principles of Planetary Climate. Cambridge University Press, Cambridge.

Pinker, R.T., Zhang, B., Dutton, E.G., May 6, 2005. Do satellites detect trends in surface solar radiation? Science 38, 850–854.

Santer, B.D., 2006. US Climate Change Science Program 2006, Temperature Trends in the Lower Atmosphere – Understanding and Reconciling Differences.

Sherwood, S.C., Bony, S., Boucher, O., Bretherton, C., Forster, P.M., Gregory, J.M., et al., 2015. Adjustments in the forcing-feedback framework for understanding climate change. AMS. http://dx.doi.org/10.1175/BAMS-D-13-00167.1.

Singer, S.F., 2011. Lack of consistency between modeled and observed temperature trends. Energy and Environment 22 (4), 375–406.

Soden, B.J., Held, I.M., 2006. An assessment of climate feedbacks in coupled ocean-atmosphere models. Journal of Climate 19, 3354–3360.

Spencer, R.W. New Satellite Upper Troposphere Product: Still No Tropical "Hotspot". http://www.drroyspencer.com/2015/05/new-satellite-upper-troposphere-product-still-no-tropical-hotspot/.

Stephens, G.L., O'Brien, D., Webster, P.J., Pilewski, P., Kato, S., Li, J-l, 2015. The albedo of Earth. Reviews of Geophysics 53. http://dx.doi.org/10.1002/2014RG000449.

Wang, Y.M., Lean, J.L., Sheeley, N.R., 2005. Modeling the Sun's magnetic field and irradiance since 1713. Astrophysical Journal 522–538.

Weart, S.R., February 2015. The Discovery of Global Warming: Simple Models of Climate Change. https://www.aip.org/history/climate/simple.htm.

CLIMATE PREDICTIONS

21

Using Patterns of Recurring Climate Cycles to Predict Future Climate Changes

D.J. Easterbrook

Western Washington University, Bellingham, WA, United States

1. INTRODUCTION

Global warming that occurred from 1978 to about 1998 pushed climate change into the forefront of potential concern. Every day the news media is filled with dire predictions of impending disasters—catastrophic melting of the Antarctic and Greenland ice sheets, drowning of major cities from sea level rise, drowning of major portions of countries, droughts, severe water shortages, no more snow, more extreme weather events (hurricanes, tornadoes), etc. With no unequivocal, cause-and-effect, tangible, physical evidence that increasing CO_2 caused this most recent global warming, adherents of this ideology have had to rely on computer models that have proven to be unreliable.

Abundant, physical, geologic evidence from the past provides a record of former periods of recurrent global warming and cooling that were far more intense than recent warming and cooling. These geologic records provide clear evidence of global warming and cooling that could not have been caused by increased CO_2. Thus, we can use these records to project global climate into the future, ie, the past is the key to the future.

Evidence-Based Climate Science, Second Edition
http://dx.doi.org/10.1016/B978-0-12-804588-6.00021-5

2. THE PAST IS THE KEY TO THE FUTURE: LESSONS FROM PAST GLOBAL CLIMATE CHANGES

2.1 Past Climate Changes

Those who advocate CO_2 as the cause of global warming have stated that never before in the Earth's history has climate changed as rapidly as in the past century, and that proves global warming is being caused by anthropogenic CO_2. Statements such as these are easily refutable by the geologic record. Fig. 21.1 shows temperature changes recorded by oxygen isotope ratios from the GISP2 ice core from the Greenland Ice Sheet. The global warming experienced during the past century pales into insignificance when compared to the magnitude of the profound climate reversals over the past 15,000 years.

The GISP2 Greenland ice core isotope data have proven to be a great source of climatic data from the geologic past. Paleo-temperatures for more than 100,000 years have been determined from nuclear accelerator measurements of thousands of oxygen isotope ratios ($^{16}O/^{18}O$) (Grootes and Stuiver, 1997), and these data have become a world standard.

Oxygen isotope ratios are a measure of paleo-temperatures at the time snow fell that was later converted to glacial ice. The age of such temperatures can be accurately measured from annual layers of accumulation of rock debris marking each summer's melting of ice and concentration of rock debris on the glacier. Paleo-temperatures from the GISP2 ice core were also reconstructed using ice core temperatures (Fig. 21.2; Cuffy and Clow, 1997; Alley, 2000). These also show exceptionally high rates of warming and cooling near the end of the Pleistocene.

The oxygen isotope and paleo-temperature data clearly show remarkable swings in climate over the past 100,000 years. In just the past 500 years, Greenland warming/cooling temperatures fluctuated back and forth about 40 times, with changes every 25—30 years (27 years on the average). None of these changes could have been caused by changes in atmospheric CO_2 because they predate the large CO_2 emissions that began about 1945. Nor can the warming of 1915—45 be related to CO_2, because it predates the soaring emissions after 1945. Thirty years of global cooling (1945—77) occurred during the big post-1945 increase in CO_2 emissions.

2.2 Magnitude and Rate of Abrupt Climate Changes

But what about the magnitude and rates of climates change? How do past temperature oscillations compare with recent global warming (1977—98) or with warming periods over the past millennia? The answer to the question of magnitude and rates of climate change can be found in the $\delta^{18}O$ and ice core temperature data (Steffensen et al., 2008).

Temperature changes in the GISP2 core over the past 25,000 years are shown in Figs. 21.1 and 21.2. The temperature curve in Fig. 21.2 is a portion of the Cuffy and Clow (1997) original curve. The horizontal axis is time and the vertical axis is temperature, based on ice core borehole temperature data. Details are discussed in their paper. Places where the curve becomes nearly vertical signify times of very rapid temperature change. Keep in mind that these are temperatures in Greenland, not global temperatures. However, correlation of the ice core temperatures with

FIGURE 21.1   Oxygen isotope ratios from the GISP2 ice core, Greenland. Note the extremely sharp rises in temperature (red) at about 14,500 years and 11,500 years, which had much greater magnitude and rate of rise than the past century. *Plotted from data by Grootes, P.M., Stuiver, M., 1997. Oxygen 18/16 variability in Greenland snow and ice with 10^3 to 10^5—year time resolution. Journal of Geophysical Research 102, 26455—26470.*

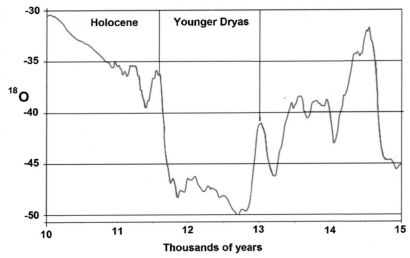

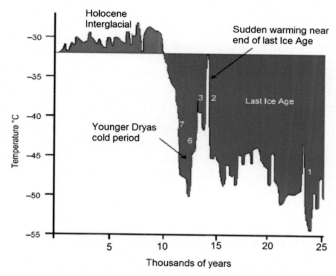

FIGURE 21.2   Greenland temperatures over the past 25,000 years recorded in the GISP2 ice core. Strong, abrupt warming is shown by nearly vertical rise of temperatures, strong cooling by nearly vertical drop of temperatures. *Modified from Cuffey, K.M., Clow, G.D., 1997. Temperature, accumulation, and ice sheet elevation in central Greenland through the last deglacial transition. Journal of Geophysical Research 102, 26383–26396.*

worldwide glacial fluctuations and correlation of modern Greenland temperatures with global temperatures confirms that the ice core record does indeed follow global temperature trends and is an excellent proxy for global temperature changes. For example, the portions of the curve from about 25,000 to 15,000 represent the last Ice Age (the Pleistocene), when huge ice sheets, thousands of feet thick, covered North America, northern Europe, and northern Russia, and alpine glaciers readvanced far downvalley.

Some of the more remarkable sudden climatic warming periods are listed later in this section and in Fig. 21.3. The numbers in Fig. 21.3 correspond to the temperature curves in Fig. 21.2.

How do the magnitude and rates of change of modern global warming/cooling compare to warming/cooling events over the past 15,000 years? We can compare the warming and cooling in the past century to approximate 100-year periods in the past 25,000 years. The scale of the curve does not allow enough accuracy to pick out exactly 100-year episodes directly from the curve, but that can be done from the annual dust layers in ice core data. Thus, not all of the periods noted here are exactly 100 years. Some are slightly more, some are slightly less, but they are close enough to allow comparison of magnitude and rates with the past century.

1. Temperature changes recorded in the GISP2 ice core from the Greenland Ice Sheet (Figs. 21.1 and 21.2) show that the global warming experienced during the past century pales into insignificance when compared to the magnitude of profound ice sheets that covered Canada and the northern United States, all of Scandinavia, and much of northern Europe and Russia.

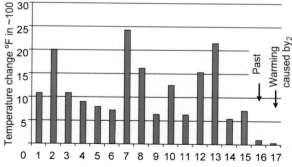

FIGURE 21.3   Magnitudes of the largest warming/cooling events over the past 25,000 years. Temperatures on the vertical axis are rise or fall of temperatures in about a century. Each column represents the rise or fall of temperature shown in Fig. 21.2. Event number 1 is about 24,000 years ago and event number 15 is about 11,000 years old. The sudden warming about 14,000 years ago caused massive melting of these ice sheets at extraordinary rates.

2. Shortly thereafter, temperatures dropped abruptly about 10°C (20°F) and temperatures then remained cold for several thousand years.
3. About 13,000 years ago, global temperatures plunged sharply ($\sim$12°C; $\sim$21°F) and a 1300-year cold period, the Younger Dryas, began.
4. 11,500 years ago, global temperatures rose sharply ($\sim$12°C; $\sim$21°F), marking the end of the Younger Dryas cold period and the end of the Pleistocene Ice Age. The end of the Younger Dryas cold period warmed by 5°C (9°F) over 30—40 years and as much as 8°C (14°F) over 40 years.

Fig. 21.2 shows comparisons of the largest magnitudes of warming/cooling events per century over the past 15,000 years. At least three warming events were 20—24 times the magnitude of warming over the past century and four were 6—9 times the magnitude of warming over the past century. The magnitude of the only modern warming that might possibly have been caused by CO_2 (1978—98) is insignificant compared to the earlier periods of warming.

2.3 Holocene Climate Changes (10,000 Years Ago to Present)

Almost all of the past 10,000 years have been warmer than the present. Figs. 21.4 and 21.5 show temperatures from the GISP2 Greenland ice core. With the exception of a brief cool period about 8200 years ago, almost all of the entire period from 10,500 to 1500 years ago was significantly warmer than the present. About 8200 years ago, the post-Ice Age interglacial warm period was interrupted by sudden global cooling that lasted for a few centuries (Figs. 21.4—21.6). During this time, alpine glaciers advanced and built moraines. The warming that followed the cool period was also abrupt. Neither the abrupt climatic cooling nor the warming that followed was preceded by atmospheric CO_2 changes.

As shown by oxygen isotope data from the GISP2 ice core, historical accounts, and various other data, the last 2000 years of the late Holocene was characterized by alternating warm and cold periods (Fig. 21.7).

2.3.1 The Roman Warm Period

The Roman Warm Period shows up prominently in the GISP2 ice core (Fig. 21.7) between about 1500 and 1800 years ago. During that time, Romans wrote of grapes and olives growing farther north in Italy that had been previously possible and of little snow or ice (Singer and Avery, 2007).

2.3.2 Dark Ages Cool Period

The Dark Ages Cool Period, which began about 1500 years ago, marks the end of 8500 years of Holocene warming and the beginning of late Holocene cold periods (Fig. 21.7). This cool period shows up conspicuously in the GISP2 ice core and in historic accounts. The Romans wrote that the Tiber River froze and snow remained on the ground for long periods (Singer and Avery, 2007).

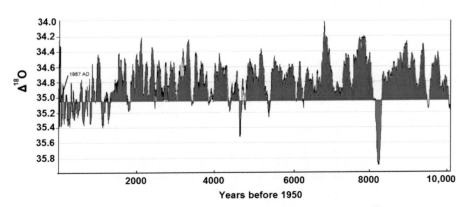

FIGURE 21.4   Greenland GISP2 oxygen isotope curve for the past 10,000 years. The vertical axis is $\delta^{18}O$, which is a temperature proxy. The red areas represent temperatures several degrees warmer than present. Blue areas are cooler times. Note the abrupt, short—term, cooling 8200 years ago and cooling from about 1500 to present. *Plotted from data by Grootes, P.M., Stuiver, M., 1997. Oxygen 18/16 variability in Greenland snow and ice with 10^3 to 10^5—year time resolution. Journal of Geophysical Research 102, 26455—26470.*

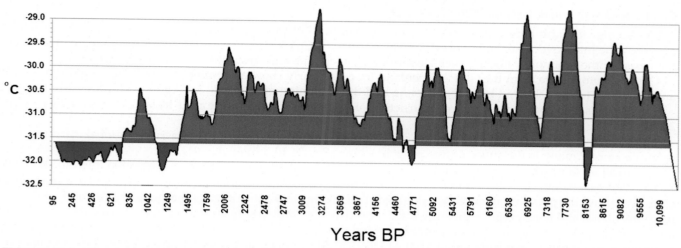

FIGURE 21.5   Greenland GISP2 temperature reconstruction for the past 10,000 years based on ice core temperatures. The paleo-temperature reconstruction is essentially the same as shown by the δ^{18}O curve (Fig. 21.4). *Modified from Cuffey, K.M., Clow, G.D., 1997. Temperature, accumulation, and ice sheet elevation in central Greenland through the last deglacial transition. Journal of Geophysical Research 102, 26383—26396.*

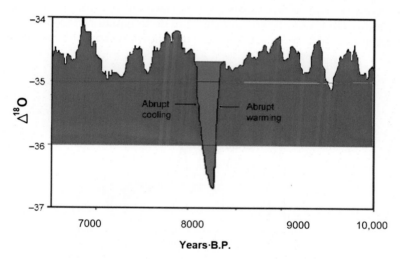

FIGURE 21.6   The 8200-year BP sudden climate change, recorded in oxygen isotope ratios in the GISP2 ice core, lasted about 200 years. *Plotted from data by Grootes, P.M., Stuiver, M., 1997. Oxygen 18/16 variability in Greenland snow and ice with 10³ to 10⁵—year time resolution. Journal of Geophysical Research 102, 26455—26470.*

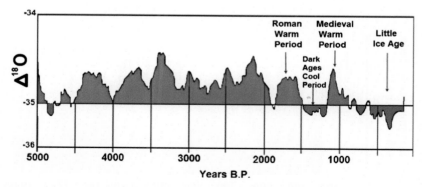

FIGURE 21.7   Greenland GISP2 oxygen isotope ratios for the past 5000 years. Red areas are warm periods and blue areas are cool periods. Virtually all of the past 5000 years was warmer than present until about 1500 years ago. *Plotted from data by Grootes, P.M., Stuiver, M., 1997. Oxygen 18/16 variability in Greenland snow and ice with 10³ to 10⁵—year time resolution. Journal of Geophysical Research 102, 26455—26470.*

2.3.3 Medieval Warm Period (900–1300 AD)

The Medieval Warm Period (MWP) is the most contentious of the late Holocene climatic oscillations because of claims by the Intergovernmental Panel on Climate Change (IPCC) and CO_2 alarmists that it didn't really happen, ie, the basis for the infamous "hockey stick" assertion of no climate changes until CO_2 increase after 1950.

Oxygen isotope data from the GISP2 Greenland ice core clearly show a prominent MWP (Fig. 21.8) between 900 and 1300 AD. It was followed by global cooling and the beginning of the Little Ice Age.

The MWP is also conspicuous on reconstruction of sea surface temperature near Iceland (Fig. 21.9; Sicre et al., 2008).

As shown by numerous studies using a wide variety of methods, the MWP was a period of global warming. One example among many is the study of tree rings in China (Fig. 21.10; Liu et al., 2011).

Historical accounts confirm the worldwide occurrence of the MWP. It was a time of warm climate from about 900 AD to 1300 AD. Its effects were evident in Europe, where grain crops flourished, alpine tree lines rose, many new cities arose, and the population more than doubled. The Vikings took advantage of the climatic amelioration to colonize Greenland, and wine grapes were grown as far north as England, where growing grapes is now not feasible, and about 500 km north of present vineyards in France and Germany. Grapes are presently grown in Germany up to elevations of about 560 m, but from about 1100 AD to 1300 AD., vineyards extended up to 780 m, implying temperatures warmer by about 1.0–1.4°C. Wheat and oats were grown around Trondheim, Norway, suggesting climates about 1°C warmer than present (Fagan, 2000).

Elsewhere in the world, prolonged droughts affected the southwestern United States and Alaska warmed. Sediments in central Japan record warmer temperatures. Sea surface temperatures in the Sargasso Sea were approximately 1°C warmer than today (Keigwin, 1996), and the climate in equatorial east Africa was drier from 1000 AD to 1270 AD. An ice core from the eastern Antarctic Peninsula shows warmer temperatures during this period.

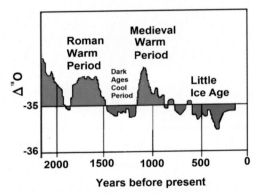

FIGURE 21.8   Oxygen isotope curve from the GISP2 Greenland ice core. (Red = warm, blue = cool.) *Plotted from* data by *Grootes, P.M., Stuiver, M., 1997. Oxygen 18/16 variability in Greenland snow and ice with 10^3 to 10^5—year time resolution. Journal of Geophysical Research 102, 26455—26470 data.*

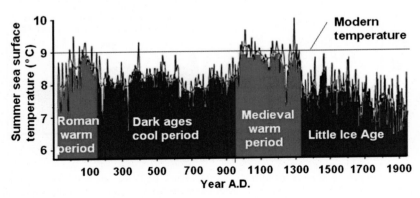

FIGURE 21.9   Summer sea surface temperatures near Iceland (Sicre et al., 2008).

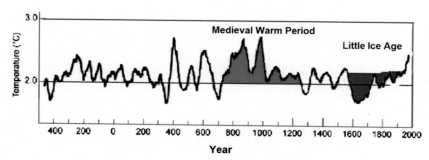

FIGURE 21.10 Temperature reconstruction from tree rings in China. (Red = warm, blue = cool.) *Modified from* Liu, Y., Cai, Q.F., Song, H.M., et al., 2011. Amplitudes, rates, periodicities and causes of temperature variations in the past 2485 years and future trends over the central-eastern Tibetan Plateau. Chinese Science Bulletin 56, 2986—2994.

Oxygen isotope studies in Greenland, Ireland, Germany, Switzerland, Tibet, China, New Zealand, and elsewhere, plus tree-ring data from many sites around the world, all confirm the existence of a global MWP. Soon and Baliunas (2003) found that 92% of 112 studies showed physical evidence of the MWP, only 2 showed no evidence, and 21 of 22 studies in the Southern Hemisphere showed evidence of Medieval warming. Evidence of the MWP at specific sites is summarized in Fagan (2007) and Singer and Avery (2007).

Evidence that the MWP was a global event is so widespread that one wonders why Mann et al. (1998) ignored it. Over a period of many decades, several thousand papers were published establishing the MWP from about 900 ADto 1300 AD. Thus, it came as quite a surprise when Mann et al. (1998), on the basis of a single tree-ring study, concluded that neither the MWP nor the Little Ice Age actually happened and that assertion became the official position of the 2001 Intergovernmental Panel on Climate Change (IPCC). The IPCC 3rd report (Climate Change, 2001) totally ignored the several 1000 publications detailing the global climate changes during the MWP and the LIA and used the Mann et al. single tree-ring study as the basis for the now-famous assertion that "Our civilization has never experienced any environmental shift remotely similar to this. Today's climate pattern has existed throughout the entire history of human civilization" (Gore, 2007). This claim was used as the main evidence that increasing atmospheric CO_2 was causing global warming, and so, as revealed in the "Climategate" scandal, advocates of the CO_2 warming theory were very concerned about the strength of data showing that the MWP was warmer than the 20th century and had occurred naturally, long before atmospheric CO_2 began to increase. The Mann et al. "hockey stick" temperature curve was at so at odds with thousands of published papers, one can only wonder how a single tree-ring study could purport to prevail over such a huge amount of data.

McIntyre and McKitrick (2003) and McKitrick and McIntyre (2005) evaluated the data in the Mann paper and concluded that the Mann curve was invalid "due to collation errors, unjustifiable truncation or extrapolation of source data, obsolete data, geographical location errors, incorrect calculation of principal components and other quality control defects". Thus, the "hockey stick" concept of global climate change is now widely considered totally invalid and an embarrassment to the IPCC.

2.3.4 The Little Ice Age

At the end of the MWP ~1300 AD, temperatures dropped sharply and the colder climate that ensued for several centuries is known as the Little Ice Age (LIA).

The cooling was devastating. Temperatures of the cold winters and cool, rainy summers were too low for growing of cereal crops, resulting in widespread famine and disease (Fagan, 2000; Grove, 2004). Glaciers in Greenland advanced and pack ice extended southward in the North Atlantic. Glaciers expanded worldwide. Temperature changes during the LIA are well shown in various paleo-temperature data (Figs. 21.7—21.10).

The population of Europe had become dependent on cereal grains as a food supply during the MWP, and when the colder climate, early snows, violent storms, and recurrent flooding swept Europe, massive crop failures occurred. Three years of torrential rains that began in 1315 led to the Great Famine of 1315—17. The Thames River in London froze over, the growing season was significantly shortened, crops failed repeatedly, and wine production dropped sharply.

Winters during the LIA were bitterly cold in many parts of the world. Advance of glaciers in the Swiss Alps in the mid-17th century (Fig. 21.11) gradually encroached on farms and buried entire villages. The Thames River and canals and rivers of the Netherlands frequently froze over during the winter (Fig. 21.12). New York Harbor froze in the winter of 1780 and people could walk from Manhattan to Staten Island. Sea ice surrounding Iceland extended for miles in every direction, closing many harbors. The population of Iceland decreased by half and the Viking colonies

FIGURE 21.11   Advance of the Rhone glacier during the Little Ice Age.

FIGURE 21.12   Cold conditions in Europe during the Little Ice Age.

in Greenland died out in the 1300s because they could no longer grow enough food there. In parts of China, warm weather crops that had been grown for centuries were abandoned. In North America, early European settlers experienced exceptionally severe winters.

Global temperatures have fluctuated about 1°F per century since the cooling of the LIA, but the warming has not been continuous. Numerous ~30-year warming periods have been interspersed with ~30-year cooling periods. Oxygen isotope data dating back to 1480 AD from the GISP2 ice core show about 40 periods of warming/cooling (Fig. 21.13). The average time span of each warm/cool period is about 27 years.

As shown by Fig. 21.13, the LIA was not a single period of global cooling, but rather multiple cold periods separated by warm periods beginning about 1300 AD and continuing into the 20th century. One of the strongest periods of cooling occurred during the Maunder Solar Minimum from 1650 AD to 1700 AD (Maunder, 1894, 1922; Eddy, 1976, 1977). Other notable cool intervals were the Dalton Solar Minimum (1790—1820 AD), the 1890 to 1915 cool period, and the 1945 to 1977 cool interval. Numerous others also occurred.

The importance of these climate fluctuations is that they show long-standing evidence of cool/warm cycles over many centuries when CO_2 could not possibly have been the cause.

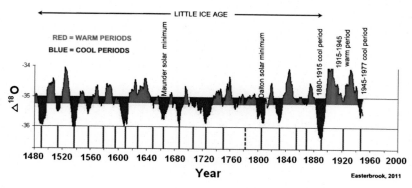

FIGURE 21.13 Oxygen isotope curve from the GISP2 Greenland ice core showing the warm/cool oscillation of temperatures over the past 500 years. The average length of each warm/cool cycle was 27 years. The cool periods correspond to solar minima.

During each warm cycle, glaciers retreated, and during each cool cycle, glaciers advanced. However, because each warm cycle was slightly warmer than the previous one, and each cool cycle not quite as cool as the previous one, glacier termini have progressively receded upvalley from their LIA maximums.

2.3.5 Climate Changes During the Past Century

The climate has changed five times since 1850 (Fig. 21.14).

The 1850—80 warm period was comparable to the two warm intervals that occurred in the 20th century. Atmospheric temperature measurements, glacier fluctuations, and oxygen isotope data from Greenland ice cores record a cool period from about 1880 to about 1915. Glaciers advanced, some nearly to the terminal positions reached during the LIA. Many cold temperature records in North America were set during this period.

Global temperatures rose steadily in the 1920s, 1930s, and early 1940s. By the mid-1940s, global temperatures were about 0.5°C (0.9°F) warmer than they had been at the turn of the century. More high-temperature records for the century were recorded in the 1930s than in any other decade of the 20th century.

Global temperatures began to cool in the mid-1940s at the point when CO_2 emissions began to soar. Many of the world's glaciers advanced during this time, and recovered a good deal of the ice lost during the 1915—45 warm period. Although CO_2 emissions soared during this interval, he climate cooled, just the opposite of what should have happened if CO_2 caused global warming.

The global cooling that prevailed from 1945 to 1977 ended abruptly in 1977 when the Pacific Ocean shifted from its cool mode to its warm mode, and global temperatures began to rise, initiating two decades of global warming. The year 1977 has been called the year of the "Great Climate Shift." During the ensuing warm period from 1978 to ~2000, alpine glaciers retreated and Arctic sea ice diminished. The abruptness of the shift in Pacific sea-surface temperatures and corresponding change from global cooling to global warming in 1977 is highly significant and strongly

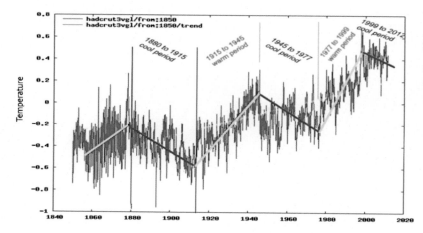

FIGURE 21.14 Global temperatures since 1850. Six global climate changes have occurred since 1850, three periods of warming and three periods of cooling. The 1850—80 and 1915—45 warming intervals were of similar magnitude and intensity to the 1977—99 warming but occurred without any significant change in CO_2.

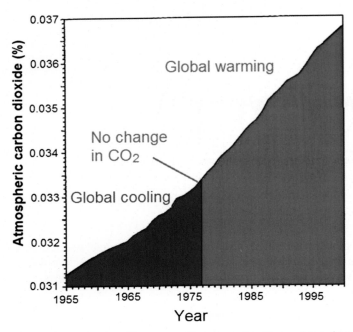

FIGURE 21.15   The abrupt 1977 climate change from cooling to warming occurred with no relationship to CO_2.

suggests a cause-and-effect relationship. The rise of atmospheric CO_2, which accelerated after 1945, shows no sudden change that could account for the "Great Climate Shift" (Fig. 21.15).

3. SIGNIFICANCE OF PAST GLOBAL CLIMATE CHANGES

The most striking thing about climate changes in the late Pleistocene is the large magnitude of the changes over very short periods of time (~20°F/ century). These are vastly greater than any climate changes that occurred in the past few centuries. Carbon dioxide *always* lags temperature and, thus, cannot have anything to do with causing these dramatic climate changes. Nor can these climate changes be caused by orbital changes of the Earth (Milankovitch cycles) because orbital changes occur far too slowly to account for such multiple, abrupt changes.

As shown by the data above, the pattern of multiple climate changes in the geologic past reveals repeated warm/cool cycles. What this means is that these recurring cycles can be projected into the future to predict possible forthcoming climates.

A critical aspect of this method is how consistent the warm/cool cycles have been in the geologic past. Since 1850, six climate changes have occurred, three warm periods and three cool periods (Fig. 21.14), averaging 30 years in length. Forty warm/cool cycles averaging 27 years each have occurred since 1480 AD (Fig. 21.13). Eighty warm/cool cycles averaging 28 years appear in the GISP2 ice core record over the past 1200 years (Fig. 21.16).

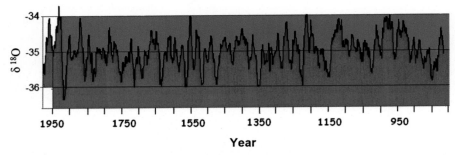

FIGURE 21.16   GISP2 oxygen ice core curve showing oscillating pattern of warm/cool periods. Each warm/cool cycle lasts an average of 28 years, the same as the PDO. *Plotted from data by Grootes, P.M., Stuiver, M., 1997. Oxygen 18/16 variability in Greenland snow and ice with 10^3 to 10^5—year time resolution. Journal of Geophysical Research 102, 26455—26470.*

The pattern of warm/cool cycles, averaging 27—30 years, appears to have been consistent over the past 1200 years. Thus, we can project this pattern into the future to predict what the climate will most likely be.

4. CORRELATION OF TEMPERATURE CYCLES AND THE PACIFIC DECADAL OSCILLATION

The Pacific Decadal Oscillation (PDO) refers to cyclical variations in sea surface temperatures in the Pacific Ocean. The PDO index is defined as the leading principal component of North Pacific monthly sea surface temperature variability (poleward of 20N for the 1900—93 period). A summary of the PDO is given in D'Aleo and Easterbrook (2011 and this volume). It was discovered in the mid-1990s by fisheries scientists studying the relationship between Alaska salmon runs, Pacific Ocean temperatures, and climate. Hare (1996) and Mantua et al. (1997) found that cyclical variations in salmon and other fisheries correlated with warm/cool changes in Pacific Ocean temperatures that followed a regular pattern. Each warm PDO phase lasted about 25—30 years and then switched to the cool mode for 25—30 years.

Fig. 21.17 shows the cold and warm modes of the PDO. During a typical PDO cold mode, cool sea surface temperatures extend from the equator northward along the coast of North America into the Gulf of Alaska. During a typical PDO warm mode, warm sea surface temperatures extend from the equator northward along the coast of North America into the Gulf of Alaska.

Fig. 21.18 shows the PDO from 1900 to 2015. The PDO was cool from 1890 to 1915, warm from 1915 to 1945, cool from 1945 to 1977, warm from 1977 to ~2000, and cool from 2001 to 2016.

Global temperatures are tied directly to sea-surface temperatures. When sea-surface temperatures are cool (cool-phase PDO), global climate cools. When sea-surface temperatures are warm (warm-phase PDO), the global climate warms, regardless of any changes in atmospheric CO_2 (Easterbrook, 2005, 2008a,b).

Fig. 21.19 shows the correlation between the PDO, advance and retreat of glaciers on Mt. Baker, Washington, and global climate. When the PDO is positive (warm), glaciers retreat and the global climate warms. When the PDO is negative (cool), glaciers advance, and the global climate warms.

During the past century, global climates have consisted of two cool periods (1880—1915 and 1945—77) and two warm periods (1915—45 and 1977—2000). In 1977, the PDO switched abruptly from its cool mode, where it had been since about 1945, into its warm mode and global climate shifted from cool to warm. This rapid switch from cool to warm has become to known as "the Great Pacific Climatic Shift" because it happened in just 1 year. Atmospheric CO_2 showed no unusual changes across this sudden climate shift (Fig. 21.15), and thus was clearly not responsible for it. Similarly, the global warming of 1915—45 could not have been caused by increased atmospheric

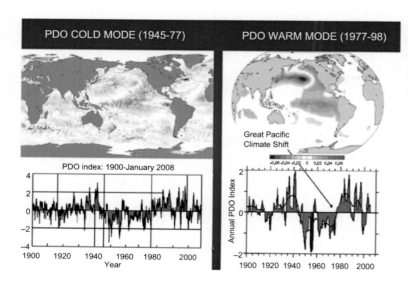

FIGURE 21.17   In 1945, the PDO (Pacific Decadal Oscillation) switched from its warm mode to its cool mode and global climate cooled from then until 1977, despite the soaring of CO_2 emissions. In 1977, the PDO switched back from its cool mode to its warm mode, initiating what is regarded as "global warming" (from 1977 to ~2000).

FIGURE 21.18 PDO from 1900 to 2015 showing warm and cool periods. *Modified from Spencer and JISAO, University of Washington.*

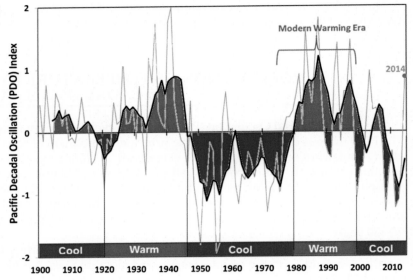

FIGURE 21.19 PDO, global temperature, and glacier advance and retreat.

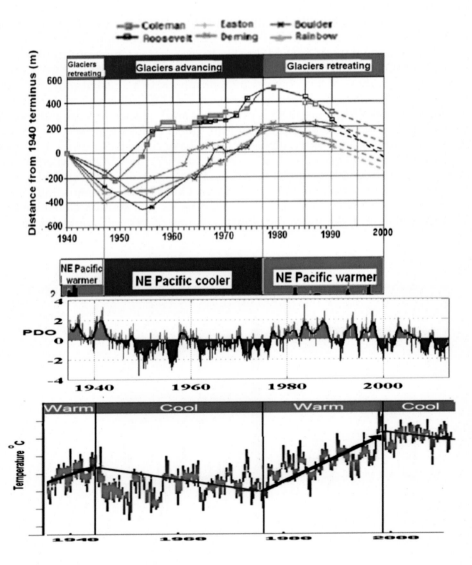

CO_2 because that time preceded the rapid rise of CO_2 emissions after 1945. When CO_2 began to increase rapidly after 1945, 30 years of global cooling ensued (1945—77), just the opposite of what should have happened if CO_2 causes global warming.

Each time this has occurred in the past century, global temperatures have remained cool for about 30 years. Thus, the current cool PDO not only explains the absence of any global warming for the past 18½ years, but also assures that cool temperatures will continue for several more decades (Easterbrook, 2001, 2006a,b, 2007, 2008c).

5. THE ATLANTIC MULTIDECADAL OSCILLATION

The Atlantic Ocean also has multidecadal warm and cool modes with periods of about 30 years, much like the PDO. With the Atlantic Multidecadal Oscillation (AMO), during warm phases, the Atlantic is warm in the tropical North Atlantic and far North Atlantic and relatively cool in the central area. During cool phases, the tropical area and far North Atlantic are cool and the central ocean is warm. For a more detailed discussion, see D'Aleo and Easterbrook (2010, 2011 and this volume).

6. WHERE IS CLIMATE HEADED DURING THE COMING CENTURY?

6.1 IPCC Predictions

What does the century have in store for global climates? According to the IPCC, the Earth is in store for climatic catastrophe this century. Computer models predict global warming of as much as 5—6°C (10—11°F) by 2100, predicated on the assumption that global warming is caused by increasing atmospheric CO_2 and that CO_2 will continue to rise.

According to the IPCC (IPCC-AR4, 2007), the ramifications of such an increase in global warming would be far reaching, even catastrophic in some areas. They predicted that by now the Arctic Ocean would be completely free of sea ice; the Greenland Ice Sheet would melt; alpine glaciers would retreat rapidly, resulting in decreased water supply in areas that depend on snowmelt; melting of Greenland and Antarctic ice would cause sea level to rise sharply, drowning major cities and flooding low coastal areas; crops would fail, resulting in widespread food shortages for people in agriculturally marginal areas; wheat/grain belts would have to shift northward; droughts would become increasingly severe in dry areas; environmental impacts would be severe, resulting in extinction of some species and drastic population decreases in other. All of these disasters are predicated on computer models that assume rising atmospheric CO_2 will cause catastrophic warming. Computer models predictions have been around long enough now that their earlier predictions can be checked against what really happened. Fig. 21.20 shows a comparison of 90 computer model predictions and actual surface and satellite temperature measurements. The computer models failed miserably to come even close to accurate predictions.

6.2 Predictions Based on Past Cyclic Climate Patterns

In 2000, I used past warming and cooling cycles over the past 1200 years and historical PDO cycles to accurately predict the end of the 1977—2000 warm period and predicted 25—30 years of global cooling.

How well has this prediction (Easterbrook, 2001) compared with actual temperature measurements since then? Fig. 21.22 shows a slight cooling trend in satellite (RSS) temperatures since 2000. Fig. 21.23 shows a slight cooling trend in HADCRUT surface temperature measurements. Fig. 21.24 shows a slight cooling trend in NOAA surface measurements in the United States. In contrast, computer models predicted up to 1°C of global warming per decade.

In light of evidence of global cooling since 2000, how can NOAA and NASA contend that virtually every year recently is the hottest ever recorded? Chapter 2 documents the wholesale corruption of surface temperature data, showing that NASA and NOAA data are no longer credible. Only the satellite data is uncorrupted and it shows no global warming for the past 18½ years.

The patterns of two kinds of data may be used to predict future climate: recurring cycles of global temperature and recurring PDO cycles. In 1999, the PDO abruptly switched from its warm mode to its cool mode and the 1978 to 1998 warm period ended, succeeded by slight cooling from 2000 to 2015. As shown in Fig. 21.21, the recurring pattern of the PDO for the past century has been alternating warm/cool periods having a duration averaging

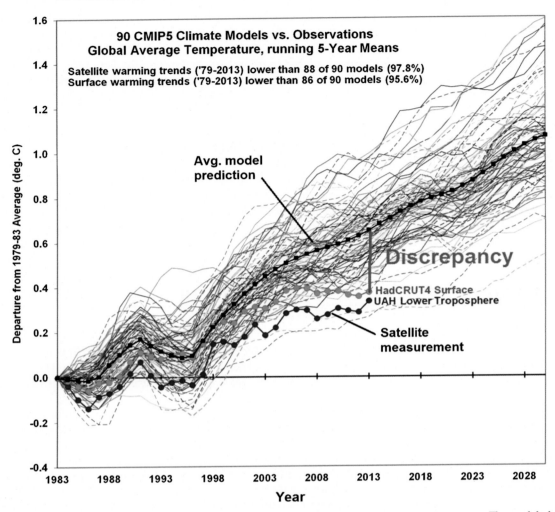

FIGURE 21.20   Comparison of computer model temperature predictions and actual temperature measurements. The models failed dismally. *Modified from Spencer.*

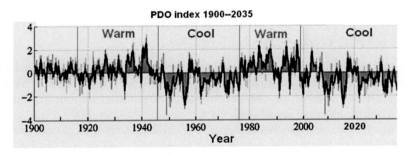

FIGURE 21.21   PDO from 1900 projected to 2035. Extending the long-term cyclic trend suggests global cooling until 2035.

25–30 years. Now that the PDO has shifted into its cool phase, if we project the previous cool period (1945–77) into the future, we should have global cooling for 25–30 years (Fig. 21.21).

The recurring pattern of temperature changes can also be used to predict future climates. Fig. 21.25 shows the HADCRT3 global surface temperature for the past century. This record can be extended 1200 years into the past

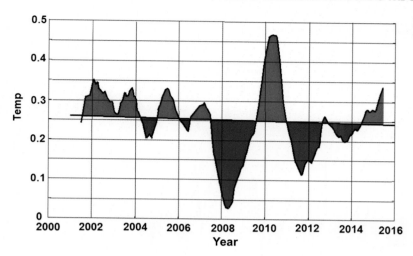

FIGURE 21.22   Satellite temperatures (RSS) from 2000 to 2015 showing a slight cooling trend.

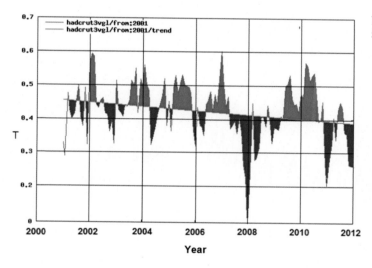

FIGURE 21.23   HADCRUT surface temperature measurements from 2000 to 2012.

using GISP2 Greenland ice core data (Fig. 21.16). These data show recurring warm/cool intervals lasting 25–30 years, so if we project the same pattern into the future, we should have global cooling for the next three decades. There have been two cycles of global cooling in the past century: 1890 to 1915 and 1945 to 1977. Because HADCRUT has essentially erased the 1945–77 cooling, the 1890–1915 cooling has been grafted onto Fig. 21.25 as the most likely prediction of future global climate.

The qualitative prediction of global cooling for the next three decades is straightforward, based on projection of the recurring cycles of PDO and global temperatures (Figs. 21.21 and 21.25). However, quantifying this prediction is more difficult because, going back in time, each of the cool periods was more intense than those that followed. Thus if the coming cool period resembles the 1890–1915 cool phase, we will have the magnitude of cooling shown on the projected temperatures in Fig. 21.25. However, the sun has just entered a Grand Solar Minimum, so if cooling resembles the last time this happened (the Dalton Solar Minimum), cooling will be more intense than that shown on Fig. 21.25. The cooling might even resemble that of the Maunder Solar Minimum, and much colder temperatures could prevail (Easterbrook, 2010; Easterbrook et al., 2013). We have no way to predict just how intense the coming cooling will be, so time will tell.

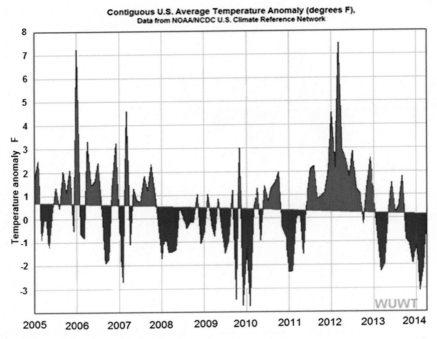

FIGURE 21.24   NOAA temperature measurements from 2005 to 2014 in the United States show a cooling trend.

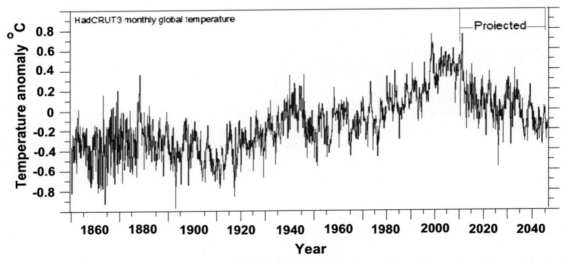

FIGURE 21.25   Projected climate for the century based on recurring climatic patterns over the past 1200 years. The projected cooling is based on repeating of the 1880 to 1915 cool period over the next three decades.

References

Alley, R.B., 2000. The Younger Dryas cold interval as viewed from central Greenland. Quaternary Science Reviews 19, 213–226.

Cuffey, K.M., Clow, G.D., 1997. Temperature, accumulation, and ice sheet elevation in central Greenland through the last deglacial transition. Journal of Geophysical Research 102, 26383–26396.

D'Aleo, J., Easterbrook, D.J., 2010. Multidecadal tendencies in ENSO and global temperatures related to multidecadal oscillations. Energy & Environment 21, 436–460.

Easterbrook, D.J., 2001. The next 25 years: global warming or global cooling? Geologic and oceanographic evidence for cyclical climatic oscillations. Geological Society of America 33, 253. Abstracts with Program.

D'Aleo, J., Easterbrook, D.J., 2011. Relationship of multidecadal global temperatures to multidecadal oceanic oscillations. Evidence-Based Climate Science 161–184.

Easterbrook, D.J., 2005. Causes and effects of abrupt, global, climate changes and global warming. Geological Society of America 37, 41. Abstracts with Program.

Easterbrook, D.J., 2006a. Causes of abrupt global climate changes and global warming predictions for the coming century. Geological Society of America 38, 77. Abstracts with Program.

Easterbrook, D.J., 2006b. The cause of global warming and predictions for the coming century. Geological Society of America 38, 235−236. Abstracts with Program.

Easterbrook, D.J., 2007. Geologic evidence of recurring climate cycles and their implications for the cause of global warming and climate changes in the coming century. Geological Society of America 39, 507. Abstracts with Programs.

Easterbrook, D.J., 2008a. Solar influence on recurring global, decadal, climate cycles recorded by glacial fluctuations, ice cores, sea surface temperatures, and historic measurements over the past millennium. In: Abstracts of American Geophysical Union Annual Meeting, San Francisco.

Easterbrook, D.J., 2008b. Implications of glacial fluctuations, PDO, NAO, and sun spot cycles for global climate in the coming decades. Geological Society of America 40, 428. Abstracts with Programs.

Easterbrook, D.J., 2008c. Correlation of climatic and solar variations over the past 500 years and predicting global climate changes from recurring climate cycles. In: Abstracts of 33rd International Geological Congress, Oslo, Norway.

Easterbrook, D.J., 2010. A Walk Through Geologic Time From Mt. Baker to Bellingham Bay, Chuckanut ed. 330 p.

Easterbrook, D.J., Ollier, C.M., Carter, R.M., 2013. Observations: the cryosphere. In: Climate Change Reconsidered II: Physical Science, pp. 645−728.

Eddy, J.A., 1976. The Maunder Minimum. Science 192, 1189−1202.

Eddy, J.A., 1977. Climate and the changing sun. Climatic Change 1, 173−190.

Fagan, B., 2000. The Little Ice Age. Basic Books, NY, 246 p.

Fagan, B., 2007. The Great Warming: Climate Change and the Rise and Fall of Civilizations. Bloomsbury Press, 283 p.

Gore, A., 2007. An Inconvenient Truth. Rodale, PA, 325 p.

Grootes, P.M., Stuiver, M., 1997. Oxygen 18/16 variability in Greenland snow and ice with 10^3 to 10^5−year time resolution. Journal of Geophysical Research 102, 26455−26470.

Grove, J.M., 2004. Little Ice Ages: Ancient and Modern. Routledge, London, UK, 718 p.

Hare, S.R., 1996. Low Frequency Climate Variability and Salmon Production (Ph.D. dissertation). School of Fisheries, University of Washington, Seattle, WA.

IPCC-AR4, 2007. Climate Change: The Physical Science Basis. Contribution of Working Group I to the Fourth Assessment Report of the Intergovernmental Panel on Climate Change. Cambridge University Press.

Keigwin, L.D., 1996. The Little Ice Age and Medieval Warm Period in the Sargasso Sea. Science 274, 1504−1508.

Liu, Y., Cai, Q.F., Song, H.M., et al., 2011. Amplitudes, rates, periodicities and causes of temperature variations in the past 2485 years and future trends over the central-eastern Tibetan Plateau. Chinese Science Bulletin 56, 2986−2994.

Mantua, N.J., Hare, S.R., Zhang, Y., Wallace, J.M., Francis, R.C., 1997. A Pacific interdecadal climate oscillation with impacts on salmon production. Bulletin of the American Meteorological Society 78, 1069−1079.

Mann, M., Bradley, R.S., Hughes, M.K., 1998. Global-scale temperature patterns and climate forcing over the past six centuries. Nature 392, 779−787.

Maunder, E.W., 1894. A prolonged sunspot minimum. Knowledge 17, 173−176.

Maunder, E.W., 1922. The prolonged sunspot minimum, 1645−1715. Journal of the British Astronomical Society 32, 140.

McIntyre, S., McKitrick, R., 2003. Corrections to the Mann et al. (1998) proxy database and Northern Hemispheric average temperature series. Energy & Environment 14, 751−771.

McKitrick, R., McIntyre, S., 2005. The M&M critique of the MBH98 Northern Hemisphere climate index: update and implications. Energy & Environment 16, 69−100.

Sicre, M., Jacob, J., Ezat, U., Rousse, S., Kissel, C., Yiou, P., Eiriksson, J., Knudsen, K.L., Jansen, E., Turon, J., 2008. Decadal variability of sea surface temperatures off North Iceland over the last 2000 years. Earth and Planetary Science Letters 268, 137−142.

Singer, S.F., Avery, D., 2007. Unstoppable Global Warming Every 1500 Years. Rowman & Littlefield Publishers, Inc, 278 p.

Soon, W., Baliunas, S., 2003. Proxy climatic and environmental changes of the past 1000 years. Climate Research 23, 89−110.

Steffensen, J.P., Andersen, K.K., Bigler, M., Clausen, H.B., Dahl-Jensen, D., Goto-Azuma, K., Hansson, M.J., Sigfus, J., Jouzel, J., Masson-Delmotte, V., Popp, T., Rasmussen, S.O., Roethlisberger, R., Ruth, U., Stauffer, B., Siggaard-Andersen, M., Sveinbjornsdottir, A.E., Svensson, A., White, J.W.C., 2008. High-resolution Greenland ice core data show abrupt climate change happens in few years. Science 321, 680−684.

Index